COMPENDIUM OF IN VIVO MONITORING IN REAL-TIME MOLECULAR NEUROSCIENCE

Volume 1: Fundamentals and Applications

COMPENDIUM OF IN VIVO MONITORING IN REAL-TIME MOLECULAR NEUROSCIENCE

Volume 1: Fundamentals and Applications

Editors

George S Wilson
University of Kansas, USA

Adrian C Michael
University of Pittsburgh, USA

NEW JERSEY • LONDON • SINGAPORE • BEIJING • SHANGHAI • HONG KONG • TAIPEI • CHENNAI

Published by

World Scientific Publishing Co. Pte. Ltd.
5 Toh Tuck Link, Singapore 596224
USA office: 27 Warren Street, Suite 401-402, Hackensack, NJ 07601
UK office: 57 Shelton Street, Covent Garden, London WC2H 9HE

Library of Congress Cataloging-in-Publication Data
Compendium of in vivo monitoring in real-time molecular neuroscience / editors, George S. Wilson, University of Kansas, USA, Adrian C. Michael, University of Pittsburgh, USA.
   volumes cm
  Includes bibliographical references and index.
  Contents: volume 1. Fundamentals and applications
  ISBN 978-9814619769 (v. 1 : hardcover : alk. paper)
  1. Neurochemistry. 2. Electrochemical analysis. I. Wilson, George S., 1939– II. Michael, Adrian C.
  QP356.3.C64 2014
  612.8'042--dc23
                                              2014029387

British Library Cataloguing-in-Publication Data
A catalogue record for this book is available from the British Library.

Copyright © 2015 by World Scientific Publishing Co. Pte. Ltd.

All rights reserved. This book, or parts thereof, may not be reproduced in any form or by any means, electronic or mechanical, including photocopying, recording or any information storage and retrieval system now known or to be invented, without written permission from the publisher.

For photocopying of material in this volume, please pay a copying fee through the Copyright Clearance Center, Inc., 222 Rosewood Drive, Danvers, MA 01923, USA. In this case permission to photocopy is not required from the publisher.

Typeset by Stallion Press
Email: enquiries@stallionpress.com

Printed in Singapore

PREFACE

The seminal work by Ralph Adams and co-workers in the early 1970s applying microelectrodes to the electrochemical detection of neuroactive substances made possible their monitoring in real time in the rodent brain. Thus, along with oxygen and ascorbate, catecholamines such as dopamine, norepinephrine, and serotonin were detected by employing an implantable electrode. The invention of the carbon fiber microelectrode by François Gonon and coworkers in the 1970s, the introduction of fast-scan cyclic voltammetry by Julian Millar and coworkers in the 1980s, and their combination and ceaseless refinement by Mark Wightman and coworkers to the present day has enhanced the sensitivity, selectivity, and temporal and spatial resolution of the *in vivo* voltammetric approach.

The repertoire of target analytes was expanded by the development of biosensors in the 1990s and early 2000s compatible with real time measurements in the brain. These include biosensors for glucose, lactate, glutamate, acetylcholine/choline, ATP, and adenosine. In addition, reactive oxygen and nitrogen species such as $\cdot O_2^-$, H_2O_2, NO, and $ONOO^-$ are electroactive and can be detected, although reliable measurements are challenged by low concentrations and transient existence.

In this volume the authors have consistently addressed the question, "Now that we have a series of monitoring tools, what can we learn using them?" Answering the question of the role of a particular analyte immediately raises the issue of how to account for time-dependent changes in concentration in response to a stimulus, the application of a pharmacological agent, changes in behavior, or in various animal models of disease. A challenge is to be able to measure more than one parameter at a time: concentrations of two different analytes, physiological parameters such as EEG activity or the frequency of cell firing. Increasingly, behavioral parameters are also correlated with changes in neurotransmitter concentration fluctuations. With the recent contributions of Kendall Lee, Paul Garris, Charles Blaha, and others, translation of electrochemical measurements to the human brain is under way.

As usual, research generates more questions than answers. More information acquired simultaneously is needed and the role of neuroactive peptides has been scarcely scratched. While implanting a sensor in brain tissue has the obvious advantage of good time and spatial resolution and provides species-specific information, it is an invasive process and we are just beginning to understand the effects of tissue alteration in the vicinity of the implant due to tissue disruption and inflammatory response. To understand chronic effects, it will be necessary to have sensors that can survive for extended periods of time in tissue. New opportunities continue to appear on the horizon: the coupling of *in vivo* monitoring to optogenetic stimulation, for example, has already cast new light on the regulation of dopamine release by cholinergic interneurons of the striatum. This is illustrated graphically on the book cover with a drawing taken from Chapter 9 by Cragg and co-workers.

George S. Wilson
Adrian C. Michael

CONTENTS

Preface   v

Chapter 1   Using Biosensors to Probe Fundamental Questions of Sleep   1
     Erik Naylor and Peter A. Petillo
   1.1 Introduction   1
   1.2 Sleep — A Beneficial Unknown   1
   1.3 Electroencephalographic Sleep Measurement   2
   1.4 Early Investigations of Sleep Physiology   4
   1.5 Biosensor Recording during Sleep   5
     1.5.1 Biosensor Basics   5
     1.5.2 Biosensor Studies during Sleep   6
       1.5.2.1 Glucose   7
       1.5.2.2 Lactate   9
       1.5.2.3 Glutamate   11
       1.5.2.4 Adenosine   13
   1.6 Biosensors and the Astrocyte Neuron Lactate Shuttle Hypothesis   15
   1.7 Conclusion   18
   Acknowlegments   20
   References   20

Chapter 2   Cortical Cholinergic Transients for Cue Detection and Attentional Mode Shifts   27
     Martin Sarter, William M. Howe and Howard Gritton
   2.1 Introduction   27
   2.2 Measurement of Second-Based Cholinergic Transients   31
   2.3 Cue Detection Mediated by Cholinergic Transients   32

2.4 Cholinergic Mediation of Cue Detection
in Trials Involving an Attentional Mode Shift 33
2.5 Gamma Synchrony as a Neurophysiological Marker
of Cholinergic Modulation of Cortical Circuitry 35
2.6 Moving Forward: Potential Neurobiological
Impact of New Electrodes and Recording Methods 37
2.7 Summary and Conclusions 38
Acknowledgment 40
References 40

Chapter 3 Real-Time Measurement of ATP and Adenosine
in the Nervous System 45
Nicholas Dale

3.1 Introduction 45
  3.1.1 The Components of Purinergic Signaling 45
    3.1.1.1 ATP receptors and release mechanisms 46
    3.1.1.2 Adenosine receptors and release mechanisms 47
    3.1.1.3 Ectonucleotidases — breakdown of ATP
and initiation of ADP and adenosine signaling 48
    3.1.1.4 The need for real-time measurements of ATP
and adenosine 49
3.2 ATP-Release, Signaling and Functions 50
  3.2.1 ATP Biosensing Principles 50
  3.2.2 The Role of ATP Signaling in Chemosensing 51
    3.2.2.1 Central CO_2 chemoreception 51
    3.2.2.2 Glucosensing in the hypothalamus 54
  3.2.3 ATP Signaling and Development 57
    3.2.3.1 Initiation of expression of
the eye-field transcription factors 57
    3.2.3.2 Regulation of development of the retina 59
3.3 Adenosine-Release Mechanisms and Functions 59
  3.3.1 Adenosine Biosensing Principles 59
  3.3.2 Activity Dependence of Adenosine Release 60
    3.3.2.1 Release mechanisms in cerebellum 61
    3.3.2.2 Release mechanisms in hippocampus 63
    3.3.2.3 Adenosine release linked to activation
of the Na^+–K^+ ATPase 63
    3.3.2.4 What have we learned? 64
  3.3.3 Sleep and Adenosine 65

3.4	Interplay between ATP and Adenosine Release: Motor Pattern Generation	67
3.5	Concluding Remarks	70
Acknowledgments		70
References		70

Chapter 4   Electrochemical Detection of Adenosine *In Vivo*   79
Ashley E. Ross and B. Jill Venton

4.1	Introduction	79
	4.1.1 Adenosine Regulation in the Central Nervous System	79
	4.1.1.1 Intra- and extracellular formation of adenosine	80
	4.1.1.2 Adenosine receptors	80
	4.1.1.3 Adenosine transporters	81
	4.1.2 Function of Adenosine in the Central Nervous System	82
	4.1.2.1 Adenosine as a neuroprotector	82
	4.1.2.2 Adenosine as a neuromodulator	83
4.2	Electrochemical Detection of Adenosine *In Vivo*	84
	4.2.1 Adenosine Characterization using FSCV	84
	4.2.1.1 Characterization of carbon-fiber microelectrodes for adenosine detection using FSCV	87
	4.2.1.2 Characterization of diamond electrodes for adenosine detection using FSCV	87
	4.2.1.3 Adenosine modulates respiratory rhythmogenesis	88
	4.2.1.4 Short electrical stimulations elicit rapid adenosine changes	88
	4.2.1.5 Adenosine transiently regulates oxygen concentrations	88
	4.2.1.6 A_1 receptors self-regulate stimulated adenosine release	89
	4.2.1.7 Stimulated adenosine release is activity-dependent in multiple brain regions	89
	4.2.1.8 FSCV to study adenosine deaminase kinetics	91
	4.2.1.9 Adenosine release during deep brain stimulation (DBS) causes microthalamotomy effect	92
	4.2.1.10 AMP hydrolysis to adenosine inhibits pain-sensing response in spinal cord	92
	4.2.1.11 Enhanced understanding of adenosine signaling using FSCV	95

4.2.2 Adenosine Characterization using Amperometric
Adenosine Micro-Biosensors 95
    4.2.2.1 Activity-dependent adenosine release in the
cerebellum is modulated by endogenous
neurotransmitters 96
    4.2.2.2 Adenosine is released during hypoxia
in the hippocampus 98
    4.2.2.3 Adenosine is released during hypoxia
in the nucleus tractus solitarii (NTS)
of the brainstem 99
    4.2.2.4 Adenosine is released during ischemia 99
    4.2.2.5 Adenosine modulates chemoreceptor responses 100
    4.2.2.6 Adenosine is linked to modulating
seizures induced by CO_2 100
    4.2.2.7 Increase in extracellular adenosine levels provides
anticonvulsant behavior during epilepsy 100
    4.2.2.8 Adenosine is involved in the sleep–wake cycle 101
    4.2.2.9 Enhanced understanding of adenosine signaling
using amperometric biosensors 102
4.3 Contributions of Electrochemical Detection
of Adenosine to the Field 102
Acknowledgments 105
References 105

Chapter 5 Real-Time *In Vivo* Neurotransmitter Measurements
Using Enzyme-Based Ceramic Microelectrode
Arrays: What We Have Learned
About Glutamate Signaling 113
Jason J. Burmeister, Erin R. Hascup,
Kevin N. Hascup, Verda Davis, Seth R. Batten,
Francois Pomerleau, Jorge E. Quintero,
Peter Huettl, Pooja M. Talauliker,
Ingrid Strömberg, Greg A. Gerhardt
5.1 Introduction 113
5.2 Materials and Methods 115
    5.2.1 Development of Ceramic-Based MEAs 115
    5.2.2 Basic Capabilities of MEAs 115
5.3 Self-Referencing Recordings using Amperometry:
How to Measure Resting Levels of Neurochemicals 116
    5.3.1 Why Perform Self-Referencing Recordings? 116

5.4	Applications to Biological Systems	119
	5.4.1 Studies in Freely Moving Rats and Mice	119
	5.4.2 Tonic Release of Neurotransmitters: Is It Neuronally-Derived?	120
	5.4.2.1 How do we know that we are measuring glutamate that is mostly derived from neurons?	120
	5.4.2.2 Resting levels of glutamate in the CNS: comparisons to microdialysis measurements	121
	5.4.3 Beyond Tonic Recordings: Phasic Release of Neurotransmitters	123
	5.4.4. Chronic Implantation: Histopathology and Biocompatibility	124
5.5	Conclusions	128
	Acknowledgments	129
	References	129

Chapter 6   Enzyme-Based Microbiosensors for Selective Quantification of Rapid Molecular Fluctuations in Brain Tissue   137
Leyda Z. Lugo-Morales and Leslie A. Sombers

6.1	Introduction	137
	6.1.1 Electrochemistry in Neuroscience	137
	6.1.2 Electrochemical Detection of Non-Electroactive Analytes	138
	6.1.3 Electrochemical Biosensors	138
6.2	Amperometric Biosensors	139
	6.2.1 Platinum (Pt)-Based Biosensors	139
	6.2.2 Methods of Enzyme Immobilization	140
	6.2.2.1 Encapsulation or entrapment	140
	6.2.2.2 Adsorption	141
	6.2.2.3 Covalent attachment and cross-linking	141
	6.2.3 Self-Referencing Amperometric Biosensors	141
	6.2.4 Applications	142
	6.2.4.1 Acetylcholine	142
	6.2.4.2 Glutamate	144
	6.2.4.3 Glucose	146
6.3	Voltammetric Biosensors	147
	6.3.1 Carbon-Fiber Microelectrodes	147
	6.3.2 Enzyme Encapsulation within Chitosan by Electrodeposition	148
	6.3.3 Enzyme-Modified Carbon-Fiber Microelectrodes for the Detection of Non-Electroactive Analytes using FSCV	148

	6.3.4 High-Resolution Sampling of Brain Glucose Concentration Fluctuations	152
6.4	Conclusions	153
	Acknowledgments	154
	References	154

Chapter 7   Monitoring and Modulating Dopamine Release and Unit Activity in Real-Time   161
Anna M. Belle and R. Mark Wightman

7.1	Introduction	161
7.2	Dopamine Neurotransmission	162
7.3	Evolution of A Waveform	164
7.4	Seeing the Brain Communicate: A Combination of Measurement Procedures	166
	7.4.1 Combining Electrochemistry with Electrophysiology	167
	7.4.2 Controlled Iontophoresis	171
7.5	Conclusions	173
	Acknowledgments	173
	References	173

Chapter 8   Quantitative Chemical Measurements of Vesicular Transmitters with Single Cell Amperometry and Electrochemical Cytometry   181
Jelena Lovrić, Xianchan Li, Andrew G. Ewing

8.1	Introduction	181
8.2	Electrochemical Cytometry	183
8.3	Mathematical Treatment of Amperometric Data: Partial Distention During Exocytosis Explains The Shape of Amperometric Events	186
	8.3.1 Simplified Discussion of the Mathematical Approach	186
	8.3.2 Modeling Supports the Hypothesis that Vesicle Pore Opening Does Not Need to be Full	189
	8.3.3 Implication that Lipids Control the Pore Once it is Open	190
8.4	Regulation of Open-and-Closed Exocytosis	190
	8.4.1 Evidence that Dynamin Regulates the Opening of the Fusion Pore	190
	8.4.2 Evidence that Actin Regulates the Closing of the Fusion Pore	192
8.5	Post-Spike Foot is Observed in Amperometry	193

	8.5.1 Observation of Post-Spike Feet	193
	8.5.2 Variation of Lipids Confirms Validity of Post-Spike Feet	197
	8.5.3 Some Exocytosis Events Appear to Open All The Way	197
8.6	Future Perspectives	198
Acknowledgments		198
References		198

Chapter 9   Coupling Voltammetry With Optogenetics to Reveal Axonal Control of Dopamine Transmission by Striatal Acetylcholine   201
Polina Kosillo, Katherine R. Brimblecombe, Sarah Threlfell and Stephanie J. Cragg

9.1	Introduction	201
	9.1.1 DA Detection with FCV	202
	9.1.2 Driving Neural Activity in Local Striatal Circuits: Electrical Stimulation *Versus* Optogenetics	203
	9.1.3 Elucidating the Role of ChIs in Striatal DA Transmission	203
9.2	Optogenetic Activation of Striatal Cholinergic Interneurons Drives Dopamine Transmission	204
	9.2.1 Methodology for Combined FCV and Optogenetics	205
	9.2.2 Key Findings	206
9.3	Critical Methodological Issues	207
	9.3.1 ChR2 Expression Method: Local Targeting *Versus* Transgenic Breeding Approaches	208
	9.3.2 Choice of AAV	210
	9.3.2.1 ChR2 expression levels	211
	9.3.2.2 Axonal expression of ChR2 and post-injection delays	211
	9.3.3 Choice of fluorescent tag	212
	9.3.4 Laser or LED	212
	9.3.5 Laser Power and Pulse Duration	214
	9.3.6 Electrochemical Analysis	214
	9.3.6.1 Photoelectric effect of CFM photo-activation	214
	9.3.6.2 Drugs as photo- and electroactive compounds	215
	9.3.7 Physiological Relevance	215
	9.3.7.1 Physiological firing frequencies	216
	9.3.7.2 Synchrony and selective activation of projection pathways	216

9.4 Summary	217
Acknowledgment	217
References	217

Chapter 10 Electrochemical Recordings During Deep Brain Stimulation in Animals and Humans: WINCS, MINCS, and Closed-Loop DBS — 225
Charles D. Blaha, Su-Youne Chang, Kevin E. Bennet and Kendall H. Lee

10.1 Introduction	225
10.1.1 DBS is a State-of-the-Art Neurosurgical Therapy	226
10.1.2 DBS Mediated DA Release in the Basal Ganglia: Pre-Clinical Studies	229
10.2 Preliminary Neurochemical Studies	229
10.2.1 STN DBS Affects Striatal DA Release in the Anesthetized Rat	231
10.2.2 Effects of Various STN DBS Parameters on Striatal DA Release in the Anesthetized Rat	232
10.3 Neurotransmitter Monitoring Systems	235
10.3.1 WINCS	235
10.3.2 MINCS	236
10.4 Recording Electrode Development	238
10.4.1 Microelectrode for Human Neurochemical Recording	238
10.5 Neurochemical Recordings in Large Animals and Neurological Patients	239
10.5.1 DA and ADO Recording in the Pig	239
10.5.2 Human Neurochemical Recordings	243
10.6 Conclusion and Future Directions	244
Acknowledgment	244
References	244

Chapter 11 Sweet Leaf: Neurochemical Advances Reveal How Cannabinoids Affect Brain Dopamine Concentrations — 251
Erik B. Oleson and Joseph F. Cheer

11.1 Introduction	251
11.2 Real-Time Monitoring of Cannabinoid Actions on Dopamine	252
11.2.1 Cannabinoids, Microdialysis, and Tonic Dopamine Concentrations	252
11.2.2 Novel Insights from FSCV	253

	11.2.3 A Current Theory on the Pharmacological Mechanism of Cannabinoid Action	254
11.3	The Endocannabinoid System	255
	11.3.1 Introduction	255
	11.3.2 Endocannabinoids, Microdialysis and Tonic Dopamine Concentrations	256
	11.3.3 Novel Insights from FSCV	257
11.4	Future Prospects	261
	Acknowledgment	263
	References	263

Chapter 12 Probing Serotonin Neurotransmission: Implications for Neuropsychiatric Disorders 269
Kevin M. Wood, David Cepeda and Parastoo Hashemi

12.1	Introduction	269
12.2	Real-Time Serotonin Monitoring	270
	12.2.1 Microdialysis	270
	12.2.2 Constant Potential Amperometry	271
	12.2.3 High Speed Chronoamperometry	272
	12.2.4 Fast-Scan Cyclic Voltammetry (FSCV)	272
12.3	The Relevance of Real-Time Serotonin Measurements to Physiology and Disease	273
	12.3.1 Endogenous *In Vivo* Processes	274
	12.3.2 Serotonin Modulation	275
	12.3.3 Serotonin Clearance	276
	12.3.4 Serotonin's Involvement in Disease	278
12.4	Conclusions	279
	References	279

Chapter 13 Voltammetric Analysis of Loss and Gain of Dopamine Function 287
Paul A. Garris, Kristen A. Keefe

13.1	Introduction	287
13.2	Modes of Dopamine Signaling: Overview	288
13.3	Dopamine Signaling in the Dopamine-Depleted Brain	289
	13.3.1 Passive Stabilization of Tonic Dopamine Signaling	289
	13.3.2 Compensatory Up-Regulation of Exocytotic Dopamine Release	292
	13.3.3 Deficits in Phasic Dopamine Signaling	293
	13.3.4 Deficits in Postsynaptic Gene Expression	295
13.4	Mechanism of AMPH Action	297

        13.4.1 Activation, not Disruption, of Phasic Dopamine Signaling   297
        13.4.2 AMPH-Induced Up-Regulation of Exocytotic Dopamine Release   300
        13.4.3 Reconciling Phasic Activation and Dopamine Efflux   302
   13.5 Summary and Conclusion   304
   Acknowledgment   304
   References   304

Chapter 14 Measurements of Dopamine Release and Uptake in Huntington's Disease Model Rodents   311
Sam V. Kaplan, Stephen C. Fowler, and Michael A. Johnson

   14.1 Introduction   311
   14.2 Potential Mechanisms of Neuronal Dysfunction and Degradation in HD   314
   14.3 Genetically Engineered HD Model Rodents   314
   14.4 Microdialysis Studies in HD Model Rodents   316
   14.5 Voltammetric Measurements of DA Release and Uptake in HD Model Rodents   317
        14.5.1 FSCV   317
        14.5.2 DA Release and Uptake Measurements in R6/2 and R6/1 HD Model Mice   318
        14.5.3 Mechanisms of Neurotransmitter Release Impairment   319
        14.5.4 DA Release and Uptake Measurements in HD Model Rats   322
            14.5.4.1 Treatment with 3-NP   322
            14.5.4.2 HDtg Rats   324
   14.6 Conclusion   328
   Acknowledgments   329
   References   329

Chapter 15 Characterizing Neuropeptide Release: From Isolated Cells to Intact Animals   335
Agatha E. Maki and Jonathan V. Sweedler

   15.1 Introduction   335
   15.2 History of Characterizing Neuropeptide Release   337
   15.3 Current Methods of Analyzing Neuropeptide Release   338

15.3.1 Cellular Methods	338
15.3.2 Intact Tissue Methods	341
15.3.3 *In vivo* Methods	342
15.4 Conclusions	344
Acknowledgment	344
References	345

Chapter 16 Advancing Chronic Intracortical Electrode Recording Function 351
Lohitash Karumbaiah, Tarun Saxena, and Ravi Bellamkonda

16.1 Introduction	351
16.2 Cortical Electrodes for Real-Time Monitoring of Neural Activity	352
16.2.1 Non-Invasive Scalp Electrodes for Measurement of EEG Activity	352
16.2.2 Minimally Invasive Subdural Electrodes for Measurement of EcoG Activity	353
16.2.3 Invasive Intracortical Electrodes for Measurement of Single Unit Activity	354
16.3 FBR to Intracortical Electrodes	356
16.3.1 Brain Response to Intracortical Electrode Implants	356
16.3.1.1 Acute response	357
16.3.1.2 Chronic response	357
16.3.2 Current Methods of Assessment	358
16.3.2.1 Histology	358
16.3.2.2 Impedance spectroscopy	358
16.3.2.3 Cyclic voltammetry	359
16.3.3 Advanced Molecular Methods	360
16.3.3.1 Non-invasive imaging of the BBB	360
16.3.3.2 Inflammatory cytokine signaling	362
16.4 Conclusions	363
References	363

Chapter 17 Measurement of Cytokines in the Brain 369
Julie A. Stenken and Michael Elkins

17.1 Introduction	369
17.2 Cytokines as Signaling Proteins	370

17.3 Clinical Interests in Cytokines in the Brain	371
17.4 Overview of Common Methods for Cytokine Measurements and Tissue Mapping	372
17.5 Cellular Sources of Cytokines	373
17.5.1 CCL2 (MCP-1)	374
17.5.2 CXCL12 (SDF-1)	374
17.5.3 CX3CL1	376
17.5.4 IL-1β	376
17.5.5 IL-6	377
17.6 Cytokine Collection from Brain ECS	377
17.6.1 Sampling Considerations: Overview of Mass Transport, Binding, and Uptake Processes	378
17.6.2 Push–Pull Perfusion	380
17.6.2.1 Device calibration	380
17.6.3 Microdialysis Sampling	381
17.6.3.1 Microdialysis sampling device calibration	382
17.6.4 Selected Applications of Cytokine Collection	383
17.6.4.1 Push–Pull perfusion applications for cytokine collection	383
17.6.4.2 Microdialysis applications for cytokine collection	384
17.7 Cytokine Production During Brain Injuries and Other Device Implantation Procedures	388
17.8 Conclusions and Future Prospects	390
Acknowledgments	391
References	391
Index	401

LIST OF CONTRIBUTORS

Seth R. Batten
Center for Microelectrode Technology, University of Kentucky
College of Medicine, Lexington, KY, USA
Chapter 5

Ravi Bellamkonda
Wallace H. Coulter Department of Biomedical Engineering
Georgia Institute of Technology and Emory School of Medicine
Atlanta, GA 30332, USA
Chapter 16
ravi@gatech.edu

Anna M. Belle
Department of Chemistry
University of North Carolina at Chapel Hill
Chapel Hill, NC 27599-3290
Chapter 7

Kevin E. Bennet
Department of Neurologic Surgery and Physiology
and Biomedical Engineering
Division of Bioengineering, Mayo Clinic
Rochester, MN 55905, USA
Chapter 10

Charles D. Blaha
Department of Psychology, University of Memphis
Memphis, TN 38152, USA
Chapter 10

Katherine R. Brimblecombe
Department of Physiology, Anatomy and Genetics
and Oxford Parkinson's Disease Centre, Sherrington Building
University of Oxford, OX1 3PT, UK
Chapter 9

Jason J. Burmeister
Center for Microelectrode Technology
JBC, Council Bluffs, IA, USA
Chapter 5

David Cepeda
Department of Chemistry
383 A. Paul Schaap Building
Wayne State University
Detroit, MI, 48202, USA
Chapter 12

Su-Youne Chang
Departments of Neurologic Surgery and
Physiology and Biomedical Engineering
Mayo Clinic, Rochester, MN 55905, USA
Chapter 10

Joseph F. Cheer
Department of Neurobiology and Anatomy
University of Maryland Baltimore
Baltimore, MD, 21201, USA
Chapter 11
jchee001@umaryland.edu

Stephanie J. Cragg
Department of Physiology, Anatomy and Genetics,
and Oxford Parkinson's Disease Centre, Sherrington Building,
University of Oxford, OX1 3PT, UK
Chapter 9
stephanie.cragg@dpag.ox.ac.uk

Nicholas Dale
School of Life Sciences
University of Warwick
Gibbet Hill Road Coventry, CV4 7AL, UK
Chapter 3
N.E.Dale@warwick.ac.uk

Verda Davis
Center for Microelectrode Technology, University of Kentucky
College of Medicine, Lexington, KY, USA
Chapter 5

Michael Elkins
Department of Chemistry and Biochemistry
University of Arkansas
Fayetteville, AR 72701, USA
Chapter 17

Andrew G. Ewing
Department of Chemical and Biological Engineering
Chalmers University of Technology
Kemivägen 10, SE-412 96 Gothenburg, Sweden
and Department of Chemistry and Molecular Biology
University of Gothenburg, Kemivägen 10
SE-412 96 Gothenburg, Sweden
Chapter 8
andrew.ewing@chem.gu.se

Stephen C. Fowler
Department of Pharmacology and Toxicology,
University of Kansas, Lawrence, KS 66045, USA
Chapter 11

Paul A. Garris
School of Biological Sciences
Illinois State University, Normal, IL 61790–4120, USA
Chapter 13
pagarri@IllinoisState.edu

Greg A. Gerhardt
Center for Microelectrode Technology, University of Kentucky
College of Medicine, Lexington, KY, USA
Chapter 5
gregg@uky.edu

Howard Gritton
Department of Biomedical Engineering
Boston University, USA
Chapter 2

Erin R. Hascup
Southern Illinois University School of Medicine
Department of Neurology
Center for Alzheimer's Disease and Related Disorders
Springfield, Illinois, USA
Chapter 5

Kevin N. Hascup
Southern Illinois University School of Medicine
Department of Neurology
Center for Alzheimer's Disease and Related Disorders
Springfield, Illinois, USA
Chapter 5

Parastoo Hashemi
Department of Chemistry
383 A. Paul Schaap Building
Wayne State University
Detroit, MI, 48202, USA
Chapter 12
phashemi@chem.wayne.edu

William M. Howe
MS: Pfizer, Inc.
Cambridge, MA 02139, USA
Chapter 2

Peter Huettl
Center for Microelectrode Technology, University of Kentucky
College of Medicine, Lexington, KY, USA
Chapter 5

Michael A. Johnson
Department of Chemistry and
R.N. Adams Institute for Bioanalytical Chemistry
University of Kansas, Lawrence, KS 66045, USA
Chapter 11
johnsonm@ku.edu

Sam V. Kaplan
Department of Chemistry and
R.N. Adams Institute for Analytical Chemistry
University of Kansas, Lawrence, KS 66045, USA
Chapter 11

Lohitash Karumbaiah
Regenerative Bioscience Center
ADS Complex, 425, River Rd.
Athens, GA 30602-2771, USA
Chapter 16

Kristen A. Keefe
Department of Pharmacology and Toxicology
University of Utah
Salt Lake City, UT 84112, USA
Chapter 13

Polina Kosillo
Department of Physiology, Anatomy and Genetics
and Oxford Parkinson's Disease Centre
Sherrington Building
University of Oxford, OX1 3PT, UK
Chapter 9

Kendall H. Lee
Departments of Neurologic Surgery and
Physiology and Biomedical Engineering
Mayo Clinic, Rochester, MN 55905, USA
Chapter 10
lee.kendall@mayo.edu

Xianchan Li
Department of Chemical and Biological Engineering
Chalmers University of Technology
Kemivägen 10, SE-412 96 Gothenburg, Sweden
Chapter 8

Jelena Lovrić
Department of Chemical and Biological Engineering
Chalmers University of Technology
Kemivägen 10, SE-412 96 Gothenburg, Sweden
Chapter 8

Leyda Z. Lugo-Morales
Department of Chemistry
North Carolina State University
Raleigh, NC. 27695, USA
Chapter 6

Agatha E. Maki
Department of Chemistry and the Beckman Institute
University of Illinois at Urbana-Champaign
Urbana, IL 61801, USA
Chapter 15

Eric Naylor
Pinnacle Technology, Inc.
2721 Oregon Street
Lawrence, Kansas 66046, USA
Chapter 1

Erik B. Oleson
Department of Psychology, University of Colorado Denver
Denver, CO 80217-3364, USA
Chapter 11

Peter A. Petillo
Pinnacle Technology, Inc.
2721 Oregon Street
Lawrence, Kansas 66046, USA
Chapter 1
alchmist@pinnaclet.com

Francois Pomerleau
Center for Microelectrode Technology, University of Kentucky
College of Medicine, Lexington, KY, USA
Chapter 5

Jorge E. Quintero
Center for Microelectrode Technology, University of Kentucky
College of Medicine, Lexington, KY, USA
Chapter 5

Ashley E. Ross
Department of Chemistry,
University of Virginia, Charlottesville
VA 22904, USA
Chapter 4

Martin Sarter
University of Michigan
Department of Psychology
4032 East Hall
530 Church Street
Ann Arbor, MI 48109-1043, USA
Chapter 2
msarter@umich.edu

Tarun Saxena
Neurological Biomaterials and Cancer
  Therapeutics Laboratory
Wallace H Coulter Department of Biomedical
  Engineering at Georgia Institute of Technology
and Emory University School of Medicine
Atlanta, GA 30332, USA
Chapter 16

Leslie A. Sombers
Department of Chemistry
North Carolina State University
Raleigh, NC. 27695, USA
Chapter 6
leslie_sombers@ncsu.edu

Julie A. Stenken
Department of Chemistry and Biochemistry
University of Arkansas
Fayetteville, AR 72701, USA
Chapter 17
jstenken@uark.edu

Ingrid Strömberg
Department of Integrative Medical Biology
Umeå University, Umeå, Sweden
Chapter 5

Jonathan V. Sweedler
Department of Chemistry and the Beckman Institute
University of Illinois at Urbana-Champaign
Urbana, IL 61801, USA
Chapter 15
sweedler@scs.illinois.edu

Pooja M. Talauliker
Center for Microelectrode Technology,
   University of Kentucky
College of Medicine, Lexington, KY, USA
Chapter 5

Sarah Threlfell
Department of Physiology, Anatomy and Genetics,
and Oxford Parkinson's Disease Centre, Sherrington Building,
University of Oxford, OX1 3PT, UK
Chapter 9

B. Jill Venton
Department of Chemistry,
   University of Virginia, Charlottesville
VA 22904, USA
Chapter 4
bjv2n@virginia.edu

Kevin M. Wood
Department of Chemistry
383 A. Paul Schaap Building
Wayne State University
Detroit, MI, 48202, USA
Chapter 12

R. Mark Wightman
Department of Chemistry
University of North Carolina at Chapel Hill
Chapel Hill, NC 27599-3290, USA
Chapter 7
rmw@unc.edu

CHAPTER 1

USING BIOSENSORS TO PROBE FUNDAMENTAL QUESTIONS OF SLEEP

Erik Naylor and Peter A. Petillo

Pinnacle Technology, Inc.

1.1 INTRODUCTION

Sleep is still largely a mystery. Though it has been extensively studied over the centuries, the underlying purpose for sleep remains unknown (Zepelin et al., 2005). In this chapter, we will present an overview of sleep states, and describe how recent studies using biosensors are changing our understanding of sleep and the associated neuroenergetics (Dash et al., 2009; Naylor et al., 2011; Dash et al., 2012; Naylor et al., 2012; Wisor et al., 2012; Dash et al., 2013). Biosensors provide a window of unprecedented detail to visualize molecular-level processes on a near real-time basis and, as such, allow correlations between known sleep physiology with underlying molecular dynamics associated with sleep function.

1.2 SLEEP — A BENEFICIAL UNKNOWN

Two questions largely define modern sleep research: (1) what analytes and physiology drive the sleep homeostatic process (Achermann and Borbely, 2011) and (2) what molecular processes underlie physiological changes associated with disrupted sleep (Van Cauter and Tasali, 2011)? Answers to these questions remain outstanding. The most common method of measuring sleep relies on observational electroencephalography (EEG) activity (Carskadon and Dement, 2011), which provides vast amounts of data on the gross neuronal activity of sleep. However, EEG does not provide any correlation with discrete molecular level events and processes. The recent body of work using biosensors in conjunction with traditional sleep measurement is now providing new insights and avenues of exploration into these critical questions by probing the relationships between the physiology of sleep with the underlying molecular neuroenergetics (Dash et al., 2009; Naylor et al., 2011; Dash et al., 2012; Naylor et al., 2012; Wisor et al., 2012; Dash et al., 2013).

Physiological control of sleep is believed to be ruled by two fundamental systems: a homeostatic control system and the circadian timing system (Achermann and Borbely, 2011). The homeostatic control system governs "sleep pressure" which continually rises when the organism is awake and decreases during sleep (Dijk and Beersma, 1989). The circadian timing system governs permissible times for sleep onset and maintains waking during appropriate times (Edgar et al., 1993). In recent years, function and control of the circadian timing system have been greatly elucidated (see reviews in Huang et al., 2011; Albrecht, 2012), but the analyte(s) and/or physiology underlying the homeostatic sleep system remain largely unknown.

Without sleep, mammals die (Everson et al., 1989). Rats subjected to continuous sleep deprivation perished after as little as two weeks (Everson et al., 1989) due to extensive bacterial infection (Benca et al., 1989; Bergmann et al., 1996; Everson, 2005). Likewise, humans afflicted with Fatal Familial Insomnia also die following a severe, extended inability to sleep (Goldfarb et al., 1992; Manetto et al., 1992). While long-term sleep deprivation is fatal, sleep deprivation lasting as little as three days in rats can result in extreme physiological effects including hyperphagia, hyperdipsia, weight loss (Everson, 1995), altered thyrotropin-releasing hormone (Everson and Nowak, 2002), increased markers of inflammation (Yehuda et al., 2009), dysfunctional temperature regulation (Bergmann et al., 1989), and alterations in bone formation (Everson et al., 2012). In addition to the well-known effects of reduced alertness and poor mood (Van Dongen et al., 2003), sleep loss in humans can also include negative impacts to immune function (Spiegel et al., 2002) along with increased risk for heart attack (Meier-Ewert et al., 2004) and diabetes (Spiegel et al., 2005). Despite a wealth of knowledge regarding the physiological effects of sleep loss, a lack of understanding of the molecular processes underlying these effects remains.

1.3 ELECTROENCEPHALOGRAPHIC SLEEP MEASUREMENT

EEG recordings of the brain are unambiguously the gold standard for differentiation of vigilance states, providing an unparalleled view of gross neuronal activity (Carskadon and Dement, 2011), and exceptional real-time measures of gross neuronal firing. Limitations of this method make it impossible to determine anything more than cortical surface activity, and as such, EEG measurements provide no insight into the activity of molecular processes such as neurotransmitter activity and neuroenergetic demands (Carskadon and Dement, 2011).

The underlying physiology of sleep was first explored during the first part of the 20th century when researchers used EEG to examine human brainwaves (Loomis et al., 1936). The EEG made it possible to record brain activity over the course of night, and, for the first time, quantify physiological differences between waking and sleep. EEG traces reflect microvolt potential differences between electrodes placed on the head. Potential differences ranging in size from 50 to 200 μV arise as a result of large-scale changes resulting from neuronal firing activity. As detailed in Figure 1.1, high frequency (> 12 Hz) EEG waves represent active, disparate neuronal activation associated with the waking state. High amplitude (> 100 μV) and low frequency (< 5 Hz) EEG patterns, observed during non-rapid eye movement (NREM) sleep, represent cortical neurons firing with a synchronized rhythm. EEG traces demonstrating a large portion of 5–7 Hz frequency waveforms with amplitudes < 100 μV are typical of rapid eye movement (REM) sleep.

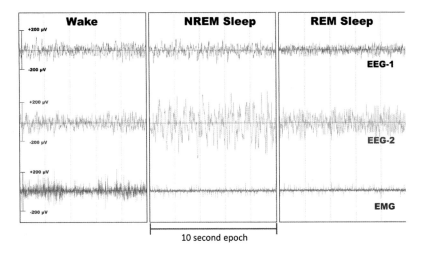

Figure 1.1 Representative waveforms for each of the three stages of rodent sleep. Wake epochs are characterized by high amplitude EMG activity coupled with high frequency (>12 Hz), low amplitude EEG activity. NREM sleep epochs are characterized by high amplitude (>100 µV), low frequency (<5 Hz) EEG activity, paired with a low amplitude EMG trace. REM sleep epochs are characterized by an abundance of 5–8 Hz frequencies (theta) in the EEG traces, whose amplitudes are <100 µV, and are coupled with a tonic EMG signal. Scoring of discrete epochs has historically been accomplished by visual recognition of these frequency and amplitude changes by a trained human scorer. Individual epochs are joined together to constitute episodes of wake, NREM sleep and REM sleep. While wake and NREM sleep episodes can be >15 min in length, REM sleep episodes are typically less than 3 min in duration. Reprinted from J. Electroanal. Chem, Vol. 656, E. Naylor, et al., Simultaneous real-time measurement of EEG/EMG and L-Glutamate in mice: A biosensor study of neuronal activity during sleep., pp. 106-113. Copyright 2011, with permission.

In addition to EEG, accurate determination of sleep state in mammals requires concurrent measurement of electromyograph muscle activity (EMG) (Steriade, 2005). EMG activity, measured by placing two independent electrodes in or around muscles, records the electrical potential arising from muscle contraction. EMG with a high amplitude reflects active movement typical of an awake animal, while a tonic EMG signal indicates quiescence or sleep (Carskadon and Dement, 2011).

Sleep onset is considered to be a gradual process, but not all sleep → wake and wake → sleep transitions take place on the same time scale. Some sleep stage transitions, such as moving from waking into NREM sleep or from NREM into REM sleep take place over a period of seconds to minutes (or longer if one suffers from insomnia) as observed on cortical EEG recordings, while some transitions, such as moving from NREM or REM sleep to waking, are virtually instantaneous. Since the molecular events that drive and control neuronal firing are not directly discernible by EEG measurements, alternate techniques are required to accurately monitor neurochemical release and uptake. Such techniques ideally target specific structures within the brain to complement electrophysiological studies and are now recognized as an important component to investigating the role of sleep (Zepelin et al., 2005).

Only recently have the neurotransmitter pathways driving the underlying physiology of state transitions been investigated (Saper et al., 2005). Prior to the introduction of biosensor technology, the only method to directly sample chemicals during the sleep process was through the use of microdialysis. While microdialysis experiments have been and continue to be important, biosensor sampling rates are generally faster, and more importantly, on the same time scale as the physiologic transitions being probed.

1.4 EARLY INVESTIGATIONS OF SLEEP PHYSIOLOGY

Microdialysis has been instrumental in the discovery and characterization of many substances involved with the sleep process (Chefer et al., 2009). Sleep-related fluctuations of many neurotransmitters such as acetylcholine (Kametani and Kawamura, 1990; Jones, 1991), serotonin (Portas et al., 1998), norepinephrine (Shouse et al., 2000), dopamine (Lena et al., 2005), GABA (Nitz and Siegel, 1997), glutamate (Watson et al., 2011), histamine (Chu et al., 2004) as well as other important compounds such as hypocretin and adenosine that may underlie the physiological changes during sleep have been identified using this technique.

The utility of microdialysis as a sampling technique is unquestioned. However, the need to physically remove molecular material from the brain both limits the number of samples that can be obtained within a given time period and may disrupt the equilibrium of these compounds within the brain (Chefer et al., 2009). Microdialysis samples during sleep are typically collected at five-minute intervals (Kametani and Kawamura, 1990; Portas et al., 1998; Chu et al., 2004; Lena et al., 2005), although intervals as small as one minute have been documented (Lopez-Rodriguez et al., 2007). As highlighted in the previous section, slow sleep transitions take place within 1–2 minutes, while more rapid changes take place in seconds. Using microdialysis, it has not been possible to reliably sample the neurochemical changes occurring during fast transitions that may be occurring on a second-by-second timescale (Naylor et al., 2011, 2012).

1.5 BIOSENSOR RECORDING DURING SLEEP

1.5.1 BIOSENSOR BASICS

For the rapid sampling needed to examine neurochemical changes and thus, neuroenergetic changes as a function of sleep state, biosensor measurements have proven more useful and reliable than other analytical techniques. Biosensors sample faster than traditional microdialysis experiments, with data typically collected at 1 Hz. Biosensors are analyte specific and can continuously record changes in concentration of an analyte over multiple days, and do not require any post-experiment sample analysis (Burmeister and Gerhardt, 2001; Rutherford et al., 2007).

An advantage and potential limitation of biosensors is the requirement of a biological recognition element *specific* to the detection analyte. This biological recognition element is typically an oxidase enzyme producing an electroactive species, which is then oxidized at the transducing element (Figure 1.2) (Lena et al., 2005; Wilson and Johnson, 2008; Naylor et al., 2011, 2012). Using the L-lactate biosensor as an example, a Pt–Ir electrode polarized at 0.6 V vs. an Ag/AgCl reference, is coated in a matrix containing the biological element L-lactate oxidase. L-lactate in the extracellular space surrounding the biosensor is converted to hydrogen peroxide (H_2O_2) by L-lactate oxidase in the sensor matrix. The enzymatically produced H_2O_2 then diffuses through the membrane layers to the Pt–Ir surface where it is oxidized to produce a measureable amperometric current. The amount of peroxide produced is directly proportional to the concentration of L-lactate, making it an excellent method for real-time detection of L-lactate within the extracellular area of the brain surrounding the biosensor. Recent advances in miniaturization and recording technology have made possible the simultaneous recording of cortical EEG alongside one or more implanted biosensors (Dash et al., 2009; Naylor et al., 2011; Dash et al., 2012; Naylor et al., 2012; Wisor et al., 2012; Dash et al., 2013).

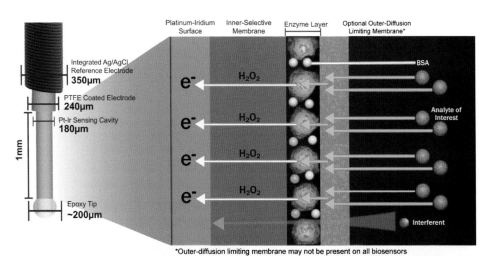

Figure 1.2 Schematic of a working biosensor, including biosensor dimensions. The analyte of interest, in the presence of water and oxygen, is converted into H_2O_2 and by-products. The enzymatically-produced H_2O_2 diffuses through the selective membrane to the Pt–Ir surface where it is oxidized at a potential of 0.6 V vs. Ag/AgCl. While it is not possible to eliminate 100% of the electroactive interferents, the inner-membrane and active removal effectively limit the amount of interferent that reaches the electrode surface such that any changes in the local concentration are insignificant and do not contribute to the observed current.

1.5.2 BIOSENSOR STUDIES DURING SLEEP

Biosensors provide high-resolution, real-time data regarding concentration changes of specific *in vivo* analytes. These characteristics are ideal to probe relationships between gross neuronal physiological activity and the rapid changes in extracellular analyte concentration. Biosensor measurements also allow quantification of neuroenergetic and neurotransmitter changes on a time scale that is concomitant with the physiology associated with most sleep transitions. In a direct comparison of glutamate concentration sampled by microdialysis to glutamate sampled by biosensors within the posterior hypothalamus/tuberomammilary nucleus region, significant increases in glutamate concentration during REM sleep in this region were recorded by a biosensor, but no such increase was noted within the microdialysis samples (John et al., 2008). Based on this finding the authors concluded, "This lack of state-related glutamate change could be due to the greater temporal and spatial resolution of the glutamate biosensor compared to microdialysis." In short, biosensors are unique and powerful analytical tools that can dissect the unknown physiological functions driving the requirement to sleep.

Studies that have used biosensors to probe sleep–wake transitions have quantified the changes in glucose, lactate, glutamate, and adenosine upon physiologic state change. These important neuroanalytes all play a role in controlling or responding to changes in sleep state. The fidelity with which biosensors are able to measure changes in these analytes as a function of time and sleep state has proven that biosensors are an important analytical tool. The question of "what can I learn from a biosensor measurement" with respect to sleep, is now at the forefront of efforts to link neuroenergetics with the underlying physiology of sleep (Wilson and Johnson, 2008).

1.5.2.1 Glucose

Glucose is the primary energy molecule delivered to the brain. Use of glucose by the brain can reach as much as 25% of the total glucose consumed by the organism (Magistretti et al., 1995). Despite this high maximum level of consumption, glucose concentrations in the brain fluctuate greatly during both waking and sleep. The first sleep investigations that used biosensors attempted to correlate glucose and sleep/wake state. These studies concluded that, when all NREM sleep epochs were averaged together, glucose concentration demonstrated an overall increase during NREM sleep periods, and an overall decrease during REM sleep as compared to waking (Netchiporouk et al., 2001).

These findings have been duplicated and refined using biosensors in both mouse (Naylor et al., 2012) and rat (Dash et al., 2013) models. These later experiments measured the change in glucose concentration at one-second intervals that were directly associated with EEG-based sleep/wake states. As reported in both studies and shown in Figure 1.3, extracellular brain glucose initially declines after waking. As the animal remained awake, glucose concentration decreased by 60–90 μM. After 10–15 minutes of continuous waking, this decline reverses, returning to or above pre-waking levels. During REM sleep, glucose concentration follows a similar pattern of declining at REM episode onset. Biosensor recordings also showed that this decline occurs very close to the onset of the REM sleep transition and persists throughout the REM episode (Naylor et al., 2012; Dash et al., 2013). However, because REM sleep in rodents seldom lasts longer than 4–5 minutes (Netchiporouk et al., 2001), this decline may not always be within a statistically detectable range.

When considered together, these findings suggest that the initial transition from a low neuronal firing state (NREM sleep) to a higher firing state (WAKE or REM) increases the overall energy demand that results in an immediate decline in the available extracellular glucose. The extra glucose demanded by active neuronal and glial cells is fulfilled by an increase in supply resulting in a return to equilibrium within several minutes.

In contrast, entry into NREM sleep in rodents produces an almost immediate rise in glucose concentrations, increasing by as much as 100 μM depending on the length of the sleep period (Naylor et al., 2012; Dash et al., 2013). Correlation

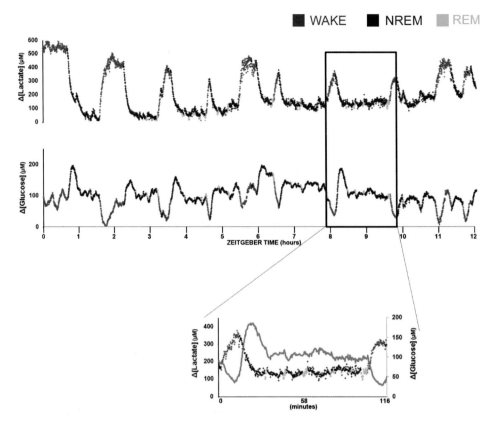

Figure 1.3 Multiple sleep/wake cycles recorded using simultaneous EEG and lactate/glucose biosensors plotted during the lights-on period. Epochs scored as wake are noted in red, non-rapid eye movement (NREM) sleep epochs are colored blue, and rapid eye movement (REM) sleep epochs are indicated in green. Concentration change for each analyte is indicated on the y-axis. The expanded inset (lower graph) corresponds to the time period on the upper graph indicted by the solid box. In the expanded graph, lactate concentration change is plotted in a manner similar to that of the large-scale graph with colors indicating sleep/wake state and glucose plotted as an overlay in orange. Reprinted from Sleep, Vol. 35(9), E. Naylor, et al., Lactate as a biomarker for sleep, pp. 1209-1222. Copyright 2012, with permission.

between this glucose rise and homeostatic sleep pressure has been investigated but, to date, no direct correspondence has been identified (Dash et al., 2013). It is likely that this increase is related to the neuroenergetics surrounding the much slower neuronal firing of NREM sleep. A decreased neuronal energy demand paired with the previously up-regulated supply necessary for waking increases the extracellular glucose concentrations for a time before cellular regulatory mechanisms such as the glucose transporter 1 (Glut1), a primary channel responsible for moving glucose across the blood brain barrier (Vannucci et al., 1997; Simpson et al., 2007), restore flow equilibrium.

As a primary energy component in the brain, molecular regulation of glucose is clearly an important component of sleep/wake transitions. Glucose use by cells changes in response to altering neuronal firing patterns. However, these transitory changes exhibit a consistent return to equilibrium if the changed state lasts more than 10 minutes.

The primary criteria for any substance to be classified as a putative sleep regulatory substance is that, "the substance and/or its receptor oscillates with sleep propensity" (Clinton et al., 2011). Another way to look at this correlation is that any substance involved with regulation of the sleep homeostat must first meet the criteria of a biomarker for sleep. The measured alterations in glucose concentration do not meet the criteria of a biomarker, nor do they correspond to known models of standard sleep homeostatic components. Therefore, glucose cannot be considered to directly affect control of the sleep homeostat. However, given the evidence that one of the well-known effects of sleep loss is an alteration of the body's response to a glucose challenge (Spiegel et al., 2005), it may be that glucose alterations during sleep transitions are also altered in cases of extended sleep deprivation and that they may be involved with regulating some of the adverse effects of sleep deprivation.

1.5.2.2 Lactate

Use of biosensors to measure lactate concentration in the brain in real time have revealed some of the most exciting and revolutionary findings surrounding the sleep process (Hu and Wilson, 1997; Shram et al., 2002; Burmeister et al., 2005). Increased brain lactate concentrations have traditionally been associated with a waking state (Shimizu et al., 1966; Cocks, 1967; Reich et al., 1972). However, in these early studies, lactate concentration was determined *post hoc* using whole-brain samples. More recent studies have greatly enhanced our understanding of lactate dynamics during sleep by pairing lactate biosensors with EEG activity (Dash et al., 2012; Naylor et al., 2012; Wisor et al., 2012) and with the simultaneous measurement of other neuroenergetically important analytes (Naylor et al., 2012). In all studies to date, extracellular lactate concentration is observed to *rise upon waking* (Figure 1.3). This rise is consistent, measurable, and calculated at a rate between 10 and 20 μM/min (Dash et al., 2012; Naylor et al., 2012; Wisor et al., 2012). This rate appears to be relatively constant regardless of whether the animal is active or not (Dash et al., 2012) or the species of rodent examined. These wake-dependent increases do not continue indefinitely, and plateau after 15 minutes of wakefulness at levels 150–200 μM higher than those observed during a sleep state (Naylor et al., 2012). This rise in lactate concentration is exceptionally robust, occurring in every instance of waking, remaining significantly elevated in

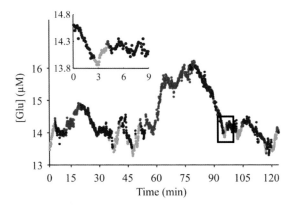

Figure 1.4 Dynamic changes in extrasynaptic glutamate concentration during waking, NREM and REM sleep. Each data point represents the average concentration of glutamate concentration across a 4s window of waking (red), NREM (blue), or REM (green). The inset shows a close-up of a NREM-to-REM transition. Reprinted from J. Neuroscience, Vol. 29(3), M.B. Dash, et al., Long-term homeostasis of extracellular glutamate in the rat cerebral cortex across sleep and waking states. pp. 620-629.

all natural wake periods greater than five minutes (Naylor et al., 2012) and continuing to remain elevated throughout extended periods of enforced sleep deprivation (Van den Noort and Brine, 1970; Kalinchuk et al., 2003; Wigren et al., 2009; Naylor et al., 2012).

Sleep onset results in a reversal of the pattern observed during waking. As detailed in Figure 1.3, extracellular lactate concentration in the brain rapidly declines, coinciding with NREM sleep onset (Dash et al., 2012; Naylor et al., 2012; Wisor et al., 2012). As with the sleep → wake transition, the changes in lactate levels observed for the wake → sleep transitions are also species independent. Lactate levels remain low throughout the NREM sleep period, due to lower levels of cellular metabolism resulting from reduced overall neuronal firing. Unlike the rate of change in glucose concentration observed with the onset of NREM sleep, the rate at which lactate concentration falls during NREM sleep appears to be proportional to the magnitude of the EEG activity at frequencies <10 Hz (Wisor et al., 2012). This change indicates that the rate of lactate decline during NREM sleep may be directly associated with slow-wave activity and, to some extent, changes in sleep pressure (Dash et al., 2012).

Despite overall depressed levels of lactate during NREM sleep, the onset of REM sleep triggers a strong, secondary rise in lactate concentration (Shram et al., 2002; Dash et al., 2012; Naylor et al., 2012; Wisor et al., 2012). This rise continues throughout the REM sleep episode. Due to the shortened nature of REM sleep in rodent models, overall changes in lactate levels do not approach those observed during waking.

Recent findings based on these studies have determined that extracellular brain lactate concentration rises and remains elevated during waking and declines and remains depressed with NREM sleep onset (Dash et al., 2012; Naylor et al., 2012; Wisor et al., 2012). These findings, replicated by multiple, independent researchers are a strong indicator of the important, yet heretofore unknown role of lactate in the wake/sleep process. No other substance to date displays such a rapid, robust and reproducible change that is directly correlated with EEG activity.

Based on these recent biosensor studies, lactate overwhelmingly meets the fluctuating criteria of a sleep homeostatic component and acts as a clear biomarker for sleep and wake. Furthermore, and given the exceptionally strong correlation between lactate and sleep/wake state, there is strong evidence that lactate is a biomarker of the sleep homeostatic process (Dash et al., 2012; Naylor et al., 2012; Wisor et al., 2012). Further studies on the true relationship between lactate and the sleep homeostat will need to focus on changes in sleep with the exogenous application or withdrawal of lactate. Although it has been shown to be a strong biomarker for sleep, further work needs to be done to better elucidate to what degree lactate may actually govern the sleep homeostat. Detailed genetic and metabolic studies must be undertaken to better define the role of lactate, or metabolites thereof, on known sleep regulatory circuits.

1.5.2.3 Glutamate

Glutamate is used by ~80% of all neurons in the cerebral cortex (Jones, 2009) and is considered the primary excitatory neurotransmitter in the brain. Measured in conjunction with sleep and waking states, microdialysis sampling of glutamate consistently reports that levels of this neurotransmitter are highest during waking throughout multiple brain regions (Azuma et al., 1996; Lena et al., 2005; Lopez-Rodriguez et al., 2007; Wigren et al., 2007; Watson et al., 2011).

The earliest reported sleep study to employ glutamate biosensors targeted three distinct brain regions (the posterior hypothalamus/tuberomammary nucleus, perifornical-lateral hypothalamus, and the cortex) to simultaneously record glutamate fluctuations in conjunction with cortical EEG recordings. Similar to earlier microdialysis findings, the authors reported significant increases of glutamate during both active waking and REM sleep in all three regions (John et al., 2008). Later biosensor studies confirmed and extended this finding to include the motor cortex in rats (Dash et al., 2009) and the prefrontal cortex in mice (Naylor et al., 2011). The strong agreement of the recent data suggests that, overall, glutamate concentration undergoes a significant increase during both waking and REM sleep and declines during NREM sleep (Figure 1.5). It has been noted that the rate of glutamate decline during NREM sleep corresponds very closely to the

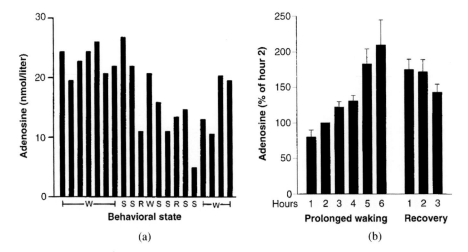

Figure 1.5 a) Adenosine concentrations in 10-min consecutive samples from an individual microdialysis probe in the basal forebrain. Labels indicate the predominant behavioral state: W, wakefulness; S, slow wave sleep [NREM sleep]; and R, REM sleep. b) Prolonged wakefulness and recovery sleep. Mean extracellular adenosine values increased in the basal forebrain during 6 hours of prolonged wakefulness [0900 to 1500; repeated measures of the analysis of variance (ANOVA) between treatments gave values of $F(8, 5) = 7.0$ and $P < 0.0001$, and the paired t-test between the second and the last hour of wakefulness gave values of $t(5) = 3.14$ and $P<0.05$]. The adenosine values decreased in the subsequent 3 hours of spontaneous recovery sleep (n = 56). Values are normalized relative to the second hour of deprivation (due to technical problems, three first-hour values were missing). Reprinted from Science, Vol. 276(5316), T. Porkka-Heiskanen, et al. Adenosine: A mediator of the sleep-inducing effects of prolonged wakefulness. pp. 1265-68. Copyright 1997, with permission.

decline in slow-wave activity (Dash et al., 2009). This decline may be related to the lower overall firing rate of glutamatergic neurons during this period and may also hint that changes in glutamate levels contribute to the sleep homeostatic load.

Biosensor studies on the relationship between glutamate and the sleep process largely correspond with microdialysis (Lena et al., 2005; John et al., 2008; Watson et al., 2011) and brain slice (Mohammed et al., 2011) studies. Extracellular glutamate is shown to rise with waking increasing at rates between 0.034 and 0.055 μM/min (Dash et al., 2009; Naylor et al., 2011; Naylor et al., 2012) before plateauing around a maximum 1.0 μM increase over baseline levels. This trend is reversed with NREM sleep onset with glutamate exhibiting a steady decline at a similar rate (0.028–0.056 μM/min) (Dash et al., 2009; Naylor et al., 2011; Naylor et al., 2012). While these levels do eventually plateau, reaching a steady-state change of ~1.0 uM, the decline is gradual and may not be fully realized until 40 minutes following a transition (Naylor et al., 2011; Naylor et al., 2012).

The overall pattern of glutamate change with sleep and wake is similar to that described by lactate suggesting that glutamate may also meet the biomarker criteria of a substance influencing homeostatic sleep regulation. However, measured glutamate concentrations were more variable and slower to respond to the physiologic state, typically taking five minutes or longer to reach the same level of change as lactate. Furthermore, the dynamic range of the overall change in glutamate concentration is at least 100× lower than that for lactate. Although glutamate is the most common neurotransmitter, it is not found ubiquitously throughout the brain (Jones, 2009). Combined with the slower rate of rise and decline exhibited by glutamate when compared to lactate within the same animal (Naylor et al., 2012), glutamate is less attractive as a candidate molecule directly responsible for homeostatic sleep regulation. Some recent studies indicate that sleep restriction alters the metabotropic glutamate receptors (Tadavarty et al., 2011; Hefti et al., 2013), suggesting that changes in glutamate physiology may underlie some of the detrimental physiological effects of sleep deprivation.

1.5.2.4 Adenosine

Neuronal activity increases energy consumption, reducing ATP ultimately to adenosine and resulting in increased extracellular levels of adenosine (Porkka-Heiskanen and Kalinchuk, 2011). This relationship has made adenosine perhaps the most studied substance to date directly implicated as a driver of the sleep process (Borbely and Tobler, 1989; Benington and Heller, 1995; Alam et al., 1999; Basheer et al., 1999; Porkka-Heiskanen et al., 2000; Dworak et al., 2010; Florian et al., 2011; Porkka-Heiskanen and Kalinchuk, 2011).

Porkka-Heiskanen et al. (1997) used EEG recording in conjunction with adenosine sample collection at 10-minute intervals in a feline model. They found brain adenosine levels fluctuated with behavioral state by continually increasing during periods of wakefulness and decreasing during sleep (Figure 1.5a). A six-hour period of forced wakefulness also resulted in a continual increase in adenosine levels with no observable plateau and then declining levels during subsequent recovery sleep periods (Figure 1.5b). More recently, experiments involving simultaneous *in vivo* microdialysis adenosine sampling and EEG recording have been undertaken using a rat model (Wigren et al., 2007; Kalinchuk et al., 2011). Both studies confirmed previous findings that extracellular adenosine continually increases in the basal forebrain during the first six hours of sleep deprivation. By extending the period of sleep deprivation to 11 hours, Kalinchuk et al. (2011) discovered adenosine concentration in the basal forebrain reaches a plateau after seven hours and that adenosine concentration in the frontal cortex also rises to significant levels after six hours of sleep deprivation. This much greater timespan before reaching a plateau as well as differing rates of concentration in different brain regions seen with adenosine stands in sharp contrast to the profile observed for lactate and glutamate under the same conditions.

Despite the perceived importance of adenosine as a driver of the sleep process, virtually all of the *in vivo* studies to date have utilized microdialysis. Current biosensors for adenosine concentration are based on either multienzyme cascades (Llaudet et al., 2003) or use fast scan cyclic voltammetry (FSCV) (Pajski and Venton, 2013) to detect changes in adenosine concentration. All biosensor and FSCV studies published to date have probed adenosine levels only in rodent brain slice preparations (Schmitt et al., 2012; Sims et al., 2013; Pajski and Venton, 2010, 2013).

The measurement of adenosine with the current generation of biosensors is also complicated by the need to control for signals from related purines, which is less of a problem for the corresponding microdialysis study. Adenosine biosensors described to date use multienzyme cascades that include nucleoside phosphorylase, xanthine oxidase, and adenosine deaminase. This enzyme cascade functions by breaking down adenosine to inosine, inosine to xanthine, and xanthine to uric acid and H_2O_2. Adenosine biosensors of this construction (ADO) are also reactive to both inosine and hypoxanthine (Llaudet et al., 2003). To control this reaction, sensors made in a similar manner but lacking adenosine deaminase (INO), are typically employed alongside standard adenosine sensors. In brain slice experiments, these two sensors are positioned close to one another, and require the signal from the INO sensor to be subtracted from current measured by the ADO sensor to determine the signal due solely to adenosine (Schmitt et al., 2012; Sims et al., 2013). The fabrication of a reliable adenosine biosensor based on a single enzymatic reaction has yet to be described. In principle, such a sensor would obviate the need to account for the presence of inosine.

In brain slices, both the basal adenosine tone and the evoked adenosine release exhibited diurnal variation, showing depressed levels during periods when the animal is typically asleep. Both of these measures were significantly higher when measured after a period of sleep deprivation, and demonstrated a reversal when treated with the inducible nitric oxide synthase (iNOS) inhibitor 1400W, suggesting that the wake-related adenosine increase is, in part, mediated by iNOS release. Because these measures were performed at discrete time points using brain slice preparations, it is not possible to quantify the rate of concentration change following discrete sleep/wake transition points as seen in studies of freely moving animals. Unlike the glucose, lactate and glutamate biosensors described above, the useable lifetime of the current generation of adenosine biosensors is limited to a few hours and a design suitable for implantation and use in a freely moving animal is not yet available.

Measurement of adenosine in brain slices has demonstrated precise increases in the extracellular hippocampal levels of adenosine in response to both normal wakefulness and sleep deprivation (Schmitt et al., 2012; Sims et al., 2013). Taken in combination with the wealth of available microdialysis data, adenosine is clearly a candidate for an analyte responsible for sleep homeostatic regulation. Adenosine not only meets the primary criteria for a homeostatic sleep component by changing with sleep and wake, but also the exogenous application of both adenosine (Virus et al., 1983) and adenosine A1 receptor agonists (Benington et al., 1995) to the basal forebrain increase sleep propensity in animals while A1 receptor blockade using cyclopentyltheophylline decreased sleep and increased wakefulness (Strecker et al., 2000).

Although biosensor studies involving adenosine are relatively recent, measurements of *ex vivo* brain adenosine concentration has demonstrated utility and will likely be continued to further investigate this critical analyte in the sleep process. Nonetheless, a truly useful adenosine biosensor for sleep recording will require a single-enzyme solution that can record for more than 24 hours from within a freely moving animal.

1.6 BIOSENSORS AND THE ASTROCYTE NEURON LACTATE SHUTTLE HYPOTHESIS

Astrocytes play a critical role in the regulation of brain metabolic responses to activity. One detailed mechanism proposed to describe the role of astrocytes in some of these responses has come to be known as the astrocyte-neuron lactate shuttle (ANLS) hypothesis (Pellerin and Magistretti, 1994; Pellerin et al., 2007; Petit et al., 2010, 2013). Although controversial, the original concept of a coupling mechanism between neuronal activity and glucose utilization that involves an activation of aerobic glycolysis in astrocytes and lactate consumption by neurons provides a heuristically valid framework for experimental studies (Pellerin et al., 2007). Furthermore, the ANLS hypothesis provides a framework wherein electrophysiological data and biosensor data may be directly compared and integrated into a composite model of sleep (Naylor et al., 2012).

Action potentials and post-synaptic processing are estimated to account for more than 70% of the total energy consumed by neurons with estimates of signal-related energy use to be as high as 30–50 μM adenosine triphosphate/g/min (Attwell and Laughlin, 2001). The primary energy analytes that are needed to help support neuronal action potentials upon waking are glucose (the primary energy source), lactate (the critical intermediate energy store), and glutamate. Multiple biosensor and microdialysis studies have independently examined changes in lactate (Shram et al., 2002), glucose (Netchiporouk et al., 2001; Johnson et al., 2007) and glutamate (John et al., 2008; Dash et al., 2009; Naylor et al., 2011) concentrations at various time points across the sleep/wake cycle, and the results are largely in agreement. Biosensor studies have added an important temporal contribution supporting the earlier investigations into brain lactate (Richter and Dawson, 1948;

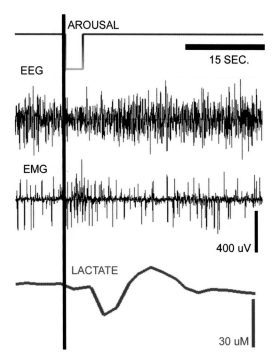

Figure 1.6 Sample trace of the lactate response to a brief arousal from NREM sleep. Sleep/wake state is summarized at the top with EEG and EMG activity plotted directly below and the lactate concentration measured during this event as a blue trace.

Shimizu et al., 1966) and glucose (Van den Noort and Brine, 1970) concentrations during sleep. It is expected that when biosensor studies on adenosine are available in freely moving animals, these results will also be in agreement with earlier studies, but with added temporal information that may help to further dissect the sleep-wake process and the sleep homeostat.

The idea that lactate, derived primarily from glucose, can act as a primary energy source during periods of neuronal activity is not a new concept (Pellerin and Magistretti, 1994; Pellerin et al., 2007; Petit et al., 2010; Suzuki et al., 2011; Petit et al., 2013). The corresponding development of biosensors optimized for *in vivo* use has allowed the codification, at the molecular level, of the underlying neuronal energy demands and their relation to physiological processes. Biosensor data reviewed herein are consistent with the predictions of the ANLS hypothesis in that extracellular lactate concentrations increase during periods of heightened neuronal activity (waking and REM sleep) and decrease during periods of neuronal quiescence (NREM sleep). The clear demarcation of physiologic state change between waking, NREM and REM sleep provides a unique model for relating *in vivo* neuronal activity with changes in neuroenergetic analytes.

The rapid sampling provided by biosensors has also allowed for analyte changes to be quantified during periods when the physiologic state is in flux. Furthermore, the observed changes in lactate and glucose concentration as a function of sleep state transition represent a critical *in vivo* demonstration of the ANLS hypothesis in a live animal model. Biosensor studies during periods of heightened neuronal activity (e.g. waking and REM sleep) have clearly established that lactate is a key energy analyte supporting increased neuronal firing. In turn, these studies have also provided clear *in vivo* evidence supporting the validity of the ANLS hypothesis.

The lactate concentration rises to significant levels within one minute of waking (Wisor et al., 2012) and entry into REM sleep (Naylor et al., 2012) then declines with sleep onset at a rate corresponding to slow-wave intensity (Dash et al., 2012; Wisor et al., 2012). Such a high degree of temporal detail allows examination of analyte activity during sleep events that are not measurable by other means. In contrast, even the most rapid microdialysis sampling methods would only reveal a rise in lactate only after it was well underway.

The transition from NREM sleep to waking and back into NREM sleep within a period of less than one minute is a common occurrence during rodent sleep. Because of the brief and unpredictable nature of these transition events, measurement of analyte activity is only possible through continuous monitoring followed by post-event identification and measurement. As shown in Figure 1.6, lactate concentration changes during a brief arousal event exhibit a significant dip immediately after onset before returning to baseline. This observation directly demonstrates that any period of increased neuronal activity results in the immediate uptake of lactate from the extracellular space. As shown by Naylor et al., *this is true for all NREM $\rightarrow$ wake transitions* (Naylor et al., 2012). This study was also the first to directly quantitate changes in lactate, glucose and glutamate during the first seconds of a transition. The initial drop in lactate concentration, corresponding to neuronal uptake, occurs within seconds of the onset of neuronal activity and before any subsequent, sustained lactate rise — an observation which directly supports the ANLS hypothesis. The Naylor study demonstrates the importance of 1Hz data, and these results further suggest the need for even faster biosensor sampling (10Hz or above) to further quantitate analyte concentration changes at physiologic transitions.

Data collected from bilateral lactate and glucose biosensors clearly demonstrate both a decrease in glucose and simultaneous increase in lactate *at every waking threshold*. The increase in lactate is always preceded by an initial decrease in lactate before the sustained increase is observed, and is true for NREM $\rightarrow$ wake and NREM $\rightarrow$ REM transitions (Naylor et al., 2012). The simultaneous recording of two biosensors (lactate + glucose) from within the same animal (Naylor et al., 2012) using probe placement in contralateral hemispheres showed changes in both lactate and glucose concentrations within the same mouse at the same time. These changes were directly related to state changes as recorded by the EEG (Figure 1.3). The increased demand for additional energy is evidenced by the significant increase in firing activity, and this demand is clearly being met, at least in part, by the rapid availability of lactate into the extracellular space.

The Naylor study (2012) demonstrates the importance of measurement of multiple, simultaneous analytes. In this experiment, two analytes were measured from contralateral hemispheres of the brain. More advanced cannula designs now allow the measurement of up to three simultaneous analytes from a single region or across multiple brain nuclei (Aillon et al., 2013). Rapid biosensor measurement of multiple analytes in this manner provides critical information about real-time neurochemical interactions providing heretofore unavailable insights regarding neurochemical pathway function during the sleep process.

1.7 CONCLUSION

One clear, important advantage of biosensors is the ability to reveal the dynamics of analytes in real time. Biosensor data probing brain analytes during discrete phases of sleep have already revealed important associations between sleep state changes and neuroenergetic components (glucose and lactate) as well as extended our understanding of sleep states and neurotransmitters (glutamate). This clarity provides researchers with a direct window into brain analyte concentration changes *as they occur* and holds the promise to revolutionize current experimental paradigms designed to answer the fundamental questions of sleep.

Given the evidence presented by lactate and glucose changes with sleep state, the first question, "to what extent, if any, might these influence the sleep homeostatic process?" can be addressed. Even during periods of increased extracellular glucose (as seen after approximately 10 minutes of sustained waking) lactate concentration remains elevated. This continued elevation in lactate suggests a *continued* high demand for oxidative energy analytes even during periods of increasing glucose supply. It has been suggested that energy demand is only partially met by available glycolytic mechanisms while the remainder, possibly as much as 60%, relies on increased lactate availability (Boumezbeur et al., 2010).

In addition to the increased firing activity already discussed, there is a great deal of structural re-organization involved with waking such as lipid bilayer expansion, protein synthesis and enlargement of synaptic sites. Recent studies suggest that the waking state increases net synaptic potentiation while sleep promotes synaptic consolidation and depression (Vyazovskiy et al., 2008; Liu et al., 2010). These are all processes requiring an enormous amount of sustained energy availability. It is not unreasonable to suggest that at least one, if not multiple, factor(s) driving the sleep homeostat consist of analytes related to this energy consumption. Both lactate and adenosine remain high during periods of wakefulness and decline during NREM sleep. However, whereas adenosine has had the benefit of over 15 years of research, the tight association of lactate and sleep cycling is more recent. Like adenosine, lactate is clearly a biomarker for sleep/wake, and meets the first criteria of putative sleep regulatory substance. Further work is now needed to establish if lactate is indeed a component of the sleep homeostatic process.

The second question of molecular processes underlying sleep-deprivation induced physiology is more complicated, but clues from the studies presented herein are now revealed. The heightened energy demands of normal waking are further increased during sleep loss. In the two biosensor studies which have examined enforced sleep loss, lactate concentration remained elevated throughout the entire period of sleep deprivation (Dash et al., 2012; Naylor et al., 2012). This continued elevation, paired with inadequate replenishment of lost reserves through sleep deficits, provides a possible pathway wherein lack of sufficient molecular energy supply may be leading to decreased neuronal branching, decreased synapse formation and/or overall neuronal depression. Researchers have found that prolonged periods of sleep deprivation result in greater lower frequency waveform activity (1–7 Hz) in both waking humans (Cajochen et al., 1999; Finelli et al., 2000; Strijkstra et al., 2003) and rats (Vyazovskiy and Tobler, 2005) suggesting that even individual neurons may be getting "tired". This tiredness may result from lacking adequate energy supplies and manifest as many of the physical symptoms of tiredness such as decreased reasoning ability, decreased reaction time and altered mood.

Further insights into the molecular processes underlying sleep-deprivation might be gleaned through chemically altering the sleep process. Specific drugs are known to profoundly affect sleep architecture. Both tricyclic antidepressants (Baumann et al., 1983) and selective serotonin uptake inhibitors (Monti and Jantos, 2005) are known to reduce REM sleep, while sodium oxybate (Vienne et al., 2012) has been shown to increase delta activity in NREM sleep. What is currently unknown is the effects these compounds exert on brain neuroenergetic pathways. If the ANLS hypothesis model is valid as suggested by the studies to date, the changes in neuroenergetics by these different drugs classes should be readily observed and fall into a predictable pattern. No biosensor studies have investigated how these compounds interact with energy analytes to influence sleep. Future studies of how the sleep process is controlled may benefit from longer periods of sleep loss in conjunction with biosensor recordings.

The use of biosensors to study, and ultimately answer, fundamental questions regarding sleep is still in its infancy. Further development of biosensor technology is desperately needed to monitor the in-depth and long-term chemical changes associated with sleep. Three areas can be highlighted for further improvement:

(1) Extensive physiological studies involving multiple, simultaneous biosensors combined with in-depth electrophysiological recordings *within the same brain region* are needed to more fully understand local neuronal activity alongside real-time extracellular analyte concentrations and relate these changes to the overall sleep state.
(2) A wider array of biosensor detection targets for neurotransmitters such as acetylcholine, ions such as Na^+ and K^+ or other neuroenergetic components such as pyruvate would provide a better palette of possible combinations for examination of sleep state physiology.
(3) The *in vivo* lifetimes of sensors must be extended. Increasing the functionality of all relevant sensors to something comparable to glucose and lactate (72+ hours) or even chronically implanted sensors lasting weeks or months is needed to more fully understand long-term brain alterations with sleep.

ACKNOWLEGMENTS

The authors wish to thank Donna A. Johnson, David A. Johnson, and Daniel Aillon for editorial assistance and useful discussions, and Molly Bellinger for assistance with graphics.

REFERENCES

Achermann P, Borbely AA (2011) Sleep homeostasis and models of sleep regulation. In: principles and practices of sleep medicine (Kryger MH et al., eds), pp 431–444. St. Louis: Elsevier Saunders.

Aillon DV, Naylor E, Barrett BS, Gwartney DV, Gabbert S, Johnson DA, Petillo PA (2013) Fast, high-resolution biosensor measurements of ethanol-induced neurotransmitter dynamics in the rat brain. In: Neuroscience 2013, San Diego, CA.

Alam MN, Szymusiak R, Gong H, King J, McGinty D (1999) Adenosinergic modulation of rat basal forebrain neurons during sleep and waking: Neuronal recording with microdialysis. J Physiol 521 Pt 3:679–690.

Albrecht U (2012) Timing to perfection: The biology of central and peripheral circadian clocks. Neuron 74:246–260.

Attwell D, Laughlin SB (2001) An energy budget for signaling in the grey matter of the brain. J Cereb Blood Flow Metab 21:1133–1145.

Azuma S, Kodama T, Honda K, Inoue S (1996) State-dependent changes of extracellular glutamate in the medial preoptic area in freely behaving rats. Neurosci Lett 214:179–182.

Basheer R, Porkka-Heiskanen T, Stenberg D, McCarley RW (1999) Adenosine and behavioral state control: Adenosine increases c-Fos protein and AP1 binding in basal forebrain of rats. Brain Res Mol Brain Res 73:1–10.

Baumann P, Gaillard JM, Perey M, Justafre JC, Le P (1983) Relationships between brain concentrations of desipramine and paradoxical sleep inhibition in the rat. J Neural Transm 56:105–116.

Benca RM, Kushida CA, Everson CA, Kalski R, Bergmann BM, Rechtschaffen A (1989) Sleep deprivation in the rat: VII. Immune function. Sleep 12:47–52.

Benington JH, Heller HC (1995) Restoration of brain energy metabolism as the function of sleep. Prog Neurobiol 45:347–360.

Benington JH, Kodali SK, Heller HC (1995) Stimulation of A1 adenosine receptors mimics the electroencephalographic effects of sleep deprivation. Brain Res 692:79–85.

Bergmann BM, Everson CA, Kushida CA, Fang VS, Leitch CA, Schoeller DA, Refetoff S, Rechtschaffen A (1989) Sleep deprivation in the rat: V. Energy use and mediation. Sleep 12:31–41.

Bergmann BM, Gilliland MA, Feng PF, Russell DR, Shaw P, Wright M, Rechtschaffen A, Alverdy JC (1996) Are physiological effects of sleep deprivation in the rat mediated by bacterial invasion? Sleep 19:554–562.

Borbely AA, Tobler I (1989) Endogenous sleep-promoting substances and sleep regulation. Physiol Rev 69:605–670.

Boumezbeur F, Petersen KF, Cline GW, Mason GF, Behar KL, Shulman GI, Rothman DL (2010) The contribution of blood lactate to brain energy metabolism in humans measured by dynamic ^{13}C nuclear magnetic resonance spectroscopy. J Neurosci 30:13983–13991.

Burmeister JJ, Gerhardt GA (2001) Self-referencing ceramic-based multisite microelectrodes for the detection and elimination of interferences from the measurement of L-glutamate and other analytes. Anal Chem 73:1037–1042.

Burmeister JJ, Palmer M, Gerhardt GA (2005) L-lactate measures in brain tissue with ceramic-based multisite microelectrodes. Biosens Bioelectron 20:1772–1779.

Cajochen C, Foy R, Dijk DJ (1999) Frontal predominance of a relative increase in sleep delta and theta EEG activity after sleep loss in humans. Sleep Res Online 2:65–69.

Carskadon MA, Dement WC (2011) Normal human sleep: An overview. In: Principles and practices of sleep medicine (Kryger MH et al., eds), pp 16–26 St. Louis: Elsevier Saunders.

Chefer VI, Thompson AC, Zapata A, Shippenberg TS (2009) Overview of brain microdialysis. In Current protocols in neuroscience/(Crawley JN et al., eds). Chapter 7: Unit 7.1.

Chu M, Huang ZL, Qu WM, Eguchi N, Yao MH, Urade Y (2004) Extracellular histamine level in the frontal cortex is positively correlated with the amount of wakefulness in rats. Neurosci Res 49:417–420.

Clinton JM, Davis CJ, Zielinski MR, Jewett KA, Krueger JM (2011) Biochemical regulation of sleep and sleep biomarkers. J Clin Sleep Med 7:S38–S42.

Cocks JA (1967) Change in the concentration of lactic acid in the rat and hamster brain during natural sleep. Nature 215:1399–1400.

Dash MB, Bellesi M, Tononi G, Cirelli C (2013) Sleep/wake dependent changes in cortical glucose concentrations. J Neurochem 124:79–89.

Dash MB, Douglas CL, Vyazovskiy VV, Cirelli C, Tononi G (2009) Long-term homeostasis of extracellular glutamate in the rat cerebral cortex across sleep and waking states. J Neurosci 29:620–629.

Dash MB, Tononi G, Cirelli C (2012) Extracellular levels of lactate, but not oxygen, reflect sleep homeostasis in the rat cerebral cortex. Sleep 35:909–919.

Dijk DJ, Beersma DG (1989) Effects of SWS deprivation on subsequent EEG power density and spontaneous sleep duration. Electroencephalogr Clin Neurophysiol 72:312–320.

Dworak M, McCarley RW, Kim T, Kalinchuk AV, Basheer R (2010) Sleep and brain energy levels: ATP changes during sleep. J Neurosci 30:9007–9016.

Edgar DM, Dement WC, Fuller CA (1993) Effect of SCN lesions on sleep in squirrel monkeys: Evidence for opponent processes in sleep-wake regulation. J Neurosci 13:1065–1079.

Everson CA (1995) Functional consequences of sustained sleep deprivation in the rat. Behav Brain Res 69:43–54.

Everson CA (2005) Clinical assessment of blood leukocytes, serum cytokines, and serum immunoglobulins as responses to sleep deprivation in laboratory rats. Am J Physiol Regul Integr Comp Physiol 289:R1054–R1063.

Everson CA, Bergmann BM, Rechtschaffen A (1989) Sleep deprivation in the rat: III. Total sleep deprivation. Sleep 12:13–21.

Everson CA, Folley AE, Toth JM (2012) Chronically inadequate sleep results in abnormal bone formation and abnormal bone marrow in rats. Exp Biol Med (Maywood) 237:1101–1109.

Everson CA, Nowak TS, Jr. (2002) Hypothalamic thyrotropin-releasing hormone mRNA responses to hypothyroxinemia induced by sleep deprivation. Am J Physiol Endocrinol Metab 283:E85–E93.

Finelli LA, Baumann H, Borbely AA, Achermann P (2000) Dual electroencephalogram markers of human sleep homeostasis: Correlation between theta activity in waking and slow-wave activity in sleep. Neuroscience 101:523–529.

Florian C, Vecsey CG, Halassa MM, Haydon PG, Abel T (2011) Astrocyte-derived adenosine and A1 receptor activity contribute to sleep loss-induced deficits in hippocampal synaptic plasticity and memory in mice. J Neurosci 31:6956–6962.

Goldfarb LG, Petersen RB, Tabaton M, Brown P, LeBlanc AC, Montagna P, Cortelli P, Julien J, Vital C, Pendelbury WW (1992) Fatal familial insomnia and familial Creutzfeldt-Jakob disease: Disease phenotype determined by a DNA polymorphism. Science 258:806–808.

Hefti K, Holst SC, Sovago J, Bachmann V, Buck A, Ametamey SM, Scheidegger M, Berthold T, Gomez-Mancilla B, Seifritz E, Landolt HP (2013) Increased metabotropic glutamate receptor subtype 5 availability in human brain after one night without sleep. Biol Psychiatry 73:161–168.

Hu Y, Wilson GS (1997) A temporary local energy pool coupled to neuronal activity: Fluctuations of extracellular lactate levels in rat brain monitored with rapid-response enzyme-based sensor. J Neurochem 69:1484–1490.

Huang W, Ramsey KM, Marcheva B, Bass J (2011) Circadian rhythms, sleep, and metabolism. J Clin Invest 121:2133–2141.

John J, Ramanathan L, Siegel JM (2008) Rapid changes in glutamate levels in the posterior hypothalamus across sleep-wake states in freely behaving rats. Am J Physiol Regul Integr Comp Physiol 295:R2041–R2049.

Johnson D, Harmon H, Naylor E, Gabbert S, AIllon D, Johnson DA, Wilson GS, Turek FW (2007) An integrated EEG/EMG/glucose system for mice and rats. Sleep 30:A24.

Jones BE (1991) Paradoxical sleep and its chemical/structural substrates in the brain. Neuroscience 40:637–656.

Jones EG (2009) The origins of cortical interneurons: Mouse versus monkey and human. Cereb Cortex 19:1953–1956.

Kalinchuk AV, McCarley RW, Porkka-Heiskanen T, Basheer R (2011) The time course of adenosine, nitric oxide (NO) and inducible NO synthase changes in the brain with sleep loss and their role in the non-rapid eye movement sleep homeostatic cascade. J Neurochem 116:260–272.

Kalinchuk AV, Urrila AS, Alanko L, Heiskanen S, Wigren HK, Suomela M, Stenberg D, Porkka-Heiskanen T (2003) Local energy depletion in the basal forebrain increases sleep. The European J Neurosci 17:863–869.

Kametani H, Kawamura H (1990) Alterations in acetylcholine release in the rat hippocampus during sleep-wakefulness detected by intracerebral dialysis. Life Sci 47:421–426.

Lena I, Parrot S, Deschaux O, Muffat-Joly S, Sauvinet V, Renaud B, Suaud-Chagny MF, Gottesmann C (2005) Variations in extracellular levels of dopamine, noradrenaline, glutamate, and aspartate across the sleep-wake cycle in the medial prefrontal cortex and nucleus accumbens of freely moving rats. J Neurosci research 81:891–899.

Liu ZW, Faraguna U, Cirelli C, Tononi G, Gao XB (2010) Direct evidence for wake-related increases and sleep-related decreases in synaptic strength in rodent cortex. J Neurosci 30:8671–8675.

Llaudet E, Botting NP, Crayston JA, Dale N (2003) A three-enzyme microelectrode sensor for detecting purine release from central nervous system. Biosens Bioelectron 18:43–52.

Loomis AL, Harvey EN, Hobart GA (1936) Electrical potentials of the human brain. J Exp Psychol 19:249–279.

Lopez-Rodriguez F, Medina-Ceja L, Wilson CL, Jhung D, Morales-Villagran A (2007) Changes in extracellular glutamate levels in rat orbitofrontal cortex during sleep and wakefulness. Arch Med Res 38:52–55.

Magistretti JP, Pellerin L, Martin J (1995) Brain energy metabolism. In: Psychopharmacology: The fourth generation of progress (Bloom FE, Kupfer DJ, eds). New York: Raven Press.

Manetto V, Medori R, Cortelli P, Montagna P, Tinuper P, Baruzzi A, Rancurel G, Hauw JJ, Vanderhaeghen JJ, Mailleux P (1992) Fatal familial insomnia: Clinical and pathologic study of five new cases. Neurology 42:312–319.

Meier-Ewert HK, Ridker PM, Rifai N, Regan MM, Price NJ, Dinges DF, Mullington JM (2004) Effect of sleep loss on C-reactive protein, an inflammatory marker of cardiovascular risk. J Am Coll Cardiol 43:678–683.

Mohammed HS, Aboul Ezz HS, Khadrawy YA, Noor NA (2011) Neurochemical and electrophysiological changes induced by paradoxical sleep deprivation in rats. Behav Brain Res 225:39–46.

Monti JM, Jantos H (2005) A study of the brain structures involved in the acute effects of fluoxetine on REM sleep in the rat. Int J Neuropsychopharmacol 8:75–86.

Naylor E, Aillon DV, Barrett BS, Wilson GS, Johnson DA, Harmon HP, Gabbert S, Petillo PA (2012) Lactate as a biomarker for sleep. Sleep 35:1209–1222.

Naylor E, Aillon DV, Gabbert S, Harmon H, Johnson DA, Wilson GS, Petillo PA (2011) Simultaneous real-time measurement of EEG/EMG and L-glutamate in mice: A biosensor study of neuronal activity during sleep. J Electroanal Chem 656:106–113.

Netchiporouk L, Shram N, Salvert D, Cespuglio R (2001) Brain extracellular glucose assessed by voltammetry throughout the rat sleep-wake cycle. Eur J Neurosci 13:1429–1434.

Nitz D, Siegel JM (1997) GABA release in the locus coeruleus as a function of sleep/wake state. Neuroscience 78:795–801.

Pajski ML, Venton BJ (2010) Adenosine release evoked by short electrical stimulations in striatal brain slices is primarily activity dependent. ACS Chem Neurosci 1:775–787.

Pajski ML, Venton BJ (2013) The mechanism of electrically stimulated adenosine release varies by brain region. Purinergic Signal 9:167–174.

Pellerin L, Bouzier-Sore AK, Aubert A, Serres S, Merle M, Costalat R, Magistretti PJ (2007) Activity-dependent regulation of energy metabolism by astrocytes: An update. Glia 55:1251–1262.

Pellerin L, Magistretti PJ (1994) Glutamate uptake into astrocytes stimulates aerobic glycolysis: A mechanism coupling neuronal activity to glucose utilization. Proc Natl Acad Sci USA 91:10625–10629.

Petit JM, Gyger J, Burlet-Godinot S, Fiumelli H, Martin JL, Magistretti PJ (2013) Genes involved in the astrocyte-neuron lactate shuttle (ANLS) are specifically regulated in cortical astrocytes following sleep deprivation in mice. Sleep 36:1445–1458.

Petit JM, Tobler I, Kopp C, Morgenthaler F, Borbely AA, Magistretti PJ (2010) Metabolic response of the cerebral cortex following gentle sleep deprivation and modafinil administration. Sleep 33:901–908.

Porkka-Heiskanen T, Kalinchuk AV (2011) Adenosine, energy metabolism and sleep homeostasis. Sleep Med Rev 15:123–135.

Porkka-Heiskanen T, Strecker RE, McCarley RW (2000) Brain site-specificity of extracellular adenosine concentration changes during sleep deprivation and spontaneous sleep: An *in vivo* microdialysis study. Neuroscience 99:507–517.

Porkka-Heiskanen T, Strecker RE, Thakkar M, Bjorkum AA, Greene RW, McCarley RW (1997) Adenosine: A mediator of the sleep-inducing effects of prolonged wakefulness. Science 276:1265–1268.

Portas CM, Bjorvatn B, Fagerland S, Gronli J, Mundal V, Sorensen E, Ursin R (1998) On-line detection of extracellular levels of serotonin in dorsal raphe nucleus and frontal cortex over the sleep/wake cycle in the freely moving rat. Neuroscience 83:807–814.

Reich P, Geyer SJ, Karnovsky ML (1972) Metabolism of brain during sleep and wakefulness. J Neurochem 19:487–497.

Richter D, Dawson RM (1948) Brain lactate in emotion. Nature 161:205.

Rutherford EC, Pomerleau F, Huettl P, Stromberg I, Gerhardt GA (2007) Chronic second-by-second measures of L-glutamate in the central nervous system of freely moving rats. J Neurochem 102:712–722.

Saper CB, Scammell TE, Lu J (2005) Hypothalamic regulation of sleep and circadian rhythms. Nature 437:1257–1263.

Schmitt LI, Sims RE, Dale N, Haydon PG (2012) Wakefulness affects synaptic and network activity by increasing extracellular astrocyte-derived adenosine. J Neurosci 32:4417–4425.

Shimizu H, Tabushi K, Hishikawa Y, Kakimoto Y, Kaneko Z (1966) Concentration of lactic acid in rat brain during natural sleep. Nature 212:936–937.

Shouse MN, Staba RJ, Saquib SF, Farber PR (2000) Monoamines and sleep: Microdialysis findings in pons and amygdala. Brain Res 860:181–189.

Shram N, Netchiporouk L, Cespuglio R (2002) Lactate in the brain of the freely moving rat: Voltammetric monitoring of the changes related to the sleep-wake states. Eur J Neurosci 16:461–466.

Simpson IA, Carruthers A, Vannucci SJ (2007) Supply and demand in cerebral energy metabolism: The role of nutrient transporters. J Cereb Blood Flow Metab 27:1766–1791.

Sims RE, Wu HH, Dale N (2013) Sleep-wake sensitive mechanisms of adenosine release in the basal forebrain of rodents: An *in vitro* study. PLOS ONE 8:e53814.

Spiegel K, Knutson K, Leproult R, Tasali E, Van Cauter E (2005) Sleep loss: A novel risk factor for insulin resistance and Type 2 diabetes. J Appl Physiol 99:2008–2019.

Spiegel K, Sheridan JF, Van Cauter E (2002) Effect of sleep deprivation on response to immunization. JAMA 288:1471–1472.

Steriade M (2005) Brain electrical activity and sensory processing during waking and sleep states. In: Principles and practices of sleep medicine, fourth edition (Kryger MH et al., eds), pp 91–100. Philadelphia: W.B. Saunders Company.

Strecker RE, Morairty S, Thakkar MM, Porkka-Heiskanen T, Basheer R, Dauphin LJ, Rainnie DG, Portas CM, Greene RW, McCarley RW (2000) Adenosinergic modulation of basal forebrain and preoptic/anterior hypothalamic neuronal activity in the control of behavioral state. Behav Brain Res 115:183–204.

Strijkstra AM, Beersma DG, Drayer B, Halbesma N, Daan S (2003) Subjective sleepiness correlates negatively with global alpha (8–12 Hz) and positively with central frontal theta (4–8 Hz) frequencies in the human resting awake electroencephalogram. Neurosci Lett 340:17–20.

Suzuki A, Stern SA, Bozdagi O, Huntley GW, Walker RH, Magistretti PJ, Alberini CM (2011) Astrocyte-neuron lactate transport is required for long-term memory formation. Cell 144:810–823.

Tadavarty R, Rajput PS, Wong JM, Kumar U, Sastry BR (2011) Sleep-deprivation induces changes in GABA(B) and mGlu receptor expression and has consequences for synaptic long-term depression. PLOS ONE 6:e24933.

Van Cauter E, Tasali E (2011) Endocrine physiology in relation to sleep and sleep disturbances. In: Principles and practices of sleep medicine (Kryger MH et al., eds), pp 291–311. St. Louis: Elsevier Saunders.

Van den Noort S, Brine K (1970) Effect of sleep on brain labile phosphates and metabolic rate. Am J Physiol 218:1434–1439.

Van Dongen HP, Maislin G, Mullington JM, Dinges DF (2003) The cumulative cost of additional wakefulness: Dose-response effects on neurobehavioral functions and sleep physiology from chronic sleep restriction and total sleep deprivation. Sleep 26:117–126.

Vannucci SJ, Gibbs EM, Simpson IA (1997) Glucose utilization and glucose transporter proteins GLUT-1 and GLUT-3 in brains of diabetic (db/db) mice. Am J Physiol 272:E267–E274.

Vienne J, Lecciso G, Constantinescu I, Schwartz S, Franken P, Heinzer R, Tafti M (2012) Differential effects of sodium oxybate and baclofen on EEG, sleep, neurobehavioral performance, and memory. Sleep 35:1071–1083.

Virus RM, Djuricic-Nedelson M, Radulovacki M, Green RD (1983) The effects of adenosine and 2'-deoxycoformycin on sleep and wakefulness in rats. Neuropharmacology 22:1401–1404.

Vyazovskiy VV, Cirelli C, Pfister-Genskow M, Faraguna U, Tononi G (2008) Molecular and electrophysiological evidence for net synaptic potentiation in wake and depression in sleep. Nat Neurosci 11:200–208.

Vyazovskiy VV, Tobler I (2005) Theta activity in the waking EEG is a marker of sleep propensity in the rat. Brain Res 1050:64–71.

Watson CJ, Lydic R, Baghdoyan HA (2011) Sleep duration varies as a function of glutamate and GABA in rat pontine reticular formation. J Neurochem 118:571–580.

Wigren HK, Rytkonen KM, Porkka-Heiskanen T (2009) Basal forebrain lactate release and promotion of cortical arousal during prolonged waking is attenuated in aging. J Neurosci 29:11698–11707.

Wigren HK, Schepens M, Matto V, Stenberg D, Porkka-Heiskanen T (2007) Glutamatergic stimulation of the basal forebrain elevates extracellular adenosine and increases the subsequent sleep. Neuroscience 147:811–823.

Wilson GS, Johnson MA (2008) In-vivo electrochemistry: What can we learn about living systems? Chem Rev 108:2462–2481.

Wisor JP, Rempe MJ, Schmidt MA, Moore ME, Clegern WC (2012) Sleep slow-wave activity regulates cerebral glycolytic metabolism. Cereb Cortex 23(8):1978–1987.

Yehuda S, Sredni B, Carasso RL, Kenigsbuch-Sredni D (2009) REM sleep deprivation in rats results in inflammation and interleukin-17 elevation. J Interferon Cytokine Res 29:393–398.

Zepelin H, Siegel JM, Tobler I (2005) Mammalian sleep. In: Principles and practices of sleep medicine, fourth edition (Kryger MH et al., eds), pp 91–100. Philadelphia: W.B. Saunders Company.

CHAPTER 2

CORTICAL CHOLINERGIC TRANSIENTS FOR CUE DETECTION AND ATTENTIONAL MODE SHIFTS

Martin Sarter[*], William M. Howe[†] and Howard Gritton[‡]

[*]University of Michigan, [†]Pfizer Inc., [‡]Boston University

2.1 INTRODUCTION

Theories on the regulation and function of the forebrain cholinergic system has been built on the experimental approaches used to manipulate this system as well as the behavioral paradigms used to illustrate the consequences of those manipulations. The Sustained Attention Task (SAT) has remained a major "workhorse" for our research on the forebrain cholinergic system and it is therefore important to first introduce the main task components. A standard task session is comprised of the random presentation of signal and non-signal trials. "Signal trials" consist of a brief (25, 50, or 500 milliseconds duration) visual cue (or signal), while "non-signal trials" have no visual cue. Following a signal or non-signal event, two response levers are extended into the operant chamber, prompting the animal to report the presence or absence of a signal *via* a press on one of two discrete manipulanda. Correct responses are defined as hits during signal trials or as correct rejections on non-signal trials. A response on the signal lever during a non-signal trial is scored as a false alarm and a press on the non-signal lever during a signal trial is recorded as a miss. Hits and correct rejections are both rewarded. A task session typically lasts 40–60 minutes and consists of 80–150 signal and non-signal trials each. We have developed versions of this task for use in rats, mice, and humans, including validation studies across all three species as well as demonstrated the SAT's usefulness for

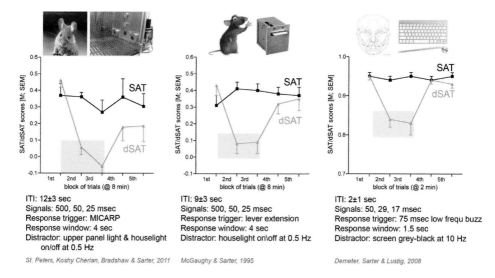

Figure 2.1 Performance by mice, rats, and humans in the standard version of the SAT (squares and black lines) and the distractor version of this task (dSAT; triangles and orange lines). The SAT has been used extensively in our research on cholinergic systems, including the measurement of cholinergic transients during performance. The SAT consists of signal and non-signal trials and generates hits and misses, and corrected rejections and false alarms, respectively. These measures are computed into one performance score, the SAT score that is depicted on the ordinate of the 3 plots (a SAT score of 1 indicates that all responses were hits and correct rejections; a score of zero indicates random lever selection). Note that SAT scores shown in Figure 2.1 were averaged over scores for 3 signal durations. For the dSAT, the distractor is presented during the second and third block of trials (shaded). As expected, human's performance is superior to that of rats and mice. However, the pattern of distractor effects is comparable among these species (for evidence indicating the validity of the SAT in these species, see McGaughy and Sarter, 1995; St. Peters et al., 2011a and Demeter et al., 2008). For the three task versions, main parameters are indicated below the plot (intertrial interval, ITI; michigan controlled access response port, MICARP).

human clinical research (McGaughy and Sarter, 1995; Demeter et al., 2008, 2011, 2013; St Peters et al., 2011a; Figure 2.1).

Three major experimental approaches have been employed to manipulate or monitor cholinergic activity during task performance. The first was the use of the immunotoxin 192 IgG-saporin, which allows for selective targeting and deletion of cholinergic neurons that innervate the cortex (e.g., Holley et al., 1994). We demonstrated that removal of 60–80% of cholinergic inputs to the anterior associative cortices robustly and permanently impaired performance on signal trials, while performance on non-signal trials was unaffected (McGaughy et al., 1996; McGaughy and Sarter, 1998). This finding was the first clear indication that the cholinergic system was a necessary component of the networks supporting attentional function, and supported the general hypothesis that the cortical cholinergic input system is necessary for the detection of instructive cues in the environment. The second major experimental approach relied on pharmacological methods to manipulate cholinergic activity. Compounds that alter or block cholinergic signaling or tone can have profound influences on cognitive performance. However, the effects of many traditionally used compounds, including antagonists of muscarinic acetylcholine receptor (mAChR; e.g., scopolamine, atropine) or the use of acetylcholinesterase inhibitors (AChEIs; e.g., physostigmine, neostigmine, or donepezil) remain difficult to interpret. For example, many muscarinic receptor antagonists also act as potent releasers of acetylcholine (ACh) due to auto-receptor binding. The net result is a rather complex combination of blockade of mAChR and concurrent stimulation of nicotinic acetylcholine receptors (nAChRs). Similarly, AChEIs drastically elevate extracellular levels of ACh, thereby stimulating both classes of acetylcholine receptors (AChRs) in a continuous and decidedly artificial manner (see also Sarter et al., 2009). Complexities aside, pharmacological experiments in animals further supported the necessity of cortical ACh for attention, cue detection (e.g., Dalley et al., 2004) and in humans, supported the putative cognitive functions of cholinergic activity (e.g., Bentley et al., 2004; Furey et al., 2000; Silver et al., 2008).

The third major experimental approach towards investigations of the functions of forebrain cholinergic systems has been the application of neurochemical methods to monitor changes in ACh release in task performing animals. Early efforts to determine brain ACh content *ex vivo* yielded important information about the consequences of brain pathology on ACh quantity (e.g., Scremin et al., 2006). However, these methods generally lacked the sensitivity, temporal resolution, and reliability to reveal the cognitive-behavioral functions of this neurotransmitter system.

The advent of microdialysis coupled with analytical methods allowed for detection of ACh in samples collected *in vivo* (Westerink et al., 1988; Damsma et al., 1988), and initiated a highly productive stream of research focusing on how ACh release correlates with behavior (e.g., Day et al., 1991; Bruno et al., 1999). Further refinement of the analytical methods permitted the exclusion of AChEIs from the dialysate (Herzog et al., 2003), and thus removed the potential confounds resulting from artificially elevated ACh levels in the sampled extracellular space (for more details see Himmelheber et al., 1998). Although these initial attempts had relatively poor temporal resolution (10–20 minutes), recent analytical advances now permit the detection of major neurotransmitters, including ACh, from much shorter collection windows (~1 minute; Song et al., 2012; Sarter et al., 2012).

Our early studies designed to monitor ACh release in the cortex of rats performing the SAT using microdialysis (described above) indicated that attentional performance produces elevated ACh levels relative to pre-task baseline. This increase in ACh release was not present in tasks that produced similar amounts of motor activity and reward density, or otherwise absent in their explicit demands on attention (Himmelheber et al., 1997, 2000; Arnold et al., 2002). Further, SAT performance-associated ACh release was shown not to correlate with levels of performance but rather demands on attention; that is, increased demands on attentional capacity lower performance levels, but are also associated with higher ACh levels than during standard, unchallenged SAT performance (Himmelheber et al., 2001; Kozak et al., 2006). Perhaps intuitively, we also found that the higher the increase in ACh release the less severe the impact the distractor had on performance (St. Peters et al., 2011b). This finding supports the view that cholinergic modulation of cortical circuitry serves to counteract the effects of the distractor, in part, by filtering the distractor from being processed and enhancing the representation of cues (see also Gill et al., 2000; Broussard et al., 2009).

Other neuronal systems recruit, interact, and influence cholinergic systems during goal directed task performance. Neuronal activity in the nucleus accumbens (NAC), for example, is necessary for performance-associated increases in cortical ACh release (Neigh et al., 2004). Furthermore, NAC glutamatergic–dopaminergic mechanisms to cooperate to enhance prefrontal ACh release (Zmarowski et al., 2005). We recently demonstrated the functional significance of NAC glutamatergic activity for SAT performance under the condition of visual distraction (St Peters et al., 2011b). We found that stimulation of N-methyl-D-aspartate (NMDA) receptors in the NAC shell, but not core promotes elevated prefrontal ACh release and rescues distractor induced impairments (St Peters et al., 2011b). These results appear to have a high level of translation validity based on our data in human subjects; human participants performing an optimized version of the SAT also show robust activation that correlates with increased attentional effort under distracting conditions (Demeter et al., 2011). In both animals (performance-associated increases in ACh release), and in humans (blood-oxygen-level-dependent (BOLD) signal), the largest increase in right prefrontal activity coincides with sustained performance (effort) simultaneous with the detrimental effects of distracting stimuli. We have also found that in schizophrenic patients' SAT performance is particularly vulnerable to the demands imposed by distracters (Demeter et al., 2013), which is consistent with other evidence and theories suggesting attenuated activation of the cholinergic system during attention-demanding performance in this disease (Kozak et al., 2007; Sarter et al., 2009).

Taken together, performance-associated ACh levels in the cortex are hypothesized to serve as a top-down signal to modulate cortical circuitry as a function of attentional effort. In this context, attentional effort is defined as a set of cognitive mechanisms that support, aid recovery, and sustain attentional performance in order to combat the effects of degrading manipulations such as distracters, disease, and fatigue (for review see Sarter et al., 2006; Sarter and Paolone, 2011).

2.2 MEASUREMENT OF SECOND-BASED CHOLINERGIC TRANSIENTS

The use of microdialysis to monitor ACh release in task-performing rats has greatly advanced our understanding of the functions of cholinergic modulation of cortical circuitry, although the interpretation of the results is restricted to the function of release on a relatively slow, or "tonic", timescale of minutes. The theoretical presence of meaningful ACh signaling events occurring at a faster temporal resolution (subsecond), while never directly demonstrated, remained an untested, yet plausible, signaling mechanism for cholinergic activity for many reasons. The unique catalytic power of acetylcholinesterase (AChE) (e.g., Quinn, 1987), and its clustering at terminals (Peng et al., 1999) offered support against the long standing view that cholinergic inputs function primarily to maintain a relatively slowly changing tonic level of ACh (for more discussion see Sarter et al., 2009). The need for methods capable of measuring cholinergic signaling with greater temporal precision and free of the constraints of off-line analytical methods was widely recognized (Huang et al., 1993; Garguilo and Michael, 1996; Xin and Wightman, 1997).

We adopted a real-time monitoring technique developed by the Gerhardt group at the University of Kentucky. This method relies on the use of ceramic-based microelectrodes with multiple platinum (Pt) recording sites for high fidelity, real-time, enzyme-based, detection of acetylcholine. For the first time, commercial based electrodes with sufficient mechanical stability and measurement properties (limit of detection for choline, sensitivity, selectivity, and linearity) made it feasible to employ fixed-potential amperometry to measure changes in current that reflect choline derived from the hydrolysis of newly released ACh, in awake animals (Burmeister et al., 2000, 2003). Collaborating with the Gerhardt group, we demonstrated the validity of this technique *in vivo*. For example, we showed that AChE inhibition in the recording field attenuates currents normally evoked by the presence of ACh, and removal of cholinergic input to the recording field correspondingly abolishes depolarization-evoked current. Collectively, these findings confirmed that current changes measured with this method reflect the oxidation of choline that is neuronally derived from the release and subsequent hydrolysis of ACh (for details see Parikh et al., 2004).

The use of choline as a reporter molecule requires more comment. Extracellular choline concentrations are relatively stable (4.85 μM; for measurement methods and details see Parikh and Sarter, 2006). Even near-complete removal of cholinergic synapses in the recording field does not appreciably alter this concentration, an observation consistent with the highly efficacious arterial plasma buffering of extracellular choline (Klein et al., 1998). As will be detailed below, the amplitude of task- and behavior-associated release events ("transients") measured by this method are not confounded by this extracellular concentration, as each transient is normalized to its own temporally proximal baseline. Thus potential variations in local choline levels that are not due to changes in ACh release are unlikely to contribute to these relative changes. Furthermore, we have characterized the role of the choline transporter (CHT) in clearing ACh-evoked transients (Parikh and Sarter, 2006), and the diffusion of choline away from the recording site does not appear to contribute to signal amplitude or clearance (see also Parikh et al., 2013).

While some methodological issues remain partly unresolved and require further research and development (e.g., long term stability of oxidizing enzymes and exclusion barriers *in vivo*), the basic validity of measures of transient cholinergic release events has consistently been supported. Indeed all neurochemical methods used for *in vivo* monitoring of synaptic mechanisms have measurement constraints and are associated with long-standing concerns about the impact of the measurement method *per se* on the regulation of the target molecule. These concerns fade into the background as the methods consistently generate orderly data that form systematic brain-behavior relationships.

2.3 CUE DETECTION MEDIATED BY CHOLINERGIC TRANSIENTS

In our first attempt at using this technique in task performing animals, we were not able to use the SAT due to technical challenges of electrical interference from operant equipment while attempting to record small pico-ampere currents (these issues were ultimately resolved and we will detail the results of recordings during SAT performance in Section 2.4 below). Therefore our initial experiments designed to measure cholinergic transients in task-performing rats utilized a relatively simple cued-appetitive response task (CART). In the CART, a visual cue is presented after a relatively long (90 ± 30 seconds) ITI. The cue predicts food delivery at one out of two randomly selected reward ports. During the ITI, animals typically were engaged in self-directed behavior (e.g., grooming). Detection was defined by a cue-evoked termination of this behavior followed by orientation and approach to the reward port(s). The major finding from these experiments was that the cues that were detected, i.e., changed ongoing grooming behavior by triggering an orienting response followed by reward seeking, evoked brief increases in cholinergic activity ("transients") selectively in the medial prefrontal cortex (mPFC). Those cues that did not trigger this behavioral shift, or "missed", did not. Additional experiments demonstrated that cue detection-evoked transients during the learning of the CART emerged as rats began to exhibit a detection response, and that shortening of the cue-reward delivery interval from a mean of six seconds to two seconds moved the onset of the transient closer to cue offset. We concluded from these experiments that cue detection, defined as a cognitive process consisting of "the entry of information concerning the presence of a signal into a system that allows the subject to report the existence of the signal by an arbitrary response indicated by the experimenter" (Posner et al., 1980), is mediated by cholinergic transients. As will be discussed next, this conclusion was further refined based on data from recordings in rats performing the SAT.

2.4 CHOLINERGIC MEDIATION OF CUE DETECTION IN TRIALS INVOLVING AN ATTENTIONAL MODE SHIFT

Building on the effects of selective removal of cholinergic inputs to the cortex, and preliminary experiments in the CART described above (McGaughy, et al., 1996; Parikh, et al., 2007), we hypothesized that in the SAT, mPFC cholinergic transients would be evoked on signal trials resulting in hits, and not on non-signal trials or signal trials ending in a miss. In agreement with this hypothesis, we found that only hits were associated with increases in cholinergic activity in the mPFC. However, we discovered that cholinergic transients occurred only on approximately 60% of detected signal trials. Exploratory analyses revealed that transients were only evoked when a hit trial was preceded by a response on the non-signal lever during the previous trial (correct rejection or miss, "incongruent hit"). When a hit followed another hit no such transient was observed (Figure 2.2; Howe et al., 2013). The trial sequence-dependent occurrence of increases in cholinergic activity on hit trials forced us to develop new predictions about the purpose and timing of cholinergic signaling during cue detection. Previous studies had demonstrated that stimulation of muscarinic receptors in the cortex selectively suppresses intracortical inputs (Hasselmo and Bower, 1992); the kind of local circuitry that might dominate non-signal trial performance or monitoring for cues based on previously learned associations and expectations (Hasselmo and Sarter, 2011). In contrast, thalamic afferent input is unaffected by muscarinic stimulation (Hasselmo and Bower, 1992). Therefore it is plausible that rather than signal detection *per se*, cholinergic transients may be necessary to facilitate a transition from expectation or monitoring for cues to cue-directed attentional processing. Under these conditions the thalamic representation of the cue is used to guide and initiate a new behavior that deviates from recent experience. Presumably network states can be maintained across subsequent trials *via* coordinated glutamate (from thalamic inputs, Parikh et al., 2008, 2010) and ACh release that activates postsynaptic muscarinic receptors and subsequently induces persistent spiking across trials in local pyramidal cells (Fransen et al, 2002, Ergorov et al., 2002, Hasselmo and Sarter, 2011), therefore a second transient is not required in the case of second hit.

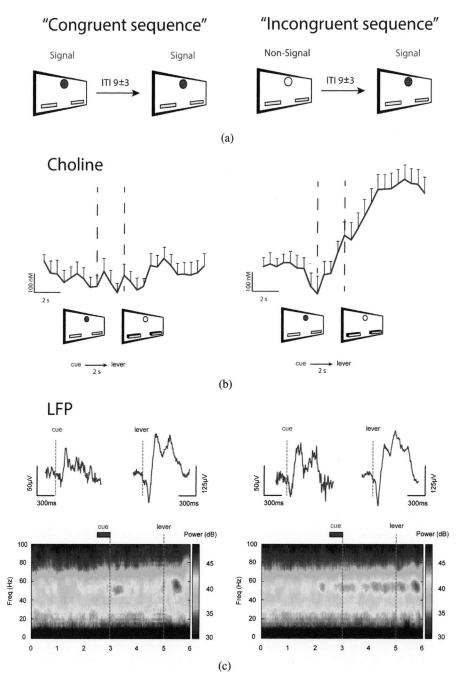

Figure 2.2 Trial sequence specific changes in mPFC signaling: (a) Cartoon illustrating trial-sequences for the two types of hits described in the main text (left: consecutive signals — congruent hits; right: signal trials separated by non-signal trials — incongruent hits). (b) Group averages of extra-cellular choline fluctuations from task-performing animals (Howe et al., 2013). Signal cues evoke cholinergic transients only if they are detected and only if they follow a non-signal trial

Figure 2.2 (*Continued*) response. (c) Mean LFPs (top) and spectral analysis (bottom) across all congruent and incongruent hit sequences from a standard 40 minutes recording session. Both detected signal cues and lever extension produce ERPs, independent of trial sequence (top). Mean session spectrograms reveal that detected cues on incongruent signal trials evoke a sustained increase in gamma frequency power that is maintained through the lever extension and reward period (bottom) temporally coinciding with cholinergic transients evoked by these trial sequences.

In addition to the rise in ACh concurrent with signal presentation, we noted that the increase in cholinergic activity on incongruent hits continued well beyond the time the animal made a response and received reward. We propose this ongoing release may have another function beyond cortical and cognitive state shifts and cue detection: namely acting as a reporter or learning signal. This speculation is in agreement with the hypothesis that ACh can increase the strength of the relationship between cue and response by augmenting long-term potentiation in both the hippocampus and cortex, effectively lowering the stimulation threshold for inducing an increase in synaptic strength (Hasselmo and Barkai, 1995; Huerta and Lisman, 1993). Further, stimulation of basal forebrain cholinergic projections has been shown to modulate gain and enhance the cortical coding of a sensory stimulus (Goard and Dan, 2009; Metherate and Ashe, 1991, 1993). In the context of animals performing the SAT, a cholinergic learning signal may serve to stabilize or increase the readiness for shifting cortical networks and refresh the processing of the response rule for cues (e.g., "if signal press left lever").

Employing biosensor technology and the SAT, we have been able to refine the role of mPFC cholinergic transients in attentional performance from a neural system generally involved in signal detection to a discrete signal observed on only select hit trials. Correct action-sequence selection, reward presentation, retrieval, or cue-detection *per se* are not sufficient for predicting the generation of transient ACh release in the SAT; rather it is only on these specific trials involving a transition from monitoring to cue-detection that cholinergic transients are observed. Preliminary data using optogenetic manipulation of prefrontal cholinergic activity supports a causal role of cholinergic transients in this process (see below, Gritton et al., 2012), and future experiments employing similar techniques can begin to further reveal how these cholinergic transients support cognitive performance outside the context of the SAT.

2.5 GAMMA SYNCHRONY AS A NEUROPHYSIOLOGICAL MARKER OF CHOLINERGIC MODULATION OF CORTICAL CIRCUITRY

Our interest in oscillations during attentional performance was driven by our desire to correlate changes in the overall state of the local prefrontal network, as captured by the local field potential (LFP), with cholinergic transients. We theorized that salient cues during SAT performance must recruit attentional networks *via* cholinergic signaling and this recruitment would be reflected in the appearance of an event related potential (ERP). We found that cue presentation on hits and lever extension on non-signal trials consistently evoked ERPs. While cholinergic signaling during SAT performance was quite specific (incongruent hits on signal trials only), ERPs were consistently evoked by stimuli in the environment that initiated a change in behavior (e.g., signal cue, lever extension, Figure 2.2). In addition, ERPs occurred during every detected cue presentation, independent of trial sequence, and suggest that while ERPs may be modulated by cholinergic signaling; their presence is not dependent on cholinergic release.

Field potential analysis also provided insight into oscillatory activity at different phases of task performance. Over the last two decades, researchers have shown an increased interest in the potential role of oscillatory activity in neural networks in serving cognitive and mnemonic processes. In particular, studies of cognition have revealed that gamma oscillations are coincident with signal discrimination, attentional shifting of behavior towards sensory stimuli, and executive processing and conflict resolution (e.g., filtering of distractors) in both humans and animals (Gray et al., 1989, Posner and Petersen, 1990, Fries et al., 2001, Fan et al., 2007). Gamma oscillations within the neocortex and hippocampus have also been shown to be sensitive to cholinergic manipulations (e.g., Fisahn et al., 1998, Rodriguez et al., 2004, Gulyas et al., 2010), therefore we speculated that cholinergic transients may be a key mediator of gamma power during attentional performance.

Initial analysis revealed that like ERPs, the emergence of gamma activity associated with behavioral orienting in rodents occurred on all detected signal trials, independent of sequence, and suggested that gamma emergence does not require ACh. We also found that neither reward presentation nor reward retrieval was associated with power increases in the gamma frequency band. Although our findings of ERPs and gamma oscillations coincident with all detected signal trials seemed discordant with trial specific recruitment of cholinergic activity, trial-sequence analysis did revealed robust differences between incongruent and congruent signal trial types. Gamma oscillations generated during cue presentation on incongruent hits were larger and more sustained when compared to consecutive hit trials (Figure 2.2). This augmented gamma signal persisted through the reward period, temporally coinciding with the sustained profile of cholinergic release identified *via* electrochemical recordings.

Traditionally, selective attention and cortical oscillations have been ascribed with the role of enhancing neuronal representations that provide information about discrete stimuli through the selection and augmentation of relevant features (gain modulation). However, recent theories of selective attention have also included the possible role of gamma oscillations in synchronizing discrete cortical circuits during signal processing that contains relevant behavioral information (Buzsaki and Draguhn, 2004, Womelsdorf and Fries, 2007, Benchenane et al., 2011). Our results support this most recent tenet and suggest that sustained periods of ACh release may promote coincident and persistent emergence of gamma oscillations in prefrontal networks. Under this condition, cholinergic signaling may facilitate synchronous high frequency oscillatory activity to promote the coordination of cortical–cortical and cortical–subcortical networks to redirect behavior from internal monitoring to cue-directed attention.

2.6 MOVING FORWARD: POTENTIAL NEUROBIOLOGICAL IMPACT OF NEW ELECTRODES AND RECORDING METHODS

Taking advantage of the latest tools for selectively probing the functions of the cholinergic system, we have been able to identify its central role in attentional performance, its tonic contribution to the top-down modulation of such performance, and finally the discovery of cue-evoked transient increases in medial prefrontal (mPF) ACh that mediate a very discrete switch in cognitive operations. Looking forward, we consider the capabilities that appear necessary for the next generation of biosensor technology to lead to further insights. The development of biosensors with the capacity for measuring multiple neurotransmitters and electrophysiological activity simultaneously is chief among desired attributes.

Cholinergic transients are hypothesized to be the product of local glutamatergic–cholinergic interactions. This hypothesis of the genesis of cholinergic transients is the net sum of a substantial body of data generated by local pharmacological manipulations of glutamatergic and cholinergic activity in anesthetized animals (Parikh et al., 2008, 2010). Preliminary results from the recording of glutamate transients in the mPFC of rats performing the SAT support this hypothesis. On hits, cue-evoked glutamate release precedes and peaks before ACh release. Further, the amplitude of glutamate release is signal duration dependent, simultaneously dictating the probability of generating a cholinergic transient and of a visual cue being detected (Howe and Sarter, 2010). This work, both in anesthetized and performing animals, has been done in separate groups where either cholinergic or glutamatergic activity was measured independently. A true test of our hypothesis concerning cholinergic–glutamatergic interactions necessitates a biosensor with the capacity for measuring glutamate and ACh at the same time, without sacrificing a independent "sentinel" channel on the same electrode assembly to control for fluctuations in current not due to the oxidation of peroxide resulting from the breakdown of either glutamate or ACh.

Our work correlating changes in the LFP to cholinergic and glutamatergic release events has furthered our understanding of the local prefrontal circuitry, underlying discrete aspects of attentional performance. We speculate that glutamate signaling coincident with cholinergic release events (e.g., Howe and Sarter, 2010) leads to the emergence of brief periods of gamma oscillations, lasting as long as the stimulus cue is present independent of trial sequence. Acetylcholine release on the other hand, promotes gamma emergence and sustains it through the decision process and the reward phase of the task. Similar to the proposed relationship between glutamatergic activity and cholinergic activity, this link to electrophysiology has been established *via* correlation across separate experiments.

Biosensor designs that offer the opportunity for simultaneous quantification of neurotransmitter release and changes in the electrophysiological signature of local networks would aid in efforts to move such work from the domain of speculation to that of causality. Further, the addition of recording sites on biosensor arrays with properties (e.g., impedance) that allow for measurements of single unit activity would also open the door to further hypothesis testing, such as our own proposing the induction of persistent spiking activity as a means of maintaining the state of prefrontal networks across successive hit trials (Hasselmo and Sarter, 2011).

2.7 SUMMARY AND CONCLUSIONS

We are approaching the ten year anniversary of utilizing enzyme-coated microelectrodes to investigate the function of the cholinergic system in task-performing animals. Clearly, this work has yielded a major revision of our classical view of the cholinergic system behaving as a diffuse neuromodulator. We now consider phasic or transient cholinergic signaling events as a cortically generated signal, as opposed to a signal primarily reflecting a property of the ascending cholinergic system. In the evolution of the cortex, inputs from ascending systems may have been increasingly integrated into cortical circuitry *via* the incorporation of heteroreceptors expressed at the terminals of these inputs. As cognitive operations simultaneously increased in complexity, these terminals began to play an increasingly pivotal role in the mediation of essential operations. As discussed above, cholinergic transients mediate the detection of cues in trials involving cortical and attentional mode shifts. Such a process requires cue-related and attentional state-related information processing across multiple telencephalic regions. The enhanced gamma power observed during "incongruent hits" (see above) suggests that the postsynaptic impact of cholinergic transients helps synchronize information processing across regions. Shifting from a state of monitoring for cues to cue-directed behavior concerns a profound change in the brain's processing state, occurs on a very brief temporal scale, as it is trial sequence-based. Cholinergic transients, as we understand them at this point, appear to fit the characteristics of a signal mediating these transitions.

Recording studies by definition generate correlational evidence. Ongoing studies using optogenetic methods to generate transients which reproduce, in terms of rise time, amplitude, and decay rate, those seen during incongruent hits, have confirmed that cholinergic transients drive cue detection behavior. Transients coinciding with signal events in the SAT also enhance detection rates. Perhaps an even more impressive proof of the causal power of cholinergic transients is that optogenetically evoked, ill-timed transients, coinciding with non-signal event windows, force false alarms (Gritton et al., in preparation).

Many questions remain unanswered, including those concerning whether the modulatory, tonic component, of ACh impacts the occurrence of cholinergic transients only indirectly *via* glutamatergic–cholinergic transient interactions (Hasselmo and Sarter, 2011), or whether there is also direct modulation of transient-generating terminals *via* mAChR or nAChR. Furthermore, we need to understand how transients are suppressed during consecutive hits (see above) and/or generated selectively during incongruent hits. It is necessary assumption that additional circuitry, including local gamma-aminobutyric acid (GABA) ergic interneurons and presynaptic autoreceptors are involved in the generation and/or suppression of transients. Can coordinated glutamatergic and cholinergic activity induce persistent spiking in local pyramidal cells and if so, what mechanism terminates that state? In what other cognitive behavioral/contexts are cholinergic transients observed, what are the release kinetics, and how does this information contribute to our understanding of their function?

Finally, our research on linking transient cholinergic activity in rats to brain activity in humans (Howe et al., 2013) illustrates the productivity of efforts to integrate basic behavioral neuroscience research with research approaches typical in cognitive neuroscience. These integrative efforts may be expanded to consider the translational significance of our new insights into cholinergic functions. Various disease processes are associated with impairments in attentional performance and the associated efficacy of cue detection and attentional shifting. Combining recordings from animal models and parallel BOLD-functional magnetic resonance imaging (fMRI) studies in patients performing the SAT may suggest that impairments in the generation of cholinergic transients essentially contribute to the cognitive symptoms of neuropsychiatric disorders, and as we acquire more insights into the cortical circuitry involved in the generation of these transients, new targets for treating these disorders may begin to emerge.

ACKNOWLEDGMENT

The authors' research was supported by NIH grants 1R01MH086530-01 and 1PO1 DA031656.

REFERENCES

Arnold HM, Burk JA, Hodgson EM, Sarter M, Bruno JP (2002) Differential cortical acetylcholine release in rats performing a sustained attention task *versus* behavioral control tasks that do not explicitly tax attention. Neuroscience 114:451–460.

Benchenane K, Tiesinga PH., et al. (2011) "Oscillations in the prefrontal cortex: a gateway to memory and attention." Curr Opin Neurobiol 21:475–485.

Bentley P, Husain M, Dolan RJ (2004) Effects of cholinergic enhancement on visual stimulation, spatial attention, and spatial working memory. Neuron 41:969–982.

Botly LC, De Rosa E (2009) Cholinergic deafferentation of the neocortex using 192 IgG-saporin impairs feature binding in rats. J Neurosci 29:4120–4130.

Broussard JI, Karelina K, Sarter M, Givens B (2009) Cholinergic optimization of cue-evoked parietal activity during challenged attentional performance. Eur J Neurosci 29:1711–1722.

Bruno JP, Sarter M, Moore AH, Himmelheber AM (1999) *In vivo* neurochemical correlates of cognitive processes: methodological and conceptual challenges. Rev Neurosci 10:25–48.

Burmeister JJ, Moxon K, Gerhardt GA (2000) Ceramic-based multisite microelectrodes for electrochemical recordings. Anal Chem 72:187–192.

Burmeister JJ, Palmer M, Gerhardt GA (2003) Ceramic-based multisite microelectrode array for rapid choline measures in brain tissue. Anal Chim Acta 481:65–74.

Buzsaki G, Draguhn A (2004) Neuronal oscillations in cortical networks. Science 304:1926–1929.

Dalley JW, Theobald DE, Bouger P, Chudasama Y, Cardinal RN, Robbins TW (2004) Cortical cholinergic function and deficits in visual attentional performance in rats following 192 IgG-saporin-induced lesions of the medial prefrontal cortex. Cereb Cortex 14:922–932.

Damsma G, Biessels PT, Westerink BH, De Vries JB, Horn AS (1988) Differential effects of 4-aminopyridine and 2,4-diaminopyridine on the *in vivo* release of acetylcholine and dopamine in freely moving rats measured by intrastriatal dialysis. Eur J Pharmacol 145:15–20.

Day J, Damsma G, Fibiger HC (1991) Cholinergic activity in the rat hippocampus, cortex and striatum correlates with locomotor activity: an *in vivo* microdialysis study. Pharmacol Biochem Behav 38:723–729.

Demeter E, Guthrie SK, Taylor SF, Sarter M, Lustig C (2013) Increased distractor vulnerability but preserved vigilance in patients with schizophrenia: evidence from a translational sustained attention task. Schizophr Res 144:136–141.

Demeter E, Hernandez-Garcia L, Sarter M, Lustig C (2011) Challenges to attention: a continuous arterial spin labeling (ASL) study of the effects of distraction on sustained attention. Neuroimage 54:1518–1529.

Demeter E, Sarter M, Lustig C (2008) Rats and humans paying attention: cross-species task development for translational research. Neuropsychology 22:787–799.

Egorov AV, Hamam BN, Fransen E, Hasselmo ME, Alonso AA (2002) Graded persistent activity in entorhinal cortex neurons. Nature 420:173–178.

Fan J, Byrne J, Worden MS, Guise KG, McCandliss BD, Fossella J, Posner MI (2007) The relation of brain oscillations to attentional networks. J Neurosci 27:6197–6206.

Fisahn A, Pike FG, Buhl EH, Paulson O (1998) Cholinergic induction of network oscillations at 40 Hz in the hippocampus *in vitro*. Nature 394:186–189.

Fransen E, Alonso AA, Hasselmo, ME (2002) Simulations of the role of the muscarinic-activated calcium-sensitive nonspecific cation current INCM in entorhinal neuronal activity during delayed matching tasks. J Neurosci, 22:1081–1097.

Fries P, Reynolds JH, Rorie AE, Desimone R (2001) Modulation of oscillatory neuronal synchronization by selective visual attention. Science 291:1560–1563.

Furey ML, Pietrini P, Alexander GE, Schapiro MB, Horwitz B (2000) Cholinergic enhancement improves performance on working memory by modulating the functional activity in distinct brain regions: a positron emission tomography regional cerebral blood flow study in healthy humans. Brain Res Bull 51:213–218.

Garguilo MG, Michael AC (1996) Amperometric microsensors for monitoring choline in the extracellular fluid of brain. J Neurosci Methods 70:73–82.

Gill TM, Sarter M, Givens B (2000) Sustained visual attention performance-associated prefrontal neuronal activity: evidence for cholinergic modulation. J Neurosci 20:4745–4757.

Giuliano C, Parikh V, Ward JR, Chiamulera C, Sarter M (2008) Increases in cholinergic neurotransmission measured by using choline-sensitive microelectrodes: enhanced detection by hydrolysis of acetylcholine on recording sites? Neurochem Int 52:1343–1350.

Goard M, Dan Y (2009) Basal forebrain activation enhances cortical coding. Nat Neurosci 12:1444–1449.

Gray CM, Konig P, Engel AK, Singer W (1989) Oscillatory responses in cat visual cortex exhibit inter-columnar synchronization which reflects global stimulus properties. Nature 338:334–337.

Gulyas AI, Szabo GG, Ulbert I, Holderith N, Monyer H, Erdelyi F, Szabo G, Freund TF, Hajos N (2010) Parvalbumin-containing fast-spiking basket cells generate the field potential oscillations induced by cholinergic receptor activation in the hippocampus. J Neurosci 30:15134–15145.

Hasselmo ME, Barkai E (1995) Cholinergic modulation of activity dependent synaptic plasticity in the piriform cortex and associative memory function in a network biophysical simulation. J Neurosci 15:6592–6604.

Hasselmo ME, Sarter M (2011) Modes and models of forebrain cholinergic neuromodulation of cognition. Neuropsychopharmacology 36:52–73.

Herzog CD, Nowak KA, Sarter M, Bruno JP (2003) Microdialysis without acetylcholinesterase inhibition reveals an age-related attenuation in stimulated cortical acetylcholine release. Neurobiol Aging 24:861–863.

Himmelheber AM, Fadel J, Sarter M, Bruno JP (1998) Effects of local cholinesterase inhibition on acetylcholine release assessed simultaneously in prefrontal and frontoparietal cortex. Neuroscience 86:949–957.

Himmelheber AM, Sarter M, Bruno JP (1997) Operant performance and cortical acetylcholine release: role of response rate, reward density, and non-contingent stimuli. Cogn Brain Res 6:23–36.

Himmelheber AM, Sarter M, Bruno JP (2000) Increases in cortical acetylcholine release during sustained attention performance in rats. Cogn Brain Res 9:313–325.

Himmelheber AM, Sarter M, Bruno JP (2001) The effects of manipulations of attentional demand on cortical acetylcholine release. Cogn Brain Res 12:353–370.

Holley LA, Wiley RG, Lappi DA, Sarter M (1994) Cortical cholinergic deafferentation following the intracortical infusion of 192 IgG-saporin: a quantitative histochemical study. Brain Res 663:277–286.

Howe WM, Sarter M (2010) Prefrontal glutamatergic-cholinergic interactions for attention: glutamatergic coding of signal salience as a function of performance levels. In: Monitoring molecules in neuroscience (Westerink B, Clinckers R, Smolders I, Sarre S, Michotte Y, eds), pp 57–59. Brussels: Vrije Universiteit.

Howe WM, Berry AS, Francois J, Gilmour G, Carp JM, Tricklebank M, Lustig C, Sarter M (2013) Prefrontal cholinergic mechanisms instigating shifts from monitoring for cues to cue-guided performance: converging electrochemical and fMRI evidence from rats and humans. J Neurosci 33:8742–8752.

Huang Z, Villarta-Snow R, Lubrano GJ, Guilbault GG (1993) Development of choline and acetylcholine Pt microelectrodes. Anal Biochem 215:31–37.

Huerta PT, Lisman JE (1993) Heightened synaptic plasticity of hippocampal CA1 neurons during a cholinergically induced rhythmic state. Nature 364:723–725.

Klein J, Köppen A, Löffelholz K (1998) Regulation of free choline in rat brain: dietary and pharmacological manipulations. Neurochem Int 32:479–485.

Kozak R, Bruno JP, Sarter M (2006) Augmented prefrontal acetylcholine release during challenged attentional performance. Cereb Cortex 16:9–17.

Kozak R, Martinez V, Young D, Brown H, Bruno JP, Sarter M (2007) Toward a neurocognitive animal model of the cognitive symptoms of schizophrenia: disruption of cortical cholinergic neurotransmission following repeated amphetamine exposure in attentional task-performing, but not non-performing, rats. Neuropsychopharmacology 32:2074–2086.

Metherate R, Ashe JH (1991) Basal forebrain stimulation modifies auditory cortex responsiveness by an action at muscarinic receptors. Brain Res 559:163–167.

Metherate R, Ashe JH (1993) Nucleus basalis stimulation facilitates thalamocortical synaptic transmission in the rat auditory cortex. Synapse 14:132–143.

McGaughy J, Kaiser T, Sarter M (1996) Behavioral vigilance following infusions of 192 IgG-saporin into the basal forebrain: selectivity of the behavioral impairment and relation to cortical AChE-positive fiber density. Behav Neurosci 110:247–265.

McGaughy J, Sarter M (1995) Behavioral vigilance in rats: task validation and effects of age, amphetamine, and benzodiazepine receptor ligands. Psychopharmacology (Berl) 117:340–357.

McGaughy J, Sarter M (1998) Sustained attention performance in rats with intracortical infusions of 192 IgG-saporin-induced cortical cholinergic deafferentation: effects of physostigmine and FG 7142. Behav Neurosci 112:1519–1525.

McGaughy J, Turchi J, Sarter M (1994) Crossmodal divided attention in rats: effects of chlordiazepoxide and scopolamine. Psychopharmacology (Berl) 115:213–220.

Neigh GN, Arnold HM, Rabenstein RL, Sarter M, Bruno JP (2004) Neuronal activity in the nucleus accumbens is necessary for performance-related increases in cortical acetylcholine release. Neuroscience 123:635–645.

Paolone G, Angelakos C, Meyer J, Robinson E, Sarter M (2013) Cholinergic control over attention in rats prone to attribute incentive salience to reward cues. J Neurosci 33:8321–8335.

Paolone G, Howe WM, Gopalakrishnan M, Decker MW, Sarter M (2010) Regulation and function of the tonic component of cortical acetylcholine release. In: Monitoring molecules in neuroscience (Westerink B, Clinckers R, Smolders S, Sarre S, Michotte Y, eds), pp. 363–365. Brussels: Vrije Universiteit Brussels.

Parikh V, Ji J, Decker MW, Sarter M (2010) Prefrontal beta2 subunit containing and alpha7 nicotinic acetylcholine receptors differentially control glutamatergic and cholinergic signaling. J Neurosci 30:3518–3530.

Parikh V, Kozak R, Martinez V, Sarter M (2007) Prefrontal acetylcholine release controls cue detection on multiple timescales. Neuropharmacology 56:141–154.

Parikh V, Man K, Decker MW, Sarter M (2008) Glutamatergic contributions to nicotinic acetylcholine receptor agonist-evoked cholinergic transients in the prefrontal cortex. J Neurosci 28:3769–3780.

Parikh V, Pomerleau F, Huettl P, Gerhardt GA, Sarter M, Bruno JP (2004) Rapid assessment of *in vivo* cholinergic transmission by amperometric detection of changes in extracellular choline levels. Eur J Neurosci 20:1545–1554.

Parikh V, Sarter M (2006) Cortical choline transporter function measured *in vivo* using choline-sensitive microelectrodes: clearance of endogenous and exogenous choline and effects of removal of cholinergic terminals. J Neurochem 97:488–503.

Parikh V, St. Peters M, Blakely RD, Sarter M (2013) The presynaptic choline transporter imposes limits on sustained cortical acetylcholine release and attention. J Neurosci 33:2326–2337.

Peng HB, Xie H, Rossi SG, Rotundo RL (1999) Acetylcholinesterase clustering at the neuromuscular junction involves perlecan and dystroglycan. J Cell Biol 145:911–921.

Posner MI, Snyder CR, Davidson BJ (1980) Attention and the detection of signals. J Exp Psychol 109:160–174.

Posner MI, Petersen SE (1990) The attention system of the human brain. Annu Rev Neurosci 13: 25–42.

Quinn DM (1987) Acetylcholinesterase: enzyme structure, reaction dynamics, and virtual transition states. Chem Rev 87:955–979.

Rodriguez R, Kallenbach U, Singer W, Munk MH (2004) Short- and long-term effects of cholinergic modulation on gamma oscillations and response synchronization in the visual cortex. J Neurosci 24:10369–10378.

Sarter M, Paolone G, Mabrouk OS, Kennedy RT (2012) Sampling from injured tissue as a blessing in disguise: tonic changes in cholinergic neurotransmission using microdialysis. In: Monitoring molecules in neuroscience, 14th International Conference. London: Imperial College.

Sarter M, Gehring WJ, Kozak R (2006) More attention must be paid: the neurobiology of attentional effort. Brain Res Rev 51:145–160.

Sarter M, Martinez V, Kozak R (2009) A neurocognitive animal model dissociating between acute illness and remission periods of schizophrenia. Psychopharmacology (Berl) 202:237–258.

Sarter M, Paolone G (2011) Deficits in attentional control: cholinergic mechanisms and circuitry-based treatment approaches. Behav Neurosci 125:825–835.

Sarter M, Parikh V, Howe WM (2009) Phasic acetylcholine release and the volume transmission hypothesis: time to move on. Nat Rev Neurosci 10:383–390.

Schiff D, Shah A, Hudson E, Nauvel T, Kalik F, Purpura P (2013) Gating of attentional effort through the central thalamus. J Neurophysiol 109:1152–1163.

Scremin OU, Li MG, Roch M, Booth R, Jenden DJ (2006) Acetylcholine and choline dynamics provide early and late markers of traumatic brain injury. Brain Res 1124:155–166.

Silver MA, Shenhav A, D'Esposito M (2008) Cholinergic enhancement reduces spatial spread of visual responses in human early visual cortex. Neuron 60:904–914.

Song P, Mabrouk OS, Hershey ND, Kennedy RT (2012) *In Vivo* neurochemical monitoring using benzoyl chloride derivatization and liquid chromatography-mass spectrometry. Anal Chem. 84:412–419.

St Peters M, Cherian AK, Bradshaw M, Sarter M (2011a) Sustained attention in mice: expanding the translational utility of the SAT by incorporating the michigan controlled access response port (MICARP). Behav Brain Res 225:574–583.

St Peters M, Demeter E, Lustig C, Bruno JP, Sarter M (2011b) Enhanced control of attention by stimulating mesolimbic-corticopetal cholinergic circuitry. J Neurosci 31:9760–9771.

Turchi J, Sarter M (1997) Cortical acetylcholine and processing capacity: effects of cortical cholinergic deafferentation on crossmodal divided attention in rats. Brain Res Cogn Brain Res 6:147–158.

Westerink BH, Hofsteede HM, Damsma G, de Vries JB (1988) The significance of extracellular calcium for the release of dopamine, acetylcholine and amino acids in conscious rats, evaluated by brain microdialysis. Naunyn Schmiedebergs Arch Pharmacol 337:373–378.

Womelsdorf T, Fries P (2007) The role of neuronal synchronization in selective attention. Curr Opin Neurobiol 17:154–160.

Xin Q, Wightman RM (1997) Transport of choline in rat brain slices. Brain Res 776:126–132.

Zmarowski A, Sarter M, Bruno JP (2005) NMDA and dopamine interactions in the nucleus accumbens modulate cortical acetylcholine release. Eur J Neurosci 22:1731–1740.

CHAPTER 3

REAL-TIME MEASUREMENT OF ATP AND ADENOSINE IN THE NERVOUS SYSTEM

Nicholas Dale
School of Life Sciences, University of Warwick

3.1 INTRODUCTION

The purines, adenosine triphosphate (ATP) and adenosine, are important biological signaling agents. Together they comprise an interdependent signaling system of surprising complexity (Figure 3.1). In this chapter I shall provide a brief overview of the fundamentals of purinergic signaling before reviewing how our use of biosensors has helped to deepen understanding of the signaling mechanisms and roles of the purines in several different physiological and developmental contexts.

3.1.1 THE COMPONENTS OF PURINERGIC SIGNALING

By virtue of the large reservoir of intracellular ATP relative to adenosine diphosphate (ADP), ATP acts as the universal store of chemical energy that powers cellular and molecular processes. However, ATP is also an important extracellular signaling molecule. The concept of ATP as a transmitter was elaborated by Burnstock in the 1970s with it being recognized firstly as a peripheral neurotransmitter (Su et al., 1971; Burnstock, 1972) and then much later as a centrally acting neurotransmitter (Edwards et al., 1992). By contrast the role of adenosine as a modulator acting on the heart was described in the 1920s (Drury and Szent-Gyorgyi, 1929). By now the purines are universally accepted as important extracellular signaling molecules in virtually every organ of the body including the brain (Burnstock, 1997, 2006; Burnstock et al., 2011). For the newcomer to the field of

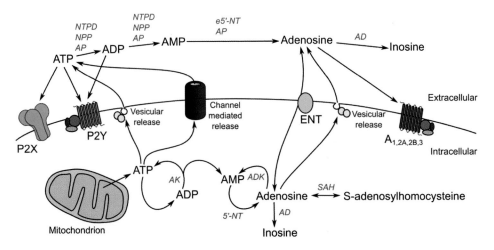

Figure 3.1 The components of purinergic signaling. The interconversion of ATP to adenosine inside and outside the cell, with release mechanisms (from vesicles and *via* large conductance channels) and receptors indicated. Enzymes responsible for interconversion steps are shown in blue: AK, adenylate kinase; ADK, adenosine kinase; AD, adenosine deaminase; AP, alkaline phosphatase; ENT, equilibrative nucleoside transporter; 5'-NT, 5'-nucleotidase; e5'-NT, ecto 5'-nucleotidase; NTPD ecto-nucleoside triphosphate dihydrophosphorylase; NPP, ecto-nucleotide pyrophosphatase; SAH, S-adenosylhomocysteine hydrolase.

purinergic signaling there are many excellent reviews especially those of Burnstock. I recently wrote a historical perspective for the Biochemical Journal which may be a good introduction for some readers (Dale, 2012). What follows is a brief, rather than comprehensive, overview to provide sufficient information and context for the experimental work that I describe later in this chapter.

3.1.1.1 ATP receptors and release mechanisms

ATP can act at both ionotropic (P2X) (Khakh, 2001) and metabotropic receptors (P2Y) (Abbracchio et al., 2006) (Figure 3.1). There are seven subtypes of P2X receptor subunits each having somewhat different properties. These subunits can combine to form a range of homotrimeric and heterotrimeric receptors (Kawate et al., 2009). The P2X receptors can only be activated by ATP and certain ATP analogs. They are non-selective cation channels that have significant permeability to Ca^{2+}. The P2X7 receptor has a curious and potentially significant property in that during continued exposure to ATP it undergoes a conformational change whereby the pore dilates to allow passage of low molecular weight compounds including ATP itself (Surprenant et al., 1996). Although originally described for the P2X7 receptor, pore dilation also occurs in P2X2 (Chaumont and Khakh, 2008) and P2X4 receptors (Khakh et al., 1999). The mechanisms of this pore dilation were controversial for some time, with some authors claiming the involvement of pannexin-1 hemichannels. This controversy has now been laid to rest by the demonstration that the P2X7 pore itself can allow permeation of large dye molecules (Browne et al., 2013).

Although there are 14 P2Y subtypes (each being a seven transmembrane segment G–protein coupled receptor) only a subset of these are considered to be ATP/ADP receptors (P2Y1, P2Y2, P2Y4, P2Y6, P2Y11, P2Y12 and P2Y13) (Abbracchio et al., 2006). Some receptors can be activated by ATP alone, some by ATP or ADP while others are ADP-preferring. For example the P2Y1 receptor can be preferentially activated by ADP and couples to G_q and G_o. The P2Y1 receptor can activate the IP_3 pathway leading to mobilization of intracellular Ca^{2+}. P2Y12 and P2Y13 are also ADP preferring receptors. Certain P2Y receptors can also be activated by uridine triphosphate (UTP) and in some cases uridine diphosphate (UDP). This raises the possibility that endogenous UTP/UDP may have a signaling role related to that of ATP/ADP. P2Y receptors can also couple to, and modulate, specific ion channels such as N-type Ca^{2+} channels and M-type K^+ channels.

ATP can be released from synaptic vesicles *via* exocytosis. A transporter called vesicular nucleotide transporter (V-NUT) selective for the nucleotides has been identified that loads ATP into vesicles (Sawada et al., 2008), and is a useful marker for cells, including astrocytes, that release ATP by this mechanism. However there are other mechanisms of ATP release (Figure 3.1) which include gap junction hemichannels (both pannexins and connexins) (Bao et al., 2004; Kang et al., 2008; Huckstepp et al., 2010a, 2010b), volume activated channels (Sabirov and Okada, 2005; Fields and Ni, 2010) and even the P2X7 receptor itself following pore dilation (Pellegatti et al., 2005).

3.1.1.2 Adenosine receptors and release mechanisms

Adenosine can claim to be one of the most important modulators in the central nervous system. There are no ionotropic channels for adenosine which act only *via* a set of four G-protein coupled receptors (A_1, A_{2A}, A_{2B} and A_3) (Fredholm et al., 2001) (Figure 3.1). The A_1 receptor in particular is widespread in brain, is implicated in neuroprotection (Rudolphi et al., 1992; Dale and Frenguelli, 2009) and control of sleep (Radulovacki et al., 1984; Porkka-Heiskanen et al., 1997; Porkka-Heiskanen and Kalinchuk, 2011), and has a range of pre- and post-synaptic inhibitory effects.

While adenosine can often arise in the extracellular space from prior release of ATP (3.1.1.3), it can be transported across membranes *via* equilibrative nucleoside transporters (ENTs) (Baldwin et al., 2004) which are abundant in the brain. ENTs equalize adenosine concentrations across the plasma membrane. Functionally they can act both to release adenosine into, and remove adenosine from the extracellular space depending on the relative concentrations of adenosine in the cytosol and extracellular space. A somewhat counter intuitive concept is that one of the main determinants of extracellular adenosine, is the intracellular enzyme adenosine kinase (ADK) (Boison, 2006). This enzyme converts adenosine to adenosine monophosphate (AMP). Thus, high activities of adenosine kinase can reduce the intracellular concentration of adenosine and maintain an inward driving force for the transport of adenosine from the extracellular space *via* the ENTs (Figure 3.1).

3.1.1.3 Ectonucleotidases — breakdown of ATP and initiation of ADP and adenosine signaling

The ectonucleotidases are extracellular enzymes that fulfil an essential role — they terminate the actions of ATP. However their actions are an important and distinctive aspect of purinergic signaling: key products of the ectonucleotidases (ADP and adenosine) have their own specific receptors and are thus biologically active. The ectonucleotidases therefore not only terminate the actions of ATP but also initiate the actions of these downstream signaling molecules (Figure 3.1).

There are four possible gene families that contribute to this conversion of ATP (Langer et al., 2008). The first is the nucleoside triphosphate dihydrophosphorylase (NTPDase) family (also known as cluster of differentiation 39 (CD39)). The NTPDases are integral membrane proteins and exist in eight variants, five of which are found on the plasma membrane, while the others are found only in intracellular compartments (NTPDase 4, 6 and 7). Some NTPDases have the potential to be secreted (Zimmermann, 1999; Zimmermann and Braun, 1999; Zimmermann, 2000; Masse et al., 2006; Robson et al., 2006; Yegutkin, 2008). All members of the NTPDase family exhibit distinct patterns of expression suggesting distinctive functional roles. The members of the NTPDase family also differ in their substrate affinities and enzymatic reactions. For example NTPDase1 will metabolize both ATP and ADP with equal efficiency, whereas NTPDase2 acts only on ATP. Thus expression of NTPDase1 results in accumulation of AMP, whereas expression of NTPDase2 will result in accumulation of ADP. The nucleotide pyrophosphatase (NPPase) family comprises of seven members. These enzymes have a single transmembrane domain and exhibit differing expression patterns and enzymatic activities (Stefan et al., 2005, 2006; Yegutkin, 2008; Masse et al., 2010). NPPases1–3 are able to hydrolyze both ATP and ADP. The alkaline phosphatase family, linked to the membrane *via* a Glycosylphosphatidylinositol (GPI) anchor, have very wide substrate specificity (ATP, ADP, AMP and even pyrophosphate) (Yegutkin, 2008). Alkaline phosphatases have particularly been associated with neural progenitor cells and their differentiation (Langer et al., 2007; Kermer et al., 2010; Diez-Zaera et al., 2011).

The final gene is the ecto 5'-nucleotidase (e5'-NT, also known as cluster of differentiation 73 (CD73)) that converts AMP to adenosine (Zimmermann, 1992; Yegutkin, 2008). This enzyme is linked to the membrane *via* a GPI anchor and is a key enzyme for the production of adenosine from AMP in the brain. It also has the curious property that it can be inhibited by ATP and ADP (Gordon et al., 1986; Slakey et al., 1986; James and Richardson, 1993), termed feed-forward inhibition, which may be of significance in providing temporal and spatial separation between the release of ATP and subsequent accumulation of adenosine (James and Richardson, 1993; Dale, 1998; Masse and Dale, 2012).

3.1.1.4 The need for real-time measurements of ATP and adenosine

When the hypothesis of chemical neurotransmission was elaborated, four essential criteria (Werman, 1966) were put forward to identify a potential neurotransmitter: the release of the neurotransmitter by physiological stimuli; the synthesis and presence of the neurotransmitter in the pre-synaptic terminal; the ability of exogenous application of the proposed neurotransmitter to mimic the biological actions of the endogenous compound; and a mechanism to terminate the actions of the neurotransmitter. As chemical neurotransmission became widely accepted, the need to demonstrate release of the neurotransmitter (not an easy task) became less urgent and was largely replaced by the demonstration that selective receptor antagonists could block the actions of the endogenous neurotransmitter.

One of the big unknowns for understanding ATP signaling mechanisms are the cellular sources for its release. ATP receptors, both P2X and P2Y, are widely distributed in the central and peripheral nervous systems, begging the question "where does the ATP come from?". Every cell contains millimolar concentrations of ATP, and is therefore a potential source of ATP (Gordon et al., 1986). Given the multiplicity of possible release mechanisms, many of which will occur outside the conventional synaptic structures, methods to measure and localize ATP release in real-time are essential.

The routes for adenosine release and the cellular sources are also poorly understood, an understanding that is hampered by the lack of ionotropic receptors capable of signaling fast synaptic transmission. Receptors for adenosine, like those for ATP, are widely distributed, yet the cellular sources for adenosine are far from obvious. For adenosine that arises from prior release of ATP, the distribution of the ectonucleotidases will be a critical factor and these may not correlate to defined anatomical structures. For the direct release of adenosine many cells have ENTs that could act to release adenosine. But this by itself is not enough: a cell's complement of intracellular enzymes, which can either create or sequester adenosine, determines whether that cell acts as a source or sink of adenosine. Once again the ability to make spatially resolved real-time measurements of adenosine has been essential to progress understanding.

3.2 ATP-RELEASE, SIGNALING AND FUNCTIONS

3.2.1 ATP BIOSENSING PRINCIPLES

Although the bioluminescent enzyme luciferase is a highly specific and sensitive method for detection of ATP release, the use of which goes back to the 1950s (Holton, 1959), it has significant disadvantages. The amount of light produced is very low. Highly sensitive cameras and comprehensive measures to eliminate stray sources of light are required for real-time spatially resolved measurement. It can be hard to obtain quantitative as opposed to qualitative data and it is a method that does not easily lend itself to *in vivo* investigations.

Our approach has been to use microelectrode biosensors that are selective for ATP (Llaudet et al., 2005). Although these biosensors are not as sensitive or selective for ATP as the luciferase method, they have the advantage of adaptability, ease of use and the ability to provide real-time quantitative measurements in many different biological contexts. The biosensors consist of two enzymes (glycerol kinase, and glycerol-3-phosphate oxidase) entrapped in a biolayer around a microelectrode (Figure 3.2). If a saturating level of glycerol is given (0.5 to 2 mM) then the cascade is purely dependent on the concentration of ATP which is consumed to produce glycerol-3-phosphate, the substrate for the second enzyme, which results in production of hydrogen peroxide (H_2O_2) in proportion to the concentration of ATP.

Although this method of ATP biosensing requires the provision of glycerol, this is not usually disadvantageous. The concentrations required are well tolerated by tissue both *in vitro* and *in vivo*, and we have never observed an effect of these concentrations of glycerol on physiological processes. Furthermore the requirement for glycerol gives an additional method of testing whether the biosensor is

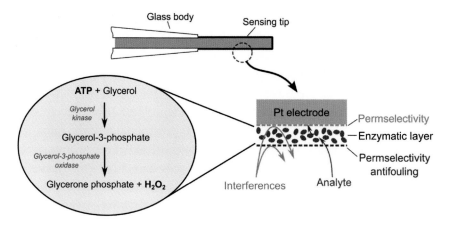

Figure 3.2 Principles and design of an ATP biosensor.

responding to ATP as opposed to some non-specific interferent — true responses to ATP will require the presence of glycerol (Llaudet et al., 2005). Biosensors based on this cascade have a lower limit of detection of around 60 nM ATP, and maintain linearity to about 100 μM. They are selective for ATP, *versus*, for example, ADP and UTP.

Other enzymatic methods for detection of ATP have been used, notably the combination of hexokinase, and glucose oxidase. The phosphorylation of glucose by hexokinase, with consequent consumption of ATP, sets up a competing reaction for the oxidation of glucose by glucose oxidase (Kueng et al., 2004; Masson et al., 2008). While this can result in a very sensitive biosensor for ATP, this cascade has the major disadvantage that it is also sensitive to variations in glucose. As glucose levels in the brain can vary from region to region and during physiological activation, this method cannot be used to detect ATP release *in vivo*.

All electrochemical biosensors have the potential to detect non-specific interferences that can oxidize on the electrode surface to generate non-specific interfering signals. To reduce the extent of this interference we grow an initial permselective polymer layer on the electrode surface onto which we then deposit the enzymatic biolayer (Figure 3.2). This permselective polymer acts to screen the electrode from interfering molecules while still allowing H_2O_2 produced in the biolayer to contact the electrode surface and generate a signal. ATP biosensors made this way have very good selectivity that allows analysis of ATP signaling with a high degree of confidence. We also utilize simultaneous recordings with "null sensors" that lack the ATP detecting enzymes (but are otherwise identical to the ATP sensors and would therefore react to interferences) as a method to further verify the specific nature of the signal recorded by the ATP biosensors.

3.2.2 THE ROLE OF ATP SIGNALING IN CHEMOSENSING

Application of ATP biosensors has led to a realization of the importance of ATP signaling in both central CO_2 and glucose chemosensitivity. There are surprising parallels between the two modes of chemosensing in the respective brain areas, which I summarize below.

3.2.2.1 Central CO_2 chemoreception

One of the first applications of the ATP biosensors was to analyze the mechanisms of respiratory chemosensitivity (Gourine et al., 2005). CO_2 is the unavoidable by-product of cellular metabolism. Humans produce about 20 moles of CO_2 per day that must be excreted by breathing. As the levels of dissolved CO_2 determine the pH of all bodily fluids, the control of partial pressure of carbon dioxide

(PCO_2) in arterial blood is therefore a vital physiological function. Both peripheral and central chemosensors measure PCO_2 and cause adaptive changes in breathing to closely regulate its level (Nattie, 1999; Nattie, 2001; Putnam et al., 2004; Richerson, 2004; Guyenet et al., 2005; Guyenet, 2008; Huckstepp and Dale, 2011). The peripheral chemosensors are found in the carotid body. One of the main locations for central CO_2 chemoreception is the ventral surface of the medulla oblongata (VMS). The VMS encompasses the retrotrapezoid nucleus which may contain chemosensitive neurons important for the regulation of breathing (Mulkey et al., 2004) and the medullary raphe which also contains chemosensitive neurons that are closely associated with blood vessels (Bradley et al., 2002; Severson et al., 2003; Corcoran et al., 2009).

Most investigators have accepted the consensus first put forward by Winterstein and Loeschcke that PCO_2 is measured only *via* consequent changes in pH (Loeschcke, 1982). However long standing, and historically neglected, evidence suggests that direct measurement of CO_2 may also be important (Eldridge et al., 1985; Shams, 1985). One reason for this is lack of a plausible molecular mechanism for the detection of CO_2 and transduction of that signal into a physiological action. By contrast the effects of pH on a range of ion channels are well accepted and understood with the result that most investigators have equated the measurement of pH with the measurement of CO_2. However, the application of ATP biosensing to the analysis of CO_2 chemoreception has fundamentally altered this view.

By using ATP biosensors, we have shown that ATP is released specifically from the classical chemosensitive areas at the ventral surface of the medulla during hypercapnia (Gourine et al., 2005) (Figure 3.3). This ATP release occurred before the adaptive changes in breathing (Gourine et al., 2005). On this basis we predicted that ATP may be released from the chemosensory cells and act as the initial trigger for neuronal activation. Application of ATP antagonists to the sites of ATP release give support to this hypothesis by considerably reducing the extent of the adaptive changes in breathing in response to hypercapnia (Gourine et al., 2005). Once released, ATP acts on neurons of the ventral respiratory column (Yao et al., 2003) including those of the retrotrapezoid nucleus (Wenker et al., 2012), to cause an increase in the frequency and depth of breathing.

The significance of this work goes beyond the simple documentation of ATP release. By pursuing the mechanisms of CO_2-dependent ATP release in isolated slices of the VMS, we identified a new chemosensory transducer that appears directly sensitive to CO_2 (Huckstepp et al., 2010b). This is connexin26 (Cx26), which is correctly located to sense changes in PCO_2 (in the leptomeninges, the astrocytes of the marginal glial layer at the ventral surface, and in cells adjacent to penetrating blood vessels) (Huckstepp et al., 2010b). CO_2-dependent ATP release in these slices depends upon Cx26 (Huckstepp et al., 2010b) and the expression of

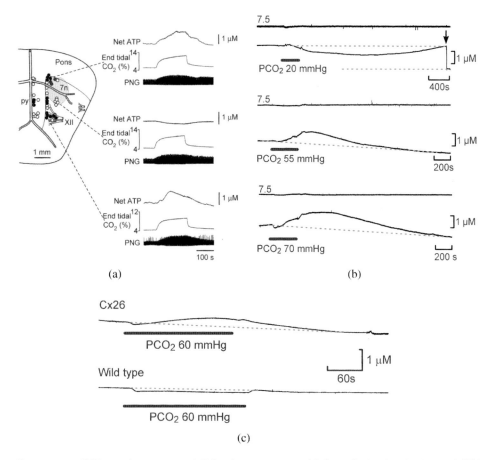

Figure 3.3 ATP signaling in central CO_2 chemoreception. (a) Spatially localized release of ATP recorded *in vivo* from ventral surface of the brain stem. 7n, facial nucleus; XII, twelfth nerve; py, pyramids. Filled circles represent sites of ATP release, open circles sites where there was no ATP release. Reproduced with permission from Gourine et al., 2005. (b) CO_2-dependent release of ATP at constant extracellular pH (top trace of each pair) from isolated slices of the medullary surface. Increase in PCO_2 causes ATP release, while a decrease reduces extracellular ATP. In the top trace, the ATP biosensor was removed from slice surface to demonstrate steady state concentration of ATP (rapid downward deflection, arrow). Reproduced with permission from Huckstepp et al., 2010b. (c) Recapitulation of CO_2-dependent ATP release in Cx26-expressing HeLa cells. Reproduced with permission from Huckstepp et al., 2010a.

Cx26 in a cell line is sufficient, by itself, to recapitulate CO_2-dependent ATP release (Huckstepp et al., 2010a) (Figure 3.3). Cx26 may well be directly sensitive to CO_2, as recordings from isolated membrane patches show that Cx26 will respond to alterations of PCO_2 at constant pH (Huckstepp et al., 2010a). Most recently, we have now shown the mechanism by which CO_2 binds directly to Cx26 to cause it to open (Meigh et al., 2013). Importantly, we were able to link Cx26 to respiratory behavior. Application of Cx26 hemichannel blockers greatly reduced the adaptive changes in breathing in response to hypercapnia. This demonstrates that Cx26 is used as a physiological chemosensory transducing molecule (Huckstepp et al., 2010b). The model emerging from these studies is that Cx26 is both the conduit for ATP release and the transducing molecule, capable of directly detecting PCO_2. For the first time this gives a mechanistic alternative or addition to the reaction theory that has dominated the field for so long (Huckstepp and Dale, 2011).

Although our work has focused on CO_2 as a chemosensory signal, Gourine and colleagues have examined the sensitivity of astrocytes of the VMS to pH. Using biosensors and Ca^{2+}-imaging they have demonstrated that changes in pH can evoke ATP release from astrocytes and consequent Ca^{2+} waves in these astrocytes (Gourine et al., 2010). By expressing channel rhodopsin-2 in astrocytes they have shown that activation of astrocytes by these light sensitive channels will enhance breathing through an ATP receptor-dependent mechanism.

The initial and continued application of ATP biosensing to CO_2 chemosensory transduction has thus had a transformative effect: identification of new cellular components of the reflex machinery; re-evaluation of the roles of some of the known components and identification of a new transducing molecule causally related to respiratory chemoreception that is directly sensitive to CO_2.

3.2.2.2 Glucosensing in the hypothalamus

The brain controls what is known as "energy homeostasis". This involves regulating: appetite and food intake; energy expenditure, by controlling metabolic rates; and energy storage, *via* fat deposition. Several interconnected areas of the hypothalamus, including the arcuate nucleus (ARC), the ventromedial hypothalamic nucleus (VMH), and the paraventricular nucleus (PVN), form a neural network that controls energy homeostasis (Morton et al., 2006). These networks monitor and integrate many signals that circulate in the body and provide information from the periphery (such as the stomach, pancreas and fat stores). These key signals not only include circulating metabolites such as glucose, free fatty acids and amino acids, but also specifically secreted hormones such as leptin, ghrelin and insulin. Many of these key nuclei controlling energy balance are in close proximity to the 3rd ventricle. The dorsal part of the lining of this ventricle contains a layer of ciliated ependymal cells. As the ventricle wall progresses more ventrally, tanycytes appear, initially interspersed with the ependymal cells, before progressively becoming more numerous and contiguous in the ventricle wall toward the median eminence (Rodriguez et al., 2005; Mullier et al., 2010). Tanycytes possess apical microvilli and have a long process that projects into the brain parenchyma.

Hypothalamic tanycytes have been divided into four major groups (Rodriguez et al., 2005). The $\beta1$ and $\beta2$ tanycytes are found very ventrally at the median eminence and send processes to the ventral surface of the brain. The α1 and α2 tanycytes lie dorsal to the β1 and β2 tanycytes. α1, α2 and β1 tanycytes have processes that project into and through the ARC and the VMH. The idea that tanycytes could somehow contribute to the control of energy homeostasis has gained widespread currency. In particular many have speculated that tanycytes could be glucosensitive. Tanycytes have all of the molecular components required for the pancreatic β cell paradigm of glucosensing (Rodriguez et al., 2005). In this model, glucose is transported into cells (usually *via* a facilitative glucose transporter) and converted to glucose-6-phosphate by hexokinase, an enzyme with a relatively low affinity for glucose that is within the physiologically important range of glucose concentrations (i.e., 4–10 mM). The glucose-6-phosphate enters the Krebs cycle and subsequent metabolism results in an increase in intracellular ATP and the ratio of ATP to ADP. This in turn causes an ATP-sensitive potassium (K-ATP) channel to close, which in the β cell, comprises Kir6.2 and an associated sulphonylurea (SUR) subunit.

We have used ATP biosensing and Ca^{2+} imaging to directly study the chemosensitive properties of tanycytes (Figure 3.4) (Dale, 2011; Frayling et al., 2011; Bolborea and Dale, 2013). Although we found that tanycytes are glucosensitive, they do not conform to the pancreatic β cell paradigm. In acutely prepared brain slices, tanycytes (α1, α2, β1) responded only rarely and weakly to changes of glucose concentration in the bathing medium. However, if glucose was applied selectively to their cell bodies *via* a puffer pipette, strong Ca^{2+} responses were observed. These responses were dependent on the release of ATP and activation of P2Y1 receptors (Figure 3.4). Interestingly, tanycytes also responded to non-metabolizable analogs of glucose (Frayling et al., 2011). This observation undermines the applicability of the pancreatic β cell paradigm as this model depends upon the ability to metabolize glucose to generate intracellular ATP. Instead we believe that tanycytes possess receptors for glucose that can also be activated by these analogs and a range of sweet tasting compounds. The precise nature of the receptor and how the signal is transduced into a Ca^{2+} wave remain to be determined.

Independent confirmation of tanycyte responsiveness to glucose comes from studies of these cells in primary culture (Orellana et al., 2012). There are some differences from the observations made in brain slices, as the cultured tanycytes appear to sense glucose in a manner that partially resembles the pancreatic β cell. Nevertheless, this study also reports the dependence on release of ATP (likely to be mediated *via* connexin 43 hemichannel gating) and subsequent activation of P2Y1 receptors (Orellana et al., 2012).

The significance of tanycyte glucosensing is an interesting area for further investigation. There is some evidence that ATP can alter feeding behavior *via* activation of P2Y1 receptors (Kittner et al., 2006; Seidel et al., 2006). Infusion of specific agonists and antagonists *via* the 3rd ventricle caused respectively an increase and

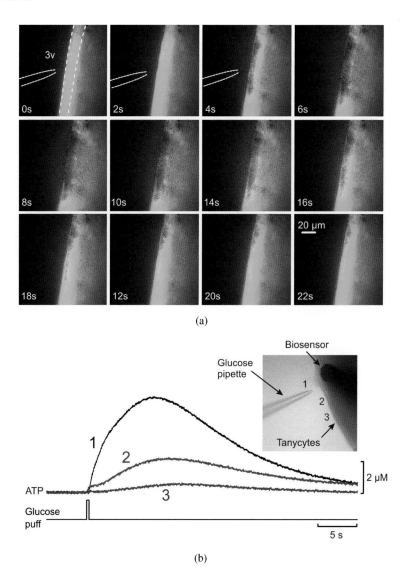

Figure 3.4 Glucosensing in hypothalamic tanycytes. (a) A Ca^{2+} wave evoked in tanycytes (delineated by dashed lines) in wall of third ventricle (3v). The position of the glucose pipette (broken lines) is indicated relative to the tanycytes. (b) ATP release from the tanycytes in response to glucose puffs (inset shows recording arrangements and positions of the puffer pipette relative to the ATP biosensor. Adapted with permission from Frayling et al., 2011.

decrease in food intake. While we cannot conclude from this that tanycytes have any role in the control of feeding, these data make it at least plausible that tanycyte signaling *via* ATP can influence activity in the hypothalamic neuronal circuitry.

A further intriguing possibility comes from the recent discovery of tanycytes as adult stem cells (Xu et al., 2005; Lee et al., 2012; Li et al., 2012; Haan et al.,

2013; Robins et al., 2013). Morphologically, tanycytes resemble radial glial cells. During development of the cortex, radial glial cells are neural progenitors and form a scaffold for the migration of newly generated neurons (Kriegstein and Alvarez-Buylla, 2009). Radial glial cells release ATP, which controls neural proliferation by acting *via* P2Y1 receptors to stimulate Ca^{2+} waves in the progenitor cells (Weissman et al., 2004). There is evidence that tanycytes are responsive to diet (Lee et al., 2012; Li et al., 2012) and differentiate to provide new neurons to the ARC (Haan et al., 2013; Robins et al., 2013). An attractive hypothesis emerging from our work is that tanycytes could integrate a number of signals (including glucose), release ATP, which if sufficiently frequent or prolonged could trigger their division and differentiation into new neurons (Bolborea and Dale, 2013). Note that the hypotheses of tanycytes as chemosensors and stem cells are not mutually exclusive but are linked *via* ATP signaling.

3.2.3 ATP SIGNALING AND DEVELOPMENT

We have used ATP biosensors to study neural development in two different contexts (Dale, 2008). The first which I describe below is in the very early embryo and our findings demonstrate a role for ATP signaling in the patterning of the early embryo and development of the future eye. The second related exemplar occurs in the developing eye and concerns the control of neural progenitor cells that give rise to the neurons of the retina. Karine Massé and I have recently reviewed the roles of purinergic signaling in early development, which may assist any reader in exploring this field further (Masse and Dale, 2012).

3.2.3.1 Initiation of expression of the eye-field transcription factors

A network of eye-field transcription factors (EFTFs), at least partially conserved through evolution, orchestrates development of the eye from anterior ectoderm at either side of the neural plate (Figure 3.5). In vertebrates, the EFTFs include genes such as *Pax6*, *Six3*, *Rx1* and *Lhx2* (Zuber et al., 2003). Until our work on ATP signaling, the mechanism by which these genes were turned on in the correct location and at the correct time were unknown. Our work looking at purinergic signaling and ATP release has helped to fill this gap, at least in the developing frog embryo (Masse et al., 2007).

We found that the enzyme NTPDase2 was a critical part of the mechanism (Figure 3.5). This enzyme converts ATP to ADP, which is the preferred substrate for the P2Y1 receptor, also involved in triggering EFTF expression. Manipulation of the expression of NTPDase2 altered the expression of the endogenous EFTFs such as *Pax6*, *Rx1*, *Otx2* and *Six3*. Simultaneous knockdown of the expression of

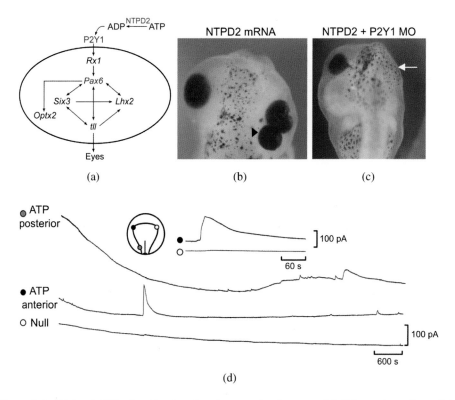

Figure 3.5 Role of ATP signaling in triggering eye development. (a) Schematic outline of the interactions of the eye field transcription factors and the role of the P2Y1 receptor and NTPDase2 in triggering this expression of this network. (b) Over-expression of NTPDase2 can trigger formation of extra eyes (triangle). (c) Morpholino antisense knock down of NTPDase2 and P2Y1 receptors can prevent the eye from developing (arrow). (d) ATP signaling in the very early embryo. A single period of ATP release is seen in the location of the future eye field. In the posterior neural plate frequent ATP release events are seen, suggesting a role for ATP signaling in the development of the nervous system (b, c and d). Adapted with permission from Masse et al., 2007.

NTPDase2 and the P2Y1 receptor could almost completely abolish *Pax6* and *Rx1* expression, greatly reduced the size of the eye and, in some cases prevented them from developing (Figure 3.5). By making biosensor recordings from the location of the eye field, we demonstrated a brief period of ATP release (Figure 3.5). The model that emerges from this work is that a collection of cells in the anterior ectoderm at the site of the future eye field, transiently releases ATP, which is converted by NTPDase2 to ADP. This ADP then activates the P2Y1 receptor and hence triggers EFTF expression (Figure 3.5). The EFTFs form a mutually activating gene network with cycles of positive feedback. *Rx1* activates *Pax6*, which then activates *Six3*, *Lhx2* and *tll*, each of which reinforces the expression of *Pax6* (Zuber et al., 2003). *Pax6* can also up-regulate its own expression. The existence of this positive feedback within the EFTF network may explain why the observed brief period of ATP release may be sufficient to trigger a very long lasting effect (Dale, 2008).

3.2.3.2 Regulation of development of the retina

The outermost layer of the developing retina, the retinal pigment epithelium (RPE), is essential for the normal development of the underlying layers (Raymond and Jackson, 1995), collectively termed the neural retina. Progenitor cells of the neural retina are in the region adjacent to the RPE (called the subventricular zone) and divide to generate the neurons and glia that comprise the fully developed retina. The rate of cell division in the ventricular zone depends upon the presence of, and close association with, the RPE. This suggests that signaling must occur between these two tissue layers.

Ca^{2+} waves occur in both the RPE (Pearson et al., 2004; Pearson et al., 2005) and developing neural retina (Sugioka et al., 1996; Pearson et al., 2002). These waves depend upon the release of ATP. In the neural retina, activation of ATP receptors by exogenous agonists evokes an increase of intracellular Ca^{2+} in, and controls the rate of division of, progenitor cells (Pearson et al., 2002; Nunes et al., 2007). Spontaneous Ca^{2+} waves occur in the RPE and can trigger Ca^{2+} waves in nearby portions of the neural retina (Pearson et al., 2004). By using ATP biosensors, we showed that the propagation of Ca^{2+} waves through the RPE depends on the release of ATP (Pearson et al., 2005). As the ATP diffuses in the extracellular space, it activates P2Y receptors on adjacent cells leading to the release of Ca^{2+} from internal stores and hence the appearance of a propagating Ca^{2+} wave. Our biosensor data suggested that the release of ATP occurred through the spontaneous opening of connexin 43 hemichannels. Specific peptide blockers of connexin 43 hemichannels prevented ATP release, and slowed cell division in the underlying ventricular zone of the neural retina (Pearson et al., 2005). Our results therefore provide a mechanism by which the RPE controls division in the neural retina through the release of ATP *via* connexin hemichannels, which subsequently activates P2Y receptors on the precursor cells in the ventricular zone.

3.3 ADENOSINE-RELEASE MECHANISMS AND FUNCTIONS

3.3.1 ADENOSINE BIOSENSING PRINCIPLES

To create adenosine-sensitive biosensors, we deposit adenosine deaminase (AD), purine nucleoside phosphorylase (PNP) and xanthine oxidase (XO) onto microelectrodes. These enzymes successively convert adenosine to inosine, hypoxanthine, xanthine and urate with the evolution of H_2O_2 which is detected electrochemically (Figure 3.6). To operate, there must be physiological levels of

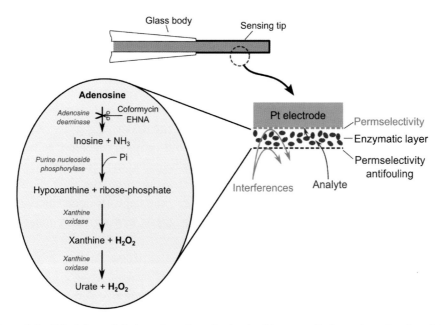

Figure 3.6 Principles and design of an adenosine/purine biosensor. Coformycin anderythro-9-(2-hydroxy-3-nonyl)adenine, EHNA, are potent blockers of adenosine deaminase and can be used to test the contribution of adenosine to the signal recorded by the biosensor.

phosphate ions (required for nucleoside phosphorylase). Obviously a sensor that comprises all three enzymes will be sensitive to all of the substrates of these enzymes. To obtain a specific adenosine measurement, differential recordings are required whereby the difference between a biosensor that has AD, PNP and XO and one that has only PNP and XO is recorded. The latter sensor will be insensitive to adenosine but will respond to all the other substrates (and interferences). This strategy can successfully abstract the specific adenosine signal from a mixture of released purines (Figure 3.7). The lower detection limit for an adenosine biosensor is around 20 nM, and these biosensors maintain a linear range to at least 20 μM (Dale, 1998; Dale et al., 2000; Llaudet et al., 2003).

3.3.2 ACTIVITY DEPENDENCE OF ADENOSINE RELEASE

There is now abundant evidence that the extracellular concentration of adenosine can increase as a result of neural activity and is a physiological feedback mechanism of considerable significance (Mitchell et al., 1993; Brundege and Dunwiddie, 1996, 1998; Pascual et al., 2005; Wall and Dale, 2007; Lovatt et al., 2012). The question as to how adenosine release/production is linked to neural activity is an important unresolved issue. The prevailing orthodoxy is that, except under pathological conditions, adenosine is not directly released from neurons, but arises instead from the catabolism of previously released ATP. Indeed there is an

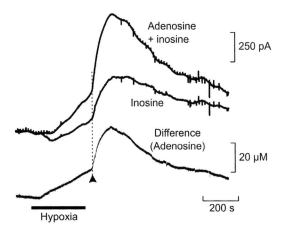

Figure 3.7 Differential recordings of adenosine release. A biosensor containing all three enzymes (adenosine deaminase, purine nucleoside phosphorylase and xanthine oxidase) will record both adenosine and inosine release. The second biosensor (Inosine) lacking adenosine deaminase will be sensitive only to inosine and downstream purines. The difference between the two sensor recordings shows the selective adenosine signal. Reproduced with permission from Frenguelli et al., 2003.

abundance of extracellular enzymes capable of mediating this conversion of ATP all the way to adenosine. Thus although there are several examples of activity-dependent ATP release, these have largely been viewed as arising from exocytotic release of ATP and subsequent conversion to adenosine.

The application of adenosine biosensors has resulted in several advances in our understanding of release mechanisms for adenosine. Biosensors provide a direct measure of adenosine release and have a number of advantages over indirect methods of inferring adenosine release (e.g., through application of adenosine receptor antagonists to monitor the effects of adenosine on synaptic transmission). For example, biosensors can be used to measure adenosine release in the absence of extracellular Ca^{2+}, or in the presence of TTX, (Wall and Dale, 2007; Klyuch et al., 2011) both of which would prevent normal synaptic transmission and thus occlude observation of the potential effects of any adenosine receptor antagonists on synaptic transmission. In the sections below, I consider our recent progress on this surprisingly complex phenomenon which demonstrates that there are several parallel mechanisms for the direct and indirect release of adenosine into the extracellular space.

3.3.2.1 Release mechanisms in cerebellum

In cerebellum, adenosine can be released in an activity-dependent manner following stimulation of the parallel fibres (Wall and Dale, 2007; Klyuch et al., 2011; Klyuch et al., 2012). Biosensor recordings reveal that adenosine release has characteristics of exocytosis: TTX- and Ca^{2+}-dependence (Wall and Dale, 2007; Klyuch et al., 2011; Klyuch et al., 2012). Adenosine release depends on spike width and can exhibit short term plasticity more typically associated with conventional synaptic potentials at cerebellar synapses (Klyuch et al., 2011). For example at very short inter-pulse intervals (50 milliseconds), adenosine release undergoes paired pulse facilitation, while at longer intervals (30–120 seconds) it undergoes paired pulse depression. These data are suggestive of an exocytotic release mechanism, which can undergo depletion and rather slow restocking.

Evidence for a direct exocytotic mechanism of adenosine release came from combining the biosensor recordings with knock-out mice lacking the key enzyme e5′-NT, which converts AMP to adenosine (Klyuch et al., 2012) (Figure 3.1). Deletion of this enzyme prevents about 90% of the conversion of extracellular ATP to adenosine in cerebellar slices. A direct pathway of activity-dependent adenosine release is maintained in these mice, demonstrating that it cannot arise from previously released ATP (Figure 3.8). By applying bafilomycin (a blocker of the H^+ ATPase that maintains a proton gradient across the vesicular membrane that is then used by transporters to concentrate transmitter inside the vesicle), we demonstrated that activity-dependent adenosine release is greatly reduced (Figure 3.8). This is the strongest evidence so far for the exocytotic release of adenosine.

However, our recordings also revealed an indirect pathway of adenosine release in the cerebellum that was dependent upon the synaptic activation of glutamate receptors. This indirect pathway was blocked in the $CD73^{-/-}$ mice demonstrating that it arose from prior release of ATP (Klyuch et al., 2012). Thus even in the

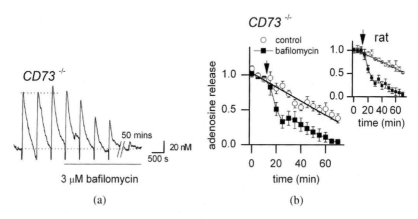

Figure 3.8 Activity-dependent release of adenosine from parallel fibres in cerebellum. (a) Adenosine biosensor recordings of adenosine release evoked by stimulation of the parallel fibres in a cerebellar slice prepared from a $CD73^{-/-}$ mouse (lacking e5′-NT). When bafilomycin was applied the amount of adenosine released was greatly reduced providing strong evidence that release was *via* exocytosis. (b) Summary time series graphs for effect of bafilomycin on adenosine release in both $CD73^{-/-}$ mice and wild-type rats. Reproduced with permission from Klyuch et al., 2012.

cerebellum there is complexity, with two distinct mechanisms of adenosine release occurring in parallel, yet having remarkably similar kinetic characteristics.

3.3.2.2 Release mechanisms in hippocampus

By studying a different brain area, the hippocampus, we found further complexity (Wall and Dale, 2013). Stimulation of the Schaffer collaterals could evoke activity-dependent ATP release in area CA1. Although kinetically, this release appeared quite similar to that which we had studied in the cerebellum, it was entirely dependent upon glutamate receptor activation. Use of $CD73^{-/-}$ mice again revealed two pathways of adenosine release, one that was fast and direct and a second, slower mechanism *via* prior release of ATP. However unlike the cerebellum the direct pathway of adenosine release was *via* the equilibrative nucleoside transporters and could be blocked by agents targeting these transporters. The indirect pathway, *via* release of ATP, came from astrocytes. Agents that interfered with astrocyte metabolism blocked the indirect pathway. Genetically modified mice, in which exocytosis of ATP has been selectively deleted from astrocytes (Pascual et al., 2005), lack this indirect pathway. Interestingly the two pathways are roughly equal in size, but transporter mediated adenosine release has a faster rise and decay time, whereas the astrocyte derived adenosine release is slower in both the rise time and decay (Figure 3.9).

3.3.2.3 Adenosine release linked to activation of the Na^+-K^+ ATPase

Around 80% of the resting energy budget of the brain is spent on pumping Na^+ out of cells to restore the ionic gradients required for electrical signaling (Attwell and Laughlin, 2001). This pumping consumes ATP and, to allow cells to exploit the conversion of ATP to ADP to perform work, the disequilibrium between ATP and ADP must be maintained. The intracellular enzyme AK is important in this respect, as it can consume two ADP molecules to produce one molecule of ATP and one molecule of AMP. The AMP is then degraded further to adenosine by the 5'-NT. Consumption of ATP therefore rapidly results in the formation of intracellular adenosine which has the potential to leak out of cells *via* facilitative transport (Figure 3.10b).

By studying AMPA-receptor evoked adenosine release we discovered a surprising requirement for an influx of Na^+ for the release of adenosine following activation of AMPA receptors (Sims and Dale, 2014). We combined Na^+-imaging and adenosine biosensing and found that an increase in intracellular Na^+ was sufficient to evoke adenosine release (Figure 3.10). We used ouabain to test the hypothesis that activation of the Na^+-K^+ATPase (NKA), by the Na^+ influx

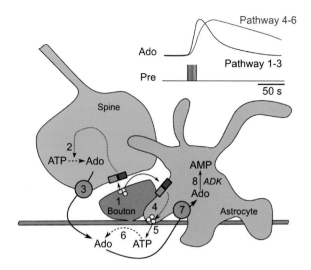

Figure 3.9 Scheme for activity-dependent release of adenosine in hippocampus. Glutamate release from the Schaffer collaterals (brown) releases glutamate(1), activates N-methyl-D-aspartate (NMDA) and α-amino-3-hydroxy-5-methyl-4-isoxazolepropionic acid (AMPA) receptors (blue and pink boxes) on the post-synaptic spine of a pyramidal neuron and NMDA and AMPA receptors on nearby astrocytic processes (4). This activation of glutamate receptors leads to an increase in intracellular adenosine(2), which is then rapidly transported out of the cell by the ENTs (3). In parallel with this, the activation of glutamate receptors on the astrocyte (4) leads to an increase in intracellular Ca^{2+} and exocytosis of ATP-containing vesicles (5). Once released, ATP is broken down to adenosine (6). Extracellular adenosine can be taken up into astrocytes *via* ENTs (7) which maintain low intracellular adenosine levels as a consequence of the activity of ADK (8). The two parallel pathways have different release kinetics, illustrated in the inset: the neuronal pathway of direct release (1–3) is rapid, while the indirect astrocytic pathway (4–6) is slower. Modified with permission from Wall and Dale, 2013.

and the resulting additional ATP consumption, was required for the AMPA-evoked adenosine release (Sims and Dale, 2014). From these results it was clear that activation of the NKA results in the efflux of adenosine (with a delay of about 90 seconds) and that if these ATPase is blocked by ouabain the adenosine efflux is inhibited (Sims and Dale, 2014) (Figure 3.10). The activation of the NKAs is thus a potentially universal mechanism that provides extracellular adenosine in proportion to the degree of neuronal activation as measured by the requirement to extrude Na^+.

3.3.2.4 What have we learned?

Activity-dependent adenosine release is surprisingly complex and varies with region in the brain. There are at least four pathways/mechanisms: 1) exocytotic — so far only described in the cerebellum; 2) fast ENT-mediated — present in the hippocampus; 3) slower astrocyte mediated release *via* conversion of ATP and 4)

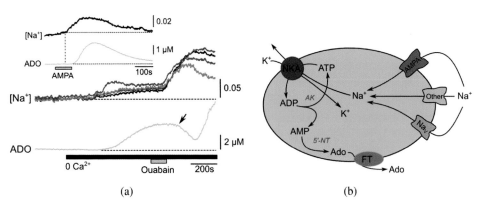

Figure 3.10 The role of the Na^+–K^+ ATPase (NKA) in activity-dependent adenosine release. (a) Inset: Adenosine release evoked by application of AMPA to basal forebrain slices is linked to an influx of intracellular Na^+, as visualized by sodium-binding benzofuranisophthalate (SBFI) fluorescence. Main panel: an increase of intracellular Na^+ evoked by application of a zero Ca^{2+} saline causes intracellular Na^+ accumulation, followed by adenosine release recorded by biosensor (ADO). Application of ouabain blocked the NKA (as shown by accumulation of intracellular Na^+). This was followed by a reduction of extracellular adenosine (arrow). (b) Hypothesized scheme to explain how a Na^+ influx through AMPA, voltage-gated Na^+ channels (Na_v) and other cation channels activates the NKA. The consumption of ATP leads *via* AK to production of AMP and thence (*via* the 5'-NT) to adenosine (Ado). The adenosine can be released to the extracellular space *via* as yet unidentified facilitative transporters (FTs). Adapted with permission from Sims and Dale, 2014.

release that is linked to the activation of the NKAs following neural signaling. Appreciation of the diversity of mechanisms underlying activity-dependent adenosine release is a major step forward. The challenge now is to understand how these different parallel mechanisms contribute to neural function, physiology and behavior.

3.3.3 SLEEP AND ADENOSINE

Sleep is subject to both circadian and homeostatic control. Sleep homoeostasis involves the tendency to spend consistent amounts of time asleep and to sleep for longer following prolonged waking. A simple explanation for sleep homoeostasis is that an endogenous somnogen can accumulate during wakefulness and dissipate during sleep. The increased accumulation of the somnogen during prolonged wakefulness then leads to a prolonged period of ensuing (rebound) sleep.

Although exogenous adenosine has long been known as a somnogen (Radulovacki et al., 1984; Radulovacki, 2005), causal evidence linking endogenous adenosine to the homoeostatic control of sleep has only arisen more recently. During wakefulness adenosine accumulates in the cholinergic basal forebrain (BFB), an important area for the control of slow wave (SW) sleep (Porkka-Heiskanen et al., 1997).

Microdialysis of the adenosine transport inhibitor nitrobenzylthioinosine (NBTI) into the BFB increased both the level of extracellular adenosine and the amount of sleep (Porkka-Heiskanen et al., 1997). In another key brain area that controls sleep, the ventrolateralpreoptic area (VLPO), adenosine excites putative sleep-active neurons *via* A_{2A} receptors (Gallopin et al., 2005). Recently, a study has demonstrated that humans with an inefficient variant of AD, which converts adenosine to inosine, report fewer awakenings during sleep and exhibit greater delta power in their SW sleep compared to those with an efficient variant of AD (Retey et al., 2005). This demonstrates the importance of adenosine and AD as regulators of human sleep.

Although adenosine is clearly an important homoeostatic regulator of sleep, the mechanisms that underlie the respective accumulation and dissipation of adenosine during wakefulness remain opaque. Nitric oxide (NO) signaling has been proposed as a trigger for enhanced adenosine release during sleep deprivation (SD) (Kalinchuk et al., 2006b). Delivery of 1400W (an inhibitor selective for inducible nitric oxide synthase (Kalinchuk et al., 2006a)), L-NAME (NG-nitro-L-arginine methyl ester, a general NOS inhibitor) and cPTIO (2-4-carboxyphenyl-4,4,5,5-tetramethylimidazoline-1-oxyl-3-oxide, a NOS scavenger (Kalinchuk et al., 2006b)) through microdialysis into the BFB inhibited the increase in extracellular adenosine concentrations in BFB that follows SD. Furthermore, both the expression of iNOS in the BFB and NO production increased during SD (Kalinchuk et al., 2011).

To advance the field further, we have developed *in vitro* methods to complement the *in vivo* studies (Sims et al., 2013). By inserting adenosine biosensors into slices of the BFB, we have studied both depolarization-evoked adenosine release and the steady state adenosine tone in rats and mice, and documented its sensitivity to the "sleep-wake status" of the animal prior to sacrifice. Adenosine release was evoked by high K^+, AMPA, NMDA and metabotropic glutamate (mGlu) receptor agonists, but not by other transmitters associated with wakefulness such as orexin, histamine or neurotensin. We found that adenosine release (both evoked and steady-state) was indeed sensitive to the sleep-wake status of the animal prior to sacrifice: the magnitude of each varied systematically with the diurnal time at which the animal was sacrificed; and SD prior to sacrifice greatly increased both evoked adenosine release and the basal tone (Sims et al., 2013) (Figure 3.11). Crucially, though the iNOS inhibitor 1400W had no effect on adenosine release in the slices derived from non-sleep deprived rats, the enhancement of both evoked adenosine release and basal tone resulting from SD was completely reversed by 1400W (Sims et al., 2013) (Figure 3.11). These data indicate that characteristics of adenosine release recorded in the BFB *in vitro*, reflect those that have been linked *in vivo* to the homeostatic control of sleep. Our results provide

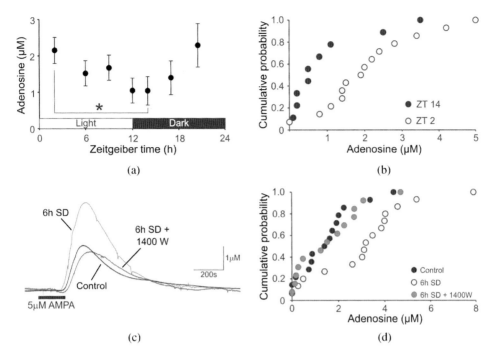

Figure 3.11 AMPA receptor evoked adenosine release recorded from slices of BFB is sensitive to diurnal time of day and SD prior to sacrifice. (a) Adenosine release reaches its highest level toward the end of dark phase (when rodents are active) and its minimum level toward the end of light phase. (b) Cumulative probability plot comparing adenosine release from individual slices at ZT2 and ZT14. (c) Examples of AMPA evoked adenosine release in the control, following six hours SD and six hours SD in the presence of the iNOS inhibitor's 1400W. (d) Cumulative probability plot comparing the effects of six hours SD and six hours SD plus 1400W to the control. Adapted with permission from Sims et al., 2013.

methodologically independent support for a key role for induction of iNOS as a trigger for enhanced adenosine release following SD and suggest that this induction may constitute a biochemical memory of this state.

In a parallel study we found that basal adenosine levels varied with the diurnal cycle in hippocampus (Schmitt et al., 2012), and that these levels could be enhanced by SD. In hippocampus we did not test the role of iNOS, but found instead that astrocytes are an important source of this adenosine in hippocampus, presumably *via* the prior release of ATP and extracellular conversion to adenosine.

3.4 INTERPLAY BETWEEN ATP AND ADENOSINE RELEASE: MOTOR PATTERN GENERATION

So far, I have largely considered the actions of ATP and adenosine separately. However in many cases these two purines act in concert. One example of this is the modulation of motor pattern generation by ATP and adenosine in the frog embryo spinal cord (Dale and Gilday, 1996). ATP increases excitability within the spinal central pattern generator, by inhibiting K^+ currents (Dale and Gilday, 1996; Brown and Dale, 2002). Exogenous ATP will lengthen swimming episodes and ATP receptor antagonists shorten them. Adenosine does the opposite by inhibiting voltage-gated Ca^{2+} currents (Dale and Gilday, 1996; Brown and Dale, 2000). Exogenous adenosine will shorten, while adenosine receptor antagonists will lengthen the episodes of swimming activity.

Many centrally generated motor patterns exhibit "centrally programmed fatigue"–a tendency to spontaneously go slow and stop in the absence of any sensory inputs. The mechanistic basis of this central fatigue has not been extensively explored, but the observation of the differential actions of ATP and adenosine led me to propose that these purines could mediate this phenomenon in the frog embryo spinal cord. Specifically, if adenosine were to arise from the prior release of ATP, its inhibitory actions in the spinal cord would only make sense if the accumulation of adenosine was slow relative to the release of ATP. The hypothesis being that the gradual accumulation of adenosine reaches a sufficient level to spontaneously terminate motor activity and that it is the rate of accumulation of adenosine that ultimately determines the length of a swimming episode. Without this hypothesized gradual accumulation, the opposing actions of adenosine and ATP would simply cancel each other out.

The most direct test of this hypothesis was to make real-time measurements of both ATP and adenosine release during motor pattern generation. Indeed, it was the desire to test this hypothesis that was the original impetus to develop biosensors for the purines. Our very first adenosine biosensor recordings allowed us to measure its accumulation during motor activity and demonstrated that it occurred progressively and slowly throughout the swimming episodes (Dale, 1998; Llaudet et al., 2003) (Figure 3.12). By contrast, when we had eventually fabricated sufficiently sensitive ATP biosensors, direct measurements of ATP showed that it was released much more quickly, and rapidly achieved a steady state plateau level, maintained throughout the episode (Llaudet et al., 2005) (Figure 3.12). The biosensor recordings contributed to the formulation of computational models, which incorporated the kinetics of ectonucleotidase action in converting ATP to adenosine. These models allowed investigation of the plausibility of this dual control system as a mediator of centrally programmed fatigue (Dale, 2002). They also allowed formulation of testable hypotheses as to the metaplasticity of motor pattern generation. This is a phenomenon whereby a swimming episode can be prolonged by a sensory stimulus that occurs early in the episode but prematurely terminated by a sensory stimulus that occurs much later in the episode. The computational models were able to reproduce this phenomenon in a reduced model system, which was then tested in the real embryo by application of ATP and adenosine receptor antagonists, which prevented the metaplasticity (Dale, 2002).

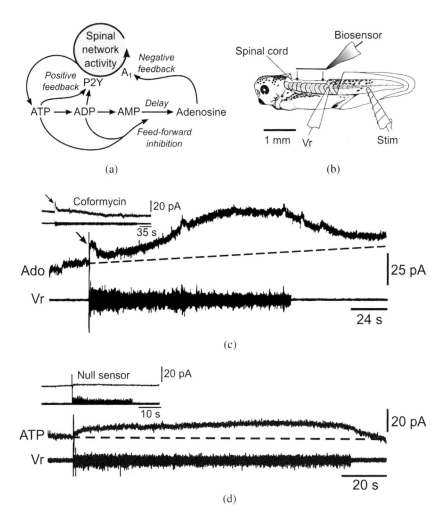

Figure 3.12 Release of adenosine and ATP during motor pattern generation in the frog embryo. (a) Schema for purinergic control of motor pattern generation. ATP released from the spinal motor network mediates positive feedback *via* P2Y receptors. The ATP is converted in the extracellular space to adenosine, with a delay, and the adenosine acts *via* A_1 receptors to mediate negative feedback. (b) Diagram of recording arrangement. Vr, ventral root recording (motor nerve); Stim, skin stimulating electrode to evoke swimming behavior. (c) Simultaneous recordings of ventral root activity and adenosine release (Ado), note the slow accumulation throughout the swimming episode. Inset: example of a control, coformycin blocks adenosine deaminase and eliminates the slowly developing signal, but leaves an initial transient signal (arrow, also arrow in main part of the panel). Thus only the slowly developing signal is due to adenosine. Reproduced with permission from Llaudet et al., 2003. (d) Simultaneous recordings of ventral root activity and ATP release. The ATP accumulates quickly and reaches a plateau level soon after the start of the swimming episode. Inset: example of a Null control. No signal is seen on the Null sensor, demonstrating that the signal seen with the ATP biosensor depends on the presence of the enzymes. Reproduced with permission from Llaudet et al., 2005.

3.5 CONCLUDING REMARKS

Direct real-time measurements of ATP and adenosine release have proven to be very valuable in analysing the signaling roles of these purines, and the mechanisms by which they are released or arise in the extracellular space. We have combined biosensor measurements with a number of other electrophysiological and imaging methods as part of a coordinated experimental program. It is perhaps this combination above all else, that makes the application of biosensors so powerful that they can be used to formulate testable hypotheses with regard to function and mechanism. Perhaps the most convincing demonstration of this is our use of ATP biosensors in the context of respiratory chemosensitivity that ultimately led to the identification of Cx26 and the direct sensing of CO_2 as a new and unexpected transducing molecule in the field of respiratory chemosensing.

ACKNOWLEDGMENTS

I thank the many colleagues and collaborators who have worked with me for more than 15 years on biosensors that have been applied to the analysis of purinergic signaling. Special thanks go to Dr Enrique Llaudet, my original co-inventor of the purine microelectrode biosensors, and to my earliest collaborators: Dr Mark Wall, Professor Bruno Frenguelli, Professor Alex Gourine and Professor Mike Spyer. I also thank the Medical Research Council (UK) and the Wellcome Trust for funding much of the work described in this chapter.

REFERENCES

Abbracchio MP, Burnstock G, Boeynaems JM, Barnard EA, Boyer JL, Kennedy C, Knight GE, Fumagalli M, Gachet C, Jacobson KA, Weisman GA (2006) International Union of Pharmacology LVIII: update on the P2Y G protein-coupled nucleotide receptors: from molecular mechanisms and pathophysiology to therapy. PharmacolRev 58:281–341.

Attwell D, Laughlin SB (2001) An energy budget for signaling in the grey matter of the brain. J Cereb Blood Flow Metab 21:1133–1145.

Baldwin SA, Beal PR, Yao SY, King AE, Cass CE, Young JD (2004) The equilibrative nucleoside transporter family, SLC29. Pflugers Arch 447:735–743.

Bao L, Locovei S, Dahl G (2004) Pannexin membrane channels are mechanosensitive conduits for ATP. FEBS Lett 572:65–68.

Boison D (2006) Adenosine kinase, epilepsy and stroke: mechanisms and therapies. Trends Pharmacol Sci 27:652–658.

Bolborea M, Dale N (2013) Hypothalamic tanycytes: potential roles in the control of feeding and energy balance. Trends Neurosci 36:91–100.

Bradley SR, Pieribone VA, Wang W, Severson CA, Jacobs RA, Richerson GB (2002) Chemosensitive serotonergic neurons are closely associated with large medullary arteries. Nat Neurosci 5:401–402.

Brown P, Dale N (2000) Adenosine A_1 receptors modulate high voltage-activated Ca^{2+} currents and motor pattern generation in the Xenopus embryo. J Physiol (Lond) 525:655–667.

Brown P, Dale N (2002) Modulation of K^+ currents in Xenopus spinal neurons by p2y receptors: a role for ATP and ADP in motor pattern generation. J Physiol 540:843–850.

Browne LE, Compan V, Bragg L, North RA (2013) P2X7 receptor channels allow direct permeation of nanometer-sized dyes. J Neurosci 33:3557–3566.

Brundege JM, Dunwiddie TV (1996) Modulation of excitatory synaptic transmission by adenosine released from single hippocampal pyramidal neurons. J Neurosci 16:5603–5612.

Brundege JM, Dunwiddie TV (1998) Metabolic regulation of endogenous adenosine release from single neurons. Neuroreport 9:3007–3011.

Burnstock G (1972) Purinergic nerves. PharmacolRev 24:509–581.

Burnstock G (1997) The past, present and future of purine nucleotides as signaling molecules. Neuropharmacology 36:1127–1139.

Burnstock G (2006) Historical review: ATP as a neurotransmitter. Trends Pharmacol Sci 27:166–176.

Burnstock G, Fredholm BB, Verkhratsky A (2011) Adenosine and ATP receptors in the brain. CurrTopMedChem 11:973–1011.

Chaumont S, Khakh BS (2008) Patch-clamp coordinated spectroscopy shows P2X2 receptor permeability dynamics require cytosolic domain rearrangements but not Panx-1 channels. Proc Natl Acad Sci U S A 105:12063–12068.

Corcoran AE, Hodges MR, Wu Y, Wang W, Wylie CJ, Deneris ES, Richerson GB (2009) Medullary serotonin neurons and central CO_2 chemoreception. Respir Physiol Neurobiol 168:49–58.

Dale N (1998) Delayed production of adenosine underlies temporal modulation of swimming in frog embryo. J Physiol (Lond) 511:265–272.

Dale N (2002) Resetting intrinsic purinergic modulation of neural activity: an associative mechanism? J Neurosci 22:10461–10469.

Dale N (2008) Dynamic ATP signaling and neural development. J Physiol 586:2429–2436.

Dale N (2011) Purinergic signaling in hypothalamic tanycytes: potential roles in chemosensing. Semin Cell Dev Biol 22:237–244.

Dale N (2012) A classic review on extracellular ATP and its signaling functions that helped to define the field's agenda for many years. Biochem J 2012:1–6.

Dale N, Gilday D (1996) Regulation of rhythmic movements by purinergic neurotransmitters in frog embryos. Nature 383:259–263.

Dale N, Frenguelli BG (2009) Release of adenosine and ATP during ischemia and epilepsy. Curr Neuropharmacol 7:160–179.

Dale N, Pearson T, Frenguelli BG (2000) Direct measurement of adenosine release during hypoxia in the CA1 region of the rat hippocampal slice. J Physiol (Lond) 526:143–155.

Diez-Zaera M, Diaz-Hernandez JI, Hernandez-Alvarez E, Zimmermann H, Diaz-Hernandez M, Miras-Portugal MT (2011) Tissue-nonspecific alkaline phosphatase promotes axonal growth of hippocampal neurons. Mol Biol Cell 22:1014–1024.

Drury AN, Szent-Gyorgyi A (1929) The physiological activity of adenine compounds with especial reference to their action upon the mammalian heart. J Physiol 68:213–237.

Edwards FA, Gibb AJ, Colquhoun D (1992) ATP receptor-mediated synaptic currents in the central nervous system. Nature 359:144–147.

Eldridge FL, Kiley JP, Millhorn DE (1985) Respiratory responses to medullary hydrogen ion changes in cats: different effects of respiratory and metabolic acidoses. J Physiol 358:285–297.

Fields RD, Ni Y (2010) Nonsynaptic communication through ATP release from volume-activated anion channels in axons. Sci Signal 3:ra73.

Frayling C, Britton R, Dale N (2011) ATP-mediated glucosensing by hypothalamic tanycytes. J Physiol 589:2275–2286.

Fredholm BB, AP IJ, Jacobson KA, Klotz KN, Linden J (2001) International Union of Pharmacology XXV: nomenclature and classification of adenosine receptors. Pharmacol Rev 53:527–552.

Frenguelli BG, Llaudet E, Dale N (2003) High-resolution real-time recording with microelectrode biosensors reveals novel aspects of adenosine release during hypoxia in rat hippocampal slices. J Neurochem 86:1506–1515.

Gallopin T, Luppi PH, Cauli B, Urade Y, Rossier J, Hayaishi O, Lambolez B, Fort P (2005) The endogenous somnogen adenosine excites a subset of sleep-promoting neurons *via* A_{2A} receptors in the ventrolateral preoptic nucleus. Neuroscience 134:1377–1390.

Gordon EL, Pearson JD, Slakey LL (1986) The hydrolysis of extracellular adenine nucleotides by cultured endothelial cells from pig aorta. Feed-forward inhibition of adenosine production at the cell surface. J Biol Chem 261:15496–15507.

Gourine AV, Llaudet E, Dale N, Spyer KM (2005) ATP is a mediator of chemosensory transduction in the central nervous system. Nature 436:108–111.

Gourine AV, Kasymov V, Marina N, Tang F, Figueiredo MF, Lane S, Teschemacher AG, Spyer KM, Deisseroth K, Kasparov S (2010) Astrocytes control breathing through pH-dependent release of ATP. Science 329:571–575.

Guyenet PG (2008) The 2008 Carl Ludwig lecture: retrotrapezoid nucleus, CO_2 homeostasis, and breathing automaticity. J Appl Physiol 105:404–416.

Guyenet PG, Stornetta RL, Bayliss DA, Mulkey DK (2005) Retrotrapezoid nucleus: a litmus test for the identification of central chemoreceptors. Exp Physiol 90:247–253.

Haan N, Goodman T, Najdi-Samiei A, Stratford CM, Rice R, El Agha E, Bellusci S, Hajihosseini MK (2013) Fgf10-expressing tanycytes add new neurons to the appetite/energy-balance regulating centers of the postnatal and adult hypothalamus. J Neurosci 33:6170–6180.

Holton P (1959) The liberation of adenosine triphosphate on antidromic stimulation of sensory nerves. J Physiol 145:494–504.

Huckstepp RT, Dale N (2011) Redefining the components of central CO_2 chemosensitivity — towards a better understanding of mechanism. J Physiol 589:5561–5579.

Huckstepp RT, Eason R, Sachdev A, Dale N (2010a) CO_2-dependent opening of connexin 26 and related beta connexins. J Physiol 588:3921–3931.

Huckstepp RT, id Bihi R, Eason R, Spyer KM, Dicke N, Willecke K, Marina N, Gourine AV, Dale N (2010b) Connexin hemichannel-mediated CO_2-dependent release of ATP in the medulla oblongata contributes to central respiratory chemosensitivity. J Physiol 588:3901–3920.

James S, Richardson PJ (1993) Production of adenosine from extracellular ATP at the striatal cholinergic synapse. J Neurochem 60:219–227.

Kalinchuk AV, Stenberg D, Rosenberg PA, Porkka-Heiskanen T (2006a) Inducible and neuronal nitric oxide synthases (NOS) have complementary roles in recovery sleep induction. Eur J Neurosci 24:1443–1456.

Kalinchuk AV, McCarley RW, Porkka-Heiskanen T, Basheer R (2011) The time course of adenosine, nitric oxide (NO) and inducible NO synthase changes in the brain with sleep loss and their role in the non-rapid eye movement sleep homeostatic cascade. J Neurochem 116:260–272.

Kalinchuk AV, Lu Y, Stenberg D, Rosenberg PA, Porkka-Heiskanen T (2006b) Nitric oxide production in the basal forebrain is required for recovery sleep. J Neurochem 99:483–498.

Kang J, Kang N, Lovatt D, Torres A, Zhao Z, Lin J, Nedergaard M (2008) Connexin 43 hemichannels are permeable to ATP. J Neurosci 28:4702–4711.

Kawate T, Michel JC, Birdsong WT, Gouaux E (2009) Crystal structure of the ATP-gated P2X(4) ion channel in the closed state. Nature 460:592–598.

Kermer V, Ritter M, Albuquerque B, Leib C, Stanke M, Zimmermann H (2010) Knockdown of tissue nonspecific alkaline phosphatase impairs neural stem cell proliferation and differentiation. Neurosci Lett 485:208–211.

Khakh BS (2001) Molecular physiology of P2X receptors and ATP signaling at synapses. Nat Rev Neurosci 2:165–174.

Khakh BS, Bao XR, Labarca C, Lester HA (1999) Neuronal P2X transmitter-gated cation channels change their ion selectivity in seconds. Nat Neurosci 2:322–330.

Kittner H, Franke H, Harsch JI, El-Ashmawy IM, Seidel B, Krugel U, Illes P (2006) Enhanced food intake after stimulation of hypothalamic P2Y1 receptors in rats: modulation of feeding behaviour by extracellular nucleotides. Eur J Neurosci 24:2049–2056.

Klyuch BP, Dale N, Wall MJ (2012) Deletion of ecto-5'-nucleotidase (CD73) reveals direct action potential-dependent adenosine release. J Neurosci 32:3842–3847.

Klyuch BP, Richardson MJ, Dale N, Wall MJ (2011) The dynamics of single spike-evoked adenosine release in the cerebellum. J Physiol 589:283–295.

Kriegstein A, Alvarez-Buylla A (2009) The glial nature of embryonic and adult neural stem cells. Annu Rev Neurosci 32:149–184.

Kueng A, Kranz C, Mizaikoff B (2004) Amperometric ATP biosensor based on polymer entrapped enzymes. Biosens Bioelectron 19:1301–1307.

Langer D, Ikehara Y, Takebayashi H, Hawkes R, Zimmermann H (2007) The ectonucleotidases alkaline phosphatase and nucleoside triphosphate diphosphohydrolase 2 are associated with subsets of progenitor cell populations in the mouse embryonic, postnatal and adult neurogenic zones. Neuroscience 150:863–879.

Langer D, Hammer K, Koszalka P, Schrader J, Robson S, Zimmermann H (2008) Distribution of ectonucleotidases in the rodent brain revisited. Cell Tissue Res 334:199–217.

Lee DA, Bedont JL, Pak T, Wang H, Song J, Miranda-Angulo A, Takiar V, Charubhumi V, Balordi F, Takebayashi H, Aja S, Ford E, Fishell G, Blackshaw S (2012) Tanycytes of the hypothalamic median eminence form a diet-responsive neurogenic niche. Nat Neurosci 15:700–702.

Li J, Tang Y, Cai D (2012) IKKbeta/NF-kappaB disrupts adult hypothalamic neural stem cells to mediate a neurodegenerative mechanism of dietary obesity and pre-diabetes. Nat Cell Biol 14:999–1012.

Llaudet E, Botting NP, Crayston JA, Dale N (2003) A three-enzyme microelectrode sensor for detecting purine release from central nervous system. Biosens Bioelectron 18:43–52.

Llaudet E, Hatz S, Droniou M, Dale N (2005) Microelectrode biosensor for real-time measurement of ATP in biological tissue. Anal Chem 77:3267–3273.

Loeschcke HH (1982) Central chemosensitivity and the reaction theory. J Physiol 332:1–24.

Lovatt D, Xu Q, Liu W, Takano T, Smith NA, Schnermann J, Tieu K, Nedergaard M (2012) Neuronal adenosine release, and not astrocytic ATP release, mediates feedback inhibition of excitatory activity. Proc Natl Acad Sci USA 109:6265–6270.

Masse K, Dale N (2012) Purines as potential morphogens during embryonic development. Purinergic Signal 8:503–521.

Masse K, Eason R, Bhamra S, Dale N, Jones EA (2006) Comparative genomic and expression analysis of the conserved NTPDase gene family in Xenopus. Genomics 87:366–381.

Masse K, Bhamra S, Eason R, Dale N, Jones EA (2007) Purine-mediated signaling triggers eye development. Nature 449:1058–1062.

Masse K, Bhamra S, Allsop G, Dale N, Jones EA (2010) Ectophosphodiesterase/nucleotide phosphohydrolase (Enpp) nucleotidases: cloning, conservation and developmental restriction. Int J Dev Biol 54:181–193.

Masson JF, Kranz C, Mizaikoff B, Gauda EB (2008) Amperometric ATP microbiosensors for the analysis of chemosensitivity at rat carotid bodies. Anal Chem 80:3991–3998.

Meigh L, Greenhalgh SA, Rodgers TL, Cann MJ, Roper DI, Dale N (2013) CO_2 directly modulates connexin 26 by formation of carbamate bridges between subunits. eLife 2:e01213.

Mitchell JB, Miller K, Dunwiddie TV (1993) Adenosine-induced suppression of synaptic responses and the initiation and expression of long-term potentiation in the CA1 region of the hippocampus. Hippocampus 3:77–86.

Morton GJ, Cummings DE, Baskin DG, Barsh GS, Schwartz MW (2006) Central nervous system control of food intake and body weight. Nature 443:289–295.

Mulkey DK, Stornetta RL, Weston MC, Simmons JR, Parker A, Bayliss DA, Guyenet PG (2004) Respiratory control by ventral surface chemoreceptor neurons in rats. Nat Neurosci 7:1360–1369.

Mullier A, Bouret SG, Prevot V, Dehouck B (2010) Differential distribution of tight junction proteins suggests a role for tanycytes in blood-hypothalamus barrier regulation in the adult mouse brain. J Comp Neurol 518:943–962.

Nattie E (1999) CO_2, brainstem chemoreceptors and breathing. Prog Neurobiol 59:299–331.

Nattie EE (2001) Central chemosensitivity, sleep, and wakefulness. Respir Physiol 129:257–268.

Nunes PHC, Calaza KD, Albuquerque LM, Fragel-Madeira L, Sholl-Franco A, Ventura AL (2007) Signal transduction pathways associated with ATP-induced proliferation of cell progenitors in the intact embryonic retina. Int J Dev Neurosci 25:499–508.

Orellana JA, Saez PJ, Cortes-Campos C, Elizondo RJ, Shoji KF, Contreras-Duarte S, Figueroa V, Velarde V, Jiang JX, Nualart F, Saez JC, Garcia MA (2012) Glucose increases intracellular free Ca^{2+} in tanycytes *via* ATP released through connexin 43 hemichannels. Glia 60:53–68.

Pascual O, Casper KB, Kubera C, Zhang J, Revilla-Sanchez R, Sul JY, Takano H, Moss SJ, McCarthy K, Haydon PG (2005) Astrocytic purinergic signaling coordinates synaptic networks. Science 310:113–116.

Pearson R, Catsicas M, Becker D, Mobbs P (2002) Purinergic and muscarinic modulation of the cell cycle and calcium signaling in the chick retinal ventricular zone. J Neurosci 22:7569–7579.

Pearson RA, Dale N, Llaudet E, Mobbs P (2005) ATP released *via* gap junction hemichannels from the pigment epithelium regulates neural retinal progenitor proliferation. Neuron 46:731–744.

Pearson RA, Catsicas M, Becker DL, Bayley P, Luneborg NL, Mobbs P (2004) Ca^{2+} signaling and gap junction coupling within and between pigment epithelium and neural retina in the developing chick. Eur J Neurosci 19:2435–2445.

Pellegatti P, Falzoni S, Pinton P, Rizzuto R, Di Virgilio F (2005) A novel recombinant plasma membrane-targeted luciferase reveals a new pathway for ATP secretion. Mol Biol Cell 16:3659–3665.

Porkka-Heiskanen T, Kalinchuk AV (2011) Adenosine, energy metabolism and sleep homeostasis. Sleep Med Rev 15:123–135.

Porkka-Heiskanen T, Strecker RE, Thakkar M, Bjorkum AA, Greene RW, McCarley RW (1997) Adenosine: a mediator of the sleep-inducing effects of prolonged wakefulness. Science 276:1265–1268.

Putnam RW, Filosa JA, Ritucci NA (2004) Cellular mechanisms involved in CO_2 and acid signaling in chemosensitive neurons. Am J Physiol Cell Physiol 287:C1493–C1526.

Radulovacki M (2005) Adenosine sleep theory: how I postulated it. Neurol Res 27:137–138.

Radulovacki M, Virus RM, Djuricic-Nedelson M, Green RD (1984) Adenosine analogs and sleep in rats. J Pharmacol Exp Ther 228:268–274.

Raymond SM, Jackson IJ (1995) The retinal pigmented epithelium is required for development and maintenance of the mouse neural retina. Curr Biol 5:1286–1295.

Retey JV, Adam M, Honegger E, Khatami R, Luhmann UF, Jung HH, Berger W, Landolt HP (2005) A functional genetic variation of adenosine deaminase affects the duration and intensity of deep sleep in humans. Proc Natl Acad Sci USA 102:15676–15681.

Richerson GB (2004) Serotonergic neurons as carbon dioxide sensors that maintain pH homeostasis. Nat Rev Neurosci 5:449–461.

Robins SC, Stewart I, McNay DE, Taylor V, Giachino C, Goetz M, Ninkovic J, Briancon N, Maratos-Flier E, Flier JS, Kokoeva MV, Placzek M (2013) alpha-Tanycytes of the adult hypothalamic third ventricle include distinct populations of FGF-responsive neural progenitors. Nat Commun 4:2049.

Robson SC, Sevigny J, Zimmermann H (2006) The E-NTPDase family of ectonucleotidases: structure function relationships and pathophysiological significance. Purinergic Signal 2:409–430.

Rodriguez EM, Blazquez JL, Pastor FE, Pelaez B, Pena P, Peruzzo B, Amat P (2005) Hypothalamic tanycytes: a key component of brain-endocrine interaction. Int Rev Cytol 247:89–164.

Rudolphi KA, Schubert P, Parkinson FE, Fredholm BB (1992) Neuroprotective role of adenosine in cerebral-ischemia. Trends Pharmacol Sci 13:439–445.

Sabirov RZ, Okada Y (2005) ATP release *via* anion channels. Purinergic Signal 1:311–328.

Sawada K, Echigo N, Juge N, Miyaji T, Otsuka M, Omote H, Yamamoto A, Moriyama Y (2008) Identification of a vesicular nucleotide transporter. Proc Natl Acad Sci USA 105:5683–5686.

Schmitt LI, Sims RE, Dale N, Haydon PG (2012) Wakefulness affects synaptic and network activity by increasing extracellular astrocyte-derived adenosine. J Neurosci 32:4417–4425.

Seidel B, Bigl M, Franke H, Kittner H, Kiess W, Illes P, Krugel U (2006) Expression of purinergic receptors in the hypothalamus of the rat is modified by reduced food availability. Brain Res 1089:143–152.

Severson CA, Wang W, Pieribone VA, Dohle CI, Richerson GB (2003) Midbrain serotonergic neurons are central pH chemoreceptors. Nat Neurosci 6:1139–1140.

Shams H (1985) Differential effects of CO_2 and H^+ as central stimuli of respiration in the cat. J Appl Physiol 58:357–364.

Sims RE, Dale N (2014) Activity-dependent adenosine release may be linked to activation of Na^+–K^+ ATPase: an *in vitro* rat study. PLoS One 9:e87481.

Sims RE, Wu HH, Dale N (2013) Sleep-wake sensitive mechanisms of adenosine release in the basal forebrain of rodents: an *in vitro* study. PLoS One 8:e53814.

Slakey LL, Cosimini K, Earls JP, Thomas C, Gordon EL (1986) Simulation of extracellular nucleotide hydrolysis and determination of kinetic constants for the ectonucleotidases. J Biol Chem 261:15505–15507.

Stefan C, Jansen S, Bollen M (2005) NPP-type ectophosphodiesterases: unity in diversity. Trends Biochem Sci 30:542–550.

Stefan C, Jansen S, Bollen M (2006) Modulation of purinergic signaling by NPP-type ectophosphodiesterases. Purinergic Signal 2:361–370.

Su C, Bevan JA, Burnstock G (1971) [^{3}H]adenosine triphosphate: release during stimulation of enteric nerves. Science 173:336–338.

Sugioka M, Fukuda Y, Yamashita M (1996) Ca^{2+} responses to ATP *via* purinoceptors in the early embryonic chick retina. J Physiol 493:855–863.

Surprenant A, Rassendren F, Kawashima E, North RA, Buell G (1996) The cytolytic P2Z receptor for extracellular ATP identified as a P2X receptor (P2X7). Science 272:735–738.

Wall MJ, Dale N (2007) Auto-inhibition of rat parallel fibre-Purkinje cell synapses by activity-dependent adenosine release. J Physiol 581:553–565.

Wall MJ, Dale N (2013) Neuronal transporter and astrocytic ATP exocytosis underlie activity-dependent adenosine release in the hippocampus. J Physiol 591:3853–3871.

Weissman TA, Riquelme PA, Ivic L, Flint AC, Kriegstein AR (2004) Calcium waves propagate through radial glial cells and modulate proliferation in the developing neocortex. Neuron 43:647–661.

Wenker IC, Sobrinho CR, Takakura AC, Moreira TS, Mulkey DK (2012) Regulation of ventral surface CO_2/H^+-sensitive neurons by purinergic signaling. J Physiol 590:2137–2150.

Werman R (1966) Criteria for identification of a central nervous system transmitter. Comp Biochem Physiol 18:745–766.

Xu Y, Tamamaki N, Noda T, Kimura K, Itokazu Y, Matsumoto N, Dezawa M, Ide C (2005) Neurogenesis in the ependymal layer of the adult rat 3rd ventricle. Exp Neurol 192:251–264.

Yao ST, Gourine AV, Spyer KM, Barden JA, Lawrence AJ (2003) Localisation of P2X2 receptor subunit immunoreactivity on nitric oxide synthase expressing neurones in the brain stem and hypothalamus of the rat: a fluorescence immunohistochemical study. Neuroscience 121:411–419.

Yegutkin GG (2008) Nucleotide- and nucleoside-converting ectoenzymes: important modulators of purinergic signaling cascade. Biochim Biophys Acta 1783:673–694.

Zimmermann H (1992) 5′-Nucleotidase: molecular structure and functional aspects. Biochem J 285:345–365.

Zimmermann H (1999) Two novel families of ectonucleotidases: molecular structures, catalytic properties and a search for function. Trends Pharmacol Sci 20:231–236.

Zimmermann H (2000) Extracellular metabolism of ATP and other nucleotides. Naunyn Schmiedebergs Arch Pharmacol 362:299–309.

Zimmermann H, Braun N (1999) Ecto-nucleotidases: molecular structures, catalytic properties, and functional roles in the nervous system. Prog Brain Res 120:371–385.

Zuber ME, Gestri G, Viczian AS, Barsacchi G, Harris WA (2003) Specification of the vertebrate eye by a network of eye field transcription factors. Development 130:5155–5167.

CHAPTER 4

ELECTROCHEMICAL DETECTION OF ADENOSINE *IN VIVO*

Ashley E. Ross and B. Jill Venton
University of Virginia

4.1 INTRODUCTION

Adenosine is an important neuromodulator and neuroprotector in the brain and is the primary breakdown product of adenosine triphosphate (ATP). Previous research about adenosine was based on slow time resolution techniques, which could not provide information on rapid signaling. Development of real-time measurements of adenosine release over the past several years has provided a wealth of information on adenosine signaling in the brain. Fast-scan cyclic voltammetry (FSCV) at carbon-fiber microelectrodes and amperometry at enzyme biosensors have been the primary techniques for real-time, *in vivo* detection of adenosine. Adenosine is now viewed not only as a slow acting neuromodulator, but as a rapidly released signaling molecule in the brain which may be important for neuromodulation or protection on a much faster time scale. Real-time measurements have led to a better understanding of how quickly adenosine can be released, the mechanisms of release in different brain regions, and how adenosine release regulates the response to acute brain injuries or conditions such as hypoxia, ischemia, hypercapnia, and epileptic seizures.

4.1.1 ADENOSINE REGULATION IN THE CENTRAL NERVOUS SYSTEM

Adenosine is a nucleoside byproduct of ATP metabolism and is specifically involved in neuromodulation (Cunha, 2008; Franco et al., 2000), neuroprotection (Cunha, 2008; Fredholm, 1997; Robertson et al., 2001; Rudolphi et al., 1992), and regulation of cerebral blood flow in the brain (O'Regan, 2005). Adenosine is also important in many other organs of the body such as the heart (Lorbar et al., 1999; Villarreal et al., 2003) and lung (Lu et al., 2012). Regulation of adenosine in the brain is not fully understood; however, many studies have examined the function of adenosine receptors, transporters, and metabolic pathways. Adenosine targeted treatments for diseases such as Parkinson's (Fuxe et al., 2001; Pinna et al., 2005) and Huntington's (Blum et al., 2003) have become increasingly more popular. Neuroprotective effects of adenosine can be exploited using adenosine receptor antagonists and agonists (Fuxe et al., 2001; Ferre et al., 2004). Adenosine is involved in several brain disorders and diseases but little is known about the immediate response of adenosine to acute disruptions in brain function or in a diseased patient.

4.1.1.1 Intra- and extracellular formation of adenosine

Intra- and extracellular adenosine formation occurs by a variety of complicated mechanisms (Figure 4.1). In general, intracellular formation of adenosine is attributed to the catabolism of cytosolic ATP to adenosine monophosphate (AMP) due to metabolic stress (Latini and Pedata, 2001) as shown in Figure 4.1. Cytosolic-5'-nucleotidase is responsible for converting AMP to adenosine. This intracellular mechanism of adenosine formation is linked to maintaining an energy balance within the cell (Latini and Pedata, 2001). ATP exists in much higher concentrations than AMP, therefore small changes in ATP concentration leads to significant changes in the amount of AMP available for adenosine production. Another method of intracellular adenosine formation proposed in Figure 4.1 is the hydrolysis of S-adenosylhomocysteine (SAH); however, this pathway is not connected to energy balance in the cell and therefore is not the main contributor to intracellular adenosine formation (Latini and Pedata, 2001).

Extracellular formation of adenosine is primarily linked to extracellular ATP metabolism (Figure 4.1). ATP is released by exocytosis as a co-transmitter and broken down to adenosine by ecto-ATPase and ecto-5'-nucleotidase (NT5E). Directly released cyclic AMP can also be extracellularly metabolized to adenosine, but this mode of formation has only been linked to slow changes in extracellular formation (not shown in Figure 4.1) (Latini and Pedata, 2001). Recently, activity-dependent release of adenosine in the extracellular space was discovered (Pajski and Venton, 2010; Wall and Dale, 2008; Klyuch et al., 2011); however, adenosine has not been identified in vesicles.

4.1.1.2 Adenosine receptors

Adenosine is regulated by four known G-protein-coupled receptors in the central nervous system (CNS) (Fredholm et al., 1994; Fredholm et al., 2005): A_1, A_{2A},

A_{2B}, and A_3. A_1 is the most abundant adenosine receptor in the brain and inhibits neurotransmission by blocking adenylyl cyclase activity. A_{2A} is the second most abundant adenosine receptor expressed in the brain and activates adenylyl cyclase activity. Both of these receptors have low nanomolar affinities for adenosine, and are, therefore, the most important receptors in the brain for adenosine signaling (Latini et al., 1996; Latini and Pedata, 2001). The excitatory receptor, A_{2B}, and the inhibitory receptor, A_3, are less abundant in the brain and are activated by micromolar concentrations of adenosine (Latini and Pedata, 2001).

A_1 receptors are widely expressed and can be found pre-synaptically, post-synaptically, and non-synaptically, whereas A_{2A} receptors are primarily in synapses (Fredholm et al., 2005; Gomes et al., 2011). However, inhibitory actions are mostly a result of pre-synaptic A_1 receptors because they are primarily found on excitatory synapses (Gomes et al., 2011; Thompson et al., 1992; Hamilton and Smith, 1991). Adenosine receptors are also located on astrocytes and glia and they may serve as a modulator of non-neuronal brain functions (Gomes et al., 2011; Calker and Biber, 2005; Haselkorn et al., 2008).

4.1.1.3 Adenosine transporters

Extracellular adenosine is regulated by bi-directional nucleoside transporters in the brain as shown in Figure 4.1 (Latini and Pedata, 2001; Baldwin et al., 2004).

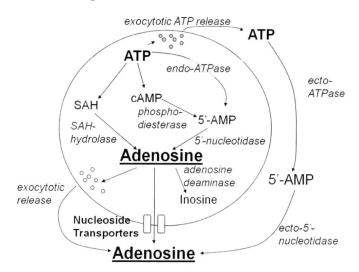

Figure 4.1 Adenosine mechanism of formation. Adenosine is formed intra- and extracellularly. Intracellularly, ATP breaks down to cAMP or SAH which further breaks down to adenosine. Adenosine can then be transported out of the cell either by nucleoside transporters or exocytosis. Extracellularly, adenosine can be formed by exocytotic release of ATP which is broken down to adenosine or by direct release of adenosine from inside the cell. Modified with permission from American Chemical Society (Pajski and Venton, 2010).

Nucleoside transporters are classified into two main categories: equilibrative nucleoside transporters (ENTs) and concentrative nucleoside transporters (Baldwin et al., 2004; Latini and Pedata, 2001). ENTs follow a concentration gradient and can transport both purines and pyrimidines, whereas concentrative transporters are driven by a sodium gradient. ENTs are the primary transporter responsible for regulation of extracellular adenosine in the brain (Latini and Pedata, 2001; Thorn and Jarvis, 1996). Equilibrative transporters which are sensitive to the selective transport inhibitor nitrobenzylthioinosine (NBTI) are called equilibrative-sensitive, es, whereas equilibrative-insensitive, ei, transporters are not sensitive to NBTI. Most research studies on adenosine transport examine esENTs because of the selective inhibitor NBTI.

4.1.2 FUNCTION OF ADENOSINE IN THE CENTRAL NERVOUS SYSTEM

Adenosine is an important neuromodulator in the brain and can regulate cerebral blood flow, the sleep–wake cycle, and excitatory neurotransmission. Modulating adenosine may help treat diseases and disorders such as stroke, ischemia, Huntington's disease, Parkinson's disease, epilepsy, and Alzheimer's disease. Due to its widespread involvement in the brain and complicated regulatory system, adenosine can be difficult to study.

4.1.2.1 Adenosine as a neuroprotector

Neuroprotection by adenosine has been linked to the activation of A_1 receptors, due to their inhibitory nature (Fredholm et al., 2005; Fredholm, 1997). A_1 receptors provide neuroprotection during ischemic attack (Sperlagh and Vizi, 2011; Pearson et al., 2006), neuroinflammation (Gomes et al., 2011; Haselkorn et al., 2008), exitotoxicity (Blum et al., 2003), and epileptic episodes (Dale and Frenguelli, 2009; Fredholm et al., 2005; Fredholm, 1997). Neuroprotection can occur presynaptically by inhibiting excitatory responses or post-synaptically by hyperpolarization of cells (Fredholm et al., 2005; de Mendonca et al., 2000; Rudolphi et al., 1992). Activation or deactivation of adenosine receptors has sparked a new avenue of pharmacological agents to be studied for therapeutics. Much is known about the long term effects of adenosine receptor manipulation; however, the rapid response of adenosine receptor manipulation, specifically during injury and acute insults, is not as well understood and will be discussed later in the chapter.

The effects on adenosine during brain injury and acute insults have been studied primarily using low temporal resolution techniques such as microdialysis coupled to high performance liquid chromatography (HPLC) (Bell et al., 1998; Deng et al., 2003; Valtysson et al., 1998) or histopathological scoring of fixed tissue collected up to months after injury (Aden et al., 2003). Adenosine was shown to be released after controlled cortical impact, a method used to study traumatic brain injury, using microdialysis coupled to HPLC (Bell et al., 1998). A peak in adenosine was observed within 20 minutes of injury and samples were taken every 10 minutes. The amount of adenosine released and how long it was elevated was determined, but the sampling time was too long to study mechanism of release. In another study, adenosine concentrations were measured during ischemia using HPLC (Valtysson et al., 1998). They determined the presence of ischemic markers, but detailed information on time course and mechanisms of release were unattainable. These studies reveal the long-term effects of adenosine during injury; however, to understand what occurs immediately following injury or during conditions such as ischemia, advances in temporal resolution were required. The recent developments of adenosine detection using FSCV and amperometry have led to further advances of adenosine characterization in the brain which will be discussed later in the chapter.

4.1.2.2 Adenosine as a neuromodulator

Distinct from its neuroprotective role, adenosine is classically recognized as a homeostatic neuromodulator in the brain and as a synaptic neuromodulator (Cunha, 2001). Like neuroprotection, neuromodulation occurs when adenosine binds to its receptor in the brain and either causes an inhibitory or excitatory response. The term "retaliatory metabolite" was given to adenosine, primarily because of its significant role in regulating cell metabolism (Cunha, 2001; Newby et al., 1985). In short, adenosine release from the cell occurs because of a concentration gradient produced when ATP is used intracellularly. Small concentration shifts in ATP produce large changes in adenosine concentration because ATP exists in much higher concentrations in the cell. When intracellular adenosine levels rise, adenosine is released by bi-directional ENTs.

In addition to slowing cell metabolism, adenosine modulates neurotransmission at the synaptic level. Adenosine can modulate dopamine (Ferre et al., 1997; Ferre et al., 1992; Quarta et al., 2004; Okada et al., 1996), serotonin (Okada et al., 2001), glutamate (Sperlagh et al., 2007), and γ-Aminobutyric acid (GABA) release (Ferre et al., 1997). Typically, adenosine acts to suppress neurotransmission, particularly in response to insults such as traumatic brain injury (Haselkorn et al., 2008; Bell et al., 1998) and stroke (Von Lubitz, 2001). Glutamate regulation by adenosine leads to reduction in excitotoxicity and can reduce damage during ischemia (Sperlagh et al., 2007). Thus, the roles of neuromodulation and neuroprotection often overlap.

Neurotransmitter modulation by adenosine is traditionally thought of as a slow process where adenosine concentration builds up minutes to hours after injury. High temporal resolution techniques have not been traditionally used to study adenosine modulation, so little is known about the extent to which adenosine modulates neurotransmission on the millisecond timescale. Over the last few years, real-time adenosine has been explored using electrochemistry (Street et al., 2011; Pajski and Venton, 2013; Pajski and Venton, 2010; Swamy and Venton, 2007; Ross and Venton, 2012; Klyuch et al., 2011; Dale and Frenguelli, 2009). These new techniques for adenosine detection have led to the understanding that adenosine is not just a slow acting neuromodulator, but also operates on the millisecond timescale. This chapter will discuss discoveries that electrochemical monitoring of adenosine have enabled and major contributions to the field.

4.2 ELECTROCHEMICAL DETECTION OF ADENOSINE *IN VIVO*

Recently, electrochemical detection of adenosine *in vivo* has become increasingly popular to monitor adenosine changes in real time. There are three major electrochemistry techniques used for detection of adenosine *in vivo*: FSCV at carbon-fiber microelectrodes, FSCV at diamond microelectrodes, and amperometry using adenosine selective enzyme sensors. FSCV at carbon-fiber microelectrodes is popular for studying the mechanism of stimulated adenosine release (Cechova and Venton, 2008; Cechova et al., 2010; Pajski and Venton, 2010), comparing adenosine release in different brain regions (Pajski and Venton, 2013), and understanding the enzymes involved in extracellular breakdown of AMP in spinal nociceptive circuits (Street et al., 2011; Street et al., 2013). FSCV at diamond microelectrodes has been used to study respiratory rhythmogenesis (Xie et al., 2006; Dong et al., 2011; Park et al., 2008). Amperometric enzyme biosensors have been used for studying the involvement of adenosine in hypercapnia (Dale, 2006), ischemia (Dale and Frenguelli, 2009), the sleep–wake cycle (Sims et al., 2013), and epilepsy (Dale and Frenguelli, 2009). Major advancements in the field regarding adenosine signaling in several biological conditions have been made using electrochemical techniques.

4.2.1 ADENOSINE CHARACTERIZATION USING FSCV

FSCV is an electrochemical technique where a triangular waveform is rapidly applied to the working electrode and the current is monitored from electroactive species undergoing redox reactions at the electrode surface. Carbon-fiber microelectrodes were developed for dopamine detection and are also typically used for adenosine detection (Stamford et al., 1984; Wightman et al., 1988; Park et al., 2010; Swamy and Venton, 2007; Pajski and Venton, 2010). The traditional triangle waveform for dopamine detection begins at $-0.4\,V$ and linearly ramps to $1.30\,V$ and back down to the holding potential until the next scan. Adenosine oxidizes at a higher potential (Dryhurst, 1977; Swamy and Venton, 2007), so $1.45\,V$ or $1.50\,V$ is used as the switching potential for adenosine detection (adenosine waveform shown in Figure 4.2a). The frequency of data collection for FSCV is

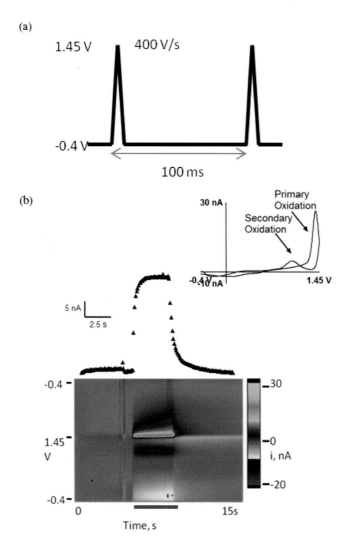

Figure 4.2 Example of the FSCV adenosine waveform and data *in vitro*. (a) The adenosine waveform used is a triangle waveform starting at a −0.4 V holding potential and linearly ramping to 1.45 V and linearly ramping back down to the holding potential. The waveform is applied at a rate of 400 V/s and is repeated every 100 milliseconds. (b) A 3-D color plot for an *in vitro* calibration of 5 μM adenosine is shown. The y-axis is voltage, x-axis is time, and the current is shown in false color. Green represents oxidative current. At four seconds, adenosine is injected into the flow cell and at eight seconds the analyte is removed, shown by the red bar. An increase in current is observed at the switching potential (1.45 V) and at 1.0 V in subsequent scans. The current *versus* time plot, at the primary oxidation peak for adenosine, is shown directly above the color plot. The CV for adenosine is in the inset. Distinct primary and secondary oxidation peaks characteristic of adenosine electrochemistry at carbon-fiber microelectrodes are marked.

normally 10 Hz, so that scans are repeated every 100 milliseconds. Fast repetition rates are beneficial for studying neurotransmitters *in vivo*. Scan rates ranging from a hundred to a thousand V/s are used for neurotransmitter detection; however, 400 V/s is the typical scan rate used for most compounds (Swamy and Venton, 2007; Venton and Wightman, 2007). High scan rates are beneficial for enhancing sensitivity, but they can produce relatively large non-faradaic background currents. The background current arises from the movement of ions in solution or the oxidation of functional groups on the surface of the electrode and is larger than the typical faradaic current. However, background currents are stable at carbon-fiber microelectrodes, which enables background subtraction, leading to a background subtracted cyclic voltammogram (CV). CVs provide a fingerprint for the analyte of interest, allowing for some selectivity between analytes.

Adenosine is an electroactive molecule, making it ideal for electrochemical detection (Dryhurst, 1977). Adenosine undergoes a series of three oxidation reactions, only two of which are typically seen at carbon-fiber microelectrodes (Swamy and Venton, 2007). The primary oxidation is at around 1.40 V and shows up on the cathodic scan due to slow kinetics. The secondary oxidation peak appears at 1.0 V. The primary oxidation for adenosine is irreversible so no reduction peak is observed. An example of a background-subtracted adenosine CV is shown in the inset of Figure 4.2b. The primary and secondary peaks for adenosine are marked. Figure 4.2b also shows a three dimensional color plot for adenosine. The plot displays voltage on the y-axis, time on the x-axis and current in false color. The green color represents oxidation of adenosine and two oxidation peaks are visible. A current *versus* time trace is often used in FSCV to monitor changes in adenosine *in vivo* (Figure 4.2, above the color plot). Current *versus* time traces are useful for studying clearance rates *in vivo* and adsorption properties *in vitro*. The trace can also be converted to a concentration *versus* time plot by using a calibration factor obtained either before or after the *in vivo* experiment.

Carbon-fiber microelectrodes are advantageous for rapid electrochemical measurements due to their ease of fabrication, affordability, and stability *in vivo*. Modifications to carbon-fiber microelectrodes are often explored to further enhance sensitivity for analytes of interest and to increase electron transfer kinetics of the oxidation and reduction reaction. Recently, carbon-fiber microelectrodes were modified with a combination of Nafion and carbon nanotubes and adenosine oxidation current with FSCV was increased four-fold (Ross and Venton, 2012). Selectivity of adenosine over ATP was increased three-fold both *in vitro* and in slices which is beneficial because the CV for ATP is similar to adenosine due to similar chemical structures and oxidation reactions.

4.2.1.1 Characterization of carbon-fiber microelectrodes for adenosine detection using FSCV

Adenosine detection using FSCV was first reported by Brajter-Toth's group in 2000 (El-Nour and Brajter-Toth, 2000). Nanostructured carbon-fiber microelectrodes were fabricated for enhanced detection of adenosine. The applied waveform was from −1.0 V to 1.50 V at a scan rate of 500 V/s. Electrochemical pretreatment of electrodes further enhanced the detection of adenosine *in vitro*. Several different methods of electrochemical treatments and buffer compositions for adenosine detection were explored, and adenosine was best detected in phosphate buffer (Brajter-Toth et al., 2000; El-Nour and Brajter-Toth, 2000) with a limit of detection (LOD) *in vitro* of 5 μM.

In 2007, our lab further characterized carbon-fiber microelectrodes for adenosine detection using FSCV (Swamy and Venton, 2007). The electrochemical reaction was analyzed and the two oxidation peaks in the adenosine CV were identified. By maximizing adsorption and minimizing noise, the LOD was 15 nM (Swamy and Venton, 2007), two orders of magnitude smaller than previously reported (Brajter-Toth et al., 2000). Several other biologically relevant molecules, chemically similar to adenosine were examined. Purines such as guanine and hypoxanthine have similar structures; however their oxidation peaks are at different potentials than adenosine; whereas, AMP and ATP have similar CVs to adenosine but carbon-fiber microelectrodes are not as sensitive to them.

4.2.1.2 Characterization of diamond electrodes for adenosine detection using FSCV

Boron-doped diamond microelectrodes have also been explored for adenosine detection with FSCV (Xie et al., 2006). Diamond is electrically insulating, so it is essential to introduce impurities into the sp^3-hybridized tetrahedral matrix of diamond, like boron, to make a conductive electrode. A thick film of boron-doped diamond is deposited onto a substrate, like metal, by chemical vapor deposition (CVD) (Park et al., 2008). Low nanomolar limits of detection were obtained with diamond microelectrodes, which is beneficial for studying adenosine levels *in vivo*. Diamond microelectrodes are advantageous compared to carbon-fiber microelectrodes due to their increased sensitivity and resistance to fouling (Park et al., 2008). Signal-to-background ratios are improved compared to carbon-fibers because of lack of functional group oxidation reactions and increased capacitance. However, despite the increase in sensitivity, they are much larger than carbon-fiber microelectrodes (80 μm) and are difficult to reproducibly insulate (Park et al., 2008).

4.2.1.3 Adenosine modulates respiratory rhythmogenesis

Diamond microelectrodes were used to study the modulation of respiratory rhythmogenesis by adenosine in the PreBötzinger complex of the rat brainstem (Xie et al., 2006). This region contains many adenosine releasing neurons responsible for regulation of respiration in the rat. After activation of adenosine A_{2A} receptors by CGS 21680 (a specific adenosine A_{2A} subtype receptor agonist), a large increase in current was observed and confirmed to be adenosine by the CV. This was the first study to use FSCV in an animal model to study adenosine and showed that adenosine may have a role in regulation of respiration (Xie et al., 2006).

4.2.1.4 Short electrical stimulations elicit rapid adenosine changes

Stimulated release has been characterized using FSCV at carbon-fiber microelectrodes by applying pulse trains at a bipolar stimulating electrode to elicit release (Pajski and Venton, 2010; Pajski and Venton, 2013; Cechova and Venton, 2008). Electrical stimulation mimics fast neuronal activation due to a discrete stimulus and causes rapid neurochemical changes in the brain. Our lab was the first to study electrically-stimulated adenosine in the caudate putamen *in vivo* using FSCV (Cechova and Venton, 2008). Stimulated adenosine release was verified by administering an ENT inhibitor, a histamine synthetic precursor, and an adenosine kinase (ADK) inhibitor. The ENT inhibitor, propentofylline, significantly decreased stimulated adenosine release, which demonstrated that short, high frequency stimulations (HFSs) caused release of adenosine from inside the cell. ABT-702, the ADK inhibitor, significantly increased stimulated adenosine release which is expected since this drug blocks adenosine metabolism. No effect was observed after the addition of L-histidine, a histamine synthetic precursor, which verified that the stimulated release was adenosine and not histamine since their CVs are fairly similar (Figure 4.3).

4.2.1.5 Adenosine transiently regulates oxygen concentrations

The effect of adenosine on oxygen levels in the brain has also been studied *in vivo* using FSCV (Cechova and Venton, 2008). Adenosine is a vasodilator and regulates cerebral blood flow so it was expected that stimulating adenosine release would affect oxygen levels in the brain (O'Regan, 2005). To measure oxygen changes, a waveform was used that scanned to a lower potential sufficient to reduce oxygen (Cechova and Venton, 2008; Venton et al., 2003). Blocking intracellular release of adenosine with propentofylline caused a decrease in the amount of oxygen detected. From this study, the effects of adenosine as a vasodilator were confirmed to occur on the millisecond timescale.

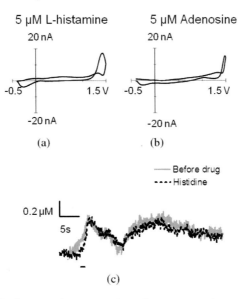

Figure 4.3 Stimulated release is adenosine and not histamine in the caudate putamen. Five μM L-histidine, a histamine precursor, was administered to verify that the stimulated release signal was adenosine. (a) *in vitro* signal of five μM L-histamine and (b) adenosine. (c) Current *versus* time trace of stimulated release *in vivo*. The signal did not change after L-histidine administration so histamine release is not being electrically stimulated in the caudate putamen. Reproduced with permission from Cechova and Venton, 2008.

4.2.1.6 A_1 receptors self-regulate stimulated adenosine release

The A_1 receptor has demonstrated autoreceptor characteristics *in vivo* using FSCV (Cechova et al., 2010). Our lab showed that adenosine A_1 receptors modulate adenosine release in the caudate putamen, which means that they self-regulate adenosine release (Cechova et al., 2010). A_1 receptor agonists decreased both stimulated dopamine and adenosine release. A_1 agonists and antagonists had an immediate effect on stimulated adenosine release, whereas the effect on dopamine release was delayed, like previously reported (Cechova et al., 2010; O'Neill et al., 2007). This study also examined how A_1 interactions with dopamine D_1 receptors modulated adenosine release. Administration of the A_1 receptor agonist, N6-cyclopentyladenosine (CPA), and the D1 receptor antagonist SCH23390 showed that the A_1–D_1 interaction modulates adenosine and dopamine release *in vivo* (Cechova et al., 2010).

4.2.1.7 Stimulated adenosine release is activity-dependent in multiple brain regions

The mechanism of stimulated adenosine release was first characterized using FSCV in brain slices from the rat caudate putamen (Pajski and Venton, 2010).

Our lab showed that stimulated adenosine release is primarily activity-dependent and compared the mechanism of release using low frequency (10 Hz) and high frequency (60 Hz) stimulation (Pajski and Venton, 2010). Blocking ENTs significantly increased both low and high frequency stimulated adenosine release, implying that intracellular formation of adenosine is not responsible for release. Low frequency stimulated adenosine release was dependent on extracellular ATP metabolism, whereas high frequency stimulated release was not. Adenosine release was dependent on calcium influx, as perfusing with either ethylene diamine tetra acetic acid (EDTA) or ethylene glycol tetra acetic acid (EGTA), to chelate calcium or calcium and magnesium respectively, decreased the signal for both low frequency and HFSs.

Activity-dependent adenosine release can be due to (1) direct vesicular release of ATP, which gets metabolized to adenosine, (2) downstream release of adenosine after exocytosis of another neurotransmitter, or (3) direct release of adenosine from the cell by exocytosis (Wall and Dale, 2008; Wall and Dale, 2007; Mitchell et al., 1993). To determine the mechanism of activity-dependent adenosine release, case 1 and case 2 were tested using FSCV (Pajski and Venton, 2010). ATP metabolism (case 1) was blocked with a combination of a nucleoside triphosphate diphosphohydrolase (NTPDase) inhibitor and an NT5E inhibitor (Pajski and Venton, 2010). Blocking ATP metabolism did not affect high frequency stimulated adenosine release; but significantly reduced low frequency stimulated release. Thus, different stimulation frequencies produced different mechanisms of adenosine release. For example, HFS occurs during stressful events, so adenosine release in response to a stressful environment is not dependent on extracellular ATP metabolism.

Case 2 was tested by decreasing dopamine release using a vesicular monoamine transporter (VMAT) inhibitor and blocking the downstream action of the ionotropic glutamate receptors, α-amino-3-hydroxy-5-methyl-4-isoxazolepropionic acid (AMPA) and N-methyl-D-aspartate (NMDA) (Pajski and Venton, 2010). Dopamine did not affect adenosine whereas blocking ionotropic glutamate receptors significantly reduced stimulated adenosine release for both low and HFSs in the caudate putamen (Pajski and Venton, 2010). Thus, adenosine release may be a downstream action of exocytotic glutamate release.

The mechanism of high frequency stimulated release was then compared in multiple brain regions: the dorsal caudate putamen, nucleus accumbens shell, CA1 region of the hippocampus, and the prefrontal cortex. Adenosine release in all regions was activity-dependent but the mechanism of release varied. In the caudate putamen and nucleus accumbens, stimulated release was dependent on ionotropic glutamate receptors whereas in the hippocampus and prefrontal cortex, adenosine release was dependent on extracellular ATP metabolism (Pajski and Venton, 2013) (Figure 4.4). This study demonstrates the complexity of adenosine release in the

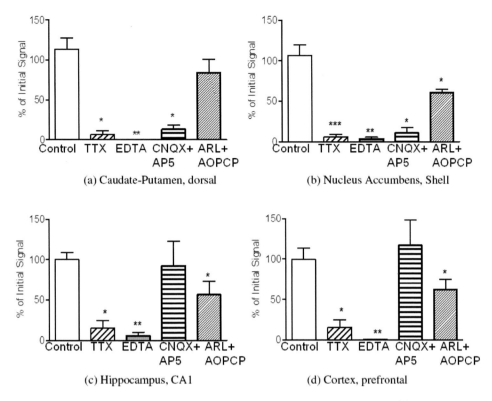

Figure 4.4 Mechanism of stimulated adenosine release varies by brain region: (a) Caudate putamen, (b) Nucleus accumbens shell, (c) CA1 region of the hippocampus and (d) Prefrontal cortex. Voltage-gated Na^+ channels were blocked by 0.5 μM tetrodotoxin (TTX) and Ca^{2+} chelated with 1 mM EDTA. In all brain regions, TTX and Ca^{2+} significantly reduced stimulated adenosine release. Adenosine release was dependent on ionotropic glutamate receptors (10 μM 6-cyano-7-nitroquinoxaline-2, 3-dione, CNQX) in the dorsal caudate putamen (a) and nucleus accumbens shell (b). Adenosine release was dependent on ATP release (blocked by 50 μM ARL 67156 + 100 μM α, β-methylene adenosine diphosphate, AOPCP) in the nucleus accumbens shell (b), hippocampus (c), and the prefrontal cortex (d). Reproduced with permission from Pajski and Venton, 2013.

brain. Although adenosine release was activity-dependent in all regions, dependence on either glutamate or ATP changed, based on the brain region.

4.2.1.8 FSCV to study adenosine deaminase kinetics

Our lab showed that FSCV at carbon-fiber microelectrodes can be used to monitor enzyme kinetics (Xu and Venton, 2010). Adenosine deaminase, the major enzyme responsible for metabolic breakdown of adenosine to inosine, was studied. Salt concentrations dramatically changed enzyme activity, and inhibitors of adenosine deaminase were not competitive in the presence of divalent cations (Xu and

Venton, 2010). Caffeine, an adenosine receptor antagonist, increased enzyme activity. This provides a fast and easy way to screen drug effects on enzyme kinetics *in vivo*.

4.2.1.9 Adenosine release during deep brain stimulation (DBS) causes microthalamotomy effect

Essential tremor patients and Parkinson's disease patients often see an immediate relief from tremors immediately following electrode implantation during DBS procedures (Chang et al., 2012). This is referred to as the microthalamotomy effect. Kendall Lee's group discovered that adenosine was elevated immediately after electrode implantation, preceding the microthalamotomy effect in the ferret thalamus (Tawfik et al., 2010). In 2012, Lee et al. implanted a DBS probe into seven human patients, and also saw the microthalamotomy effect along with release of adenosine in all patients (Chang et al., 2012). Future studies could probe whether adenosine is a cause or effect of the microthalamotomy effect during implantation in DBS.

The Lee group also discovered that HFS in ferret thalamic slices causes non-activity-dependent adenosine release, indicating a non-neuronal source such as astrocytes (Tawfik et al., 2010). The waveform used for adenosine detection involved scanning from −0.4 to 1.5 V at a scan rate of 400 V/s (Figure 4.5a). Figure 4.5b shows adenosine is released within 10–15 seconds after HFS. A clear primary and secondary peak is observed, indicating the release is from adenosine. The primary and secondary peaks have similar current *versus* time profiles (Figure 4.5c), but the secondary peak did not last as long. Figure 4.5d shows the background current before and after stimulation, and a small change is indicated. The CV confirming adenosine release is shown in Figure 4.5e, and the primary and secondary peaks are marked. No adenosine pharmacology experiments were performed, but they could help verify adenosine release in the future.

4.2.1.10 AMP hydrolysis to adenosine inhibits pain-sensing response in spinal cord

AMP hydrolyzes to adenosine in spinal lamina II neurons responsible for nociceptive (pain-sensing) responses (Street et al., 2011; Street et al., 2013). Three enzymes were discovered to hydrolyze AMP to adenosine in these neurons: prostatic acid phosphatase (PAP), NT5E, and tissue-nonspecific alkaline phosphatase (TNAP). Figure 4.6 shows example data for adenosine release in spinal lamina II neurons. AMP was applied to the slices and subsequent adenosine formation was measured at carbon-fiber microelectrodes (Figure 4.6a) (Street et al., 2011). CVs

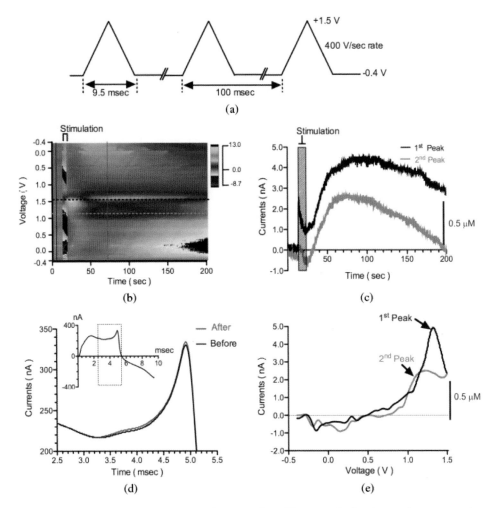

Figure 4.5 Adenosine release occurs after HFS (125 Hz, 200 µA, for 5 seconds, 100 µs pulse width) in thalamic slices. (a) FSCV waveform for adenosine detection involved scanning from −0.4 to 1.5 V at 400 V/s. (b) Color plot for adenosine shows adenosine release immediately following stimulation, as indicated by the black box at top of color plot. Release lasted for two minutes after stimulation. (c) Current *versus* time trace of the primary and secondary peak of adenosine after stimulation. (d) Background current changes over time before and after stimulation. (e) CV for adenosine where primary and secondary peaks are denoted by the black arrows. Reproduced with permission from Tawfik et al., 2010.

for adenosine in brain slices were compared to *in vitro* (Figure 4.6b), and both exhibit clear primary and secondary oxidation products. Color plots for all mouse types are shown in Figures 4.6c–4.6f. Knockouts of PAP, NT5E, and double knockout (dKO) were all compared. Adenosine release was significantly reduced in both NT5E and dKO at pH 7.4 (Figure 4.6h) and was significantly reduced in all knockouts at pH 5.6 (Figure 4.6j).

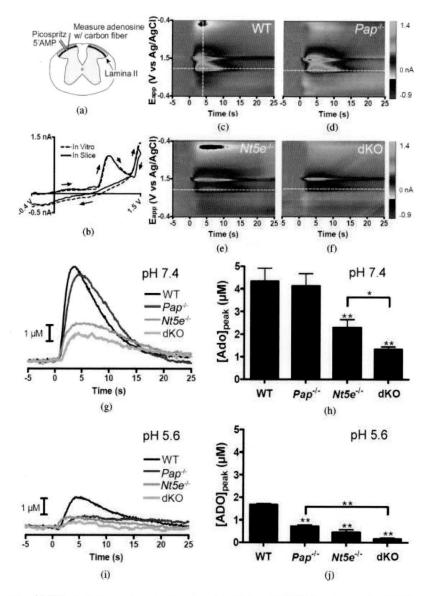

Figure 4.6 AMP hydrolysis to adenosine is reduced in PAP and NT5E knockout mice. Wild-type (WT), PAP$^{-/-}$, NT5E$^{-/-}$, and dKO mice were compared. (a) Schematic of experimental set up. 5'-AMP was picospritzed onto spinal lamina II neurons and degradation to adenosine was detected at carbon-fiber microelectrodes with FSCV. (b) Example CV of adenosine for both *in vitro* and in slice data. Two oxidation peaks characteristic for adenosine are observed. Color plot for WT (c), PAP$^{-/-}$ (d), NT5E$^{-/-}$ (e), and dKO (f) show differences in concentration of adenosine released. Differences in pH were tested and pH 7.4 (g) produced more current for all mouse types compared to pH 5.6 (i). Average adenosine concentrations for each mouse type in pH 7.4 (h) and pH 5.6 (j) show less adenosine in more acidic pH. Adenosine was significantly reduced in NT5E$^{-/-}$ and dKO (h and j) and in PAP$^{-/-}$ (j). Reproduced with permission from Street et al., 2011.

Street et al. also discovered spontaneous adenosine transients which occurred in spinal lamina II slices (Street et al., 2011). dKO mice of PAP and NT5E had reduced adenosine transient frequency, likely because of limited AMP hydrolysis. This was the first time spontaneous transient adenosine release had been observed in the CNS.

4.2.1.11 Enhanced understanding of adenosine signaling using FSCV

FSCV detection of adenosine with carbon-fiber microelectrodes has led to a wealth of information on the timescale of adenosine signaling and mechanisms of release. FSCV experiments have shown that the mechanism of transient adenosine release is always activity-dependent (Pajski and Venton, 2010), although the specific dependence on glutamate signaling or ATP breakdown can vary by region (Pajski and Venton, 2013). This is different than slower signaling which is primarily due to reverse transport (Sweeney, 1996; Latini and Pedata, 2001). Rapid neuromodulation by adenosine was also discovered during DBS electrode implantation for Parkinson's disease patients (Chang et al., 2012). Parkinson's disease is a dopamine dysregulation disease so treatments have primarily focused on dopamine therapies such as L-3,4-dihydroxyphenylalanine (L-DOPA) (Bourne, 2001; Zigmond et al., 1990; Cools, 2006; Funkiewiez et al., 2006), but now may be targeted towards adenosine regulation of dopamine release. FSCV has also been used to study the mechanism of adenosine formation during pain-sensing regulation (Street et al., 2011). A new enzyme was discovered in spinal cord neurons which is responsible for adenosine formation from AMP. The most attractive attributes of FSCV with carbon-fiber microelectrodes are rapid detection and low limits of detection. Previously, adenosine was monitored on the minute to hour time scale using microdialysis coupled to HPLC, so fast changes in neuronal activity went undetected (Koos et al., 1997; Sciotti and Vanwylen, 1993). Utilization of fast sensing techniques such as FSCV has changed our understanding of the time course of adenosine signaling in the brain. Not only is adenosine a slow acting neuromodulator, but it is important on the millisecond timescale in biological functions such as respiration (Xie et al., 2006). Understanding adenosine modulation on a rapid timescale will be beneficial for further developing adenosine therapeutics for disease treatments.

4.2.2 ADENOSINE CHARACTERIZATION USING AMPEROMETRIC ADENOSINE MICRO-BIOSENSORS

The development of small amperometric adenosine sensors in the early 2000s (Llaudet et al., 2003) has facilitated many studies on the release mechanisms of adenosine (Klyuch et al., 2012a) and its roles in brain diseases (Dale and Frenguelli, 2009; Dale, 2006). Adenosine selective enzyme sensors are made by entrapping xanthine oxidase, purine nucleoside phosphorylase, and adenosine deaminase on a platinum microelectrode using a derivatized pyrrole polymer (Llaudet et al., 2003). This enzyme cascade causes adenosine to break down to inosine then hypoxanthine and then to uric acid and hydrogen peroxide (Dale et al., 2000). The hydrogen peroxide concentration is proportional to purine concentration and hydrogen peroxide is detected at the electrode using amperometry. The diameter of the biosensors ranges from 7 μm to 50 μm and the linear concentration range is similar to carbon-fiber microelectrodes (50 nM to 20 μM) (Dale and Frenguelli, 2012). A null sensor which contains no enzymes is often used in parallel with the adenosine sensor so that background subtraction of non-adenosine activity can be achieved. A sensor which lacks the first enzyme to break down adenosine to inosine, can be used to verify adenosine is detected and not one of its metabolites (Dale and Frenguelli, 2012). The enzyme sensors are typically held at 500 mV using a potentiostat so that hydrogen peroxide can be oxidized (Llaudet et al., 2003; Dale et al., 2000). Adenosine biosensors are most often used in brain slices and have provided a wealth of information on adenosine biological function. Enzyme biosensors are commercially available from Sarissa-biomedical.

4.2.2.1 Activity-dependent adenosine release in the cerebellum is modulated by endogenous neurotransmitters

Stimulated release has been characterized using amperometric adenosine biosensors. Activity-dependent adenosine release occurred after electrical stimulation of cerebellar slices (Wall and Dale, 2007). Because the release was modulated by receptors which act on parallel fiber-Purkinje cell synapses, parallel fibers were considered the most likely release sites (Wall and Dale, 2007). Parallel fiber-Purkinje cells are inhibited by both $GABA_B$ and metabotropic glutamate 4 (mGlu4) receptors (Neale et al., 2001; Batchelor and Garthwaite, 1992). Administration of either a $GABA_B$ receptor agonist, baclofen or mGluR agonist, L-2-amino-4-phosphonobutyric acid (L-AP4), inactivated parallel fiber-Purkinje cell transmission and decreased adenosine release; therefore, adenosine release requires activation of parallel fiber-Purkinje cells and is modulated by GABA and glutamate in this region (Wall and Dale, 2007).

Blocking K^+ channels with 4-aminopyridine (4-AP) increased adenosine release in the cerebellum (Klyuch et al., 2011). After applying 60 μM 4-AP, a single pulse stimulus could be used to evoke adenosine release (Klyuch et al., 2011). The adenosine metabolism blocker erythro-9-(2-hydroxy-3-nonyl) adenine (EHNA), was used to verify adenosine release. Electrophysiological experiments

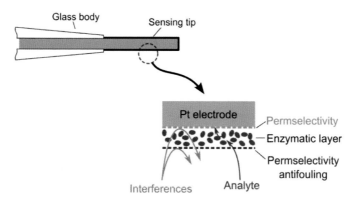

Figure 4.7 Schematic of biosensor. The enzyme layer is coated between two permselective layers which enhance biocompatibility and block interferences. For adenosine, the enzymatic layer consists of xanthine oxidase, purine nucleoside phosphorylase and adenosine deaminase in a derivatized pyrrole polymer. The sensing tip is insulated, is between 7–50 μm in diameter, and protrudes 0.5 to 2 mm in length. Reproduced with permission from Dale and Frenguelli, 2012.

confirmed that 4-AP increased the width of the action potential by increasing Ca^{2+} influx into the cell, allowing more adenosine to be released (Klyuch et al., 2011). Interestingly, adenosine release without 4-AP was glutamate receptor independent whereas adenosine release after 4-AP was AMPA receptor dependent. Release in the presence of 4-AP was also found to be activity-dependent by blocking Ca^{2+} influx and using TTX to block voltage-gated sodium channels (Klyuch et al., 2011).

In 2012, Klyuch et al. studied how both mechanisms (train and single-spike) of adenosine release in the cerebellum were controlled by the metabotropic receptors A_1, $GABA_B$, and mGlu4 (Klyuch et al., 2012b). They discovered that metabotropic receptors (A_1, $GABA_B$, mGlu4) modulate adenosine release, including the adenosine release which is independent of ionotropic glutamate activation (train evoked) (Klyuch et al., 2012b). The adenosine receptor A_1, together with $GABA_B$ and mGlu4, potentiated adenosine release though A_1 is most important in adenosine modulation. Blocking metabotropic receptors caused train stimulus evoked adenosine release which was previously found to be independent of ionotropic glutamate activation in 2011 (Klyuch et al., 2011). Metabotropic receptor inhibition also attenuated single spike adenosine release which is dependent on ionotropic glutamate receptor activation (Klyuch et al., 2012b). Adenosine release in the cerebellum is highly complex and is regulated by several different types of receptors.

Further characterization of activity-dependent adenosine release in the cerebellum revealed that a portion of release is not due to extracellular breakdown of ATP (Klyuch et al., 2012a). Activity-dependent adenosine release in the cerebellum of mice lacking the CD73 (cluster of differentiation 73) gene which encodes NT5E (which converts AMP to adenosine) revealed ATP independence (Klyuch et al., 2012a). The adenosine release that was independent of ATP was modulated by mGlu4 activation. These findings further support the complexity of adenosine signaling within the brain and that direct exocytosis may be possible.

4.2.2.2 Adenosine is released during hypoxia in the hippocampus

Adenosine release during hypoxia has been an accepted phenomenon for several years (van Wylen et al., 1986; Wallman-Johansson and Fredholm, 1994; Koos et al., 1997). Hypoxia is a physiological condition in which adequate oxygen is not delivered to the tissue. Previous studies of adenosine during hypoxia heavily relied on low temporal resolution techniques such as HPLC (Koos et al., 1997; Pedata et al., 1993). More recently, adenosine biosensors have been used to understand the time course of adenosine signaling during hypoxia (Dale et al., 2000; Frenguelli et al., 2003; Frenguelli et al., 2007; Otsuguro et al., 2011).

In 2000, Dale et al. discovered and validated adenosine release during hypoxia in the hippocampus (Dale et al., 2000). Hypoxic episodes of different durations were studied and adenosine increased steadily during both 5 minute and 10 minute episodes (Dale et al., 2000). Doubling the hypoxic episode time resulted in doubled adenosine release, confirming the relationship between hypoxia and adenosine. The increase in adenosine correlated with a decrease in synaptic transmission (Dale et al., 2000). Removal of extracellular Ca^{2+} significantly increased adenosine release whereas increased Ca^{2+} reduced adenosine release. This suggests that adenosine release is not dependent on Ca^{2+} but the amount is regulated by extracellular Ca^{2+} (Dale et al., 2000). Later, Pearson et al. showed that repeated exposures to hypoxia in the hippocampus resulted in weakened depression of synaptic transmission and adenosine release (Pearson et al., 2001). Release is partially restored after recovery or exogenous adenosine application. These findings suggest a depletable but replenishable pool of adenosine (Pearson et al., 2001).

Smaller amperometric enzyme sensors were used in 2003 by Frenguelli et al. to confirm adenosine release during hypoxia (Frenguelli et al., 2003). The smaller sensors provided a more accurate depiction of adenosine release during hypoxia because they were inserted in the tissue, instead of being placed on top (Frenguelli et al., 2003). The improved sensor also lead to the discovery of the post-hypoxia purine efflux (PPE) (Frenguelli et al., 2003). PPE is a large increase in adenosine production as the brain returns to normoxia. The function of PPE is unknown, but it was mostly attributed to reoxygenation of the brain causing changes in nitric oxide, adenosine, or intracellular pH (Frenguelli et al., 2003).

4.2.2.3 Adenosine is released during hypoxia in the nucleus tractus solitarii (NTS) of the brainstem

In vivo measurements in the NTS region of the brainstem confirmed adenosine release as a result of hypoxia (Gourine et al., 2002). Adenosine release was much smaller in the rostral region of the ventrolateral medulla (VLM) than in the NTS. Adenosine release in both regions was not correlated with a decrease in respiration, so adenosine release was not believed to modulate the hypoxia-induced decrease in respiration. Adenosine was measured on the surface of the region and within the region, and greater amounts of adenosine were detected on the surface, unlike previous studies in this region without hypoxia (Dale et al., 2002). The different concentrations of adenosine point to different structures within this region releasing adenosine. Overall, this study demonstrated adenosine release in a region other than the hippocampus during hypoxia and that adenosine release was not the cause of decreased respiration during hypoxia.

4.2.2.4 Adenosine is released during ischemia

Amperometric enzyme biosensors have also been used to study adenosine release during ischemia or oxygen/glucose deprivation (Pearson et al., 2003; Dale and Frenguelli, 2009; Pearson et al., 2006). Ischemia is inadequate blood flow to tissue that causes a decrease in oxygen. Ischemic attack always results in hypoxia; however, hypoxia is not always caused by ischemia. Ischemia is often simulated experimentally by reducing oxygen and glucose. Similar to hypoxia, ischemic conditions caused adenosine to rise at the same time excitatory synaptic transmission decreased (Pearson et al., 2006). The decrease in synaptic transmission was not as dramatic in the presence of an A_1 receptor antagonist and the rate of recovery was also enhanced from 60 minutes to 15 minutes (Pearson et al., 2006). This study provided evidence that adenosine is neuroprotective during ischemia.

Adenosine release precedes ATP release during ischemia in the hippocampus (Frenguelli et al., 2007). Amperometric enzyme biosensors were used to compare ATP and adenosine release during ischemia in the hippocampus. ATP release did not occur until after adenosine release and ATP release required extracellular Ca^{2+} while adenosine release was enhanced by removal of extracellular Ca^{2+}. Ionotropic glutamate receptor inhibition increased ATP release but only slightly enhanced adenosine release. Because adenosine was released before ATP and was larger, adenosine is not a result of ATP metabolism. Long periods of ischemia resulted in an anoxic depolarization and ATP was only released at this point.

Boison et al. used amperometric adenosine sensors *in vivo* to show increased adenosine in a transgenic mice model (fb–adk–def) with reduced ADK in the forebrain (Shen et al., 2011). Transgenic mice were better protected during ischemic-induced stroke, likely due to elevated adenosine levels.

4.2.2.5 Adenosine modulates chemoreceptor responses

In 2002, the first *in vivo* measurements were made using amperometric adenosine sensors (Dale et al., 2002) on the dorsal surface of the brainstem and the NTS of the medulla oblongata. Adenosine release was measured in response to a defense reaction, induced by stimulating the hypothalamus defense area (HDA). The defense reaction involves an increase in respiration, heart rate, and blood pressure. The NTS is responsible for regulating many cardiovascular and respiratory responses. In the NTS, adenosine modulated baroreceptor and chemoreceptor reflexes involved in responding to pressure and chemical stimuli, respectively. Thus, adenosine release in the brainstem correlates with cardiovascular defense responses.

4.2.2.6 Adenosine is linked to modulating seizures induced by CO_2

Respiration rate and blood flow is regulated by carbon dioxide (CO_2) levels and *vice versa*. Hypercapnia, an increase in CO_2, causes a decrease in neuron excitability, whereas hypocapnia, a decrease in CO_2, increases neuron excitability (Dulla et al., 2005). Decreasing CO_2 is a common method for inducing seizure while increasing CO_2 is often used in sedation. The level of CO_2 controls tissue pH, and pH has also been linked to neuronal excitability. Dulla et al. showed that increasing CO_2 decreased pH and increased adenosine in the hippocampus (Dulla et al., 2005). The increase in adenosine correlated with the inhibition of synaptic transmission. These effects were modulated by A_1 receptors, ATP receptors, and ecto-ATPase. Decreases in adenosine levels during hypocapnia were linked to hyperventilation-induced epileptic seizures. During hypercapnia, ATPase blockers did not change adenosine release, and a cocktail of A_1 and ATP receptor antagonists were needed to completely inhibit release. The combination of these findings leads to the conclusion that adenosine and ATP work in concert with one another as neuromodulators during hypercapnia (Dulla et al., 2005).

4.2.2.7 Increase in extracellular adenosine levels provides anticonvulsant behavior during epilepsy

Basal synaptic adenosine levels are regulated by astrocytic ADK which regulates seizure activity (Etherington et al., 2009). One of the main pathways for regulating adenosine is phosphorylation of adenosine to AMP by ADK. Etherington et al. showed that inhibition of ADK increased extracellular adenosine which inhibited excitatory neurotransmission in the hippocampus. Seizure activity evoked by either Mg^{2+} free artificial cerebrospinal fluid (aCSF) or HFS was greatly attenuated by the increase in extracellular adenosine (Etherington et al., 2009). ADK is primarily located in glial fibrillary acidic protein (GFAP)-positive astrocytes, so astrocytes are thought to be important during seizure activity. Because ADK regulates adenosine during evoked seizures and adenosine induces anticonvulsant activity, ADK regulation may be a novel therapy for epilepsy.

4.2.2.8 Adenosine is involved in the sleep–wake cycle

Sleep can be broadly divided into two categories: slow wave sleep and rapid eye movement sleep. Sleep is divided into these two main categories based on four criteria: posture, changes in electroencephalogram (EEG), increase in response threshold to external stimuli, and rapid reversibility (Bjorness and Greene, 2009). These four criteria are often used to determine sleep/wake activity in mammals. Behaviors during sleep are controlled by either homeostatic mechanisms or circadian rhythms. Adenosine is involved primarily during slow wave sleep (Bjorness and Greene, 2009; Bjorness et al., 2009). Schmitt et al. discovered that wakefulness regulates extracellular adenosine levels through an astrocytic-SNARE-dependent mechanism in hippocampal slices (Schmitt et al., 2012). To understand this, the authors compared extracellular adenosine concentrations using amperometric biosensors in WT mice and transgenic mice, which have a double negative transgene specific to disrupt SNARE complex formation (dnSNARE). Both sets of mice were sleep deprived and their adenosine levels compared. Extracellular adenosine was significantly larger in sleep deprived WT mice than in dnSNARE mice. This study showed that sleep deprivation increased adenosine and that this release was dependent on neurotransmission from astrocytes in the hippocampus (Schmitt et al., 2012).

Dale's group recently characterized the mechanism of adenosine release during the sleep–wake cycle in the basal forebrain (BFB) (Sims et al., 2013). The BFB is involved in the ascending arousal system (Murillo-Rodriguez et al., 2004). Previous studies have linked nitric oxide (NO) signaling to increased adenosine release during sleep (Kalinchuk et al., 2006). The relationship between sleep and adenosine was studied in three rodent models: rats, mice, and hamsters (Sims et al., 2013) and results were generally similar in all the rodents. Adenosine levels were measured at seven different stages of the diurnal cycle, spanning the light and dark stages, by sacrifice and brain slice preparation at those time points. Extracellular adenosine release increased after administration of ionotropic and group I mGlu receptor agonists but did not change from orexin, histamine, and neurotensin administration, suggesting not all depolarizing stimuli contribute to adenosine release. Adenosine release varied during the diurnal cycle. Adenosine release was greatest at the end of the dark cycle and less during the light cycle.

Dale's group also showed that the rise in adenosine in the BFB due to sleep deprivation was dependent on NO produced from inducible nitric oxide synthase (iNOs) (Sims et al., 2013). Blocking iNOs, scavenging NO, or using a non-specific inhibitor of NO caused decreased adenosine during sleep deprivation. Figure 4.8 shows comparisons of adenosine release for controls (non-sleep deprived), sleep deprived, sleep deprived with iNOs inhibition (with 1400W), and non-sleep deprived with iNOs inhibition. The diurnal dependent variation of adenosine may be independent of iNOs because adenosine did not change during control conditions (no sleep deprivation, Figures 4.8a and 4.8b) after iNOs inhibition. Basal tone of adenosine followed a similar pattern to adenosine release (Figure 4.8d). However, the cortex did not follow the same pattern as the BFB (Figures 4.8e and 4.8f). Overall, adenosine release is dependent on NOs during sleep deprivation and changes based on the diurnal sleep cycle.

4.2.2.9 Enhanced understanding of adenosine signaling using amperometric biosensors

The development and application of amperometric biosensors selective for adenosine has provided information on the role of adenosine signaling in several diseases. Experiments with biosensors have shown that adenosine is released rapidly in response to hypoxia (Frenguelli et al., 2003), ischemia (Dale and Frenguelli, 2009) and hypercapnia (Dale, 2006). Adenosine also regulates the sleep–wake cycle (Sims et al., 2013), chemoreception (Dale et al., 2002), and epileptic seizures (Dale and Frenguelli, 2009; Etherington et al., 2009). Different mechanisms of release occur during different brain pathologies, demonstrating that adenosine has many modes of formation. The selectivity of biosensors is their most attractive attribute, so that the molecule being detected is known. Implementing amperometric biosensors in other preparations and brain regions will allow further characterization of the neuromodulatory role of adenosine in the brain.

4.3 CONTRIBUTIONS OF ELECTROCHEMICAL DETECTION OF ADENOSINE TO THE FIELD

The advent of rapid real-time monitoring of adenosine *in vivo* has led to new knowledge of the time course of adenosine signaling in the brain. Previous adenosine research relied upon slow temporal techniques such as microdialysis and

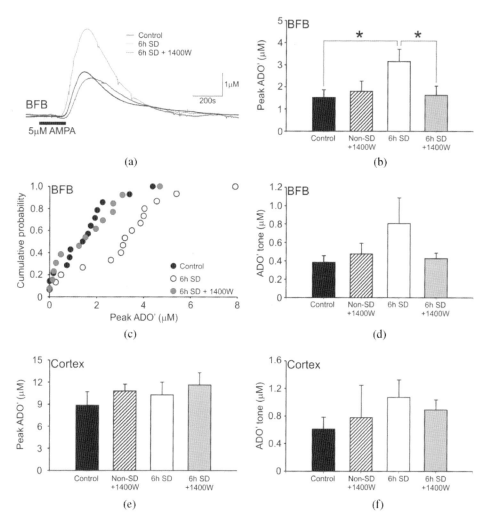

Figure 4.8 Increased adenosine after sleep deprivation is modulated by iNOs production. Adenosine release was measured after six hours for control animals (not sleep deprived), sleep deprived, and sleep deprived with 1400 W which inhibits iNOs production. (a) Shows raw data for all three conditions described above in the BFB. The data is normalized to the 10 μM adenosine calibration. (b) Adenosine release was significantly greater in sleep deprived animals. Animals which were not sleep deprived or where iNOs production was inhibited showed less adenosine release. (c) Cumulative probability distribution of control individual adenosine responses, sleep deprived, and sleep deprived with iNOs inhibition. (d) Basal tone was compared in all four conditions, and the patterned followed that of adenosine release. (e) In the cortex, there were no significant differences between adenosine release or (f) basal adenosine tone. Reproduced with permission from Sims et al., 2013.

HPLC (Porkka-Heiskanen et al., 2000; Pedata et al., 2001). These techniques established a fundamental understanding of the importance of adenosine in the brain, but mechanistic information was more difficult to obtain, especially about rapid signaling. Information on the rapid timescale of adenosine signaling was still unknown; however the long term effects of adenosine were revealed (Pedata et al., 2001). Over the last several years, FSCV with carbon-fiber microelectrodes (Swamy and Venton, 2007) and amperometric adenosine biosensors (Llaudet et al., 2003) have gained increased popularity. These studies have revealed that there is rapid signaling of adenosine, a time scale that was previously unknown (Cechova and Venton, 2008; Pajski and Venton, 2010; Dale and Frenguelli, 2012; Frenguelli et al., 2003; Dale et al., 2002). More research is needed to fully understand the complicated pathways of adenosine formation and signaling.

Rapid adenosine monitoring has led to the overall understanding that adenosine is not just a slow-acting, retaliatory metabolite. Adenosine can be released and cleared on the order of seconds, much like neurotransmitters. Electrochemical measurements have proven that some adenosine is activity-dependent (Klyuch et al., 2012b; Pajski and Venton, 2010) and that activity-dependent release can be either direct release or *via* downstream effects of other neurotransmitters. The mechanism of release is dependent on the brain region and on the stimulation (Pajski and Venton, 2013). For example, adenosine can be stimulated by electrical pulses, elevated K^+, and AMPA receptor activation. Thus, adenosine release is complicated and characterization of other major brain regions would be beneficial to understand the full scope of adenosine signaling in the brain. Lastly, because some evidence points to a direct release of adenosine, more studies to examine whether adenosine is directly released from vesicles are warranted.

A slow buildup of adenosine during brain injuries and non-physiological conditions such as hypoxia had been demonstrated, but studies with electrochemical sensors have shown that there are also rapid adenosine events during injury and hypoxia. Adenosine can be released immediately following hypoxia (Frenguelli et al., 2003), ischemia (Frenguelli et al., 2007) and during defense reactions (Dale et al., 2002). The function of rapid adenosine signaling and downstream effects have not been fully characterized. Future studies to measure receptor activation will provide information about the functional consequences of rapid adenosine signaling.

Only a few disease models have been studied to examine the effects on rapid adenosine release. Because adenosine is ubiquitous, i.e., present throughout the brain, real-time sensing of adenosine could potentially reveal adenosine signaling during many behaviors or diseases. Understanding the mechanism of adenosine release during diseases, such as Parkinson's disease or stroke, may lead to development of novel drugs to manipulate adenosine regulation for therapeutic uses. The field of real-time monitoring of adenosine is still relatively young but it is growing as an understanding of the role of adenosine signaling in many brain functions increases. The use of sensors to understand basic adenosine neurobiology could assist in therapeutic development in the future.

ACKNOWLEDGMENTS

The authors would like to thank the NIH (R01NS076875) and NSF (CHE0645587522) for funding the Venton lab adenosine work. We would also like to thank Drs. Mark Zylka, Nicholas Dale, and Kendall Lee for providing figures for this chapter.

REFERENCES

Aden U, Halldner L, Lagercrantz H, Dalmau I, Ledent C, Fredholm BB (2003) Aggravated brain damage after hypoxic ischemia in immature adenosine A(2A) knockout mice. Stroke 34:739–744.

Baldwin SA, Beal PR, Yao SY, King AE, Cass CE, Young JD (2004) The equilibrative nucleoside transporter family, SLC29. Pflugers Arch 447:735–743.

Batchelor AM, Garthwaite J (1992) $GABA_B$ receptors in the parallel fibre pathway of rat cerebellum. Eur J Neurosci 4:1059–1064.

Bell MJ, Kochanek PM, Carcillo JA, Mi ZC, Schiding JK, Wisniewski SR, Clark RSB, Dixon CE, Marion DW, Jackson E (1998) Interstitial adenosine, inosine, and hypoxanthine are increased after experimental traumatic brain injury in the rat. J Neurotrauma 15:163–170.

Bjorness TE, Greene RW (2009) Adenosine and sleep. Curr Neuropharmacol 7:238–245.

Bjorness TE, Kelly CL, Gao T, Poffenberger V, Greene RW (2009) Control and function of the homeostatic sleep response by adenosine A1 receptors. J Neurosci 29:1267–1276.

Blum D, Hourez R, Galas MC, Popoli P, Schiffmann SN (2003) Adenosine receptors and Huntington's disease: implications for pathogenesis and therapeutics. Lancet Neurol 2:366–374.

Bourne JA (2001) SCH 23390: the first selective dopamine D1-like receptor antagonist. CNS Drug Rev 7:399–414.

Brajter-Toth A, El-Nour KA, Cavalheiro ET, Bravo R (2000) Nanostructured carbon fiber disk electrodes for sensitive determinations of adenosine and uric acid. Anal Chem 72:1576–1584.

Calker D, Biber K (2005) The role of glial adenosine receptors in neural resilience and the neurobiology of mood disorders. Neurochem Res 30:1205–1217.

Cechova S, Elsobky AM, Venton BJ (2010) A1 receptors self-regulate adenosine release in the striatum: evidence of autoreceptor characteristics. Neuroscience 171:1006–1015.

Cechova S, Venton BJ (2008) Transient adenosine efflux in the rat caudate-putamen. J Neurochem 105:1253–1263.

Chang SY, Kim I, Marsh MP, Jang DP, Hwang SC, Van Gompel JJ, Goerss SJ, Kimble CJ, Bennet KE, Garris PA, Blaha CD, Lee KH (2012) Wireless fast-scan cyclic voltammetry to monitor adenosine in patients with essential tremor during deep brain stimulation. Mayo Clin Proc 87:760–765.

Cools R (2006) Dopaminergic modulation of cognitive function-implications for L-DOPA treatment in Parkinson's disease. Neurosci Biobehav Rev 30:1–23.

Cunha RA (2001) Adenosine as a neuromodulator and as a homeostatic regulator in the nervous system: different roles, different sources and different receptors. Neurochem Int 38:107–125.

Cunha RA (2008) Adenosine neuromodulation and neuroprotection. In: Handbook of neurochemistry and molecular neurobiology (Lajtha A, Vizi ES, eds), pp 255–273. US: Springer.

Dale N (2006) The acid nature of CO_2-evoked adenosine release in the CNS. J Physiol 574:633.

Dale N, Frenguelli BG (2009) Release of adenosine and ATP during ischemia and epilepsy. Curr Neuropharmacol 7:160–179.

Dale N, Frenguelli BG (2012) Measurement of purine release with microelectrode biosensors. Purinergic Signal 8:27–40.

Dale N, Gourine AV, Llaudet E, Bulmer D, Thomas T, Spyer KM (2002) Rapid adenosine release in the nucleus tractus solitarii during defence response in rats: real-time measurement *in vivo*. J Physiol 544:149–160.

Dale N, Pearson T, Frenguelli BG (2000) Direct measurement of adenosine release during hypoxia in the CA1 region of the rat hippocampal slice. J Physiol 526:143–155.

de Mendonca A, Sebastiao AM, Ribeiro JA (2000) Adenosine: does it have a neuroprotective role after all? Brain Res Brain Res Rev 33:258–274.

Deng Q, Watson CJ, Kennedy RT (2003) Aptamer affinity chromatography for rapid assay of adenosine in microdialysis samples collected *in vivo*. J Chromatogr A 1005:123–130.

Dong H, Wang S, Galligan JJ, Swain GM (2011) Boron-doped diamond nano/microelectrodes for biosensing and *in vitro* measurements. Front Biosci (Schol Ed) 3:518–540.

Dryhurst G (1977) Electrochemistry of biological molecules. New York: Academic Press.

Dulla CG, Dobelis P, Pearson T, Frenguelli BG, Staley KJ, Masino SA (2005) Adenosine and ATP link PCO_2 to cortical excitability *via* pH. Neuron 48:1011–1023.

El-Nour KA, Brajter-Toth A (2000) Sensitivity of electrochemically nanostructured carbon fiber ultramicroelectrodes in the determination of adenosine. Electroanalysis 12:305–310.

Etherington LA, Patterson GE, Meechan L, Boison D, Irving AJ, Dale N, Frenguelli BG (2009) Astrocytic adenosine kinase regulates basal synaptic adenosine levels and seizure activity but not activity-dependent adenosine release in the hippocampus. Neuropharmacology 56:429–437.

Ferre S, Ciruela F, Canals M, Marcellino D, Burgueno J, Casado V, Hillion J, Torvinen M, Fanelli F, de Benedetti P, Goldberg SR, Bouvier M, Fuxe K, Agnati LF, Lluis C,

Franco R, Woods A (2004) Adenosine A(2A)-dopamine D-2 receptor-receptor heteromers. Targets for neuro-psychiatric disorders. Parkinsonism Relat Disord 10:265–271.

Ferre S, Fredholm BB, Morelli M, Popoli P, Fuxe K (1997) Adenosine-dopamine receptor-receptor interactions as an integrative mechanism in the basal ganglia. Trends Neurosci 20:482–487.

Ferre S, Fuxe K, von EG, Johansson B, Fredholm BB (1992) Adenosine-dopamine interactions in the brain. Neuroscience 51:501–512.

Franco R, Ferre S, Agnati L, Torvinen M, Gines S, Hillion J, Casado V, Lledo P, Zoli M, Lluis C, Fuxe K (2000) Evidence for adenosine/dopamine receptor interactions: indications for heteromerization. Neuropsychopharmacology 23:S50–S59.

Fredholm BB (1997) Adenosine and neuroprotection. Neuroprotective agents and cerebral ischaemia. Int Rev Neurobiol 40:259–280.

Fredholm BB, Abbracchio MP, Burnstock G, Daly JW, Harden TK, Jacobson KA, Leff P, Williams M (1994) Nomenclature and classification of purinoceptors. Pharmacol Rev 46:143–156.

Fredholm BB, Chen JF, Cunha RA, Svenningsson P, Vaugeois JM (2005) Adenosine and brain function. Int Rev Neurobiol 63:191–270.

Frenguelli BG, Llaudet E, Dale N (2003) High-resolution real-time recording with microelectrode biosensors reveals novel aspects of adenosine release during hypoxia in rat hippocampal slices. J Neurochem 86:1506–1515.

Frenguelli BG, Wigmore G, Llaudet E, Dale N (2007) Temporal and mechanistic dissociation of ATP and adenosine release during ischaemia in the mammalian hippocampus. J Neurochem 101:1400–1413.

Funkiewiez A, Ardouin C, Cools R, Krack P, Fraix V, Batir A, Chabardes S, Benabid AL, Robbins TW, Pollak P (2006) Effects of levodopa and subthalamic nucleus stimulation on cognitive and affective functioning in Parkinson's disease. Mov Disord 21:1656–1662.

Fuxe K, Stromberg I, Popoli P, Rimondini-Giorgini R, Torvinen M, Ogren SO, Franco R, Agnati LF, Ferre S (2001) Adenosine receptors and Parkinson's disease. Relevance of antagonistic adenosine and dopamine receptor interactions in the striatum. Adv Neurol 86:345–353.

Gomes CV, Kaster MP, Tome AR, Agostinho PM, Cunha RA (2011) Adenosine receptors and brain diseases: neuroprotection and neurodegeneration. Biochim Biophys Acta 1808:1380–1399.

Gourine AV, Llaudet E, Thomas T, Dale N, Spyer KM (2002) Adenosine release in nucleus tractus solitarii does not appear to mediate hypoxia-induced respiratory depression in rats. J Physiol 544:161–170.

Hamilton BR, Smith DO (1991) Autoreceptor-mediated purinergic and cholinergic inhibition of motor nerve terminal calcium currents in the rat. J Physiol 432:327–341.

Haselkorn ML, Shellington D, Jackson E, Vagni V, Feldman K, Dubey R, Gillespie D, Bell M, Clark R, Jenkins L, Schnermann J, Homanics G, Kochanek P (2008) Adenosine A1 receptor activation as a brake on neuroinflammation after experimental traumatic brain injury in mice. J Neurotrauma 25:901–910.

Kalinchuk AV, Lu Y, Stenberg D, Rosenberg PA, Porkka-Heiskanen T (2006) Nitric oxide production in the basal forebrain is required for recovery sleep. J Neurochem 99:483–498.

Klyuch BP, Dale N, Wall MJ (2012a) Deletion of ecto-5'-nucleotidase (CD73) reveals direct action potential-dependent adenosine release. J Neurosci 32:3842–3847.

Klyuch BP, Dale N, Wall MJ (2012b) Receptor-mediated modulation of activity-dependent adenosine release in rat cerebellum. Neuropharmacology 62:815–824.

Klyuch BP, Richardson MJE, Dale N, Wall MJ (2011) The dynamics of single spike-evoked adenosine release in the cerebellum. J Physiol (Lond) 589:283–295.

Koos BJ, Kruger L, Murray TF (1997) Source of extracellular brain adenosine during hypoxia in fetal sheep. Brain Res 778:439–442.

Latini S, Pazzagli M, Pepeu G, Pedata F (1996) A2 adenosine receptors: their presence and neuromodulatory role in the central nervous system. Gen Pharmac 27:925–933.

Latini S, Pedata F (2001) Adenosine in the central nervous system: release mechanisms and extracellular concentrations. J Neurochem 79:463–484.

Llaudet E, Botting NP, Crayston JA, Dale N (2003) A three-enzyme microelectrode sensor for detecting purine release from central nervous system. Biosens Bioelectron 18:43–52.

Lorbar M, Fenton RA, Duffy AJ, Graybill CA, Dobson JG Jr. (1999) Effect of aging on myocardial adenosine production, adenosine uptake and adenosine kinase activity in rats. J Mol Cell Cardiol 31:401–412.

Lu Q, Newton J, Hsiao V, Shamirian P, Blackburn MR, Pedroza M (2012) Sustained adenosine exposure causes lung endothelial barrier dysfunction *via* nucleoside transporter-mediated signaling. Am J Respir Cell Mol Biol 47:604–613.

Mitchell JB, Lupica CR, Dunwiddie TV (1993) Activity-dependent release of endogenous adenosine modulates synaptic responses in the rat hippocampus. J Neurosci 13:3439–3447.

Murillo-Rodriguez E, Blanco-Centurion C, Gerashchenko D, Salin-Pascual RJ, Shiromani PJ (2004) The diurnal rhythm of adenosine levels in the basal forebrain of young and old rats. Neuroscience 123:361–370.

Neale SA, Garthwaite J, Batchelor AM (2001) Metabotropic glutamate receptor subtypes modulating neurotransmission at parallel fibre-Purkinje cell synapses in rat cerebellum. Neuropharmacology 41:42–49.

Newby AC, Worku Y, Holmquist CA (1985) Adenosine formation. evidence for a direct biochemical link with energy metabolism. Adv Myocardiol 6:273–284.

O'Neill C, Nolan BJ, Macari A, O'Boyle KM, O'Connor JJ (2007) Adenosine A1 receptor-mediated inhibition of dopamine release from rat striatal slices is modulated by D1 dopamine receptors. Eur J Neurosci 26:3421–3428.

O'Regan M (2005) Adenosine and the regulation of cerebral blood flow. Neurol Res 27:175–181.

Okada M, Mizuno K, Kaneko S (1996) Adenosine A1 and A2 receptors modulate extracellular dopamine levels in the rat striatum. Neurosci Lett 212:53–56.

Okada M, Nutt DJ, Murakami T, Zhu G, Kamata A, Kawata Y, Kaneko S (2001) Adenosine receptor subtypes modulate two major functional pathways for hippocampal serotonin release. J Neurosci 21:628–640.

Otsuguro K, Wada M, Ito S (2011) Differential contributions of adenosine to hypoxia-evoked depressions of three neuronal pathways in isolated spinal cord of neonatal rats. Br J Pharmacol 164:132–144.

Pajski ML, Venton BJ (2010) Adenosine release evoked by short electrical stimulations in striatal brain slices is primarily activity-dependent. ACS Chem Neurosci 1:775–787.

Pajski ML, Venton BJ (2013) The mechanism of electrically stimulated adenosine release varies by brain region. Purinergic Signal 9:167–174.

Park J, Aragona BJ, Kile BM, Carelli RM, Wightman RM (2010) *In vivo* voltammetric monitoring of catecholamine release in subterritories of the nucleus accumbens shell. Neuroscience 169:132–142.

Park J, Quaiserova-Mocko V, Patel BA, Novotny M, Liu A, Bian X, Galligan JJ, Swain GM (2008) Diamond microelectrodes for *in vitro* electroanalytical measurements: current status and remaining challenges. Analyst 133:1724.

Pearson T, Currie AJ, Etherington LA, Gadalla AE, Damian K, Llaudet E, Dale N, Frenguelli BG (2003) Plasticity of purine release during cerebral ischemia: clinical implications? J Cell Mol Med 7:362–375.

Pearson T, Damian K, Lynas RE, Frenguelli BG (2006) Sustained elevation of extracellular adenosine and activation of A1 receptors underlie the post-ischaemic inhibition of neuronal function in rat hippocampus *in vitro*. J Neurochem 97:1357–1368.

Pearson T, Nuritova F, Caldwell D, Dale N, Frenguelli BG (2001) A depletable pool of adenosine in area CA1 of the rat hippocampus. J Neurosci 21:2298–2307.

Pedata F, Corsi C, Melani A, Bordoni F, Latini S (2001) Adenosine extracellular brain concentrations and role of A2A receptors in ischemia. Ann N Y Acad Sci 939:74–84.

Pedata F, Latini S, Pugliese AM, Pepeu G (1993) Investigations into the adenosine outflow from hippocampal slices evoked by ischemia-like conditions. J Neurochem 61:284–289.

Pinna A, Volpini R, Cristalli G, Morelli M (2005) New adenosine A2A receptor antagonists: actions on parkinson's disease models. Eur J Pharmacol 512:157–164.

Porkka-Heiskanen T, Strecker RE, McCarley RW (2000) Brain site-specificity of extracellular adenosine concentration changes during sleep deprivation and spontaneous sleep: an *in vivo* microdialysis study. Neuroscience 99:507–517.

Quarta D, Borycz J, Solinas M, Patkar K, Hockemeyer J, Ciruela F, Lluis C, Franco R, Woods AS, Goldberg SR, Ferre S (2004) Adenosine receptor-mediated modulation of dopamine release in the nucleus accumbens depends on glutamate neurotransmission and N-methyl-D-aspartate receptor stimulation. J Neurochem 91:873–880.

Robertson CL, Bell MJ, Kochanek PM, Adelson PD, Ruppel RA, Carcillo JA, Wisniewski SR, Mi ZC, Janesko KL, Clark RSB, Marion DW, Graham SH, Jackson EK (2001) Increased adenosine in cerebrospinal fluid after severe traumatic brain injury in infants and children: association with severity of injury and excitotoxicity. Crit Care Med 29:2287–2293.

Ross AE, Venton BJ (2012) Nafion-CNT coated carbon-fiber microelectrodes for enhanced detection of adenosine. Analyst 137:3045–3051.

Rudolphi KA, Schubert P, Parkinson FE, Fredholm BB (1992) Adenosine and brain ischemia. Cerebrovasc Brain Metab Rev 4:346–369.

Schmitt LI, Sims RE, Dale N, Haydon PG (2012) Wakefulness affects synaptic and network activity by increasing extracellular astrocyte-derived adenosine. J Neurosci 32:4417–4425.

Sciotti VM, Vanwylen DGL (1993) Increases in interstitial adenosine and cerebral blood-flow with inhibition of adenosine kinase and adenosine deaminase. J Cereb Blood Flow Metab 13:201–207.

Shen HY, Lusardi TA, Williams-Karnesky RL, Lan JQ, Poulsen DJ, Boison D (2011) Adenosine kinase determines the degree of brain injury after ischemic stroke in mice. J Cereb Blood Flow Metab 31:1648–1659.

Sims RE, Wu HHT, Dale N (2013) Sleep–wake sensitive mechanisms of adenosine release in the basal forebrain of rodents: an *in vitro* study. Plos One 8:e53814.

Sperlagh B, Vizi ES (2011) The role of extracellular adenosine in chemical neurotransmission in the hippocampus and basal ganglia: pharmacological and clinical aspects. Curr Top Med Chem 11:1034–1046.

Sperlagh B, Zsilla G, Baranyi M, Illes P, Vizi ES (2007) Purinergic modulation of glutamate release under ischemic-like conditions in the hippocampus. Neuroscience 149:99–111.

Stamford JA, Kruk ZL, Millar J, Wightman RM (1984) Striatal dopamine uptake in the rat: *in vivo* analysis by fast cyclic voltammetry. Neurosci Lett 51:133–138.

Street SE, Kramer NJ, Walsh PL, Taylor-Blake B, Yadav MC, King IF, Vihko P, Wightman RM, Millan JL, Zylka MJ (2013) Tissue-nonspecific alkaline phosphatase acts redundantly with PAP and NT5E to generate adenosine in the dorsal spinal cord. J Neurosci 33:11314–11322.

Street SE, Walsh PL, Sowa NA, Taylor-Blake B, Guillot TS, Vihko P, Wightman RM, Zylka MJ (2011) PAP and NT5E inhibit nociceptive neurotransmission by rapidly hydrolyzing nucleotides to adenosine. Mol Pain 7:80.

Swamy BEK, Venton BJ (2007) Subsecond detection of physiological adenosine concentrations using fast-scan cyclic voltammetry. Anal Chem 79:744–750.

Sweeney MI (1996) Adenosine release and uptake in cerebellar granule neurons both occur *via* an equilibrative nucleoside carrier that is modulated by G proteins. J Neurochem 67:81–88.

Tawfik VL, Chang SY, Hitti FL, Roberts DW, Leiter JC, Jovanovic S, Lee KH (2010) Deep brain stimulation results in local glutamate and adenosine release: investigation into the role of astrocytes. Neurosurgery 67:367–375.

Thompson SM, Haas HL, Gahwiler BH (1992) Comparison of the actions of adenosine at pre-synaptic and post-synaptic receptors in the rat hippocampus *invitro*. J Physiol (Lond) 451:347–363.

Thorn JA, Jarvis SM (1996) Adenosine transporters. Gen Pharmac 27:613–620.

Valtysson J, Persson L, Hillered L (1998) Extracellular ischaemia markers in repeated global ischaemia and secondary hypoxaemia monitored by microdialysis in rat brain. Acta Neurochir 140:387–395.

van Wylen DG, Park TS, Rubio R, Berne RM (1986) Increases in cerebral interstitial fluid adenosine concentration during hypoxia, local potassium infusion, and ischemia. J Cereb Blood Flow Metab 6:522–528.

Venton BJ, Michael DJ, Wightman RM (2003) Correlation of local changes in extracellular oxygen and pH that accompany dopaminergic terminal activity in the rat caudate-putamen. J Neurochem 84:373–381.

Venton BJ, Wightman RM (2007) Pharmacologically induced, subsecond dopamine transients in the caudate-putamen of the anesthetized rat. Synapse 61:37–39.

Villarreal F, Zimmermann S, Makhsudova L, Montag AC, Erion MD, Bullough DA, Ito BR (2003) Modulation of cardiac remodeling by adenosine: *in vitro* and *in vivo* effects. Mol Cell Biochem 251:17–26.

Von Lubitz DK (2001) Adenosine in the treatment of stroke: yes, maybe, or absolutely not? Expert Opin Investig Drugs 10:619–632.

Wall M, Dale N (2008) Activity-dependent release of adenosine: a critical re-evaluation of mechanism. Curr Neuropharmacol 6:329–337.

Wall MJ, Dale N (2007) Auto-inhibition of rat parallel fibre-Purkinje cell synapses by activity-dependent adenosine release. J Physiol 581:553–565.

Wallman-Johansson A, Fredholm BB (1994) Release of adenosine and other purines from hippocampal slices stimulated electrically or by hypoxia/hypoglycemia. Effect of chlormethiazole. Life Sci 55:721–728.

Wightman RM, Amatore C, Engstrom RC, Hale PD, Kristensen EW, Kuhr WG, May LJ (1988) Real-time characterization of dopamine overflow and uptake in the rat striatum. Neuroscience 25:513–523.

Xie S, Shafer G, Wilson CG, Martin HB (2006) *In vitro* adenosine detection with a diamond-based sensor. Dia Rel Mater 15:225–228.

Xu Y, Venton BJ (2010) Rapid determination of adenosine deaminase kinetics using fast-scan cyclic voltammetry. Phys Chem Chem Phys 12:10027–10032.

Zigmond MJ, Abercrombie ED, Berger TW, Grace AA, Stricker EM (1990) Compensations after lesions of central dopaminergic neurons: some clinical and basic implications. Trends Neurosci 13:290–295.

CHAPTER 5

REAL-TIME *IN VIVO* NEUROTRANSMITTER MEASUREMENTS USING ENZYME-BASED CERAMIC MICROELECTRODE ARRAYS: WHAT WE HAVE LEARNED ABOUT GLUTAMATE SIGNALING

Jason J. Burmeister[*], Erin R. Hascup[†], Kevin N. Hascup[†], Verda Davis[‡], Seth R. Batten[‡], Francois Pomerleau[‡], Jorge E. Quintero[‡], Peter Huettl[*], Pooja M. Talauliker[‡], Ingrid Strömberg[§], Greg A. Gerhardt[‡]

[*]Center for Microelectrode Technology, [†]Southern Illinois University School of Medicine, [‡]University of Kentucky, [§]Umeå University

5.1 INTRODUCTION

Since the pioneering work of Adams and coworkers (Kissinger et al., 1973; McCreery et al., 1974; Wightman et al., 1976) in the 1970s, many research groups have strived to develop microelectrode technologies for the direct chemical measurements of neurotransmitters in the extracellular space of the brain (Wightman et al., 1976; Hu et al., 1994; Lowry et al., 1998; Burmeister et al., 2000; Dale et al., 2000; Oldenziel et al., 2006b; Wassum et al., 2008; Zesiewicz et al., 2013). Interestingly, the world has been recording neuronal electrical activity of the brain since the 1930s, but this is a resulting consequence of the chemical communication of the brain, which makes up approximately 90% of brain communication. Over the last decade, our group has worked on the development of a more universal microelectrode array (MEA) technology for second-by-second measurements of neurotransmitters, neuromodulators, and markers of brain energy/metabolism. Historically, microdialysis has been a "gold standard" for *in vivo* neurochemical measurements during drug effects and behavior, because of the ability to analyze virtually any substance that crosses the microdialysis membrane. Unfortunately, the routine sampling rate of this technique (5–20 minutes) is a significant limitation for measuring the rapid dynamics (milliseconds–seconds) of fast signaling neurotransmitters, such as L-glutamate in the brain (Timmerman and Westerink, 1997; Kennedy et al., 2002; Watson et al., 2006). Although recent advancements have allowed for sub-minute microdialysis measures, this methodology has not been extensively employed because of technical limitations of the methods (Lada et al., 1997; Tucci et al., 1997; Rossell et al., 2003). In addition, chronic microdialysis experiments are generally limited to 6–8 hours in duration, although in some instances may be longer (Clapp-Lilly et al., 1999; Kennedy et al., 2002). However, there has been the need to record for many days and possibly even weeks to determine the long-term, chronic effects of drugs. Moreover, for the typical microdialysis probe, the large size (1–4 mm length, 0.2–0.3 mm diameter) produces apparent tissue damage that may confound the interpretation of the results (Clapp-Lilly et al., 1999). Thus, microelectrode technologies with improved spatial-temporal resolution have been developed to solve some of the limitations of microdialysis methods.

The most widely used microelectrode methods have been those employing carbon fiber-based microelectrodes. These have been developed and used for measuring rapid changes in central nervous system (CNS) levels of dopamine and other electroactive monoamines in anesthetized, and more recently in awake, behaving animals (Suaud-Chagny et al., 1993; Westerink, 2004). These techniques, however, have had limited success for monitoring other non-electroactive neurotransmitters such as glutamate and acetylcholine mainly because of the inherent lack of self-referencing capabilities, limits of detection, or temporal resolution.

Our laboratory and others have made great strides toward developing biosensors capable of measuring these non-electroactive neurotransmitters (Hu et al., 1994; Oldenziel et al., 2006a; Tian et al., 2009). We have developed an advanced technology using enzyme-based ceramic MEAs for chronic neurotransmitter measurements in rodents and non-human primates (Rutherford et al., 2007; Hascup et al., 2008; Stephens et al., 2010). A variety of neurochemicals can be measured using the MEA technology in conjunction with constant voltage amperometry or chronoamperometry using the Fast Analytical Sensing Technology (FAST16mkIII; Quanteon, LLC, Nicholasville, KY) recording system. While the experiments and examples discussed in this chapter can be applied to other neurochemicals, we have the most extensive experience measuring the major excitatory neurotransmitter in the CNS, the L-glutamate (Rutherford et al., 2007; Hascup et al., 2008; Hascup et al., 2010). Therefore, in this chapter we focus on glutamatergic neurotransmission and what we have learned over the last decade about this important and elusive neurotransmitter.

5.2 MATERIALS AND METHODS

5.2.1 DEVELOPMENT OF CERAMIC-BASED MEAS

Over the last decade our group has used the basic concepts of photolithographic formation of microelectrodes on silicon substrates (Wise et al., 1976) to develop ceramic alumina or aluminum oxide (Al_2O_3) substrate-based MEAs. Relatively inert FDA (Food and Drug Administration) approved materials are used in fabrication including Al_2O_3, polyimide for encapsulation and platinum (Pt) recording (electrophysiological or chemical) sites and platinum/iridium (Pt/Ir) stimulation sites. MEAs are mass fabricated using photolithographic techniques and cut out using computer-controlled diamond saw technology. Hundreds of MEAs with different or identical designs can be simultaneously patterned on a single ceramic wafer. Operational biocompatibility is based on the ability to record single-unit activity of neurons or electrochemical signals of a test molecule such as peroxide and post-experimentation integrity of the MEAs for up to six months, *in vivo* in awake rats. High purity, non-conducting Al_2O_3 substrates (99.9%, Coors, Golden, CO) are polished to 50–125 μm in thickness (2.5 × 2.5 cm or 5.0 × 5.0 cm wafers). The metal Pt recording sites or stimulation sites, connecting lines and bonding pads are patterned using a double mask design. One or both sides of the MEA are coated with 1.25 μm polyimide to insulate the connecting lines. Pt or Pt/Ir (0.25 μm) can be additionally sputtered onto the recording or stimulation sites. MEAs with 4–8 recording sites have been fabricated with recording sites ranging from 10 × 10 μm to 15 × 330 μm. Various recording site sizes and arrangements have been selected to conform to specific brain layers (Burmeister et al., 2002; Opris et al., 2011; Stephens et al., 2011). The front and back of the ceramic substrate can be patterned to increase recording site density and create isolated front and back site pairs. A practical advantage of multiple recording sites over a single electrode is that even if one site fails useful data can be received from the other recording sites.

5.2.2 BASIC CAPABILITIES OF MEAS

These conformal MEAs have been used for many experiments in both rodents and non-human primates and have been applied to chronic recordings of multiple single-unit neural activity, electrical stimulation and electrochemical recordings of neurotransmitters in awake rats, non-human primates and possibly in future applications to human neurosurgery (Day et al., 2006; Quintero et al., 2007; Dash et al., 2009; Zhang et al., 2009; Konradsson-Geuken et al., 2010; Stephens et al., 2010; Choi et al., 2012; Hampson et al., 2012; Onifer et al., 2012; Opris et al., 2012b; Opris et al., 2012a; Howe et al., 2013; Zhou et al., 2013). Single-unit neuronal activity has been recorded for up to six months in the hippocampus of freely moving rats. Hippocampal pyramidal cells in the rat have been stimulated and recorded. Single-unit recordings from non-human primate frontal cortex and hippocampus in awake animals have been performed using a specially designed deep recording electrode for the primate brain. Neurochemical recordings of glutamate signaling in the motor and frontal cortex of anesthetized monkeys as well as recordings of glutamate levels in the prefrontal cortex, hippocampus and putamen of awake rats and non-human primates have been performed. New Pt/Ir surfaces formed on the more standard Pt surfaces have been characterized for improved electrical stimulation and enhanced recording site surface area. In conjunction with Ad-Tech® Medical Instrument Corporation, flexible shank electrodes have been developed for future studies in non-human primates and patients with epilepsy, brain tumors, Parkinson's disease, deep brain stimulation (DBS) surgery or traumatic brain injury (TBI). In addition, a four shank electrode design with 32 active sites for recording and stimulation in the hippocampus of awake monkeys is currently being used and further developed for electrochemical and electrophysiological studies in the hippocampus of awake non-human primates (work in progress).

5.3 SELF-REFERENCING RECORDINGS USING AMPEROMETRY: HOW TO MEASURE RESTING LEVELS OF NEUROCHEMICALS

5.3.1 WHY PERFORM SELF-REFERENCING RECORDINGS?

The MEA photolithographic mass fabrication process allows exceptional flexibility and control of the geometric layout of multiple recording sites. Multiple recording sites within close proximity allow for the use of a technique we call self-referencing (Burmeister and Gerhardt, 2001). Historically, electrochemical recordings have focused on ways to minimize the "non-faradaic" or non-specific background signals such as solvent dipole reorientation, adsorption, desorption, movement of electrolyte ions at the recording site surfaces from measurements such that the major signal measured is due to the analyte of interest. Unfortunately, in biological systems there are many contributors to changes in an electrode's background, or "non-faradaic" response, such as pH shifts and changes in Ca^{2+}. Self-referencing has been an extremely powerful tool for removing the effects of chemical interferents that might contribute to a portion of the analyte signal and for permitting the subtraction of noise, which is present on both the analyte and control sites, from the analyte signal. In addition, one of the biggest advantages of self-referencing is that it makes it possible to measure both tonic levels and phasic changes of the neurochemical, rather than only the changes from a baseline or transient neurotransmitter changes (Burmeister and Gerhardt, 2001). Self-referencing subtraction cannot be achieved using single microelectrode and is a unique recording feature of multisite MEAs. One of the key aspects of using this approach is differentially coating electrodes to have control and analyte-sensing sites in close, defined proximity, but physically coating MEAs in this fashion is a challenge. To address this issue, we are developing a micro-coating apparatus capable of differentially coating side-by-side pairs that is impossible to do by hand. Also, we are developing a two sided MEA with identical recording sites on both the front and back of the microelectrode. The front and back design allows for one side to be coated for glutamate detection while the reverse side is coated as a control for self-referencing. The advantage then is that either the micro-coating apparatus or the hand enzyme coating may be employed with the front and back design (Figure 5.1).

Self-referencing subtraction is possible when one recording site is capable of recording the compound of interest while an adjacent site cannot (See Figures 5.2 and 5.5 and Burmeister and Gerhardt, 2001; Day et al., 2006; Rutherford et al., 2007). For example, first one uses a highly selective enzyme, like glutamate oxidase

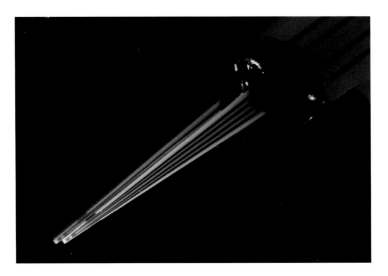

Figure 5.1 Conformal front/back MEA. The front sites are shown along with a reflection of the back sites. The recording sites are 150 × 20 μm and are arranged in pairs; substrate/MEA thickness is 125 μm and tapers from 100 μm wide at the tip to 160 μm at the top of the upper recording sites. Ag/AgCl (silver/silver chloride) reference electrodes are utilized for all recordings (not shown).

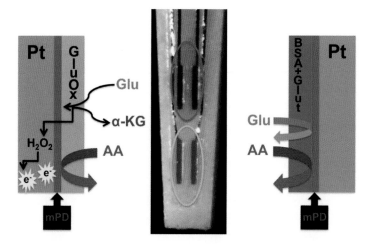

Figure 5.2 Enzyme-based coatings permit the measurement of molecules that could not otherwise be oxidized on an electrode surface. Layers (poly-(meta-phenylenediamine), mPD) to minimize the oxidation of unwanted molecules are applied to the surface of the MEA. Glu: glutamate, α-KG: alpha ketoglutarate, AA: ascorbic acid, GluOX: glutamate oxidase, BSA: bovine serum albumin, Glut: glutaraldehyde.

(GluOx) which is extremely selective for L-glutamate over other amino acids (Kusakabe et al., 1983). Then GluOx is entrapped in a protein matrix such as bovine serum albumin (BSA) that is cross-linked with glutaraldehyde to retain the high enzyme-selectivity and activity after immobilization onto the MEA surface ((Burmeister et al., 2013) green area shown in Figure 5.2).

Local glutamate molecules contact the immobilized GluOx where they are converted to the reporter molecule, hydrogen peroxide (H_2O_2) plus α-ketoglutarate. Because converted H_2O_2 is directly proportional to local glutamate, H_2O_2 is oxidized and quantified at the Pt recording sites to yield a current that is also proportional to local glutamate concentration. Although interfering species can be removed by self-referencing subtraction, the use of a selective layer improves MEA performance for glutamate detection. The sulfonated ionomer Nafion® is used to repel the anions such as uric acid, ascorbic acid, and 3, 4-dihydroxyphenyl acetic acid (DOPAC) by charge repulsion. Cationic interfering chemicals including dopamine, serotonin, and norepinephrine are not blocked and remain detectable when using Nafion®. Recently, we have further improved the selectivity of the microelectrodes by using mPD (Hinzman et al., 2010; Stephens et al., 2011). Once electro-polymerized onto the recording sites, mPD produces a size exclusion layer, which repels molecules larger than the size of peroxide, thus effectively blocking dopamine, serotonin, and norepinephrine as well as uric acid, ascorbic acid, and DOPAC. However, some small molecules like nitric oxide or other unknown electrochemically active substances could still be detected when using mPD as an exclusion layer. This has been a limitation for many groups using only a single microelectrode. Self-referencing subtraction is particularly useful to detect and remove interfering signals that are present on the glutamate detecting and sentinel recording sites as well as for removing non-faradaic background current. When the sentinel site response (blue area in Figure 5.2) is subtracted from the glutamate site response (green area in Figure 5.2), a nearly specific measure of glutamate is the result (e.g., see Figure 5.5; Day et al., 2006; Hascup et al., 2008). Glutamate concentrations as low as 0.1 μM (S/N = 3, n = 20) with a spatial resolution of 150 μm × 20 μm and a tip width of 100 μm (for the MEA design shown in Figure 5.1 with 30 μm between the adjacent recording sites and 100 μm between the upper and lower pairs) have been measured using this methodology. Recently, the correlation between responses from the glutamate and sentinel sites has been used as a metric to further confirm that self-reference subtracted signals are indeed specific to the glutamate site alone. Chronic recordings of glutamate have been made in numerous studies using MEAs implanted in rats and mice for up to 17 days (Rutherford et al., 2007; Hascup et al., 2008; Dash et al., 2009; Hascup et al., 2010). A software package that is customized to *in vivo* electrochemical signaling analysis allows longer recording periods (24 hours +) in awake animals to be analyzed for spontaneous, transient glutamate signals as well as potassium-induced phasic changes of glutamate (Hascup et al., 2011a; Hinzman et al., 2012).

5.4 APPLICATIONS TO BIOLOGICAL SYSTEMS

5.4.1 STUDIES IN FREELY MOVING RATS AND MICE

There are several advantages in recording glutamate and other neurotransmitters in freely moving animals. Anesthesia has been shown to alter mostly tonic glutamate levels and in some circumstances phasic glutamate release dynamics (Liachenko et al., 1998, 1999; Rutherford et al., 2007). Recordings in freely moving animals allow for monitoring glutamate neurotransmission without the effects of anesthesia. Behavioral studies can be coupled to neurochemical measurements to investigate changes in glutamate that are related to inherent or learned behaviors. Glutamate MEAs have been shown to yield reliable glutamate recordings for at least one week in freely moving mice and rats (Rutherford et al., 2007; Hascup et al., 2008). Compared to most microdialysis studies, MEA glutamate recordings are capable of multiple independent or continuous measures that can be more readily continued over repeated days (see Hascup et al., 2012). With direct application to behavioral paradigms or disease state models where tonic and/or phasic changes of glutamate are likely over the course of the experiment, freely moving MEA glutamate measures have applications in a variety of studies relating but not limited to addiction, sensitization, circadian rhythms, epilepsy, and TBI studies.

5.4.2 TONIC RELEASE OF NEUROTRANSMITTERS: IS IT NEURONALLY-DERIVED?

5.4.2.1 How do we know that we are measuring glutamate that is mostly derived from neurons?

A major issue that has surrounded *in vivo* measures of glutamate has also surrounded the origin of the signal (see (Timmerman and Westerink, 1997). Both neuronal and glial processes are involved in the regulation of tonic glutamate levels and many prior microdialysis studies support the view that resting levels sampled by microdialysis methods may not be derived from neurons. Recent pharmacological studies in awake rats using glutamate MEAs have shed light on the origins of tonic and phasic glutamate signals (Rutherford et al., 2007; Hascup et al., 2010). Modulation of resting glutamate levels in rat prefrontal cortex; result is shown in Figure 5.3.

Resting glutamate levels were significantly lower (~40–50%) than vehicle with local application of the Na^+ channel blocker tetrodotoxin (TTX) (3 μl, 100 μM) or the Ca^{2+} channel blocker ω-conotoxin (1 μl, 3 μM). These results support the fact that the resting levels of glutamate measured by the MEAs in the rat prefrontal cortex are derived from neurons because the inhibitory effects of both a sodium channel blocker and a calcium channel blocker are known to directly affect the release of neurotransmitters from neurons.

Second, the contribution of Group II metabotropic glutamate receptors (mGluR$_{2/3}$), which act as inhibitory auto-receptors and reside primarily on

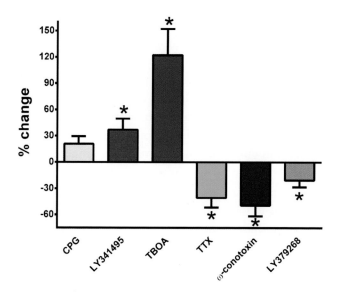

Figure 5.3. Local application of compounds to identify mechanisms responsible for resting glutamate levels in awake rats. Adapted with permission from Hascup et al., 2010.

pre-synaptic neurons in the prefrontal cortex, was examined. An ~35% increase in resting levels was observed after delivery of the mGluR$_{2/3}$ antagonist LY341495 (3 μl, 100 μM) while an ~20% decrease was found with the mGluR$_{2/3}$ agonist LY379268 (3 μl, 100 μM). These results support the idea that there is a significant neuronal glutamate contribution to resting levels as well as due to the modulation of resting levels detected following local administration of these drugs that can directly affect output of glutamate from nerve terminals.

Third, when the non-selective glutamate transporter inhibitor D, L-threo-β-benzyloxyaspartate (TBOA; 1 μl, 100 μM) was locally applied there was an ~120% increase in resting levels, supporting the idea that extracellular glutamate levels are regulated by glutamate transporters located largely on glia (see Danbolt et al., 2001) and indicating that there is significant spontaneous efflux of glutamate that is detected following glutamate uptake inhibition.

Microdialysis results suggest that the cystine/glutamate exchanger is involved in glutamate homeostasis (Melendez et al., 2005). In contrast, when using glutamate MEAs in the frontal cortex of rats, local application of the cystine/glutamate antiporter inhibitor (S)-4-carboxyphenylglycine (CPG; 3 μl, 50 μM) did not significantly alter resting glutamate. In fact, a slight increase in glutamate levels was observed. Thus, we conclude that the MEAs record glutamate levels that are significantly composed of a neuronal component. However, the lack of a complete blockade of the signals suggests that there is another extracellular "pool" of glutamate, which is also detected by the MEAs, that is likely derived from mechanisms relating to the control of extracellular levels by glia. Recent studies in a rat model of TBI support that the resting levels of glutamate 48 hours after a TBI are not affected by the aforementioned drugs, such as the calcium channel blocker, which can inhibit resting levels, and support that, especially under certain brain trauma and/or other pathological conditions, MEAs can measure signals that are not solely derived from neuronal release of glutamate (Magistretti, 2009; Pellerin and Magistretti, 2012). Additional studies are needed to explore regulation of glutamate by non-neuronal sources that may represent new targets for drug development.

5.4.2.2 Resting levels of glutamate in the CNS: comparisons to microdialysis measurements

Microdialysis sampling with offline high performance liquid chromatography (HPLC) quantification is considered by many to be the 'gold standard' for *in vivo* measures of resting glutamate. The relationship between MEA and microdialysis/HPLC-measured glutamate levels therefore need to be understood. We have shown that MEA measured resting levels can be quantitated for at least 7–10 days *in vivo*. However, one recent report suggested that the high selectivity of GluOx may be altered when glutaraldehyde is used as the cross-linking agent and thereby resulting in an increase in perceived resting glutamate levels when using wire-type microelectrodes (Vasylieva et al., 2013). However, studies using ceramic-based MEAs support that GluOx is selective over other amino acids when using our coating protocol of entrapping GluOx in a glutaraldehyde cross-linked BSA matrix (Burmeister et al., 2013). For one, we only observed minimal responses to select amino acids with glutaraldehyde cross-linked glutamate oxidase-coated MEAs, and furthermore, functional glutamate MEA biosensors employed for *in vivo* experiments use multiple means to minimize the potential interference from electroactive molecules (e.g., exclusion layers and self-referencing subtraction). The perceived lack of selectivity of some enzyme-coated MEAs has been suggested to result in higher levels of resting glutamate than recordings using wire electrodes and some microdialysis measures (Vasylieva et al., 2013). We have recorded glutamate resting levels in a variety of brain regions, species, and animal strains as seen in Figure 5.4. In addition, Figure 5.4 also shows resting levels of glutamate recorded by microdialysis in numerous brain areas and species under a variety of conditions. Glutamate levels *in vivo* are regulated primarily by transporters located on glia allowing for extra-synaptic levels of or "spill over" of glutamate into the extracellular fluid. These sodium- and chloride-dependent transporters are low affinity, high capacity systems (see Danbolt, 2001). Resting levels of glutamate reported with either microdialysis or MEAs are within the range of glutamate affinities for glutamate transporters, which are 5 to 100 micromolar (Danbolt, 2001). This low affinity for glutamate and the extra-synaptic location of the transporters supports extracellular levels of glutamate that can likely exist close to synaptic areas in the low to mid micromolar range.

In general, microdialysis levels tend to be lower than our measured values, but they clearly overlap. We argue that the MEA measures of glutamate will be somewhat higher than most microdialysis measures because: (1) the microelectrodes produce less damage to brain tissue than microdialysis (Clapp-Lilly et al., 1999; Borland et al., 2005; Hascup et al., 2009), (2) they are closer to the synaptic sources because of the thinner diffusion layer compared to microdialysis, and (3) they are not limited by the extraction fraction of microdialysis. In addition, one might expect that extracellular levels of glutamate vary from region to region due to the variable numbers of glia compared to synapses (glial synapse ratios) and the diversity of glutamate receptors, especially the mGluR subtypes (Ferraguti and Shigemoto, 2006; Azevedo et al., 2009). Thus, our extensive studies with MEAs, taken together with microdialysis measures, support that resting levels of glutamate vary based on the species and strain, the brain area studied, and the animal model. In fact, findings in a recent report support our measurements of resting glutamate levels and the diversity and levels of glutamate measurements within the brain. Using low-flow push–pull perfusion sampling to obtain microelectrode-like spatial

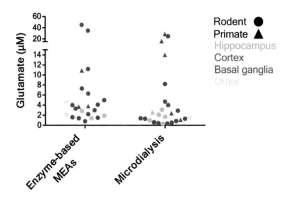

Figure 5.4 Glutamate concentrations *in vivo* reported by our group are comparable to glutamate values obtained with microdialysis reported in the literature. Values reported are from different brain regions and different species. Note: change in upper segment of y-axis scale. MEAs: (Day et al., 2006; Quintero et al., 2007; Rutherford et al., 2007; Hascup et al., 2008; Thomas et al., 2009; Hascup et al., 2010; Hinzman et al., 2010; Stephens et al., 2010; Hascup et al., 2011b; Hascup et al., 2011a; Mattinson et al., 2011; Stephens et al., 2011; Hascup et al., 2012; Hinzman et al., 2012; Matveeva et al., 2012a; Matveeva et al., 2012b; Onifer et al., 2012; Thomas et al., 2012; Nevalainen et al., 2013); Microdialysis: (Ueda and Tsuru, 1995; Yang et al., 1995; Battaglia et al., 1997; Biggs et al., 1997; Timmerman et al., 1999; Reinstrup et al., 2000; Schulz et al., 2000; Giovannini et al., 2001; Segovia et al., 2001; Galvan et al., 2003; Cavus et al., 2005; Clinckers et al., 2005; Calcagno et al., 2006; Segovia et al., 2006; Ballini et al., 2008; Hernandez et al., 2008; Lupinsky et al., 2010; Slaney et al., 2011; Li et al., 2012; Vasylieva et al., 2013).

resolution, Slaney and colleagues describe heterogeneity in basal glutamate concentration of the midbrain with some areas having low glutamate levels (2 μM) and neighboring ones having higher glutamate levels (24 μM). (Slaney et al., 2012). Moreover, MEA measures of resting glutamate (tonic levels) are at least 40%–50% derived from neuronal sources (Hascup et al., 2010) in contrast with the lack of sodium- and calcium-dependence seen with many microdialysis recordings (Timmerman and Westerink, 1997). Thus, our extensive studies support that resting glutamate levels vary because of the biology of the glutamate system and not because of an inherent poor selectivity of MEA recording methods.

5.4.3 BEYOND TONIC RECORDINGS: PHASIC RELEASE OF NEUROTRANSMITTERS

Phasic release of neurotransmitters from synchronized or burst firing of neuronal activity can result from naturally occurring behaviors (Opris et al., 2012b). Such behavioral events can lead to the release of a multitude of neurotransmitters similar to artificial stimuli like potassium-induced depolarization or electrical stimulation. Recently, using MEA recordings we have shown that tonic (resting) and phasic (transient release) glutamate signals are dissociable events. For example,

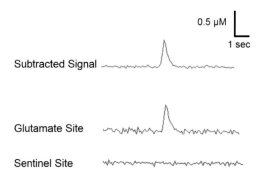

Figure 5.5 Spontaneous transient recorded in the hippocampus dentate gyrus (DG) of an awake rat. No stimulation was used to evoke the spontaneous transient.

initial MEA studies conducted in rats under anesthesia at various ages revealed that tonic levels of glutamate are often minimally affected by normal aging. However, local applications of high potassium result in phasic glutamate release with dynamic age-related decreases in pre-synaptic release and glial-mediated clearance of glutamate (Nickell et al., 2005; Hascup et al., 2007; Stephens et al., 2011). Lacking the rapid temporal resolution of the glutamate MEAs, previous results using *in vivo* microdialysis have not consistently shown age-related changes in glutamate signaling. In awake rats we have demonstrated that potassium-evoked release and resting levels of glutamate can be measured (unpublished observations Hascup et al., 2007; Rutherford et al., 2007). Also in awake rats, we have recorded spontaneous transients of glutamate in nucleus accumbens, striatum, prefrontal cortex and hippocampus. Figure 5.5 shows an example of a glutamate transient detected in the hippocampus of an awake rat that is clearly seen on the glutamate detecting site of the MEA but not observed on the sentinel site that cannot measure glutamate.

Using MEAs placed in the hippocampus, isoflurane-anesthetized rats or mice, we have recorded both tonic and phasic glutamate release in the DG, CA3 and CA1 sub-regions of the tri-synaptic pathway. Average spike amplitudes and average number of transients are seen in Figure 5.6 for recordings in the DG, CA3 and CA1 regions that clearly demonstrate the capabilities of measuring these spontaneous transient glutamate events in the anesthetized rat.

5.4.4. CHRONIC IMPLANTATION: HISTOPATHOLOGY AND BIOCOMPATIBILITY

Our group and others would like to follow neurochemical signaling for months and possibly years in brain regions of a behaving rat or mouse. Presently, MEAs can reliably measure glutamate in awake animals for 7–10 days. However,

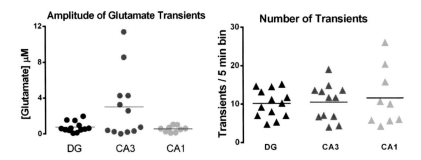

Figure 5.6 Average amplitudes and average number of spontaneous glutamate transients recorded in the various sub-regions of the rat hippocampus.

single-unit neuronal recordings have been successfully recorded for at least six months using our ceramic-based MEAs placed in the rat hippocampus (unpublished observations). In addition, the reporter molecule H_2O_2 has been reliably detected *in vivo* for at least 90 days (unpublished observation). Changes in the enzyme layer are likely limiting *in vivo* glutamate recordings to 7–10 days as histology studies and single-unit recordings support a minimal change in the microenvironment surrounding the MEA in the CNS (Hascup et al., 2009). As with orthopedic prostheses composed of ceramic Al_2O_3, where there is enhanced integration into host tissue, i.e., when the roughness value of the implant surface is equivalent to the dimensions of bone-forming cells (50–400 μm), our ceramic-based MEAs appear to integrate well into neuronal tissues (Hulbert et al., 1972). However, ceramic-based neural devices are relatively new. Thus, published studies on their CNS biocompatibility are few (Rutherford et al., 2007; Hascup et al., 2009). MEAs chronically implanted up to six months in rats result in minimal damage to the surrounding brain tissue in the prefrontal cortex or hippocampus (see Figure 5.7; Talauliker, 2010).

MEA implantation results in minimal observed changes in astrocytes (glial fibrillary acidic protein, GFAP) and microglia (ionized calcium binding adaptor molecule 1, Iba1) (Rutherford et al., 2007; Hascup et al., 2009) — the two major cell types participating in the inflammatory reaction to tissue injury (Biran et al., 2005, 2007). Because glutamate transporters located on the cell membrane of astrocytes are the primary mechanism by which glutamate is cleared from the extracellular environment, glial changes in the recording environment of the glutamate MEA could present a challenge as both tonic and phasic levels of glutamate could be affected (Danbolt, 2001). While in the days and weeks following implantation microglia accumulate around the electrode (Turner et al., 1999; Turner et al., 2001); microglia appear to stabilize after one week post MEA implantation. A physical barrier that limits microelectrode access to healthy neurons by astrocyte and microglia encapsulation has been reported with other types

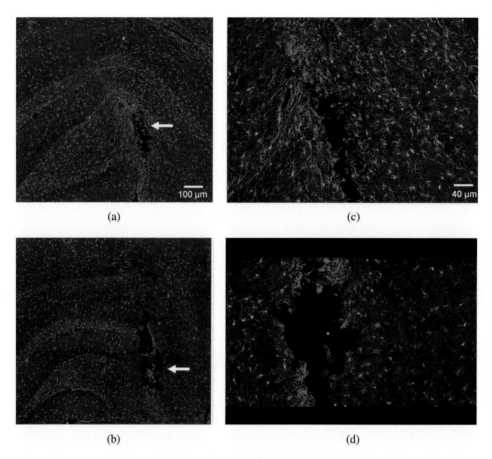

Figure 5.7 Hippocampal coronal sections co-labeled for GFAP and Iba1 at weeks 8 and 16 after chronic MEA implantation. GFAP (red) and Iba1 (green) labeled tissue shown at 8 weeks (a, c) and 16 weeks (b, d) after surgery. A well-defined glial scar is visible at eight weeks but becomes contracted and more compact by four months. Scale a, b: 100 μm, Scale c, d: 40 μm. White arrows indicate track location.

of microelectrodes (silicon, Pt wire references) (Rousche and Normann, 1998; Turner et al., 1999; Szarowski et al., 2003; Biran et al., 2005; Polikov et al., 2005; Spataro et al., 2005; Biran et al., 2007; Leung et al., 2008). This does not appear to be the case with ceramic-based MEAs where single-unit activity recordings are possible for up to six months *in vivo*. In marked contrast to the effects that occur from other chronically implanted devices (i.e., silicon microelectrodes or microdialysis probes) (Szarowski et al., 2003), with ceramic-based MEAs a relatively small glial scar is formed following chronic implantation in the prefrontal cortex (Hascup et al., 2009). This finding has been further confirmed with recent studies of chronic hippocampal implants examining GFAP staining in the hippocampus. While it significantly increased following MEA implantation, GFAP staining

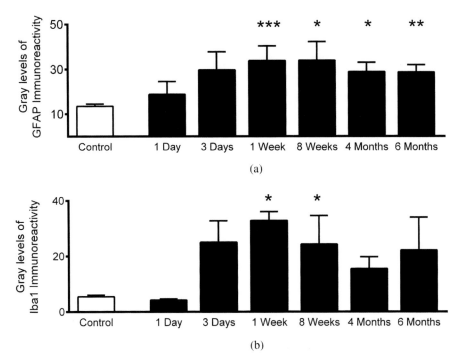

Figure 5.8 Quantitative analysis of mean GFAP and Iba1 immunoreactivity around MEA tracks in the rat hippocampus. Gray levels of GFAP (Panel a) and Iba1 (Panel b) immunoreactivity around the MEA track were compared for time points ranging from one day to six months following implantation in the rat hippocampus, relative to control tissue. Statistical comparisons were conducted using a one way ANOVA followed by Dunnet's test. GFAP immunoreactivity was significantly higher by one week (33.8 ± 6.8) and remained elevated up to six months. Iba1 levels were also significantly elevated by one week (32.9 ± 3.2) but diminished after eight weeks. Significance was defined as *: $p<0.05$, **: $p<0.01$, ***: $p<0.001$.

subsided considerably by four months (Figure 5.7; Talauliker, 2010). Hippocampal coronal sections from eight weeks post-surgery demonstrated intense GFAP immunoreactivity around the implant track (Figures 5.7a and 5.7c). The extended processes of reactive astrocytes contributing to the glial scar were readily distinguishable from normal astrocytes. At four months (16 weeks) post-surgery, the width of the glial sheath decreased and became more compact (Figures 5.7b–5.7d).

Quantitative estimation of total GFAP immunoreactivity at all data points showed that increased staining intensity was detected at day 1 following MEA implantation (Figure 5.8). A significant increase in intensity levels was measured at 1 week ($p<0.001$). Beyond eight weeks a decrease in GFAP levels was noted though overall immunoreactivity remained significantly elevated ($p<0.05$) for the duration of the implant.

How MEA surfaces are altered with chronic implantation must be determined to improve the long term performance. High quality *in vivo* glutamate signals are dependent upon MEA glutamate sensitivity, selectivity, total electroactive area and reaction kinetics. The characterization of the intrinsic physical and electrochemical properties of the Pt recording surfaces is necessary to improve recording capabilities including topography, electrochemical behavior, chemical reactivity and molecular composition. Immobilized GluOx converts L-glutamate into H_2O_2, which is subsequently oxidized at a polarized electrode recording surface. Because of its biocompatibility and high catalytic activity for amperometric measurements of H_2O_2, Pt is widely used in planar while Pt or Pt/Ir is often used for wire-based amperometric microelectrodes (Gerhardt et al., 2000; Hu et al., 1994; Graciela et al, 2011). Although it is generally thought that the reaction mechanism of H_2O_2 oxidation on Pt surfaces is independent of electrode geometry, this similarity may not translate to equivalent recording performance per unit area on MEAs (Talauliker et al., 2011). In our recent review of the current scientific literature, the performance of MEAs rivaled or exceeded that of other electrode subtypes in terms of sensitivity per unit area (Talauliker et al., 2011). The ultra-high purity Pt required for the MEA fabrication process seems to yield recording sites with high effective surface area and activity for H_2O_2 oxidation than most wire-type microelectrodes. A series of studies has recently been performed to characterize the intrinsic physical properties and surface topography of an implantable ceramic-based neural device for intracerebral monitoring of neurotransmitter dynamics (Talauliker et al., 2011). More studies to better understand the MEA surfaces before and after implantation are needed to improve the long-term performance of implantable devices. The combination of high purity recording surface and precise layering yields MEAs with excellent glutamate response over a wide range (0.2 to >200 μM) and fast temporal resolution (~1 second) (Burmeister and Gerhardt, 2001).

5.5 CONCLUSIONS

There are some notable conclusions that have resulted from numerous studies using the MEA technology *in vivo* to study glutamate neuronal systems and the regulation of glutamate signaling. First, our studies support that the MEAs are durable and can be mass fabricated. Second glutamate resting levels are largely derived from neuronal elements and are highly affected by drugs that alter neuronal and glial regulation of glutamate. Third, the ability to perform self-referencing and the fast MEA response allows measures of both tonic (resting) and phasic (transients) glutamate signaling to be measured. Fourth, investigations in rodents and non-human primates show that glutamate signaling is heterogeneous and that resting levels measured using glutamate MEAs are comparable to microdialysis, especially if we compare to zero flow methods. Finally, the MEAs produce little damage to CNS allowing chronic glutamate measurements for up to 10 days and possibly months in the future. MEA technology has provided an exciting platform for understanding the chemical messages of the nervous system, and with continued refinements and advancements, the future possibilities seem limitless.

ACKNOWLEDGMENTS

Supported by NIDA; R21DA033796-01, DARPA; N66001-09-C-2080 and NIH; National Center for Research Resources and the National Center for Advancing Translational Sciences through grant UL1RR033173-01. The content is solely the responsibility of the authors and does not necessarily represent the official views of the NIH.

Disclosure of competing interest: GAG is principal owner of Quanteon LLC. JEQ, PH, FP have served as consultants to Quanteon LLC.

REFERENCES

Azevedo FA, Carvalho LR, Grinberg LT, Farfel JM, Ferretti RE, Leite RE, Jacob Filho W, Lent R, Herculano-Houzel S (2009) Equal numbers of neuronal and non-neuronal cells make the human brain an isometrically scaled-up primate brain. J Comp Neurol 513:532–541.

Ballini C, Corte LD, Pazzagli M, Colivicchi MA, Pepeu G, Tipton KF, Giovannini MG (2008) Extracellular levels of brain aspartate, glutamate and GABA during an inhibitory avoidance response in the rat. J Neurochem 106:1035–1043.

Battaglia G, Monn JA, Schoepp DD (1997) *In vivo* inhibition of veratridine-evoked release of striatal excitatory amino acids by the group II metabotropic glutamate receptor agonist LY354740 in rats. Neurosci Lett 229:161–164.

Biggs CS, Fowler LJ, Whitton PS, Starr MS (1997) Extracellular levels of glutamate and aspartate in the entopeduncular nucleus of the rat determined by microdialysis: Regulation by striatal dopamine D2 receptors *via* the indirect striatal output pathway? Brain Res 753:163–175.

Biran R, Martin DC, Tresco PA (2005) Neuronal cell loss accompanies the brain tissue response to chronically implanted silicon microelectrode arrays. Exp Neurol 195:115–126.

Biran R, Martin DC, Tresco PA (2007) The brain tissue response to implanted silicon microelectrode arrays is increased when the device is tethered to the skull. J Biomed Mater Res A 82:169–178.

Borland LM, Shi G, Yang H, Michael AC (2005) Voltammetric study of extracellular dopamine near microdialysis probes acutely implanted in the striatum of the anesthetized rat. J Neurosci Methods 146:149–158.

Burmeister J, Pomerleau F, Palmer M, Day B, Huettl P, Gerhardt G (2002) Improved ceramic-based multisite microelectrode for rapid measurements of L-glutamate in the CNS. J Neurosci Methods 119:163–171.

Burmeister JJ, Gerhardt GA (2001) Self-referencing ceramic-based multisite microelectrodes for the detection and elimination of interferences from the measurement of L-glutamate and other analytes. Anal Chem 73:1037–1042.

Burmeister JJ, Moxon K, Gerhardt GA (2000) Ceramic-based multisite microelectrodes for electrochemical recordings. Anal Chem 72:187–192.

Burmeister JJ, Davis VA, Quintero JE, Pomerleau F, Huettl P, Gerhardt GA (2013) Glutaraldehyde cross-linked glutamate oxidase-coated microelectrode arrays: Selectivity and resting levels of glutamate in the CNS. ACS Chem Neurosci 4:721–728.

Calcagno E, Carli M, Invernizzi RW (2006) The 5-HT(1A) receptor agonist 8-OH-DPAT prevents prefrontocortical glutamate and serotonin release in response to blockade of cortical NMDA receptors. J Neurochem 96:853–860.

Cavus I, Kasoff WS, Cassaday MP, Jacob R, Gueorguieva R, Sherwin RS, Krystal JH, Spencer DD, Abi-Saab WM (2005) Extracellular metabolites in the cortex and hippocampus of epileptic patients. Ann Neurol 57:226–235.

Choi HB, Gordon GRJ, Zhou N, Tai C, Rungta RL, Martinez J, Milner TA, Ryu JK, McLarnon JG, Tresguerres M, Levin LR, Buck J, MacVicar BA (2012) Metabolic communication between astrocytes and neurons *via* bicarbonate-responsive soluble adenylyl cyclase. Neuron 75:1094–1104.

Clapp-Lilly KL, Roberts RC, Duffy LK, Irons KP, Hu Y, Drew KL (1999) An ultrastructural analysis of tissue surrounding a microdialysis probe. J Neurosci Methods 90:129–142.

Clinckers R, Gheuens S, Smolders I, Meurs A, Ebinger G, Michotte Y (2005) *In vivo* modulatory action of extracellular glutamate on the anticonvulsant effects of hippocampal dopamine and serotonin. Epilepsia 46:828–836.

Dale N, Pearson T, Frenguelli BG (2000) Direct measurement of adenosine release during hypoxia in the CA1 region of the rat hippocampal slice. J Physiol 526:143–155.

Danbolt NC (2001) Glutamate uptake. Prog Neurobiol 65:1–105.

Dash MB, Douglas CL, Vyazovskiy VV, Cirelli C, Tononi G (2009) Long-term homeostasis of extracellular glutamate in the rat cerebral cortex across sleep and waking states. J Neurosci 29:620–629.

Day BK, Pomerleau F, Burmeister JJ, Huettl P, Gerhardt GA (2006) Microelectrode array studies of basal and potassium-evoked release of L-glutamate in the anesthetized rat brain. J Neurochem 96:1626–1635.

Ferraguti F, Shigemoto R (2006) Metabotropic glutamate receptors. Cell Tissue Res 326:483–504.

Galvan A, Smith Y, Wichmann T (2003) Continuous monitoring of intracerebral glutamate levels in awake monkeys using microdialysis and enzyme fluorometric detection. J Neurosci Methods 126:175–185.

Gerhardt GA, Burmeister JJ, Meyers RA (2000) Voltammetry *in vivo* for chemcial analysis of the nervous system. In: Encyclopedia of analytical chemistry, pp 710–731. Chichester: John Wiley & Sons.

Giovannini MG, Rakovska A, Benton RS, Pazzagli M, Bianchi L, Pepeu G (2001) Effects of novelty and habituation on acetylcholine, GABA, and glutamate release from the frontal cortex and hippocampus of freely moving rats. Neuroscience 106:43–53.

Hampson RE, Gerhardt GA, Marmarelis V, Song D, Opris I, Santos L, Berger TW, Deadwyler SA (2012) Facilitation and restoration of cognitive function in primate prefrontal cortex by a neuroprosthesis that utilizes minicolumn-specific neural firing. J Neural Eng 9:056012.

Hascup ER, af Bjerken S, Hascup KN, Pomerleau F, Huettl P, Stromberg I, Gerhardt GA (2009) Histological studies of the effects of chronic implantation of ceramic-based microelectrode arrays and microdialysis probes in rat prefrontal cortex. Brain Res 1291:12–20.

Hascup ER, Hascup KN, Stephens M, Pomerleau F, Huettl P, Gratton A, Gerhardt GA (2010) Rapid microelectrode measurements and the origin and regulation of extracellular glutamate in rat prefrontal cortex. J Neurochem 115:1608–1620.

Hascup ER, Hascup KN, Pomerleau F, Huettl P, Hajos-Korcsok E, Kehr J, Gerhardt GA (2012) An allosteric modulator of metabotropic glutamate receptors (mGluR$_2$), (+)-TFMPIP, inhibits restraint stress-induced phasic glutamate release in rat prefrontal cortex. J Neurochem 122:619–627.

Hascup KN, Hascup ER, Pomerleau F, Huettl P, Gerhardt GA (2008) Second-by-second measures of L-glutamate in the prefrontal cortex and striatum of freely moving mice. J Pharmacol Exp Ther 324:725–731.

Hascup KN, Hascup ER, Stephens ML, Glaser PE, Yoshitake T, Mathe AA, Gerhardt GA, Kehr J (2011a) Resting glutamate levels and rapid glutamate transients in the prefrontal cortex of the flinders sensitive line rat: A genetic rodent model of depression. Neuropsychopharmacology 36:1769–1777.

Hascup KN, Bao X, Hascup ER, Hui D, Xu W, Pomerleau F, Huettl P, Michaelis ML, Michaelis EK, Gerhardt GA (2011b) Differential levels of glutamate dehydrogenase 1 (GLUD1) in Balb/c and C57BL/6 mice and the effects of overexpression of Glud1 gene on glutamate release in striatum. ASN Neuro 3:e00057.

Hascup KN, Rutherford EC, Quintero JE, Day BK, Nickell JR, Pomerleau F, Huettl P, Burmeister J, Gerhardt GA, Michael AC, Borland LM (2007) Second-by-second measures of L-glutamate and other neurotransmitters using enzyme-based microelectrode arrays. In: Electrochemical methods for neuroscience (Michael AC, Borland LM, eds), pp 407–450. Boca Raton, FL: CRC Press.

Hernandez LF, Segovia G, Mora F (2008) Chronic treatment with a dopamine uptake blocker changes dopamine and acetylcholine but not glutamate and GABA concentrations in prefrontal cortex, striatum and nucleus accumbens of the awake rat. Neurochem Int 52:457–469.

Hinzman JM, Thomas TC, Quintero JE, Gerhardt GA, Lifshitz J (2012) Disruptions in the regulation of extracellular glutamate by neurons and glia in the rat striatum two days after diffuse brain injury. J Neurotrauma 29:1197–1208.

Hinzman JM, Thomas TC, Burmeister JJ, Quintero JE, Huettl P, Pomerleau F, Gerhardt GA, Lifshitz J (2010) Diffuse brain injury elevates tonic glutamate levels and potassium-evoked glutamate release in discrete brain regions at two days post-injury: An enzyme-based microelectrode array study. J Neurotrauma 27:889–899.

Howe WM, Berry AS, Francois J, Gilmour G, Carp JM, Tricklebank M, Lustig C, Sarter M (2013) Prefrontal cholinergic mechanisms instigating shifts from monitoring for cues to cue-guided performance: converging electrochemical and fMRI evidence from rats and humans. J Neurosci 33:8742–8752.

Hu Y, Mitchell KM, Albahadily FN, Michaelis EK, Wilson GS (1994) Direct measurement of glutamate release in the brain using a dual enzyme-based electrochemical sensor. Brain Res 659:117–125.

Hulbert SF, Morrison SJ, Klawitter JJ (1972) Tissue reaction to three ceramics of porous and non-porous structures. J Biomed Mater Res 6:347–374.

Kennedy RT, Thompson JE, Vickroy TW (2002) *In vivo* monitoring of amino acids by direct sampling of brain extracellular fluid at ultralow flow rates and capillary electrophoresis. J Neurosci Methods 114:39–49.

Kissinger PT, Hart JB, Adams RN (1973) Voltammetry in brain tissue — a new neurophysiological measurement. Brain Res 55:209–213.

Konradsson-Geuken A, Wu HQ, Gash CR, Alexander KS, Campbell A, Sozeri Y, Pellicciari R, Schwarcz R, Bruno JP (2010) Cortical kynurenic acid bi-directionally modulates prefrontal glutamate levels as assessed by microdialysis and rapid electrochemistry. Neuroscience 169:1848–1859.

Kusakabe H, Midorikawa Y, Fujishima T, Kuninaka A, Yoshino H (1983) Purification and properties of a new enzyme, L-glutamate oxidase, from *Streptomyces sp* X-199-6 grown on wheat bran. Agric Biol Chem 47:1323–1328.

Lada MW, Vickroy TW, Kennedy RT (1997) High temporal resolution monitoring of glutamate and aspartate *in vivo* using microdialysis on-line with capillary electrophoresis with laser-induced fluorescence detection. Anal Chem 69:4560–4565.

Leung BK, Biran R, Underwood CJ, Tresco PA (2008) Characterization of microglial attachment and cytokine release on biomaterials of differing surface chemistry. Biomaterials 29:3289–3297.

Li J, von Pfostl V, Zaldivar D, Zhang X, Logothetis N, Rauch A (2012) Measuring multiple neurochemicals and related metabolites in blood and brain of the rhesus monkey by using dual microdialysis sampling and capillary hydrophilic interaction chromatography-mass spectrometry. Anal Bioanal Chem 402:2545–2554.

Liachenko S, Tang P, Somogyi GT, Xu Y (1998) Comparison of anaesthetic and non-anaesthetic effects on depolarization-evoked glutamate and GABA release from mouse cerebrocortical slices. Br J Pharmacol 123:1274–1280.

Liachenko S, Tang P, Somogyi GT, Xu Y (1999) Concentration-dependent isoflurane effects on depolarization-evoked glutamate and GABA outflows from mouse brain slices. Br J Pharmacol 127:131–138.

Lowry JP, Miele M, O'Neill RD, Boutelle MG, Fillenz M (1998) An amperometric glucose-oxidase/poly(o-phenylenediamine) biosensor for monitoring brain extracellular glucose: *in vivo* characterisation in the striatum of freely-moving rats. J Neurosci Methods 79:65–74.

Lupinsky D, Moquin L, Gratton A (2010) Interhemispheric regulation of the medial prefrontal cortical glutamate stress response in rats. J Neurosci 30:7624–7633.

Magistretti PJ (2009) Role of glutamate in neuron-glia metabolic coupling. Am J Clin Nutr 90:875S–880S.

Mattinson CE, Burmeister JJ, Quintero JE, Pomerleau F, Huettl P, Gerhardt GA (2011) Tonic and phasic release of glutamate and acetylcholine neurotransmission in subregions of the rat prefrontal cortex using enzyme-based microelectrode arrays. J Neurosci Methods 202:199–208.

Matveeva EA, Davis VA, Whiteheart SW, Vanaman TC, Gerhardt GA, Slevin JT (2012a) Kindling-induced asymmetric accumulation of hippocampal 7S SNARE complexes correlates with enhanced glutamate release. Epilepsia 53:157–167.

Matveeva EA, Price DA, Whiteheart SW, Vanaman TC, Gerhardt GA, Slevin JT (2012b) Reduction of vesicle-associated membrane protein 2 expression leads to a kindling-resistant phenotype in a murine model of epilepsy. Neuroscience 202:77–86.

Mazzonea GL, Nistri A (2011) Electrochemical detection of endogenous glutamate release from rat spinal cord organotypic slices as a real-time method to monitor excitotoxicity, J Neurosci Methods 197:128–132.

McCreery RL, Dreiling R, Adams RN (1974) Voltammetry in brain tissue: the fate of injected 6-hydroxydopamine. Brain Res 73:15–21.

Melendez RI, Hicks MP, Cagle SS, Kalivas PW (2005) Ethanol exposure decreases glutamate uptake in the nucleus accumbens. Alcohol Clin Exp Res 29:326–333.

Nevalainen N, Lundblad M, Gerhardt GA, Stromberg I (2013) Striatal glutamate release in L-DOPA-induced dyskinetic animals. PLoS One 8:e55706.

Nickell J, Pomerleau F, Allen J, Gerhardt GA (2005) Age-related changes in the dynamics of potassium-evoked L-glutamate release in the striatum of Fischer 344 rats. J Neural Transm 112:87–96.

Oldenziel WH, Dijkstra G, Cremers TI, Westerink BH (2006a) *In vivo* monitoring of extracellular glutamate in the brain with a microsensor. Brain Res 1118:34–42.

Oldenziel WH, Dijkstra G, Cremers TI, Westerink BH (2006b) Evaluation of hydrogel-coated glutamate microsensors. Anal Chem 78:3366–3378.

Onifer SM, Quintero JE, Gerhardt GA (2012) Cutaneous and electrically evoked glutamate signaling in the adult rat somatosensory system. J Neurosci Methods 208:146–154.

Opris I, Hampson RE, Stanford TR, Gerhardt GA, Deadwyler SA (2011) Neural activity in frontal cortical cell layers: evidence for columnar sensorimotor processing. J Cogn Neurosci 23:1507–1521.

Opris I, Hampson RE, Gerhardt GA, Berger TW, Deadwyler SA (2012a) Columnar processing in primate pFC: evidence for executive control microcircuits. J Cogn Neurosci 24:2334–2347.

Opris I, Fuqua JL, Huettl PF, Gerhardt GA, Berger TW, Hampson RE, Deadwyler SA (2012b) Closing the loop in primate prefrontal cortex: inter-laminar processing. Front Neural Circuits 6:88.

Pellerin L, Magistretti PJ (2012) Sweet sixteen for ANLS. J Cereb Blood Flow Metab 32:1152–1166.

Polikov VS, Tresco PA, Reichert WM (2005) Response of brain tissue to chronically implanted neural electrodes. J Neurosci Methods 148:1–18.

Quintero JE, Day BK, Zhang Z, Grondin R, Stephens ML, Huettl P, Pomerleau F, Gash DM, Gerhardt GA (2007) Amperometric measures of age-related changes in glutamate regulation in the cortex of rhesus monkeys. Exp Neurol 208:238–246.

Reinstrup P, Stahl N, Mellergard P, Uski T, Ungerstedt U, Nordstrom CH (2000) Intracerebral microdialysis in clinical practice: baseline values for chemical markers during wakefulness, anesthesia, and neurosurgery. Neurosurgery 47:701–709.

Rossell S, Gonzalez LE, Hernandez L (2003) One-second time resolution brain microdialysis in fully awake rats. Protocol for the collection, separation and sorting of nanoliter dialysate volumes. J Chromatogr B Analyt Technol Biomed Life Sci 784:385–393.

Rousche PJ, Normann RA (1998) Chronic recording capability of the utah intracortical electrode array in cat sensory cortex. J Neurosci Methods 82:1–15.

Rutherford EC, Pomerleau F, Huettl P, Stromberg I, Gerhardt GA (2007) Chronic second-by-second measures of L-glutamate in the central nervous system of freely moving rats. J Neurochem 102:712–722.

Schulz MK, Wang LP, Tange M, Bjerre P (2000) Cerebral microdialysis monitoring: determination of normal and ischemic cerebral metabolisms in patients with aneurysmal subarachnoid hemorrhage. J Neurosurg 93:808–814.

Segovia G, Del Arco A, Prieto L, Mora F (2001) Glutamate-glutamine cycle and aging in striatum of the awake rat: effects of a glutamate transporter blocker. Neurochem Res 26:37–41.

Segovia G, Yague AG, Garcia-Verdugo JM, Mora F (2006) Environmental enrichment promotes neurogenesis and changes the extracellular concentrations of glutamate and GABA in the hippocampus of aged rats. Brain Res Bull 70:8–14.

Slaney TR, Mabrouk OS, Porter-Stransky KA, Aragona BJ, Kennedy RT (2012) Chemical gradients within brain extracellular space measured using low-flow push–pull perfusion sampling *in vivo*. ACS Chem Neurosci 4:321–329.

Slaney TR, Nie J, Hershey ND, Thwar PK, Linderman J, Burns MA, Kennedy RT (2011) Push–pull perfusion sampling with segmented flow for high temporal and spatial resolution *in vivo* chemical monitoring. Anal Chem 83:5207–5213.

Spataro L, Dilgen J, Retterer S, Spence AJ, Isaacson M, Turner JN, Shain W (2005) Dexamethasone treatment reduces astroglia responses to inserted neuroprosthetic devices in rat neocortex. Exp Neurol 194:289–300.

Stephens ML, Pomerleau F, Huettl P, Gerhardt GA, Zhang Z (2010) Real-time glutamate measurements in the putamen of awake rhesus monkeys using an enzyme-based human microelectrode array prototype. J Neurosci Methods 185:264–272.

Stephens ML, Quintero JE, Pomerleau F, Huettl P, Gerhardt GA (2011) Age-related changes in glutamate release in the CA3 and dentate gyrus of the rat hippocampus. Neurobiol Aging 32:811–820.

Suaud-Chagny MF, Cespuglio R, Rivot JP, Buda M, Gonon F (1993) High sensitivity measurement of brain catechols and indoles *in vivo* using electrochemically treated carbon-fiber electrodes. J Neurosci Methods 48:241–250.

Szarowski DH, Andersen MD, Retterer S, Spence AJ, Isaacson M, Craighead HG, Turner JN, Shain W (2003) Brain responses to micro-machined silicon devices. Brain Res 983:23–35.

Talauliker PM (2010) Characterization and optimization of microelectrode arrays for glutamate measurements in the rat hippocampus. In: Anatomy & Neurobiology. Lexington, Kentucky USA: Kentucky.

Talauliker PM, Price DA, Burmeister JJ, Nagari S, Quintero JE, Pomerleau F, Huettl P, Hastings JT, Gerhardt GA (2011) Ceramic-based microelectrode arrays: recording surface characteristics and topographical analysis. J Neurosci Methods 198:222–229.

Thomas TC, Grandy DK, Gerhardt GA, Glaser PE (2009) Decreased dopamine D4 receptor expression increases extracellular glutamate and alters its regulation in mouse striatum. Neuropsychopharmacology 34:436–445.

Thomas TC, Hinzman JM, Gerhardt GA, Lifshitz J (2012) Hypersensitive glutamate signaling correlates with the development of late-onset behavioral morbidity in diffuse brain-injured circuitry. J Neurotrauma 29:187–200.

Tian FM, Gourine AV, Huckstepp RTR, Dale N (2009) A microelectrode biosensor for real-time monitoring of L-glutamate release. Anal Chim Acta 645:86–91.

Timmerman W, Westerink BH (1997) Brain microdialysis of GABA and glutamate: what does it signify? Synapse 27:242–261.

Timmerman W, Cisci G, Nap A, de Vries JB, Westerink BH (1999) Effects of handling on extracellular levels of glutamate and other amino acids in various areas of the brain measured by microdialysis. Brain Res 833:150–160.

Tucci S, Rada P, Sepulveda MJ, Hernandez L (1997) Glutamate measured by 6-seconds resolution brain microdialysis: capillary electrophoretic and laser-induced fluorescence detection application. J Chromatogr B Biomed Sci Appl 694:343–349.

Turner JA, Lee JS, Martinez O, Medlin AL, Schandler SL, Cohen MJ (2001) Somatotopy of the motor cortex after long-term spinal cord injury or amputation. IEEE Trans Neural Syst Rehabil Eng 9:154–160.

Turner JN, Shain W, Szarowski DH, Andersen M, Martins S, Isaacson M, Craighead H (1999) Cerebral astrocyte response to micromachined silicon implants. Exp Neurol 156:33–49.

Ueda Y, Tsuru N (1995) Simultaneous monitoring of the seizure-related changes in extracellular glutamate and gamma-aminobutyric acid concentration in bilateral hippocampi following development of amygdaloid kindling. Epilepsy Res 20:213–219.

Vasylieva N, Maucler C, Meiller A, Viscogliosi H, Lieutaud T, Barbier D, Marinesco S (2013) Immobilization method to preserve enzyme specificity in biosensors: consequences for brain glutamate detection. Anal Chem 85:2507–2515.

Wassum KM, Tolosa VM, Wang J, Walker E, Monbouquette HG, Maidment NT (2008) Silicon wafer-based platinum microelectrode array biosensor for near real-time measurement of glutamate *in vivo*. Sensors 8:5023–5036.

Watson CJ, Venton BJ, Kennedy RT (2006) *In vivo* measurements of neurotransmitters by microdialysis sampling. Anal Chem 78:1391–1399.

Westerink RH (2004) Exocytosis: using amperometry to study pre-synaptic mechanisms of neurotoxicity. Neurotoxicology 25:461–470.

Wightman RM, Strope E, Plotsky PM, Adams RN (1976) Monitoring of transmitter metabolites by voltammetry in cerebrospinal fluid following neural pathway stimulation. Nature 262:145–146.

Yang CS, Tsai PJ, Lin NN, Liu L, Kuo JS (1995) Elevated extracellular glutamate levels increased the formation of hydroxyl radical in the striatum of anesthetized rat. Free Radic Biol Med 19:453–459.

Zesiewicz T, Shaw J, Allison K, Staffetti J, Okun M, Sullivan K (2013) Update on treatment of essential tremor. Curr Treat Options Neurol 15:410–423.

Zhang H, Lin SC, Nicolelis MA (2009) Acquiring local field potential information from amperometric neurochemical recordings. J Neurosci Methods 179:191–200.

Zhou N, Rungta RL, Malik A, Han H, Wu DC, MacVicar BA (2013) Regenerative glutamate release by pre-synaptic NMDA receptors contributes to spreading depression. J Cereb Blood Flow Metab 33:1582–1594.

CHAPTER 6

ENZYME-BASED MICROBIOSENSORS FOR SELECTIVE QUANTIFICATION OF RAPID MOLECULAR FLUCTUATIONS IN BRAIN TISSUE

Leyda Z. Lugo-Morales and Leslie A. Sombers
North Carolina State University

6.1 INTRODUCTION

6.1.1 ELECTROCHEMISTRY IN NEUROSCIENCE

The ability to reliably monitor molecules in live brain tissue is critical to the field of neuroscience because brain function is governed by subsecond chemical neurotransmission events. This presents a significant analytical challenge — zeptomole quantities of neuroactive molecules must be detected in a chemically complex sample with minimal perturbation of the tissue. Thus, fundamental progress in neuroscience requires small, sensitive and selective analytical tools to monitor multiple molecules in real-time. Electrochemical techniques provide instantaneous information, and thus they are preferred for real-time *in vivo* measurements of electroactive substances with high spatiotemporal resolution (Michael and Borland, 2007). The original *in vivo* electrochemical experiments were conducted by Adams and colleagues (Kissinger et al., 1973). Today, some forty years later, electrochemistry is a benchmark tool in neuroscience. This progress was made possible by the miniaturization of standard electrodes to the micrometer scale, advances in the field of microelectronics, and the development of carbon-fiber microelectrodes by Wightman (Dayton et al., 1980; Dayton and Wightman, 1980) and others (Armstrong-James and Millar, 1979).

6.1.2 ELECTROCHEMICAL DETECTION OF NON-ELECTROACTIVE ANALYTES

A subset of neurotransmitters is electroactive, and thus these molecules can be easily detected at microelectrodes using electrochemistry. These include the catecholamines (dopamine, epinephrine, and norepinephrine) (Park et al., 2009; Herr et al., 2012; Park et al., 2012) and serotonin (Hashemi et al., 2009; Hashemi et al., 2012). Dopamine (DA) has been the focus of a considerable amount of research since its presence in the brain was confirmed (Carlsson and Waldeck, 1958; Carlsson, 1959). Over the past decade, significant advances in understanding the biological role and functional consequences associated with this electroactive neurochemical have been made using electrochemistry. It is highly implicated in the mediation of incentive salience (wanting) and reinforcement learning (Day et al., 2007; Gan et al., 2010; Flagel et al., 2011; Howe et al., 2013a). Additionally, significant advances have been made in understanding the dysfunction of DA neurons in diseases such as Parkinson's (Bergstrom and Garris, 2003; Lundblad et al., 2009; Bergstrom et al., 2011; Nevalainen et al., 2011) and depression (Hascup et al., 2011).

Despite an overwhelming number of publications focused on DA, its precise function in the brain is not well understood. An important point to consider is that neurotransmission in the brain involves a massive number of highly interconnected neuronal networks and modulatory systems. Other neurotransmitters in the brain such as acetylcholine (ACh), gamma-aminobutyric acid (GABA), and glutamate, are thought to play crucial roles in the modulation of DA dynamics (Johnson and North, 1992; Johnson et al., 1992; Chergui et al., 1993; Karreman et al., 1996; Overton and Clark, 1997; Zheng and Johnson, 2002; Maskos, 2008; Sombers et al., 2009; Parker et al., 2011; Barrot et al., 2012; Cachope et al., 2012; Exley et al., 2012; Threlfell et al., 2012; Wickham et al., 2013). Additionally, other equally important brain chemicals that are not neurotransmitters, such as glucose and lactate, play a critical role in brain energetics in both healthy and disease states (Pellerin and Magistretti, 1994; Dwyer, 2002; Hall et al., 2012). These molecules are non-electroactive and thus real-time electroanalysis is not straightforward. Biosensors are used to overcome this limitation by converting non-electroactive analyte into a secondary electroactive product that works as a reporter molecule.

6.1.3 ELECTROCHEMICAL BIOSENSORS

Since the first biosensor device made by Clark and Lyons (Clark and Lyons, 1962), a whole field has developed to enable the electrochemical detection of molecules that are non-electroactive in nature (Zhang, 2007). The fundamental principle common to all biosensors is the use of a biological recognition element that is able to selectively recognize the non-electroactive molecule of interest. Enzymes are proteins that are highly selective for a specific substrate. Thus they are commonly exploited as signal transducers in the fabrication of biosensors. The detection of the non-electroactive substance relies on the generation of an electroactive molecule during the enzymatic oxidation of the analyte. This electroactive molecule (usually hydrogen peroxide, H_2O_2) works as the reporter molecule. It can be electrochemically oxidized at an electrode surface, and the current generated can be used as an index to quantify the species of interest. This is only possible if the current measured for the reporter molecule is directly proportional to the amount of enzymatically oxidized non-electroactive analyte.

6.2 AMPEROMETRIC BIOSENSORS

Constant potential amperometry is a controlled potential technique in which a constant potential is applied to the working electrode (with respect to a reference electrode). The fixed potential is sufficient to oxidize or reduce all electroactive substances (including the analyte) that undergo an electron transfer reaction at the electrode surface. The high sampling rate used in constant-potential amperometry allows recordings to be made in the subsecond timescale. Most commercially available biosensors operate using this electrochemical strategy.

6.2.1 PLATINUM (Pt)-BASED BIOSENSORS

Pt is typically used as the electrode material of choice in the fabrication of amperometric biosensors, because it provides an electro-catalytic surface for the oxidation of H_2O_2 (Hall et al., 1998; Wilson and Johnson, 2008). Although all electrodes can exhibit biofouling when used *in vivo* due to non-specific adsorption of biological molecules to the surface, metal electrodes are particularly prone to this problem. This results in an unstable electrochemical background and negatively impacts electron transfer kinetics. Furthermore, amperometry cannot differentiate between multiple electroactive analytes contributing to the measured current. Thus, these electrodes are modified with chemically selective polymeric coatings to reduce fouling and facilitate analyte discrimination in tissue (Figure 6.1). This has a double purpose, serving to optimize the dynamic linear range for the analyte and to minimize the effects of oxygen variations on electrode response (Chen et al., 2002; Ward et al., 2002; Burmeister et al., 2005). However, electrode performance becomes dependent on coating integrity, and detection is slowed due to increased diffusion coefficients for analytes traveling through multiple coatings (Hu and Wilson, 1997; Chen et al., 2002; Hascup et al., 2007; Wilson and

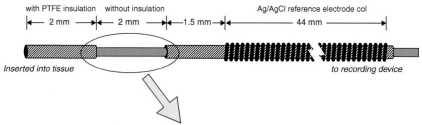
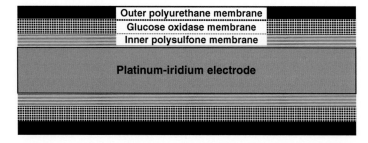

Figure 6.1 A schematic drawing of an implantable glucose sensor. The membrane coatings enable selective measurements and can also serve to restrict the diffusion of the analyte (enzyme substrate) to the electrode surface without limiting the diffusion of the enzyme co-substrate, often oxygen. Reproduced with permission from Ward et al., 2002. Copyright 2002 Elsevier.

Johnson, 2008). Sensitivity is also decreased in the presence of a diffusional barrier (Chen et al., 2002; Wilson and Johnson, 2008).

6.2.2 METHODS OF ENZYME IMMOBILIZATION

Enzyme immobilization methods for the fabrication of biosensors have been broadly classified into the following categories (Buerk, 1993; Pantano and Kuhr, 1995; Mulchandani and Rogers, 1998; Wilson and Gifford, 2005; Brena and Batista-Viera, 2006):

6.2.2.1 Encapsulation or entrapment

A polymer or membrane film can be employed to contain the enzyme near the electrode surface, minimizing its diffusion. The selected polymer must be permeable to the substrate of interest as well as the product of the enzymatic reaction, allowing for diffusion without significantly compromising the response time of the sensor. Redox polymers (Mitala and Michael, 2006), electronically conductive polymers (Gerard et al., 2002; Amatore et al., 2006; Ren et al., 2006), ion exchange polymers (Mizutani et al., 1995), and biopolymers (Yi et al., 2005) can all be used for the purpose of enzyme immobilization, depending on the enzyme and application. This immobilization strategy is ideal to minimize enzyme denaturation (breakdown of the enzyme's three-dimensional structure).

6.2.2.2 Adsorption

This is the simplest of the immobilization methods because the enzyme is directly adsorbed onto the electrode surface without the use of coupling reagents. Despite this, adsorption is considered to be the least adequate method for immobilization because the lifetime of the biosensor is abbreviated in comparison to biosensors prepared by other techniques. This is due to the very weak non-covalent interactions between the enzyme and the support (Pantano and Kuhr, 1995; Mulchandani and Rogers, 1998).

6.2.2.3 Covalent attachment and cross-linking

These two methods can be employed alone or in conjunction with other enzyme immobilization methods. They consist of using chemical reagents to bind the enzyme to the sensing substrate, or to form covalent bonds between enzyme molecules, creating a network.

6.2.3 SELF-REFERENCING AMPEROMETRIC BIOSENSORS

A common approach to address the non-selective nature of amperometric recordings is to incorporate a control recording site into the sensor design (see Figure 6.2). In one design, a non-enzyme-modified "sentinel" electrode is located 100–300 μm away from the enzyme-coated electrode (Parikh et al., 2004; Burmeister et al., 2005). The sentinel electrode is used to verify the signal in a self-referencing subtraction paradigm that is the electrochemical equivalent of a double beam experiment in spectroscopy (Burmeister and Gerhardt, 2001; Burmeister et al., 2002; Hascup et al., 2007). However, this paradigm can introduce inaccuracies because individual electrodes exhibit varying sensitivities for different chemical agents, and an unanticipated and significant chemical heterogeneity exists at brain recording sites separated by 75 μm (Wightman et al., 2007). Another drawback is crosstalk. This can be electrical or diffusional, occurring between the recording sites, precluding their simultaneous use *in vivo*. Thus, the enzyme-coated and sentinel electrodes are often used on separate days (Kiyatkin and Lenoir, 2012) or in different groups of animals (Kiyatkin et al., 2013), introducing significant variability. Nonetheless, this technology has been successfully used to evaluate how transient changes in localized neurochemical concentration underlie the mediation of specific cognitive operations, decision-making processes, and disease states (*vide infra*).

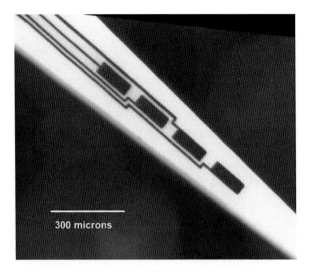

Figure 6.2 A ceramic-based multisite microelectrode for selective measurements of non-electroactive analytes. These electrodes are designed to eliminate non-specific signal using a self-referencing subtraction paradigm. Individual sensors are modified with the appropriate enzyme and layers of exclusion polymers. Sentinels (enzyme-null electrodes) are fabricated in an identical fashion, but lacking the enzyme. The difference in signal reflects changes in the level of the analyte of interest. Reproduced with permission from Burmeister et al., 2005. Copyright 2004 Elsevier.

6.2.4 APPLICATIONS

6.2.4.1 Acetylcholine

Choline is generated in the brain when newly released ACh is broken down in the extracellular space by the endogenous enzyme, acetylcholinesterase. Thus, choline-sensitive microelectrodes that use changes in choline as an index to measure neuronal ACh release have been developed by several groups. Michael and colleagues developed such a microsensor by immobilizing horseradish peroxidase, choline oxidase, and a cross-linkable redox polymer poly(vinylpyridine) with pendant osmium-based redox groups onto a carbon-fiber electrode, and coating with a Nafion® overlayer (Garguilo and Michael, 1994). Amperometry was used to reliably detect choline, even in the presence of ascorbate, with a detection limit of ~10 μM choline. The sensor was used to investigate the rate of ACh clearance from the extracellular fluid through cholinesterase activity (Garguilo and Michael, 1996). However, it is important to note that choline is not generated solely by the enzymatic breakdown of ACh. Choline is also a precursor to ACh, and it is used in other biological processes such as membrane formation. Thus, Gerhardt and colleagues have developed ceramic-based multisite microelectrode arrays that utilize a dual-enzyme strategy that is sensitive to both ACh and choline, to account

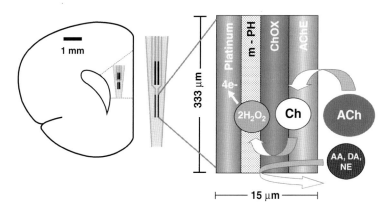

Figure 6.3 Microelectrode arrays that use a dual-enzyme detection strategy can be used to selectively detect the rapid release of ACh. In this design, sets of electrodes were coated with either choline oxidase (choline-sensitive sites) or both choline oxidase and acetylcholinesterase (ACh/choline-sensitive sites). Here, the coatings on the sites sensitive to both analytes are shown. The two enzymes are covalently bound to the electrode surface. ACh is enzymatically hydrolyzed to choline. The subsequent oxidation of choline generates H_2O_2, which diffuses through the m-polyphenylene diamine (mPD) exclusion layer, and is finally oxidized at the Pt electrode surface to generate a current. Reproduced with permission from Bruno et al., 2006. Copyright 2004 John Wiley and Sons.

for the signal from these other endogenous choline sources (Bruno et al., 2006). These electrodes work in a similar fashion to the self-referencing technique described above; however, sets of recording sites are coated with enzyme. One set of electrodes is sensitive to choline only. These are coated with choline oxidase and other immobilization components. A second set of recording sites is coated with both choline oxidase and acetylcholinesterase (as well as the other immobilization components). Thus, this set is sensitive to both choline and ACh (Figure 6.3). The signal collected from the choline-sensitive site is subtracted from that collected at the ACh/choline-sensitive site, to generate the signal that corresponds to the rapid release of ACh. These sensors have been used to measure ACh release in the prefrontal cortex of anesthetized rats following local depolarization with high K^+, or with infusion of scopolamine, demonstrating the potential of this approach for direct measurements of cholinergic transmission *in vivo* (Bruno et al., 2006).

The cholinergic projections to the cortex have traditionally been described as a relatively slow-acting neuromodulator system. In the first reported use of ceramic-based multisite microelectrode arrays in awake and freely-moving animals, Sarter and colleagues challenged this description by electrochemically investigating cholinergic activity in rats performing cued appetitive response tasks (Parikh et al., 2007). The work measured transient increases in choline (on the scale of seconds) that resulted from the hydrolysis of newly released ACh, through the use of amperometry at Pt recording sites coated with Nafion® and cross-linked choline oxidase. These increases in cholinergic concentration were seen in the medial prefrontal cortex when the animal detected cues predicting subsequent reward delivery that evoked an attentional shift to the reward ports. The data reveal that the ACh transients mediate a cognitive operation, rather than simply indicating sensory processing (of the cue). Port approach, reward delivery, and retrieval did not confound the signals. The transient increases in cholinergic concentration were superimposed over more slowly changing (on the scale of minutes) levels of cholinergic concentration.

In a later work, additional measurements of transient ACh fluctuations were carried out in the thalamic input layer of the right prelimbic cortex of rats performing a sustained attention task (Howe et al., 2013b). Cholinergic transients peaked at ~250 nM above baseline, and occurred only during specific trial sequences that required a shift from a state of perceptual attention (monitoring for cues) to the generation of a cue-evoked response. These data further delineate the specific role of the cortical cholinergic input system in integrating external cues with internal representations to initiate and guide behavior.

6.2.4.2 Glutamate

Electrophysiological techniques have been used extensively to characterize the effects of excitatory signaling in the brain. Recently, however, several groups have taken a major step forward by directly evaluating glutamatergic transmission. Kehr and colleagues have investigated differences in excitatory signaling between genetic rodent models of depression using the self-referencing, ceramic-based microelectrode array technology in Flinders sensitive line and control Flinders resistant line rats (Hascup et al., 2011). A current hypothesis proposes that excessive glutamatergic neurotransmission contributes to depression. This study measured rapid glutamate transients (on the time scale of tens of seconds) (see Figure 6.4) in the prefrontal cortex of awake rats, and found that the average transient duration and amplitude were increased in 12 to 15 months old Flinders sensitive line rats (~2 μM glutamate, released over ~35 seconds) compared with age-matched controls (~0.5 μM glutamate, released over ~15 seconds); however, these differences were not observed in younger animals. The data suggest that this time point could mark a critical transition period in this animal model.

In another ground breaking work, Maidment and colleagues have used amperometric Nafion®/polypyrrole/glutamate oxidase-modified Pt biosensors in a self-referencing, microelectrode array to shed light on the role glutamatergic signaling plays during decision-making processes that guide self-initiated action sequences (Wassum et al., 2012). Glutamate transients were recorded in the basolateral amygdala of rats performing a self-paced sequence task for sucrose reward. In this

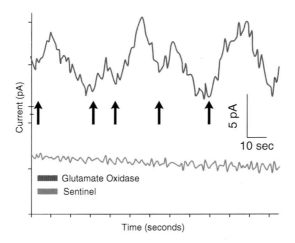

Figure 6.4 Rapid glutamate transients. An example amperometric trace reflecting spontaneous glutamate signals collected in the prefrontal cortex of a 12 to 15 months old Flinders sensitive line rat. Reproduced with permission from Hascup et al., 2011. Copyright Macmillan Publishers Ltd.

study, the current output from each electrode in the array was subtracted from the baseline current for that electrode. The current change from baseline on the Nafion®/polypyrrole-coated electrode was then subtracted from the current change from baseline collected on the Nafion®/polypyrrole/glutamate oxidase-coated sensors to obtain a low-noise change-from-baseline glutamate current signal. The majority of these transients were ~2–5 seconds in duration with an average rise time that closely matched the response time of the sensors *in vitro* (~0.8 seconds). Thus, the extracellular glutamate concentration may be changing more rapidly than these measurements suggest. The transients were action potential dependent and appeared to originate from neuronal projections from the orbitofrontal cortex. Interestingly, they were time-locked to lever-pressing actions and varied according to task engagement. They were considerably faster than those recorded in the prefrontal cortex using similar sensors (Figure 6.4), which were not correlated with any behavioral event (Hascup et al., 2011).

The behavioral paradigm required rats to perform a fixed sequence of two different lever press actions to earn sucrose pellets (Wassum et al., 2012). The first lever was continuously available. When pressed, a second lever was inserted into the chamber. Pressing it resulted in sucrose delivery and retraction of the second lever. Electrochemical data was processed as described above, and the resulting signals collected in the 10 seconds before and after a lever press event were averaged across trials for each rat, and then across rats. Glutamate transients (the largest were ~6–7 μM) occurred more frequently when animals were engaged in the lever-pressing task, and both transient amplitude and frequency were positively correlated with press rate. These data directly demonstrate that glutamate transients in the basolateral amygdala are related to the decision-making processes that occur before these lever press actions. However, it is important to note that glutamate transients tended to precede pressing at either lever, and they were also prevalent at the conclusion of the action sequence, during reward consumption. Thus, transient glutamate activity was not solely related to the initiating decision. When the value of the sucrose reward was reduced by satiety, lever-pressing activity was reduced and a decrease in task-related glutamate transients was recorded.

6.2.4.3 Glucose

Glucose sensors have a significant presence on the global biosensor market, due in part to the prevalence of diabetes and its projected increase in the coming years. Despite this, significant gaps exist in our understanding of how brain energy utilization is coupled to neural function. Kiyatkin and colleagues have begun to directly investigate this in the nucleus accumbens and substantia nigra pars reticulata of awake rats using commercially available Nafion®/polyurethane/glucose oxidase-modified platinum/iridium (Pt/Ir) biosensors coupled to amperometry (Kiyatkin and Lenoir, 2012). This work evaluates local fluctuations in extracellular glucose levels induced by various stimuli (audio, tail pinch, social interaction, and intravenous cocaine). Electrochemical data were sampled at one Hz (mean current over one second), and averaged in 60 seconds (slow) or four seconds (rapid) bins. To exclude non-specific contributions to the signal, sensors that were not coated were also used. These were fabricated in a manner identical to the glucose sensors, except for the absence of glucose oxidase (GOx) enzyme. Mean changes in current recorded at glucose-null sensors were subtracted from the mean changes in current recorded at the glucose sensors to generate a current differential. The individual measurements used to generate the differential were performed in separate animals, or in single animals on multiple test days.

The authors estimate the basal levels of glucose to be ~540 μM in the nucleus accumbens, and ~407 μM in the substantia nigra pars reticulata. However, these may be underestimates because the multiple layers on the sensor limit glucose diffusion to the electrode surface. Glucose levels fluctuated in response to stimuli or cocaine administration, with latencies on the seconds scale. Overall, the changes were relatively small in magnitude (5–10% of baseline glucose concentrations). However, the magnitude and duration were variable, and dependent on both brain region and the nature of the stimulus. This finding is consistent with the appropriate distribution of available energetic resources to areas with differing metabolic requirements, to most effectively fuel neuronal responses. The short latency (seconds) to glucose signal onset may be surprising because time is required for vascular transit and diffusion into tissue; however, this time scale is in agreement with the latency for a local increase in oxygen after electrical stimulation (Devor et al., 2011). This suggests that the underlying vascular machinery is optimized to quickly deliver resources in response to local energy demands. Finally, the authors also found that extracellular glucose levels were relatively resistant to large, passive increases in blood glucose levels, but strongly increased during general anesthesia. Indeed, it is important to note that general anesthesia reduces brain activity, and has been reported to generate a hyperglycemic state (Maggi and Meli, 1986; Sanchez-Pozo et al., 1988).

6.3 VOLTAMMETRIC BIOSENSORS

Fast-scan cyclic voltammetry (FSCV) is an alternate electrochemical approach that has become a benchmark tool for *in vivo* measurements of electrochemically active analytes in the brain, because it combines high sensitivity and millisecond temporal resolution with chemical discrimination of multiple analytes (Robinson et al., 2008; Roberts et al., 2013). In contrast to amperometry, which operates at a fixed potential, voltammetric current is collected as the electrode potential is periodically scanned at >100 V/sec across an appropriate potential window and back. The effective sampling rate is 10 Hz, as a complete cyclic voltammogram consisting of ~1000 data points is collected every 100 milliseconds. Each voltammogram provides an electrochemical signature, allowing individual analytes to be simultaneously quantified and distinguished from one another, not because of an interference excluding membrane, but because the analytes have unique electrochemistry that can be differentiated from the electrochemistry of the interferents. Combining the benefits of FSCV with enzyme-modified carbon-fiber microelectrodes is a novel approach that eliminates the requirement to subtract non-specific measurements collected at separate enzyme-null electrodes (spatial averaging), as well as the use of polymeric exclusion layers that slow electrode response time. This allows for the selective detection of non-electroactive species with optimal spatial and temporal resolution.

6.3.1 CARBON-FIBER MICROELECTRODES

Carbon materials are a preferred sensing substrate for *in vivo* applications due to their wide potential window, biocompatibility, inert nature, and low cost (Netchiporouk et al., 1995; Huffman and Venton, 2009; Mauko et al., 2009). Most importantly, carbon resists biofouling, and offers a surface that can be renewed with each successive voltammetric scan (Takmakov et al., 2010). Additionally, the carbon fiber surface is rich in oxygen containing functional groups which play a significant role in facilitating electron-transfer reactions (Pantano and Kuhr, 1991;

McCreery, 2008; Roberts et al., 2010; Takmakov et al., 2010). This provides a means by which chemical or electrochemical modification of the carbon surface can be used to enhance the sensitivity of detection (Hopper and Kuhr, 1994; Roberts et al., 2010). Carbon-fiber microelectrodes are fabricated by aspirating a single fiber (generally between 5 and 30 μm in diameter) into a borosilicate capillary which serves as the insulating material. Using a micropipette puller, the capillary containing the fiber is heated and pulled, generating two glass insulated carbon-fiber microelectrodes. Finally, the fiber length at the sensing tip is adjusted to about 100 μm, using a blade and optical microscope.

6.3.2 ENZYME ENCAPSULATION WITHIN CHITOSAN BY ELECTRODEPOSITION

Chitosan is a natural polyaminosaccharide derived from chitin, which is a major constituent of crustacean shells, exoskeletons of insects, and the cell wall of fungi. The chitosan biopolymer is a cationic polyelectrolyte with a pKa of ~6.5, obtained by partial deacetylation of chitin to create a material with a variety of molecular weights and chemical characteristics (Krajewska, 2004; Yi et al., 2005). It is non-toxic, physiologically inert, and has a strong affinity for proteins, facilitating enzyme encapsulation (Muzzarelli et al., 1976). Most importantly, the solubility of chitosan can be controlled by electrochemically manipulating the pH of the solution (Krajewska, 2004).

Encapsulation of an enzyme at an electrode within chitosan polymer can be achieved by way of electrodeposition (Yi et al., 2005; Zangmeister et al., 2006). In this process, an electrochemically conditioned electrode is dipped in an aqueous solution of chitosan and enzyme and connected to a DC power supply. When the cathodic potential is sufficient to reduce the protons (H^+) in solution to hydrogen gas (H_2), a shift in pH is induced in close proximity to the electrode surface. This causes the polymer to locally deprotonate and deposit as a water-insoluble film on the electrode surface, thereby encapsulating enzyme molecules.

6.3.3 ENZYME-MODIFIED CARBON-FIBER MICROELECTRODES FOR THE DETECTION OF NON-ELECTROACTIVE ANALYTES USING FSCV

We have used the electrodeposition of chitosan to confine GOx at the surface of carbon-fiber microelectrodes, creating a simple, biocompatible, and inexpensive microbiosensor (Figure 6.5) (Lugo-Morales et al., 2013). GOx is a well-characterized oxidoreductase enzyme that produces H_2O_2 in the presence of its substrate and co-substrate, glucose and oxygen, respectively. Our group recently developed

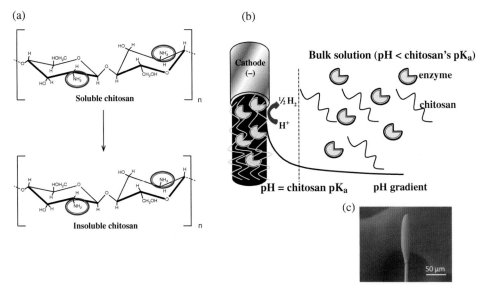

Figure 6.5 Carbon-fiber microbiosensors. (a) A chitosan polymer is used to entrap GOx at the electrode surface. Chitosan solubility is dependent on pH. (b) Schematic illustrating the electrodeposition of chitosan. (c) Scanning electron micrograph of a carbon-fiber microelectrode with an electro-deposited chitosan membrane encapsulating GOx.

and characterized a fast-scan voltammetric detection scheme for H_2O_2 (Sanford et al., 2010), and we have used it at these novel microbiosensors to detect enzymatically-generated H_2O_2 with electrochemical selectivity. This approach precludes the requirement for polymeric coatings and/or the use of a null electrode to afford selective measurements, as H_2O_2 is distinguished by its characteristic cyclic voltammogram, even in the presence of other analytes or interferents. These novel microbiosensors have been fully characterized *in vitro* and used to quantitatively and selectively monitor glucose fluctuations in live brain tissue with unprecedented chemical and spatial resolution (Lugo-Morales et al., 2013). Furthermore, this novel biosensing strategy is widely applicable to the immobilization of any H_2O_2 producing enzyme, enabling rapid monitoring of many non-electroactive enzyme substrates.

In vitro characterization was carried out in a flow-injection apparatus, in which a bolus of glucose was reproducibly introduced to the electrode surface (Figure 6.6) (Lugo-Morales et al., 2013). The background-subtracted voltammograms show a single peak at ~ + 1.2 V (*versus* Ag/AgCl, using a four-pole Bessel filter, 2.5 KHz), indicative of the voltammetric signature for H_2O_2 (Sanford et al., 2010). Physiological glucose concentrations generated a linear response with a limit of detection of ~13 μM and sensitivity of ~19 nA/mM glucose. Enzyme modification did not slow detection, as the subsecond response time to a supra-physiological

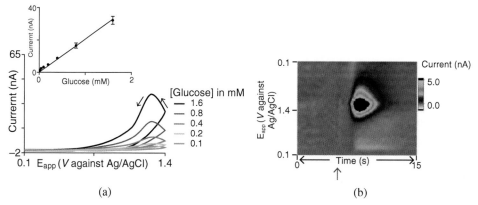

Figure 6.6 Voltammetric detection of glucose at GOx/chitosan-modified microelectrodes. (a) Background-subtracted voltammograms for H_2O_2 enzymatically generated in response to glucose. Inset, linear response to physiological glucose concentrations. (b) Representative color plot comprised of 150 voltammograms collected over 15 seconds. The current (color scale) generated in response to glucose (arrow) is plotted with respect to the applied potential (V, y-axis) and collection time (s, x-axis). Reproduced with permission from Lugo-Morales et al., 2013. Copyright 2013 American Chemical Society.

concentration of H_2O_2 (50 μM) was not significantly different between enzyme-modified and bare microelectrodes. As a point of comparison, amperometric studies have been performed *in vivo* using commercially-produced GOx/polymer-coated Pt/Ir microelectrodes with response times for glucose detection ranging from ~5–10 seconds (Gifford et al., 2005; Kiyatkin and Lenoir, 2012; Finnerty et al., 2013; Kiyatkin et al., 2013). In our work the chitosan thickness was estimated to be ~5.7 μm using a previously described model (Kawagoe and Wightman, 1994), resulting in an overall microbiosensor diameter of ~20 μm. Additionally, these sensors are stable over time and selective. Performance was not significantly affected by fluctuations of oxygen concentration in the physiological range despite the lack of an outer diffusional barrier to limit glucose flux.

The advantages of the voltammetric approach are readily evident when comparing the performance of our GOx/chitosan-modified carbon-fiber microelectrodes using FSCV and conventional amperometry. Figure 6.7 presents voltammetric data collected in response to a 1 mM sample of glucose (a) followed by a mixture (b) of 1 mM glucose and 250 μM ascorbic acid (AA), a principal electroactive interferent. The data are displayed in color plots, each containing 300 background-subtracted voltammograms collected over 30 seconds. The ordinate is applied potential, the abscissa is time, and the current (nA) is displayed in false color. Enzymatically generated H_2O_2 was detected in response to the glucose sample, as evidenced by the current collected at ~ + 1.2 V (Figure 6.7a, top, arrow). The concentration *versus* time trace extracted from the data is also presented (bottom).

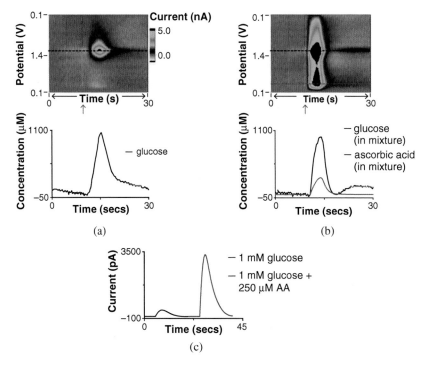

Figure 6.7 Comparing the performance of the GOx/chitosan-modified microelectrodes using FSCV and conventional amperometry. (a, b) Voltammetric data. Representative color plots (top panels) and respective analyte concentration *versus* time traces (bottom panel) collected in response to a bolus injection (arrow) of (a) 1 mM glucose (black) or (b) a sample containing 1 mM glucose (black) in the presence of AA (250 μM, blue). Concentration *versus* time traces for both analytes were extracted from the data using principal component regression. Glucose and AA were readily distinguished and simultaneously quantified at the single recording site. (c) Amperometric data. Our sensors are different from conventional biosensors in that they rely on voltammetry for selectivity, and do not incorporate separate measures to ensure the exclusion of non-specific signal. Thus, when the same solutions were interrogated with the microbiosensor held at a constant-potential of + 1.0 V, the two analytes could not be distinguished and the current from the oxidation of AA was a contributor to the signal (pink). Reproduced with permission from Lugo-Morales et al., 2013. Copyright 2013 American Chemical Society.

In contrast, the mixture generated anodic current at two distinct potentials, as shown in Figure 6.7b (top, arrow). These currents are due to the oxidation of H_2O_2 (~ + 1.2 V) and AA (~ + 0.4 V). AA does not interfere with the detection of glucose, despite the lack of polymeric membranes to promote chemical selectivity, and these analytes are simultaneously quantifiable at the single recording site. Conventional amperometric detection was used to detect the same samples with these sensors, and these data are presented in Figure 6.7c. A color plot is not presented, as all data were collected at + 1.0 V. The current collected in response to the glucose sample is shown in black, and that collected in response to the mixture

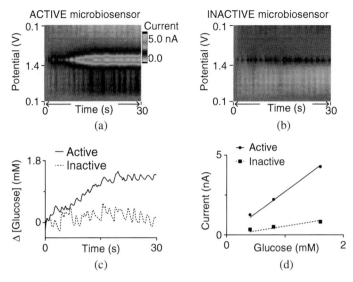

Figure 6.8 Rapid glucose fluctuations recorded in the CPu. Representative voltammetric recordings simultaneously obtained at an active (a) and at a heat-inactivated (b) carbon-fiber microbiosensor in response to an intravenous infusion of glucose. (c) Glucose concentration *versus* time traces extracted from the voltammetric data. (d) Post-calibration demonstrated that the active biosensor was significantly more sensitive to glucose than the inactivated electrode, which demonstrated some residual enzymatic activity. Reproduced with permission from Lugo-Morales et al., 2013. Copyright 2013 American Chemical Society.

is shown in pink. When responding to the sample containing both glucose and AA, current was generated by both analytes, because these sensors are different from conventional biosensors in that they do not incorporate separate measures to ensure the exclusion of non-specific signal. These data clearly demonstrate how the electrochemical selectivity inherent to the voltammetric approach enables multiple analytes to be simultaneously detected and quantified at a single microelectrode, without requiring polymeric coatings and/or self-referencing schemes to exclude interferents. If the additional analytes (or interferents) are electroactive and have electrochemistry that is different from that of H_2O_2, then chemical resolution is possible.

6.3.4 HIGH-RESOLUTION SAMPLING OF BRAIN GLUCOSE CONCENTRATION FLUCTUATIONS

The voltammetric strategy has enabled us to overcome the limitations typically associated with enzyme-modified electrodes, and to accomplish high-resolution sampling of glucose fluctuations at discrete locations in the dorsal striatum. Figure 6.8 shows example voltammetric recordings simultaneously collected at a

GOx/chitosan-modified electrode implanted in the caudate putamen (CPu) (Figure 6.8a), and at a heat-inactivated, but otherwise identical microbiosensor (Figure 6.8b) implanted in the contralateral CPu in an anesthetized rat. The color plots each contain 300 background-subtracted voltammograms recorded over 30 second epochs. Infusion of 30–50% glucose (120–300 mg in 0.3–1.0 mL saline over 30 seconds, intravenous) caused a significant increase in extracellular brain glucose concentration that rose over approximately 10 seconds and was only detected using the active microbiosensor (Lugo-Morales et al., 2013). This slow rise in extracellular glucose concentration measured 388 ± 168 μM, and was significantly elevated when compared to data collected following saline infusion (n = 5, t = 2.878, *P = 0.0451). Close examination of the gradual rise in glucose concentration also revealed superimposed, rapid glucose fluctuations. The dynamics of these rapid events did not significantly differ from those measured after systemic saline injection. Voltammograms extracted from these *in vivo* data statistically correlate with those collected *in vitro* during post-calibration. A small intermittent signal was collected on the heat-inactivated microelectrode. Post-calibration indicated that this electrode possessed some residual enzymatic activity; however, we cannot rule out the possibility that the residual signal is due to endogenous H_2O_2 signaling in this brain region (Spanos et al., 2013).

6.4 CONCLUSIONS

The brain is a cacophony of chemical signaling that to date has remained largely unexplored. Advances in microbiosensor technology will enable investigations on how these individual voices underlie various aspects of complex brain function. We have reviewed several examples of researchers taking important steps toward this goal, but much remains to be done. Self-referencing amperometric biosensors have enabled Sarter and colleagues to begin to tease out the role that ACh plays in attention (Parikh et al., 2007; Howe et al., 2013b); however, this molecule plays a critical role in many brain nuclei. Indeed, there is significant evidence to suggest that ACh modulates the mesolimbic DA system (Maskos, 2008; Cachope et al., 2012; Exley et al., 2012; Threlfell et al., 2012; Wickham et al., 2013). Similarly, researchers have recently elucidated rapid glutamate transients in the rat prefrontal cortex (Hascup et al., 2011) and basolateral amygdala (Wassum et al., 2012), investigating their role in depression and motivated behavior, respectively. However, glutamate is a ubiquitously present neurotransmitter that is central to innumerable brain functions, and many questions remain unanswered. As a final example, ~25% of the glucose used by the body is consumed by brain functions; however, the detailed molecular mechanisms by which the brain utilizes glucose are still not completely understood (Belanger et al.,

2011). Kiyatkin et al. have evaluated local fluctuations in extracellular glucose levels induced by various stimuli (Kiyatkin and Lenoir, 2012), and our research group has accomplished high-resolution sampling of subsecond glucose fluctuations at discrete locations in the dorsal striatum (Lugo-Morales et al., 2013). However, studies investigating the mechanisms linking glucose dynamics to neural activity, or linking impaired glucose utilization to the onset of neuropathies are badly needed. Continued advances in microsensor technology will help to advance these and innumerable other studies that are fundamentally important to a variety of fields.

ACKNOWLEDGMENTS

Research in the Sombers laboratory on this topic has been carried out with grants from the National Institutes of Health, the National Science Foundation, and NCSU Department of Chemistry. In addition, we gratefully acknowledge our coworkers, past and present, for the studies cited in this review.

REFERENCES

Amatore C, Arbault S, Bouton C, Coffi K, Drapier JC, Ghandour H, Tong Y (2006) Monitoring in real-time with a microelectrode the release of reactive oxygen and nitrogen species by a single macrophage stimulated by its membrane mechanical depolarization. Chembiochem 7:653–661.

Armstrong-James M, Millar J (1979) Carbon-fiber microelectrodes. J Neurosci Methods 1:279–287.

Barrot M, Sesack SR, Georges F, Pistis M, Hong S, Jhou TC (2012) Braking dopamine systems: a new GABA master structure for mesolimbic and nigrostriatal functions. J Neurosci 32:14094–14101.

Belanger M, Allaman I, Magistretti PJ (2011) Brain energy metabolism: focus on astrocyte-neuron metabolic cooperation. Cell Metab 14:724–738.

Bergstrom BP, Garris PA (2003) "Passive stabilization" of striatal extracellular dopamine across the lesion spectrum encompassing the presymptomatic phase of Parkinson's disease: a voltammetric study in the 6-OHDA-lesioned rat. J Neurochem 87:1224–1236.

Bergstrom BP, Sanberg SG, Andersson M, Mithyantha J, Carroll FI, Garris PA (2011) Functional reorganization of the presynaptic dopaminergic terminal in parkinsonism. Neuroscience 193:310–322.

Brena B, Batista-Viera F (2006) Immobilization of enzymes. In: Immobilization of enzymes and cells (Guisan J, ed), pp 15–30. New Jersey: Humana Press.

Bruno JP, Gash C, Martin B, Zmarowski A, Pomerleau F, Burmeister J, Huettl P, Gerhardt GA (2006) Second-by-second measurement of acetylcholine release in prefrontal cortex. Eur J Neurosci 24:2749–2757.

Buerk DG (1993) Biosensors: theory and applications. Lancaster, PA: Technomic Publishing Company.

Burmeister JJ, Gerhardt GA (2001) Self-referencing ceramic-based multisite microelectrodes for the detection and elimination of interferences from the measurement of L-glutamate and other analytes. Anal Chem 73:1037–1042.

Burmeister JJ, Palmer M, Gerhardt GA (2005) L-lactate measures in brain tissue with ceramic-based multisite microelectrodes. Biosens Bioelectron 20:1772–1779.

Burmeister JJ, Pomerleau F, Palmer M, Day BK, Huettl P, Gerhardt GA (2002) Improved ceramic-based multisite microelectrode for rapid measurements of L-glutamate in the CNS. J Neurosci Methods 119:163–171.

Cachope R, Mateo Y, Mathur BN, Irving J, Wang HL, Morales M, Lovinger DM, Cheer JF (2012) Selective activation of cholinergic interneurons enhances accumbal phasic dopamine release: setting the tone for reward processing. Cell Rep 2:33–41.

Carlsson A (1959) Detection and assay of dopamine. Pharmacol Rev 11:300–304.

Carlsson A, Waldeck B (1958) A fluorimetric method for the determination of dopamine (3-hydroxytyramine). Acta Physiol Scand 44:293–298.

Chen XH, Matsumoto N, Hu YB, Wilson GS (2002) Electrochemically mediated electrodeposition/electropolymerization to yield a glucose microbiosensor with improved characteristics. Anal Chem 74:368–372.

Chergui K, Charlety PJ, Akaoka H, Saunier CF, Brunet JL, Buda M, Svensson TH, Chouvet G (1993) Tonic activation of NMDA receptors causes spontaneous burst discharge of rat midbrain dopamine neurons *in vivo*. Eur J Neurosci 5:137–144.

Clark LC Jr, Lyons C (1962) Electrode systems for continuous monitoring in cardiovascular surgery. Ann N Y Acad Sci 102:29–45.

Day JJ, Roitman MF, Wightman RM, Carelli RM (2007) Associative learning mediates dynamic shifts in dopamine signaling in the nucleus accumbens. Nat Neurosci 10:1020–1028.

Dayton MA, Wightman RM (1980) Carbon-fiber microelectrodes as *in vivo* voltammetric probes of endogenous chemical transmitter concentration. Clin Res 28:A798.

Dayton MA, Brown JC, Stutts KJ, Wightman RM (1980) Faradaic electrochemistry at micro-voltammetric electrodes. Anal Chem 52:946–950.

Devor A, Sakadzic S, Saisan PA, Yaseen MA, Roussakis E, Srinivasan VJ, Vinogradov SA, Rosen BR, Buxton RB, Dale AM, Boas DA (2011) "Overshoot" of O_2 is required to maintain baseline tissue oxygenation at locations distal to blood vessels. J Neurosci 31:13676–13681.

Dwyer DS, ed (2002) Glucose metabolism in the brain. San Diego: Elsevier Science.

Exley R, McIntosh JM, Marks MJ, Maskos U, Cragg SJ (2012) Striatal alpha 5 nicotinic receptor subunit regulates dopamine transmission in dorsal striatum. J Neurosci 32:2352–2356.

Finnerty NJ, Bolger FB, Palsson E, Lowry JP (2013) An investigation of hypofrontality in an animal model of schizophrenia using real-time microelectrochemical sensors for glucose, oxygen, and nitric oxide. ACS Chem Neurosci 4:825–831.

Flagel SB, Clark JJ, Robinson TE, Mayo L, Czuj A, Willuhn I, Akers CA, Clinton SM, Phillips PE, Akil H (2011) A selective role for dopamine in stimulus-reward learning. Nature 469:53–57.

Gan JO, Walton ME, Phillips PE (2010) Dissociable cost and benefit encoding of future rewards by mesolimbic dopamine. Nat Neurosci 13:25–27.

Garguilo MG, Michael AC (1994) Quantitation of choline in the extracellular fluid of brain tissue with amperometric microsensors. Anal Chem 66:2621–2629.

Garguilo MG, Michael AC (1996) Amperometric microsensors for monitoring choline in the extracellular fluid of brain. J Neurosci Methods 70:73–82.

Gerard M, Chaubey A, Malhotra BD (2002) Application of conducting polymers to biosensors. Biosens Bioelectron 17:345–359.

Gifford R, Batchelor MM, Lee Y, Gokulrangan G, Meyerhoff ME, Wilson GS (2005) Mediation of *in vivo* glucose sensor inflammatory response *via* nitric oxide release. J Biomed Mater Res A 75A:755–766.

Hall CN, Klein-Flugge MC, Howarth C, Attwell D (2012) Oxidative phosphorylation, not glycolysis, powers presynaptic and postsynaptic mechanisms underlying brain information processing. J Neurosci 32:8940–8951.

Hall SB, Khudaish EA, Hart AL (1998) Electrochemical oxidation of hydrogen peroxide at platinum electrodes. Part 1. An adsorption-controlled mechanism. Electrochim Acta 43:579–588.

Hascup KN, Hascup ER, Stephens ML, Glaser PE, Yoshitake T, Mathe AA, Gerhardt GA, Kehr J (2011) Resting glutamate levels and rapid glutamate transients in the prefrontal cortex of the Flinders Sensitive Line rat: a genetic rodent model of depression. Neuropsychopharmacology 36:1769–1777.

Hascup KN, Rutherford EC, Quintero JE, Day BK, Nickell JR, Pomerleau F, Huettl P, Burmeister J, Gerhardt GA (2007) Second-by-second measures of glutamate and other neurotransmitters using enzyme-based microelectrode arrays. In: Electrochemical methods for neuroscience (Michael AC, Borland LM, eds), pp 407–450. Boca Raton: CRC Press.

Hashemi P, Dankoski EC, Petrovic J, Keithley RB, Wightman RM (2009) Voltammetric detection of 5-hydroxytryptamine release in the rat brain. Anal Chem 81:9462–9471.

Hashemi P, Dankoski EC, Lama R, Wood KM, Takmakov P, Wightman RM (2012) Brain dopamine and serotonin differ in regulation and its consequences. Proc Natl Acad Sci U S A 109:11510–11515.

Herr NR, Park J, McElligott ZA, Belle AM, Carelli RM, Wightman RM (2012) *In vivo* voltammetry monitoring of electrically evoked extracellular norepinephrine in subregions of the bed nucleus of the stria terminalis. J Neurophysiol 107:1731–1737.

Hopper P, Kuhr WG (1994) Characterization of the chemical architecture of carbon-fiber microelectrodes. 3. Effect of charge on the electron-transfer properties of ECL reactions. Anal Chem 66:1996–2004.

Howe MW, Tierney PL, Sandberg SG, Phillips PE, Graybiel AM (2013a) Prolonged dopamine signalling in striatum signals proximity and value of distant rewards. Nature 500:575–579.

Howe WM, Berry AS, Francois J, Gilmour G, Carp JM, Tricklebank M, Lustig C, Sarter M (2013b) Prefrontal cholinergic mechanisms instigating shifts from monitoring for cues to cue-guided performance: converging electrochemical and fMRI evidence from rats and humans. J Neurosci 33:8742–8752.

Hu YB, Wilson GS (1997) Rapid changes in local extracellular rat brain glucose observed with an *in vivo* glucose sensor. J Neurochem 68:1745–1752.

Huffman ML, Venton BJ (2009) Carbon-fiber microelectrodes for *in vivo* applications. Analyst 134:18–24.

Johnson SW, North RA (1992) Opioids excite dopamine neurons by hyperpolarization of local interneurons. J Neurosci 12:483–488.

Johnson SW, Seutin V, North RA (1992) Burst firing in dopamine neurons induced by N-methyl-D-aspartate: role of electrogenic sodium pump. Science 258:665–667.

Karreman M, Westerink BH, Moghaddam B (1996) Excitatory amino acid receptors in the ventral tegmental area regulate dopamine release in the ventral striatum. J Neurochem 67:601–607.

Kawagoe KT, Wightman RM (1994) Characterization of amperometry for *in vivo* measurement of dopamine dynamics in the rat brain. Talanta 41:865–874.

Kissinger PT, Hart JB, Adams RN (1973) Voltammetry in brain tissue — a new neurophysiological measurement. Brain Res 55:209–213.

Kiyatkin EA, Lenoir M (2012) Rapid fluctuations in extracellular brain glucose levels induced by natural arousing stimuli and intravenous cocaine: fueling the brain during neural activation. J Neurophysiol 108:1669–1684.

Kiyatkin EA, Wakabayashi KT, Lenoir M (2013) Physiological fluctuations in brain temperature as a factor affecting electrochemical evaluations of extracellular glutamate and glucose in behavioral experiments. ACS Chem Neurosci 4:652–665.

Krajewska B (2004) Application of chitin- and chitosan-based materials for enzyme immobilizations: a review. Enzyme Microb Technol 35:126–139.

Lugo-Morales LZ, Loziuk PL, Corder AK, Toups JV, Roberts JG, McCaffrey KA, Sombers LA (2013) Enzyme-modified carbon-fiber microelectrode for the quantification of dynamic fluctuations of nonelectroactive analytes using fast-scan cyclic voltammetry. Anal Chem 85:8780–8786.

Lundblad M, af Bjerken S, Cenci MA, Pomerleau F, Gerhardt GA, Stromberg I (2009) Chronic intermittent L-DOPA treatment induces changes in dopamine release. J Neurochem 108:998–1008.

Maggi CA, Meli A (1986) Suitability of urethane anesthesia for physiopharmacological investigations in various systems. Part 1: General considerations. Experientia 42:109–114.

Maskos U (2008) The cholinergic mesopontine tegmentum is a relatively neglected nicotinic master modulator of the dopaminergic system: relevance to drugs of abuse and pathology. Br J Pharmacol 153:S438–S445.

Mauko L, Ogorevc B, Pihlar B (2009) Response behavior of amperometric glucose biosensors based on different carbon substrate transducers coated with enzyme-active layer: a comparative study. Electroanal 21:2535–2541.

McCreery RL (2008) Advanced carbon electrode materials for molecular electrochemistry. Chem Rev 108:2646–2687.

Michael AC, Borland LM (2007) Electrochemical methods for neuroscience. Boca Raton: CRC Press/Taylor & Francis.

Mitala JJ, Michael AC (2006) Improving the performance of electrochemical microsensors based on enzymes entrapped in a redox hydrogel. Anal Chim Acta 556:326–332.

Mizutani F, Yabuki S, Hirata Y (1995) Amperometric L-lactate-sensing electrode based on a polyion complex layer containing lactate oxidase — Application to Serum and Milk Samples. Anal Chim Acta 314:233–239.

Mulchandani A, Rogers KR, eds (1998) Enzyme and microbial biosensors: techniques and protocols. New Jersey: Humana Press.

Muzzarelli RA, Barontini G, Rocchetti R (1976) Immobilized enzymes on chitosan columns: alpha-chymotrypsin and acid phosphatase. Biotechnol Bioeng 18:1445–1454.

Netchiporouk LI, Shulga AA, Jaffrezicrenault N, Martelet C, Olier R, Cespuglio R (1995) Properties of carbon-fiber microelectrodes as a basis for enzyme biosensors. Anal Chim Acta 303:275–283.

Nevalainen N, af Bjerken S, Lundblad M, Gerhardt GA, Stromberg I (2011) Dopamine release from serotonergic nerve fibers is reduced in L-DOPA-induced dyskinesia. J Neurochem 118:12–23.

Overton PG, Clark D (1997) Burst firing in midbrain dopaminergic neurons. Brain Res Brain Res Rev 25:312–334.

Pantano P, Kuhr WG (1991) Characterization of the chemical architecture of carbon-fiber microelectrodes. 1. Carboxylates. Anal Chem 63:1413–1418.

Pantano P, Kuhr WG (1995) Enzyme-modified microelectrodes for *in vivo* neurochemical measurements. Electroanal 7:405–416.

Parikh V, Kozak R, Martinez V, Sarter M (2007) Prefrontal acetylcholine release controls cue detection on multiple timescales. Neuron 56:141–154.

Parikh V, Pomerleau F, Huettl P, Gerhardt GA, Sarter M, Bruno JP (2004) Rapid assessment of *in vivo* cholinergic transmission by amperometric detection of changes in extracellular choline levels. Eur J Neurosci 20:1545–1554.

Park J, Kile BM, Wightman RM (2009) *In vivo* voltammetric monitoring of norepinephrine release in the rat ventral bed nucleus of the stria terminalis and anteroventral thalamic nucleus. Eur J Neurosci 30:2121–2133.

Park J, Wheeler RA, Fontillas K, Keithley RB, Carelli RM, Wightman RM (2012) Catecholamines in the bed nucleus of the stria terminalis reciprocally respond to reward and aversion. Biol Psychiatry 71:327–334.

Parker JG, Wanat MJ, Soden ME, Ahmad K, Zweifel LS, Bamford NS, Palmiter RD (2011) Attenuating GABA(A) receptor signaling in dopamine neurons selectively enhances reward learning and alters risk preference in mice. J Neurosci 31: 17103–17112.

Pellerin L, Magistretti PJ (1994) Glutamate uptake into astrocytes stimulates aerobic glycolysis: a mechanism coupling neuronal activity to glucose utilization. Proc Natl Acad Sci U S A 91:10625–10629.

Ren GL, Xu XH, Liu Q, Cheng J, Yuan XY, Wu LL, Wan YZ (2006) Electrospun poly(vinyl alcohol)/glucose oxidase biocomposite membranes for biosensor applications. React Funct Polym 66:1559–1564.

Roberts JG, Moody BP, McCarty GS, Sombers LA (2010) Specific oxygen-containing functional groups on the carbon surface underlie an enhanced sensitivity to dopamine at electrochemically pretreated carbon fiber microelectrodes. Langmuir 26:9116–9122.

Roberts JG, Lugo-Morales LZ, Loziuk PL, Sombers LA (2013) Real-time chemical measurements of dopamine release in the brain. In: Leading methods in dopamine research (Kabbani N, ed), pp 275–294. New York: Humana Press.

Robinson DL, Hermans A, Seipel AT, Wightman RM (2008) Monitoring rapid chemical communication in the brain. Chem Rev 108:2554–2584.

Sanchez-Pozo A, Alados JC, Sanchez-Medina F (1988) Metabolic changes induced by urethane-anesthesia in rats. Gen Pharmacol 19:281–284.

Sanford AL, Morton SW, Whitehouse KL, Oara HM, Lugo-Morales LZ, Roberts JG, Sombers LA (2010) Voltammetric detection of hydrogen peroxide at carbon-fiber microelectrodes. Anal Chem 82:5205–5210.

Sombers LA, Beyene M, Carelli RM, Wightman RM (2009) Synaptic overflow of dopamine in the nucleus accumbens arises from neuronal activity in the ventral tegmental area. J Neurosci 29:1735–1742.

Spanos M, Gras-Najjar J, Letchworth JM, Sanford AL, Toups JV, Sombers LA (2013) Quantitation of hydrogen peroxide fluctuations and their modulation of dopamine dynamics in the rat dorsal striatum using fast-scan cyclic voltammetry. ACS Chem Neurosci 4:782–789.

Takmakov P, Zachek MK, Keithley RB, Walsh PL, Donley C, McCarty GS, Wightman RM (2010) Carbon microelectrodes with a renewable surface. Anal Chem 82:2020–2028.

Threlfell S, Lalic T, Platt NJ, Jennings KA, Deisseroth K, Cragg SJ (2012) Striatal dopamine release is triggered by synchronized activity in cholinergic interneurons. Neuron 75:58–64.

Ward WK, Jansen LB, Anderson E, Reach G, Klein JC, Wilson GS (2002) A new amperometric glucose microsensor: *in vitro* and short-term *in vivo* evaluation. Biosens Bioelectron 17:181–189.

Wassum KM, Tolosa VM, Tseng TC, Balleine BW, Monbouquette HG, Maidment NT (2012) Transient extracellular glutamate events in the basolateral amygdala track reward-seeking actions. J Neurosci 32:2734–2746.

Wickham R, Solecki W, Rathbun L, McIntosh JM, Addy NA (2013) Ventral tegmental area alpha-6-beta-2 nicotinic acetylcholine receptors modulate phasic dopamine release in the nucleus accumbens core. Psychopharmacology (Berl) 229:73–82.

Wightman RM, Heien ML, Wassum KM, Sombers LA, Aragona BJ, Khan AS, Ariansen JL, Cheer JF, Phillips PE, Carelli RM (2007) Dopamine release is heterogeneous within microenvironments of the rat nucleus accumbens. Eur J Neurosci 26: 2046–2054.

Wilson GS, Gifford R (2005) Biosensors for real-time *in vivo* measurements. Biosens Bioelectron 20:2388–2403.

Wilson GS, Johnson MA (2008) *In vivo* electrochemistry: what can we learn about living systems? Chem Rev 108:2462–2481.

Yi H, Wu LQ, Bentley WE, Ghodssi R, Rubloff GW, Culver JN, Payne GF (2005) Biofabrication with chitosan. Biomacromolecules 6:2881–2894.

Zangmeister RA, Park JJ, Rubloff GW, Tarlov MJ (2006) Electrochemical study of chitosan films deposited from solution at reducing potentials. Electrochim Acta 51:5324–5333.

Zhang HJX, Wang J (2007) Electrochemical sensors, biosensors and their biomedical applications. Burlington, MA: Academic Press.

Zheng F, Johnson SW (2002) Group I metabotropic glutamate receptor-mediated enhancement of dopamine cell burst firing in rat ventral tegmental area *in vitro*. Brain Res 948:171–174.

CHAPTER 7

MONITORING AND MODULATING DOPAMINE RELEASE AND UNIT ACTIVITY IN REAL-TIME

Anna M. Belle and R. Mark Wightman
University of North Carolina at Chapel Hill

7.1 INTRODUCTION

A vital step in creating effective treatments for addiction and learning disorders is to understand the neuronal circuitry that drives voluntary reward-directed behaviors and how drugs may hinder, help or hijack this system. Intracranial self stimulation (ICSS) is a well established behavioral paradigm in which an animal learns to press a lever to stimulate a region in their own brain and is considered a robust method to study learning and reward (Olds and Milner, 1954). Use of this technique combined with selective lesions and pharmacological manipulation of selected brain regions has indicated that the nucleus accumbens (NAc) is an important junction for the transmission of information during ICSS (Saddoris et al., 2013).

The role of dopamine in circuits that control learning, goal-oriented behaviors, and addiction has been of interest to neuroscientists for many decades. It is known that dopaminergic neurons, many of which terminate in the NAc, change their firing patterns in response to rewards and cues that predict reward (Mirenowicz and Schultz, 1994), and that dopamine neurotransmission is vital for reinforcing a behavior (Di Chiara and Imperato, 1988). However, these discoveries were made using electrophysiology to monitor the firing pattern of neurons or microdialysis to look at prolonged changes in extracellular dopamine concentration, not dopamine concentration changes on the timescale of cell firing. Dopaminergic neurons generally fire several times per second with periodic bursts of increased frequency

(Grace and Bunney, 1984), so a technique that can distinguish changes in dopamine concentration on this timescale is essential.

Fast-scan cyclic voltammetry (FSCV) can measure changes in local dopamine concentrations up to 60 times per second (Kile et al., 2012), making it the perfect technique to monitor dopamine release in freely-moving animals (Garris et al., 1997; Rebec et al., 1997; Robinson and Wightman, 2007). Thanks to technical advances over the past decade FSCV can now be combined with electrophysiological recordings and iontophoresis, allowing researchers to monitor and manipulate pre- and post-synaptic activity during behavior. The technical considerations and resulting discoveries from the combination of these techniques are discussed here.

7.2 DOPAMINE NEUROTRANSMISSION

Neurons are brain cells responsible for the integration and transmission of information throughout the brain; they receive, process, and transmit information to and from discrete populations of neurons within specific circuits in the brain. Classic neurotransmission occurs when the activation of receptors on dendrites of a cell begins a cascade of intracellular processes that often include the generation and propagation of an action potential, a voltage difference across the neuronal membrane. The frequency of the firing of action potentials of a cell is referred to as its unit activity, and the modulation of unit activity is the way in which information is encoded. When an action potential propagates along the neuron's axon to its terminals, it triggers the release of neurotransmitter into the extracellular space. There the neurotransmitters can bind to specific receptors on a proximal neuron. The modulation of target neurons *via* neurotransmitter release is central to the regulation of an organism's behavior.

Dopamine release in the NAc plays an extensive role in governing motivated behaviors (Salamone and Correa, 2012) and we have shown that its release coincides with learned associations for rewarding stimuli and drugs of abuse (Phillips et al., 2003; Robinson and Wightman, 2007; Owesson-White et al., 2009; Beyene et al., 2010; Day et al., 2010). There is a high density of dopamine cell bodies in the ventral tegmental area (VTA) that send their axonal projections to many regions of the brain including the striatum. The ventral striatum is divided into two subregions: the NAc core and NAc shell. In both NAc subregions, dopamine terminals form synapses onto spines found on the dendrites of medium spiny neurons (MSNs) (Yung et al., 1995). MSNs comprise 95% of the cell bodies in the NAc and release gamma-aminobutyric acid (GABA), an inhibitory neurotransmitter, upon firing (Chang and Kitai, 1985). Neurons with cell bodies located in regions other than the VTA, including other NAc MSNs, also synapse onto these MSNs, making their activation a complex process known to be essential for movement, learning, and motivation, and that is facilitated by dopamine release in the region.

When dopamine is released from terminals it can then bind to any of the five types of dopamine receptors (D1–D5). The predominant dopamine receptors in the NAc, are the D1 receptor (D1R) and D2 receptor (D2R). Both of these receptors are G-protein linked, but D1Rs activate G-proteins that stimulate adenylyl cyclase, while D2Rs activate G-proteins that inhibit adenylyl cyclase. D1Rs also have a lower binding affinity for dopamine than D2Rs. This difference in binding affinities supports the idea that normal basal levels of dopamine in the brain constantly ensure the activation of the majority of D2Rs, while sudden phasic increases in dopamine release activate D1Rs (Dugast et al., 1997; Berke and Hyman, 2000).

D2Rs are found on both pre- and post-synaptic terminals. Pre-synaptic receptors are commonly referred to as D2-autoreceptors (Roth, 1979) and are found to have inhibitory effects on dopamine release from terminals. D2-autoreceptors have been linked to short term modulation of dopamine release (Kita et al., 2007), making them an ideal target when attempting to manipulate evoked dopamine release quickly. Additionally, there are biochemical (Helmreich et al., 1982; Martin et al., 1982; Claustre et al., 1985), electrophysiological (Skirboll et al., 1979), and behavioral (Bradbury et al., 1984) data suggesting that dopamine agonists exhibit greater potency at D2-autoreceptors than at post-synaptic D2Rs. While both pre- and post-synaptic D2 receptors exhibit similar pharmacology (Elsworth and Roth, 1997), they differ in the G-proteins they use to inhibit adenylyl cyclase (Montmayeur et al., 1993; Guiramand et al., 1995). Functionally, this means more binding may be required at post-synaptic receptors to elicit comparable adenylyl cyclase inhibition. The net effect of dopamine release on MSNs is dependent on a balance between the binding of D1Rs, post-synaptic D2Rs, and D2-autoreceptors. Indeed, it appears that the more we know about dopamine neurotransmission the more subtle and yet complex its signaling capabilities seem.

While the different effects of D1R and D2R activation on MSNs has long been recognized, it was recently discovered that D1Rs and D2Rs are in fact segregated onto two different MSN populations (Valjent et al., 2009; Gerfen and Surmeier, 2011). In the dorsal striatum approximately half of MSNs contain D1Rs and project axons to the output nuclei of the basal ganglia (substantia nigra and internal capsule of the globus pallidus). These D1 MSNs are termed the "direct pathway." Other MSNs contain exclusively D2Rs and project to the external capsule of the globus pallidus. These D2 MSNs are termed the "indirect pathway" because they synapse with neurons that also project back to the output nuclei. Although this circuitry was originally characterized in the dorsal striatum, it is also found in the NAc (Ikemoto, 2007).

7.3 EVOLUTION OF A WAVEFORM

A number of electrochemical techniques can be used to detect neurotransmitters *in vivo* (Robinson et al., 2008) but cyclic voltammetry has the unique ability to provide a distinct current trace (cyclic voltammogram) for the oxidation and/or reduction of electroactive compounds. This allows identification and quantification of compounds, a critical feature that prevents incorrect conclusions about the role of a substance in a particular behavior (Wightman and Robinson, 2002). FSCV has become the electrochemical technique of choice for monitoring changes in neurotransmitter levels in the brain because it adds high temporal resolution to the chemical selectivity and high sensitivity of FSCV (Millar et al., 1985). The detection of monoamines with FSCV has been optimized over the years to allow detection of low nanomolar concentrations of dopamine (Keithley et al., 2011). However, the detection limit is not the only factor that must be considered when deciding on a waveform. By varying the rate and range of potentials applied to the carbon-fiber microelectrode, the detection limit of a neuroactive species and the temporal response of the electrode can be optimized.

The first studies in freely-moving rats detected dopamine with a waveform that held the carbon-fiber at −0.4 V, increased linearly up to 1.0 V and back to −0.4 V in 9.3 milliseconds once every 100 milliseconds. This potential waveform is less frequently used today for the detection of neurotransmitters in freely-moving animals because of its lack of sensitivity. However, many important discoveries were made with its use because of its temporal resolution. These include studies of rapid dopamine release in response to reward delivery and how the magnitude of the dopamine changes as an animal learns a behavior (Garris et al., 1999; Kilpatrick et al., 2000). This high speed recording of dopamine release during behavioral tasks supported the hypothesis that dopamine release is required for learning a rewarding task, but it does not signal the reward itself.

The lack of sensitivity of the original waveform used for dopamine detection limited FSCV to the measurement of "extracellular dopamine changes after electrical stimulation of cell bodies in the substantia nigra pars compacta, rather than spontaneous or gradual changes in extracellular dopamine" (Budygin et al., 2001). Fortunately, extension of the original waveform increased dopamine sensitivity nine-fold *in vivo* allowing detection of 5 nM, a level seen for some naturally occurring changes in dopamine (Heien et al., 2003). The improved detection limit was achieved by holding the electrode at −0.6 V and linearly increasing voltage to 1.4 V and back to −0.6 V in 10 milliseconds once every 100 milliseconds. Increased sensitivity resulted from two consequences of this extended waveform: adsorptive pre-concentration of dopamine occurs when the electrode is held at a more negative potential between scans, and the reactive carbon-fiber surface is constantly cleaned as a consequence of over-oxidation of its functional groups. This extended waveform decreased the temporal response of the technique by 1.2 seconds, however, and decreased the selectivity of the electrode for the oxidation and reduction of dopamine over other species. Thus this waveform is most useful for experiments carried out in regions where only a single electroactive compound is present and when the detection limit of dopamine is of higher priority than the electrode's temporal response. This waveform allowed monitoring of naturally occurring increases in dopamine release in the NAc (Phillips et al., 2003) and allowed detection of basal dopamine level changes over 90 second periods (Heien et al., 2005).

The waveform now used for dopamine detection *in vivo* combines the detection limit of the extended waveform and the temporal response of the original waveform. The scan starts at –0.4 V and linearly increases to 1.3 V at the same rate as the extended waveform, taking 8.5 milliseconds to complete (Heien et al., 2004). This is repeated every 100 milliseconds. This "dopamine waveform" allows detection of 8 nM dopamine without the sacrifice in temporal response seen with pre-adsorption at –0.6 V. This is the waveform of choice in experiments where FSCV is used to monitor dopamine or norepinephrine and pH changes in the brain of freely-moving animals.

With the development of methods to quantitatively measure the dynamics of several analytes simultaneously (Heien et al., 2004) and to display thousands of cyclic voltammograms simultaneously to qualitatively monitor the temporal dynamics of each analyte (Michael et al., 1998), we were able to begin to ask and answer increasingly complex questions about dopamine signaling dynamics in behavior. One of these questions was how the dynamics of dopamine signaling affected the balance between cerebral blood flow and metabolism in a region. A waveform was designed that allows for the reduction of oxygen at the carbon-fiber when biased to a negative potential (–1.4 V). By scanning up to 1.0 V we can also see the oxidation of dopamine and any changes in hydrogen ion concentrations within the same scan, but the sensitivity of this waveform for monoamines is greatly reduced due to the reduced magnitude of the positive scan and a holding potential of 0.0 V (Zimmerman and Wightman, 1991; Kennedy et al., 1992). Monitoring oxygen, pH, and dopamine simultaneously has allowed the discoveries that dopamine seems to have little effect on the balance between cerebral blood flow (oxygen changes) and metabolism (pH changes) in both anesthetized (Zimmerman and Wightman, 1991; Venton et al., 2003) and awake animals (Cheer et al., 2006; Ariansen et al., 2012). The information about metabolic dynamics during neurotransmission that this waveform allows us to monitor, also allowed us to be the first to show with the temporal and spatial resolution of our microelectrodes, that the endocannabinoid CB1 receptor mediates changes in the balance between cerebral blood flow and metabolism in the brain of awake animals

(Cheer et al., 2006). This finding has implications for the development of new treatments for cerebral vascular disorders and is an excellent example of the sorts of questions FSCV can answer about the brain.

7.4 SEEING THE BRAIN COMMUNICATE: A COMBINATION OF MEASUREMENT PROCEDURES

Development of an FSCV method that is optimized for monitoring neuronal processes was a task that analytical chemists, with an interest in neuroscience could confidently undertake. However, the design of small animal behavioral experiments to probe the role of dopamine dynamics encompassed concepts far beyond those normally addressed by chemists. We employed a number of unconventional stimuli such as the popping of bubble wrap outside of cages to monitor the startle response of rats (Robinson and Wightman, 2004). We made the unanticipated discovery that the presence of female rats induced subsecond dopamine increases in male rats. We also discovered that these dopamine transients were unrelated to sexual differentiation (Rebec et al., 1997; Robinson et al., 2001; Robinson et al., 2002). Nevertheless, the initial attempts at behavioral experiments turned out to be excellent proof-of-concept for this new measurement technique in freely-moving rats.

In 2000, the Wightman lab partnered with the lab of Dr. Regina Carelli, a psychologist with demonstrated proficiency in small animal behavior. Carelli was known for using electrophysiology to monitor cell firing in behaviors involving reward-based learning and drugs of abuse (Carelli and Deadwyler, 1997; Carelli, 2000, 2002). Extracellular electrophysiological recordings monitor voltage changes in the space around a cell that result from potential changes arising from propagation of an action potential. Carelli and coworkers had focused on the electrophysiology responses of dopaminergic neurons in rats and thus it was relatively straight forward to re-examine the same behaviors while monitoring changes in dopamine release with FSCV (Carelli, 2002). The combined expertise brought by psychologists and chemists working in parallel allowed significant breakthroughs in understanding the dynamics of dopamine transmission in behavior and addiction. The temporal resolution of FSCV resolves distinct dopamine increases coinciding with the cue, indicating imminent lever availability from those that occur at the lever presentation, two seconds later (Stuber et al., 2005a; Day et al., 2007). In addition, the temporal resolution of FSCV showed that the dopamine release occurred before the reward was presented once the animal knew to expect lever presentation at a fixed interval (Cheer et al., 2007). This directly supported earlier theories that hypothesized that dopamine is responsible for reward prediction or its expectation, instead of being a component of the hedonistic aspect (Schultz, 1998).

The early experiments used a fixed time between lever availability so it was unclear whether dopamine was playing a role in the measurement of elapsed time or was signaling that the lever was present. We now use a variable time-out period between lever presentations to avoid teaching the animal that the timing of cue presentation is predictable (Owesson-White et al., 2008). FSCV used in a variety of reward-based behavioral paradigms established that in the NAc, dopamine release occurs during behavioral tasks with the same rapid dynamics as changes in cell firing (Phillips et al., 2003; Roitman et al., 2004; Stuber et al., 2005b; Stuber et al., 2005a; Owesson-White et al., 2008; Jones et al., 2010; Sugam et al., 2012).

7.4.1 COMBINING ELECTROCHEMISTRY WITH ELECTROPHYSIOLOGY

In a 2004 review, Carelli and Wightman reviewed the microcircuitry in the NAc involved in drug addiction (Carelli and Wightman, 2004). The review contained a single figure in which data from two separate experiments in two separate animals were overlaid to show cell firing and dopamine changes both rapidly occurred at the administration of cocaine. The modulation of cell firing in NAc MSNs during behavioral tasks appeared to be dopamine-dependent (Yun et al., 2004) but the relationship between the two was (and still is!) not fully understood. While the Wightman lab had previously ventured into simultaneous electrochemistry and single-unit recording experiments in anesthetized preparations (Ewing et al., 1983; Kuhr et al., 1987), the two techniques were carried out at separate electrodes positioned 500 μm apart (Ewing et al., 1983) or in completely different brain regions (Kuhr et al., 1987). The development of the electronics and methods to combine the two techniques on a single electrode in a freely-moving animal was first accomplished by the lab in 2005 (Cheer et al., 2005) and has been an ongoing area of research (Owesson-White et al., 2009; Cacciapaglia et al., 2011; Takmakov et al., 2011; Belle et al., 2013).

The key to this combined technique is the ability to use a carbon-fiber microelectrode both for electrochemistry and electrophysiology. This concept was first developed by Julian Millar and coworkers in anesthetized animals (Armstrong-James and Millar, 1979, 1988; Stamford et al., 1993). Several advantages arise from using the same electrode for both measurements, including the ability to measure action potentials from the neurons that are being influenced by the local neurotransmitter release sensed at the electrode (Su et al., 1990), the minimization of tissue damage, and the decreased complexity of surgeries. Rapidly alternating between these two techniques without glitches or artifacts requires complex circuitry and modified experimental parameters to freely-moving systems (Garris et al., 1997), a project that required the collaboration of chemists, neurobiologists, and electrical engineers (Cheer et al., 2005; Takmakov et al., 2011).

A switch is used to alternate the carbon-fiber between a current amplifier for electrochemical detections and a voltage follower for single-unit recording. The head stage where this first stage of signal modification is completed is located on the rats' head within a few centimeters of the electrode, minimizing the noise in the amplified signal. Figure 7.1 shows the timing and output when switching between the two circuits. When the voltage follower is connected, voltage changes occurring at the carbon-fiber are recorded. Alternatively when the current amplifier is connected to the carbon-fiber, it is ramped through a potential window and the resulting currents recorded. The original design of this experiment included a solid-state switch that unfortunately would sometimes allow electrochemical voltages to leak over into the single-unit recordings (Cheer et al., 2005). While these voltage changes looked distinct from unit activity, they often overwhelmed the recordings making it difficult to elicit firing changes in the unit of interest. Two improvements were made to help prevent this leakage.

The first modification was to the electrical circuit. There is now a new solid-state relay chosen for its low leakage, low charge transfer, low and matched input capacitance, low resistance (10 Ω), and fast (<1 μs) switching time (Takmakov et al., 2011). These characteristics prevent excess charging currents during the switch and current leakage between the two circuits. The second modification that helped prevent leakage of voltage between the two systems was the timing of the voltammetric scan to the carbon-fiber. The unit recording interval always has a 20 milliseconds gap every 180 milliseconds during which the 8.5 milliseconds voltammetric scan occurs. In initial applications of this technique, the potential scan occurred at the very end of this 20 milliseconds interval. This was done to maximize the adsorption of dopamine to the electrode while it was held at a negative potential; however, this resulted in current fluctuations when the electrode was switched over to electrophysiology. These fluctuations would manifest as large voltage changes in the single-unit data. To correct this, a potential of −0.4 V is applied for pre-concentration of dopamine and current stabilization of the electrode for only the first five milliseconds of the 20 milliseconds interval. The electrode is then ramped to 1.3 V and back to −0.4 V where it remains for ~five milliseconds before the mode is switched. This timing is shown in Figure 7.1. The ability to look at these changes concurrently and with respect to external stimuli or behavioral patterns allows us to begin to directly tease apart the effects of dopamine neurotransmission by monitoring dopamine release from terminals and the unit activity of the cell with dopamine receptors in the vicinity. However, the required modifications of the dopamine waveform result in some sacrifices in electrochemical detection. The sampling rate is doubled every 200 milliseconds and the dopamine detection limit is only 62% of the traditional dopamine waveform (Cheer et al., 2005).

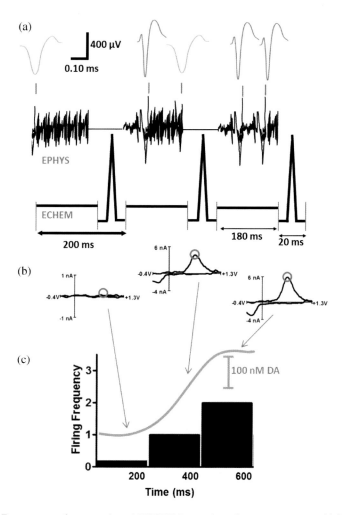

Figure 7.1 Data output from combined FSCV/electrophysiology experiments. (a) The output of the carbon-fiber electrode was connected to a voltage follower to monitor cell firing, seen as voltage changes over time (upper circuitry). The lines above the voltage read out indicate regions of the 180 milliseconds recording interval that have been expanded above. The dark grey spikes are firing of the MSN of interest. The light grey spikes are an example of an excluded event. Every 200 milliseconds the electrode was switched from the electrophysiology circuit to the lower voltammetry circuitry for 20 milliseconds. In this position, the amplifier controlled the electrode potential and the current was monitored. When the lower circuit was completed, the electrode was held at −0.4 V over 8.5 milliseconds, scanned to +1.3 V and then back to −0.4 V. During this time, there were no voltage changes seen in the electrophysiology (when circuit was open but program still recorded data). (b) The cyclic voltammograms resulting from changes of each of the three sequential potential scans. (c) The simultaneous presentation of electrophysiological and electrochemical data. The histogram shows the frequency with which the cell fired during the 200 milliseconds bind, and trace superimposed over the graph shows changes in the concentration of dopamine detected over the same time period.

This dual technique's first discovery was to show that in the same location dopamine release and changes in cell firing were both synchronized to lever pressing in trained animals. MSNs that showed increases OR decreases at lever press (and electrical stimulation), showed a simultaneous increase in dopamine (Cheer et al., 2005). This study also showed that cell firing (but not dopamine release) changed with the administration of GABAergic antagonists; supporting the idea that dopamine altered the probability of a cell firing but was not the neurotransmitter directly responsible for NAc MSNs firing. Later work demonstrated a positive correlation between the concentration of dopamine and magnitude of cell firing change and observed that dopamine release was seen in all locations with MSNs responsive to behavioral stimuli (Owesson-White et al., 2009). MSNs that were unresponsive during the behavior were all in locations with no dopamine release during the behavior (Owesson-White et al., 2009; Cacciapaglia et al., 2012). The difference between correlation of dopamine release and cell firing in the two subregions of the NAc, the core and the shell, was also investigated. Dopamine in the core was closely time-locked to the reinforced response of the lever press. In the shell, dopamine was released over a longer duration and did not coincide as greatly with lever pressing (Owesson-White et al., 2009). The same subregion specific dopamine dynamics were seen for natural rewards as well (Cacciapaglia et al., 2012). These results demonstrate the heterogeneity of dopamine release and suggest this release is positioned to selectively modulate specific MSNs.

After establishing the link between dopamine release and cell firing during ICSS (Cheer et al., 2005, 2007; Owesson-White et al., 2009), we investigated whether this link was also seen with natural rewards (Cacciapaglia et al., 2011). Rats were trained to press a lever for a sucrose pellet instead of a direct electrical stimulation. The lever was made available at a variable interval and preceded by a two seconds cue. Locations with a surge in dopamine concentration at the onset of the cue and (to a lesser extent) with lever press, showed one of four alterations in MSN firing: inhibition at cue presentation, excitation at cue presentation, inhibition at lever press, and excitation at lever press. To see if this surge of dopamine was in fact responsible for the coincidental changes in cell firing, N-methyl-D-aspartate (NMDA) receptors were blocked in the VTA attenuating the burst firing of dopamine cells in the region (Chergui et al., 1993). This decreased dopamine release in the NAc. MSNs that showed an excitation to the cue onset or lever press became non-phasic, while cells that were inhibited at either the cue or the lever press were unaffected by the diminished dopamine release. This showed dopamine's ability to selectively modulate discrete pathways within the NAc and suggested that this was a selective modulation of the direct (D1 MSNs) or indirect (D2 MSNs) pathway.

7.4.2 CONTROLLED IONTOPHORESIS

Iontophoresis is a technique that uses current to induce the migration of a solution of ions through a glass pipette. It was developed in the early 1950s by W. L. Natsuk, a student of A. L. Hodgkin (Nastuk, 1953). While attempting to understand how ions contributed to the actions of acetylcholine at the neuromuscular junction, he noticed that acetylcholine would naturally leak out of a glass pipette pulled to a fine tip and that application of current to the pipette solution ejected even more acetylcholine onto the junction (Hicks, 1984). From there, iontophoresis increased in popularity and extensive studies on the technique were carried out (Krnjevic et al., 1963a, 1963b; Crawford and Curtis, 1964; Curtis and Nastuk, 1964; Bradley and Candy, 1970; Bloom, 1974; Simmonds, 1974; Freedman et al., 1975; Purves, 1977, 1979). The technique's popularity for studying receptor dynamics *in vivo* is due to the fact that drugs can be quickly, selectively, and locally delivered to the site of action with minimal disruption of tissue. Systemic drug administration is only useful for drugs that can pass the blood-brain barrier without metabolic degradation. Even in cases where the drug can cross the blood-brain barrier, the drug affects the entire brain making it difficult to study discrete brain region effects (Bloom, 1974). Additionally, systemic drug administration can alter animal behavior, making it difficult to look at the drug effects in the brain during behavior (Hernandez and Cheer, 2012). These problems are avoided by using iontophoresis to study the pharmacology of the brain.

A drawback to iontophoresis was the inability to monitor or quantify the amount of drug delivered. This made it impossible to differentiate a null response to drug application from a clogged glass pipette. Additionally, too little drug delivered could result in a false negative whereas excessive application could lead to non-specific effects. Applied pump currents are commonly used to compare ejections (Pierce and Rebec, 1995; Kiyatkin and Rebec, 1996, 1999b), but the same pump current ejects different drug concentrations from barrel to barrel (Herr et al., 2008). Modifications to the design of Millar and co-workers, which coupled iontophoresis barrels to carbon-fiber microelectrodes, allow detection of electroactive compounds ejected with iontophoresis at the neighboring electrode (Armstrong-James et al., 1981).

Using these coupled iontophoresis probes, electro-osmosis was found to contribute significantly to the observed drug delivery (Herr et al., 2008). Electro-osmosis is a phenomenon caused by the attraction between the cations in solution and the ionizable silanol groups on the glass capillary surface. When a positive current is applied to the capillary, the cations along the wall migrate toward the anode, inducing a bulk movement of solution, termed electro-osmotic flow. The variability in iontophoretic ejections is associated with variability of electro-osmotic flow from barrel to barrel, while electrophoretic mobility (ionic migration) for a given species is consistent. Using an electroactive neutral molecule as an internal standard to monitor the variability in electro-osmotic flow, and subsequently the amount of drug delivered from different barrels, allows us to control this variability. These insights into iontophoresis enable quantitative delivery of electroactive and electroinactive drugs by monitoring the coejection of an electroactive molecule from the same barrel (Herr et al., 2008). When the relative mobilities of the coejected substances are known, monitoring the concentration of the electroactive molecule with the carbon-fiber electrode provides an indirect measure of the relative concentration of the coejected non-electroactive substance (Herr et al., 2008; Herr et al., 2010).

While FSCV had previously been coupled to iontophoresis to monitor modulation of cell firing by electroactive compounds (Kiyatkin and Rebec, 1996, 1997, 1999a, 1999b & 2000; Rebec, 1998; Kiyatkin et al., 2000), it was not until 2010 that the technique was used *in vivo* to alter release from dopamine terminals in the region of drug application (Herr et al., 2010). These papers established that controlled iontophoresis could be used to quickly (<60 seconds) modulate dopamine release by affecting D2-autoreceptors and the dopamine transporter in anesthetized (Herr et al., 2010) and freely-moving animals (Belle et al., 2013). Monitoring iontophoresis ejections eliminates the fear of ionic and electrical artifacts from the ejection current altering the neuronal environment as well. With controlled iontophoresis coupled to the simultaneous electrochemistry/electrophysiology technique we can watch for these problems during an experiment. While changes to the neuronal environment are not seen from ejection of saline or the electroactive marker molecule, the ejection current is seen to affect the electrode, electrical connections can be altered from a different barrel on the probe used to prevent any data collection problems (Belle et al., 2013).

With the discovery of the discrete location of D1Rs and D2Rs on separate populations of MSNs (Valjent et al., 2009; Gerfen and Surmeier, 2011) came a huge breakthrough in how neuronal networks in the brain work together in reward, behavior and addiction. In the NAc core, only 6% of MSNs have both D1Rs and D2Rs, while 53% of MSNs have D1Rs exclusively (D1 MSNs) and 41% of MSNs have D2Rs exclusively (D2 MSNs) (Valjent et al., 2009). While the role of each receptor during learning and reward has been investigated (Surmeier et al., 2007), direct electrophysiological evidence has been lacking, partly because of the difficulties in differentiating D1 MSNs and D2 MSNs *in vivo* (Venance and Glowinski, 2003).

Coupling controlled iontophoresis to the same microelectrode in a combined FSCV and single-unit recording experiment allows for pharmacological identification of MSNs based on their responses to D1R and D2R antagonists (Belle et al., 2013). The firing rates of NAc MSNs in awake animals were monitored before, during and after a 15 seconds iontophoretic ejection of specific dopamine receptor antagonist. Changes in response to these antagonists were seen both immediately and on a prolonged timescale (as an overall change in the firing rate of a neuron after application). Looking at prolonged changes, 40% of MSNs increased their firing rate after local application of a D2R antagonist, 46% of MSNs exhibited a decreased firing rate after local application of the D1R antagonist, and only 11% of MSNs responded to both antagonists. These results are in agreement with previously reported distributions for dopamine receptor subtypes on MSNs (Valjent et al., 2009) supporting the method as a way to discriminate between and selectively modulate D1 MSNs and D2 MSNs *in vivo*.

7.5 CONCLUSIONS

As our knowledge of the subtle and complex signaling required for the brain to encode reward-directed behaviors increases, the ability to selectively modulate and monitor dopamine release and MSN firing in freely-moving animals engaged in behavioral tasks becomes even more essential. The development of a technique to monitor subsecond dopamine fluctuations in freely-moving animals has allowed the study of naturally and electrically evoked dopamine during reward-motivated tasks and provided the ability to watch slower, tonic changes in dopamine. Furthermore, combining this ability with concurrent observation of cell firing patterns *in vivo*, is essential for understanding neurotransmission. The technique has shown that MSNs alter their firing during a behavioral task, receive dopaminergic inputs during the behavior and that dopamine release is required for excitations of MSNs during behavior. This combined technique is a passive way to 'listen in' on neurotransmission and the ability to selectively and locally modulate this neuronal conversation will allow an even greater understanding of the purpose of dopaminergic pathways in learning and reward.

ACKNOWLEDGMENTS

The authors thank the staff of the Electronics Facility at the University of North Carolina at Chapel Hill, especially Collin McKinney and Matt Verber for advances in the circuitry required for these experiments and Nathan Rodeberg for review of manuscript. This research was supported by NIH (DA010900).

REFERENCES

Ariansen JL, Heien MLAV, Hermans A, Phillips PEM, Hernadi I, Bermudez M, Schultz W, Wightman RM (2012) Monitoring extracellular pH, oxygen, and dopamine during reward delivery in the striatum of primates. Front Behav Neurosci 6:36.

Armstrong-James M, Millar J (1979) Carbon fibre microelectrodes. J Neurosci Methods 1:279–287.

Armstrong-James M, Millar J (1988) High-speed cyclic voltammetry and unit recording with carbon fibre microelectrodes. In: Measurement of neurotransmitter release *in vivo* (Marden C, ed), pp209–224. New York: John Wiley and Sons Ltd.

Armstrong-James M, Fox K, Kruk ZL, Millar J (1981) Quantitative ionophoresis of catecholamines using multibarrel carbon fibre microelectrodes. J Neurosci Methods 4:385–406.

Belle AM, Owesson-White C, Herr NR, Carelli RM, Wightman RM (2013) Controlled iontophoresis coupled with fast-scan cyclic voltammetry/electrophysiology in awake, freely moving animals. ACS Chem Neurosci 5:761–71.

Berke JD, Hyman SE (2000) Addiction, dopamine, and the molecular mechanisms of memory. Neuron 25:515–532.

Beyene M, Carelli RM, Wightman RM (2010) Cue-evoked dopamine release in the nucleus accumbens shell tracks reinforcer magnitude during intracranial self-stimulation. Neuroscience 169:1682–1688.

Bloom FE (1974) To spritz or not to spritz: the doubtful value of aimless iontophoresis. Life Sci 14:1819–1834.

Bradbury AJ, Cannon JG, Costall B, Naylor RJ (1984) A comparison of dopamine agonist action to inhibit locomotor activity and to induce stereotyped behaviour in the mouse. Eur J Pharmacol 105:33–47.

Bradley PB, Candy JM (1970) Iontophoretic release of acetylcholine, noradrenaline, 5-hydroxytryptamine and D-lysergic acid diethylamide from micropipettes. Br J Pharmacol 40:194–201.

Budygin EA, Phillips PE, Robinson DL, Kennedy AP, Gainetdinov RR, Wightman RM (2001) Effect of acute ethanol on striatal dopamine neurotransmission in ambulatory rats. J Pharmacol Exp Ther 297:27–34.

Cacciapaglia F, Wightman RM, Carelli RM (2011) Rapid dopamine signaling differentially modulates distinct microcircuits within the nucleus accumbens during sucrose-directed behavior. J Neurosci 31:13860–13869.

Cacciapaglia F, Saddoris MP, Wightman RM, Carelli RM (2012) Differential dopamine release dynamics in the nucleus accumbens core and shell track distinct aspects of goal-directed behavior for sucrose. Neuropharmacology 62:2050–2056.

Carelli RM (2000) Activation of accumbens cell firing by stimuli associated with cocaine delivery during self-administration. Synapse 35:238–242.

Carelli RM (2002) Nucleus accumbens cell firing during goal-directed behaviors for cocaine *vs.* 'natural' reinforcement. Physiol Behav 76:379–387.

Carelli RM, Deadwyler SA (1997) Cellular mechanisms underlying reinforcement-related processing in the nucleus accumbens: electrophysiological studies in behaving animals. Pharmacol Biochem Behav 57:495–504.

Carelli RM, Wightman RM (2004) Functional microcircuitry in the accumbens underlying drug addiction: insights from real-time signaling during behavior. Curr Opin Neurobiol 14:763–768.

Chang HT, Kitai ST (1985) Projection neurons of the nucleus accumbens: an intracellular labeling study. Brain Res 347:112–116.

Cheer JF, Wassum KM, Wightman RM (2006) Cannabinoid modulation of electrically evoked pH and oxygen transients in the nucleus accumbens of awake rats. J Neurochem 97:1145–1154.

Cheer JF, Heien ML, Garris PA, Carelli RM, Wightman RM (2005) Simultaneous dopamine and single-unit recordings reveal accumbens GABAergic responses: implications for intracranial self-stimulation. Proc Natl Acad Sci USA 102:19150–19155.

Cheer JF, Aragona BJ, Heien ML, Seipel AT, Carelli RM, Wightman RM (2007) Coordinated accumbal dopamine release and neural activity drive goal-directed behavior. Neuron 54:237–244.

Chergui K, Charlety PJ, Akaoka H, Saunier CF, Brunet JL, Buda M, Svensson TH, Chouvet G (1993) Tonic activation of NMDA receptors causes spontaneous burst discharge of rat midbrain dopamine neurons *in vivo*. Eur J Neurosci 5:137–144.

Claustre Y, Fage D, Zivkovic B, Scatton B (1985) Relative selectivity of 6,7-dihydroxy-2-dimethylaminotetralin, N-n-propyl-3-(3-hydroxyphenyl) piperidine, N-n-propylnorapomorphine and pergolide as agonists at striatal dopamine autoreceptors and postsynaptic dopamine receptors. J Pharmacol Exp Ther 232:519–525.

Crawford JM, Curtis DR (1964) The excitation and depression of mammalian cortical neurones by amino acids. Br J Pharmacol Chemother 23:313–329.

Curtis DR, Nastuk WL (1964) Micro-electrophoresis. In: Physical techniques in biological research, pp144–190. New York: Academic Press.

Day JJ, Roitman MF, Wightman RM, Carelli RM (2007) Associative learning mediates dynamic shifts in dopamine signaling in the nucleus accumbens. Nat Neurosci 10:1020–1028.

Day JJ, Jones JL, Wightman RM, Carelli RM (2010) Phasic nucleus accumbens dopamine release encodes effort- and delay-related costs. Biol Psychiatry 68:306–309.

Di Chiara G, Imperato A (1988) Drugs abused by humans preferentially increase synaptic dopamine concentrations in the mesolimbic system of freely moving rats. Proc Natl Acad Sci USA 85:5274–5278.

Dugast C, Brun P, Sotty F, Renaud B, Suaud-Chagny MF (1997) On the involvement of a tonic dopamine D2-autoinhibition in the regulation of pulse-to-pulse-evoked dopamine release in the rat striatum *in vivo*. Naunyn Schmiedebergs Arch Pharmacol 355:716–719.

Elsworth JD, Roth RH (1997) Dopamine synthesis, uptake, metabolism, and receptors: relevance to gene therapy of parkinson's disease. Exp Neurol 144:4–9.

Ewing AG, Alloway KD, Curtis SD, Dayton MA, Wightman RM, Rebec GV (1983) Simultaneous electrochemical and unit recording measurements: characterization of the effects of D-amphetamine and ascorbic acid on neostriatal neurons. Brain Res 261:101–108.

Freedman R, Hoffer BJ, Woodward DJ (1975) A quantitative microiontophoretic analysis of the responses of central neurones to noradrenaline: interactions with cobalt, manganese, verapamil and dichloroisoprenaline. Br J Pharm 54:529–539.

Garris PA, Christensen JR, Rebec GV, Wightman RM (1997) Real-time measurement of electrically evoked extracellular dopamine in the striatum of freely moving rats. J Neurochem 68:152–161.

Garris PA, Kilpatrick M, Bunin MA, Michael D, Walker QD, Wightman RM (1999) Dissociation of dopamine release in the nucleus accumbens from intracranial self-stimulation. Nature 398:67–69.

Gerfen CR, Surmeier DJ (2011) Modulation of striatal projection systems by dopamine. Annu Rev Neurosci 34:441–466.

Gratton A, Wise RA (1994) Drug- and behavior-associated changes in dopamine-related electrochemical signals during intravenous cocaine self-administration in rats. J Neurosci 14:4130–4146.

Grace AA, Bunney BS (1984) The control of firing patterns in nigral dopamine neurons: burst firing. J Neurosci 4:2877–2890.

Guiramand J, Montmayeur J-P, Ceraline J, Bhatia M, Borrelli E (1995) Alternative splicing of the dopamine D2 receptor directs specificity of coupling to G-proteins. J Biol Chem 270:7354–7358.

Heien ML, Johnson MA, Wightman RM (2004) Resolving neurotransmitters detected by fast-scan cyclic voltammetry. Anal Chem 76:5697–5704.

Heien ML, Phillips PE, Stuber GD, Seipel AT, Wightman RM (2003) Overoxidation of carbon-fiber microelectrodes enhances dopamine adsorption and increases sensitivity. Analyst 128:1413–1419.

Heien ML, Khan AS, Ariansen JL, Cheer JF, Phillips PE, Wassum KM, Wightman RM (2005) Real-time measurement of dopamine fluctuations after cocaine in the brain of behaving rats. Proc Natl Acad Sci USA 102:10023–10028.

Helmreich I, Reimann W, Hertting G, Starke K (1982) Are presynaptic dopamine autoreceptors and postsynaptic dopamine receptors in the rabbit caudate nucleus pharmacologically different? Neuroscience 7:1559–1566.

Hernandez G, Cheer JF (2012) Effect of CB1 receptor blockade on food-reinforced responding and associated nucleus accumbens neuronal activity in rats. J Neurosci 32:11467–11477.

Herr NR, Kile BM, Carelli RM, Wightman RM (2008) Electroosmotic flow and its contribution to iontophoretic delivery. Anal Chem 80:8635–8641.

Herr NR, Daniel KB, Belle AM, Carelli RM, Wightman RM (2010) Probing presynaptic regulation of extracellular dopamine with iontophoresis. ACS Chem Neurosci 1:627–638.

Hicks TP (1984) The history and development of microiontophoresis in experimental neurobiology. Prog Neurobiol 22:185–240.

Ikemoto S (2007) Dopamine reward circuitry: two projection systems from the ventral midbrain to the nucleus accumbens-olfactory tubercle complex. Brain Res Rev 56:27–78.

Jones JL, Day JJ, Aragona BJ, Wheeler RA, Wightman RM, Carelli RM (2010) Basolateral amygdala modulates terminal dopamine release in the nucleus accumbens and conditioned responding. Biol Psychiatry 67:737–744.

Keithley RB, Takmakov P, Bucher ES, Belle AM, Owesson-White CA, Park J, Wightman RM (2011) Higher sensitivity dopamine measurements with faster-scan cyclic voltammetry. Anal Chem 83:3563–3571.

Kennedy RT, Jones SR, Wightman RM (1992) Simultaneous measurement of oxygen and dopamine: coupling of oxygen consumption and neurotransmission. Neuroscience 47:603–612.

Kile BM, Walsh PL, McElligott ZA, Bucher ES, Guillot TS, Salahpour A, Caron MG, Wightman RM (2012) Optimizing the temporal resolution of fast-scan cyclic voltammetry. ACS Chem Neurosci 3:285–292.

Kilpatrick MR, Rooney MB, Michael DJ, Wightman RM (2000) Extracellular dopamine dynamics in rat caudate-putamen during experimenter-delivered and intracranial self-stimulation. Neuroscience 96:697–706.

Kita JM, Parker LE, Phillips PE, Garris PA, Wightman RM (2007) Paradoxical modulation of short-term facilitation of dopamine release by dopamine autoreceptors. J Neurochem 102:1115–1124.

Kiyatkin EA, Rebec GV (1996) Dopaminergic modulation of glutamate-induced excitations of neurons in the neostriatum and nucleus accumbens of awake, unrestrained rats. J Neurophysiol 75:142–153.

Kiyatkin EA, Rebec GV (1997) Iontophoresis of amphetamine in the neostriatum and nucleus accumbens of awake, unrestrained rats. Brain Res 771:14–24.

Kiyatkin EA, Rebec GV (1999a) Striatal neuronal activity and responsiveness to dopamine and glutamate after selective blockade of D1 and D2 dopamine receptors in freely moving rats. J Neurosci 19:3594–3609.

Kiyatkin EA, Rebec GV (1999b) Modulation of striatal neuronal activity by glutamate and GABA: iontophoresis in awake, unrestrained rats. Brain Res 822:88–106.

Kiyatkin EA, Rebec GV (2000) Dopamine-independent action of cocaine on striatal and accumbal neurons. Eur J Neurosci 12:1789–1800.

Kiyatkin EA, Kiyatkin DE, Rebec GV (2000) Phasic inhibition of dopamine uptake in nucleus accumbens induced by intravenous cocaine in freely behaving rats. Neuroscience 98:729–741.

Krnjevic K, Laverty R, Sharman DF (1963a) Iontophoretic release of adrenaline, noradrenaline and 5-hydroxytryptamine from micropipettes. Br J Pharmacol Chem 20:491–496.

Krnjevic K, Mitchell JF, Szerb JC (1963b) Determination of iontophoretic release of acetylcholine from micropipettes. J Physiol 165:421–436.

Kuhr WG, Wightman RM, Rebec GV (1987) Dopaminergic neurons: simultaneous measurements of dopamine release and single-unit activity during stimulation of the medial forebrain bundle. Brain Res 418:122–128.

Martin GE, Williams M, Haubrich DR (1982) A pharmacological comparison of 6,7-dihydroxy-2-dimethylaminotetralin (TL-99) and N-n-propyl-3-(3-hydroxyphenyl) piperidine with (3-PPP) selected dopamine agonists. J Pharmacol Exp Ther 223:298–304.

Michael D, Travis ER, Wightman RM (1998) Color images for fast-scan CV measurements in biological systems. Anal Chem 70:586A–592A.

Millar J, Stamford JA, Kruk ZL, Wightman RM (1985) Electrochemical, pharmacological and electrophysiological evidence of rapid dopamine release and removal in the rat caudate nucleus following electrical stimulation of the median forebrain bundle. Eur J Pharmacol 109:341–348.

Mirenowicz J, Schultz W (1994) Importance of unpredictability for reward responses in primate dopamine neurons. J Neurophysiol 72:1024–1027.

Montmayeur JP, Guiramand J, Borrelli E (1993) Preferential coupling between dopamine D2 receptors and G-proteins. Mol Endocrinol 7:161–170.

Nastuk WL (1953) Membrane potential changes at a single muscle endplate produced by transitory application of acetylcholine with an electrically controlled microjet. Fed Proc 12:102.

Olds J, Milner P (1954) Positive reinforcement produced by electrical stimulation of septal area and other regions of rat brain. J Comp Physiol Psychol 47:419–427.

Owesson-White CA, Cheer JF, Beyene M, Carelli RM, Wightman RM (2008) Dynamic changes in accumbens dopamine correlates with learning during intracranial self-stimulation. Proc Natl Acad Sci USA 105:11957–11962.

Owesson-White CA, Ariansen J, Stuber GD, Cleaveland NA, Cheer JF, Wightman RM, Carelli RM (2009) Neural encoding of cocaine-seeking behavior is coincident with phasic dopamine release in the accumbens core and shell. Eur J Neurosci 30:1117–1127.

Phillips PE, Stuber GD, Heien ML, Wightman RM, Carelli RM (2003) Subsecond dopamine release promotes cocaine seeking. Nature 422:614–618.

Pierce RC, Rebec GV (1995) Iontophoresis in the neostriatum of awake, unrestrained rats: differential effects of dopamine, glutamate and ascorbate on motor- and nonmotor-related neurons. Neuroscience 67:313–324.

Purves RD (1977) The release of drugs from iontophoretic pipettes. J Theor Biol 66:789–798.

Purves RD (1979) The physics of iontophoretic pipettes. J Neurosci Methods 1:165–178.

Rebec GV (1998) Real-time assessments of dopamine function during behavior: single-unit recording, iontophoresis, and fast-scan cyclic voltammetry in awake, unrestrained rats. Alcohol Clin Exp Res 22:32–40.

Rebec GV, Christensen JR, Guerra C, Bardo MT (1997) Regional and temporal differences in real-time dopamine efflux in the nucleus accumbens during free-choice novelty. Brain Res 776:61–67.

Robinson DL, Wightman RM (2004) Nomifensine amplifies subsecond dopamine signals in the ventral striatum of freely-moving rats. J Neurochem 90:894–903.

Robinson DL, Wightman RM (2007) Rapid dopamine release in freely moving rats. In: Electrochemical methods for neuroscience (Michael AC, Borland LM, eds), pp 17–36. Boca Raton, FL: CRC Press.

Robinson DL, Heien ML, Wightman RM (2002) Frequency of dopamine concentration transients increases in dorsal and ventral striatum of male rats during introduction of conspecifics. J Neurosci 22:10477–10486.

Robinson DL, Hermans A, Seipel AT, Wightman RM (2008) Monitoring rapid chemical communication in the brain. Chem Rev 108:2554–2584.

Robinson DL, Phillips PE, Budygin EA, Trafton BJ, Garris PA, Wightman RM (2001) Sub-second changes in accumbal dopamine during sexual behavior in male rats. Neuroreport 12:2549–2552.

Roitman MF, Stuber GD, Phillips PE, Wightman RM, Carelli RM (2004) Dopamine operates as a subsecond modulator of food seeking. J Neurosci 24:1265–1271.

Roth RH (1979) Dopamine autoreceptors: pharmacology, function and comparison with post-synaptic dopamine receptors. Commun Psychopharmacol 3:429–445.

Saddoris MP, Sugam JA, Cacciapaglia F, Carelli RM (2013) Rapid dopamine dynamics in the accumbens core and shell: learning and action. Front Biosci (Elite Ed) 5:273–288.

Salamone JD, Correa M (2012) The mysterious motivational functions of mesolimbic dopamine. Neuron 76:470–485.

Schultz W (1998) Predictive reward signal of dopamine neurons. J Neurophysiol 80:1–27.

Simmonds MA (1974) Quantitative evaluation of responses to microiontophoretically applied drugs. Neuropharmacol 13:401–406.

Skirboll LR, Grace AA, Bunney BS (1979) Dopamine auto- and postsynaptic receptors. Electrophysiological evidence for differential sensitivity to dopamine agonists. Science 206:80–82.

Stamford JA, Palij P, Davidson C, Jorm CM, Millar J (1993) Simultaneous "real-time" electrochemical and electrophysiological recording in brain slices with a single carbon-fibre microelectrode. J Neurosci Methods 50:279–290.

Stuber GD, Wightman RM, Carelli RM (2005a) Extinction of cocaine self-administration reveals functionally and temporally distinct dopaminergic signals in the nucleus accumbens. Neuron 46:661–669.

Stuber GD, Roitman MF, Phillips PE, Carelli RM, Wightman RM (2005b) Rapid dopamine signaling in the nucleus accumbens during contingent and noncontingent cocaine administration. Neuropsychopharmacology 30:853–863.

Su MT, Dunwiddie TV, Gerhardt GA (1990) Combined electrochemical and electrophysiological studies of monoamine overflow in rat hippocampal slices. Brain Res 518:149–158.

Sugam JA, Day JJ, Wightman RM, Carelli RM (2012) Phasic nucleus accumbens dopamine encodes risk-based decision-making behavior. Biol Psychiatry 71:199–205.

Surmeier DJ, Ding J, Day M, Wang Z, Shen W (2007) D1 and D2 dopamine-receptor modulation of striatal glutamatergic signaling in striatal medium spiny neurons. Trends Neurosci 30:228–235.

Takmakov P, McKinney CJ, Carelli RM, Wightman RM (2011) Instrumentation for fast-scan cyclic voltammetry combined with electrophysiology for behavioral experiments in freely moving animals. Rev Sci Instrum 82:074302.

Valjent E, Bertran-Gonzalez J, Herve D, Fisone G, Girault JA (2009) Looking BAC at striatal signaling: cell-specific analysis in new transgenic mice. Trends Neurosci 32:538–547.

Venance L, Glowinski J (2003) Heterogeneity of spike frequency adaptation among medium spiny neurones from the rat striatum. Neuroscience 122:77–92.

Venton BJ, Michael DJ, Wightman RM (2003) Correlation of local changes in extracellular oxygen and pH that accompany dopaminergic terminal activity in the rat caudate-putamen. J Neurochem 84:373–381.

Wightman RM, Robinson DL (2002) Transient changes in mesolimbic dopamine and their association with 'reward'. J Neurochem 82:721–735.

Yun IA, Wakabayashi KT, Fields HL, Nicola SM (2004) The ventral tegmental area is required for the behavioral and nucleus accumbens neuronal firing responses to incentive cues. J Neurosci 24:2923–2933.

Yung KK, Bolam JP, Smith AD, Hersch SM, Ciliax BJ, Levey AI (1995) Immunocytochemical localization of D1 and D2 dopamine receptors in the basal ganglia of the rat: light and electron microscopy. Neuroscience 65:709–730.

Zimmerman JB, Wightman RM (1991) Simultaneous electrochemical measurements of oxygen and dopamine *in vivo*. Anal Chem 63:24–28.

CHAPTER 8

QUANTITATIVE CHEMICAL MEASUREMENTS OF VESICULAR TRANSMITTERS WITH SINGLE CELL AMPEROMETRY AND ELECTROCHEMICAL CYTOMETRY

Jelena Lovrić[*], Xianchan Li[*], Andrew G. Ewing[*,†]

[*]Chalmers University of Technology, [†]University of Gothenburg

Understanding the mechanism of exocytosis is among the most important molecular problems in neuroscience today. Even though much is known about this process, the extent that the vesicles are emptied upon fusion is a topic that is being debated. Electrochemical techniques, such as amperometry and patch-clamp measurements, provide unique chemical and biological insights into the mechanism of the exocytotic process at the fundamental level. In this chapter we present a new technique called electrochemical cytometry and summarize key current evidence which leads to the conclusion that during exocytosis the majority of vesicles partially open to release transmitters and then close directly again. These insights suggest that open-and-closed exocytosis is the normal mode of neuronal communication.

8.1 INTRODUCTION

Exocytosis is a vital process in neuronal communication. Signals are transferred through neurons by means of a depolarization of the cell membrane known as the action potential. The cells can form a specialized structure, termed as synapse,

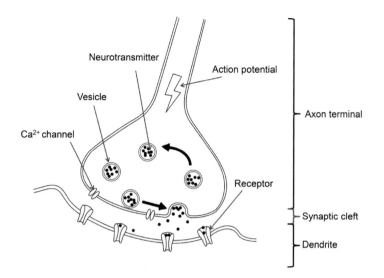

Figure 8.1 Schematic representation of the structure of the chemical synapse.

through which the signal is transferred from one cell to the next (Figure 8.1). Since the electrical signal cannot pass the synaptic gap, this transfer is achieved in a different way, i.e., releasing transmitters. Prior to release, neurotransmitter molecules are contained within vesicles in the nerve terminal. The arrival of the action potential causes the vesicles to fuse with the membrane and release their contents into the synaptic cleft. The transmitters diffuse across the synapse and bind to the specific receptors on the postsynaptic membrane and thus a signal is transduced. Until now, several types of neurotransmitters, such as amino acids, monoamines, peptides and others have been found but their function in the brain is not always clear. The initial release event is named exocytosis and involves the fusion of a vesicle with the cell plasma membrane. During this process, a fusion pore, which connects the inside of the vesicle and the exterior of the cell, is formed. In this case, the transmitter molecules previously trapped by a lipid bilayer are released into the extracellular space. Mechanistic studies of this process are central to understanding neurotransmission. Therefore, several different methods have been developed to detect, measure, and characterize the release of transmitter molecules with biological or artificial models (Leszczyszyn et al., 1990; Wightman et al., 1991; Chow et al., 1992; Detoledo et al., 1993; Chen et al., 1995; Wightman et al., 1995; Finnegan et al., 1996; Anderson et al., 1999; Colliver et al., 2000; Westerink et al., 2000; Sombers et al., 2005; Amatore et al., 2006, 2007; Uchiyama et al., 2007).

Methods to observe and quantify individual exocytotic events have traditionally revolved around electron microscopy and patch-clamp capacitance measurements (Leszczyszyn et al., 1990; Matthews, 1996). However, in 1990, Wightman and coworkers showed that they could monitor individual exocytotic events of easily oxidized messengers occurring on the millisecond time scale by use of amperometric measurements at microelectrodes (Wightman et al., 1991; Chow et al., 1992). Furthermore, they were able to quantify the number of molecules released. This work began with adrenal chromaffin cells (Wightman et al., 1991; Dernick et al., 2005). Typically, a carbon-fiber electrode is placed near the cell membrane and held at a constant potential which is sufficient to oxidize the transmitter (typically <1.0 V *versus* Ag/AgCl reference electrode). The carbon-fiber is held at a constant potential exceeding the redox potential of the desired analyte. As the analyte comes into contact with the carbon-fiber surface, it is oxidized, thereby releasing electrons, and generating a current. The charge (Q) from current transients on the amperometric trace generated by the oxidation can be related to the mole amount of transmitters (N) oxidized using Faraday's law ($Q = nNF$), where n is the number of electrons exchanged in the oxidation reaction($2e^-$ for most monoamines) and F is Faraday's constant (96,485 C/mol) (Cans and Ewing, 2011). Patch-clamp and amperometry have also been combined to create a hybrid technique called patch-amperometry (Dernick et al., 2005), where both the membrane capacitance and the released substance can be measured. This provides an advantage in providing measurements of release combined with information about the vesicle holding the transmitter. This must be done on the exposed portions of the cell and will only work for messengers that are electrochemically active. Overall, electrochemical methods are uniquely able to quantify the amount of chemical messenger release, and many investigations with this approach have been carried out yielding important information about exocytosis and neurobiology.

A new concept of transient release has been recently discussed and named "extended kiss-and-run" or what could also be called open-and-closed exocytosis. In this chapter, we present a new method, called electrochemical cytometry, to determine the total transmitter content in vesicles containing electroactive molecules. This will then be compared to single cell amperometry to provide evidence for partial release and argue that open-and-closed exocytosis is normal exocytosis. As part of this argument, we present models of amperometric data and evidence for a "post exocytosis foot" indicating that the vesicle is closing again. We do not attempt to be comprehensive surrounding exocytosis methods and do not discuss many of the other elegant electrochemical methods used to measure this process.

8.2 ELECTROCHEMICAL CYTOMETRY

We have developed a new technology, which we term electrochemical cytometry, to separate, lyse and quantitatively measure the transmitter content of individual secretory vesicles isolated from cells or tissues *via* end-column electrochemical

Figure 8.2 Investigation of the content of isolated vesicles with electrochemical cytometry. Vesicles are isolated from cells off-line by differential centrifugation. A suspension of vesicles is then electrokinetically injected onto a fused-silica capillary that terminates onto a PDMS-based microfluidic platform. As individual vesicles exit the separation capillary, they are flushed with lysis solution from neighboring channels in a sheath-flow format. Their membranes are rapidly lysed when they interact with this solution and the electrode, and their contents are subsequently detected by end-column amperometry at the cylindrical carbon-fiber microelectrode. Adapted with permission from Omiatek et al., 2013.

detection (CE-EC). Although conventional CE-EC has been used to quantitatively detect transmitters present in single cells, it could not differentiate between the presence of cytosolic and vesicular transmitter in the resultant electropherograms (Woods et al., 2005). With the electrochemical cytometry device, it is possible to directly investigate the content of single vesicles using a hybrid capillary-microfluidic-electrochemical platform (Figure 8.2) (Omiatek et al., 2009, 2010 & 2013). In this approach, secretory vesicles are first isolated off-line from cell or tissue homogenates using differential centrifugation. Then they are isolated as individual entities in a fused silica capillary by electrophoresis. The capillary terminates into a polydimethylsiloxane (PDMS)-based microfluidic chamber where lysis buffer is continuously delivered and focused on to a cylindrical carbon-fiber microelectrode in a sheath-flow format. Once the individual intact vesicles exit the capillary, they are lysed and amperometrically detected at the microelectrode.

The electrochemical cytometry device has been successfully characterized for both the sensitivity and selectivity of the method (coulometric efficiency >95%) using large unilamellar vesicles (diameter ~200 nm), a model for the secretory vesicle, both containing and lacking an electroactive analyte, dopamine (Omiatek et al., 2009). This provides a potential platform to quantitatively probe the transmitter content of nanometer-sized vesicles. Indeed, quantification of large dense core vesicle (LDCV) content has been accomplished for PC12 cell vesicles

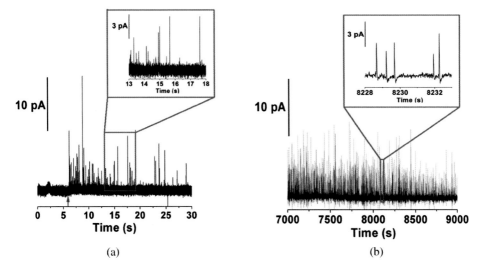

Figure 8.3 Amperometric quantification of catecholamine amounts in PC12 cells. (a). Representative amperometric trace resulting from exocytotic release at intact cells. The arrow indicates elevated K$^+$ application used to induce exocytosis. (b). Representative electropherogram for the detection of individual vesicles by electrochemical cytometry. Vesicles were isolated off-line by differential centrifugation from a matched population of PC12 cells investigated in (a). Vesicles in (b) were injected at 111 Vcm^{-1} for five seconds and separated in an applied field of 333 Vcm^{-1}. Individual events are shown in the expanded axes in (a) and (b) in order to illustrate typical peak characteristics observed for the analyses. The amperometric signals from single-cell experiments were filtered at 2 kHz bandpass, and at 15 Hz bandpass for the electrochemical cytometry experiments, providing different baselines for these different experimental approaches. Adapted with permission from Omiatek et al., 2010.

(Omiatek et al., 2010). A 2000-s portion of a representative electropherogram from the amperometric electro-oxidation of catecholamine in vesicles extracted from PC12 cells is compared with that from measurements of single cell release in which exocytosis is induced by a chemical secretagogue and detected at the cell surface by amperometry (Figure 8.3). An expanded view of the axis (five seconds) is also shown in order to illustrate the typical peak characteristics. The electrochemical cytometry platform provides a high-throughput approach for the measurement of vesicular transmitter content within the cell (~5,000 vesicles per ~1 nL injection *versus* >100 vesicle release events with conventional single cell amperometry experiments for a single stimulation per cell). As each current transient on an amperometric trace can be nearly attributed to an individual vesicle release, the measured charge (Q) from current transients on the amperometric trace can be related to the mole amount of transmitter (N) detected per vesicle using Faraday's law ($Q = nNF$), as described above. When the average amount of neurotransmitter in single vesicles extracted from PC12 cells, obtained from electrochemical cytometry, is compared to the amount expelled during a single release event in a single PC12 cell obtained from conventional single cell amperometry, the conclusion is that only 40% of a vesicle's content is released during the average exocytosis event, a premise that quantitatively contradicts classical hypotheses of all-or-none quantal release.

In a most recent study, we have quantified catecholamine content in individual synaptic vesicles directly sampled from mouse brain tissue (Figure 8.4) (Omiatek et al., 2013). An average of 33,000 catecholamine molecules were measured per vesicle. This amount is considerably greater than typically measured during quantal release at cultured neurons, which suggests again that exocytotic release is not all-or-none. Moreover, vesicular catecholamine content was significantly increased (115% increase, two hours after L-3,4-dihydroxyphenylalanine (L-DOPA) administration), which supports that synaptic vesicles *in vivo* are not saturated with transmitters. Conversely, *in vivo* administration of reserpine decreased the catecholamine content in vesicles. Further, electrochemical cytometry was applied to investigate the effects of the psychostimulant amphetamine on catecholamine content in mammalian cells. The number of catecholamine molecules was rapidly decreased by 50%, one hour after *in vivo* administration, supporting the "weak base hypothesis" that amphetamine displaces transmitter from synaptic vesicles and reduces quantal size. These findings indicate that electrochemical cytometry is an effective approach to quantify transmitter molecules in individual vesicles in animal models relevant for pharmacology. Also, the concept of partial vesicular release raises the possibility for vesicular transmitter content to become a new pharmacological target.

8.3 MATHEMATICAL TREATMENT OF AMPEROMETRIC DATA: PARTIAL DISTENTION DURING EXOCYTOSIS EXPLAINS THE SHAPE OF AMPEROMETRIC EVENTS

8.3.1 SIMPLIFIED DISCUSSION OF THE MATHEMATICAL APPROACH

Considering the high temporal resolution of the amperometric technique, it is possible to characterize the properties of the fusion pore, a nanometric liquid channel that connects the vesicle interior with the extracellular matrix during exocytosis, in some detail. The fusion pore allows neurotransmitters stored in the vesicles to diffuse into the synaptic cleft and the flux of chemical messengers through fusion pore is resolved as a pre-spike foot. This is characterized as a pedestal like feature of the spike, showing a slight increase in current before the larger current peak (Cans and Ewing, 2011) (more detail is given in section 8.5.1). The fraction of amperometric spikes with pre-spike features as well as the fraction of electroactive

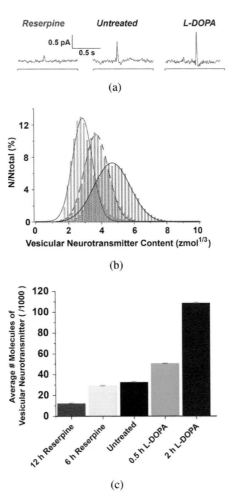

Figure 8.4 Pharmacological manipulation of striatal vesicle neurotransmitter content as measured by electrochemical cytometry of vesicles from mouse brain. (a). Representative current spikes for mice receiving reserpine treatment (left, 12 hours, 20 mg/kg), untreated (center), or L-DOPA treatment (right, 2 hours, 50 mg/kg). (b). Normalized frequency histograms for reserpine-treated mice (red, 12 hours, 20 mg/kg dose, distribution mean = 2.9 $zmol^{1/3}$) and L-DOPA-treated mice (blue, 2 hours, 50 mg/kg dose, distribution mean = 4.7 $zmol^{1/3}$) striatal vesicular dopamine content *versus* controls (black, 3.7 $zmol^{1/3}$). Data are plotted as cube root transforms. Bin size = 0.2 $zmol^{1/3}$. Fits were obtained from a Gaussian distribution of the data. (c). Cumulative analysis of striatal vesicles from mice with respect to drug and time after treatment. The time after pharmacological treatment is associated with regulation of neurotransmitter content in mouse synaptic vesicles. Mean numbers of molecules of vesicular dopamine for mice dosed with 20 mg/kg reserpine for 12 hours and 6 hours treatments were 18,000 ± 100 (N = 7,602) and 29,000 ± 400 (N = 3,029) molecules, respectively. Mean amount of vesicular dopamine from untreated mice striatal tissue was 33,000 ± 100 (N = 20,331) molecules. Mean numbers of molecules of vesicular dopamine for mice dosed with 50 mg/kg L-DOPA for 0.5 hour and 2 hours were 51,000 ± 200 (N = 23,700) and 71,000 ± 400 (N = 22,047) molecules, respectively. Mean values are significantly different from each other (one-way ANOVA, $p < 0.001$). Adapted with permission from Omiatek et al., 2013.

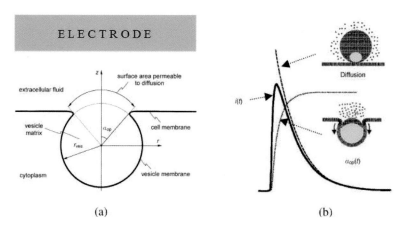

Figure 8.5 Mathematical modeling of amperometric measured transmitter release. (a). Cross-section of an ideal vesicle fusion with the definition of the aperture angle as a time dependent function α_{op} (t) (b). Schematic convolution of the aperture function α_{op} (t) and of diffusion giving rise to an amperometric spike, i(t). Legend: current spike (solid line), membrane fusion (dotted line), diffusion (dashed line). Adapted with permission from Amatore et al., 2010.

neurotransmitters released through a fusion pore depend on the physicochemical properties of the membrane such as curvature, rigidity, lipid composition, etc. (Mosharov and Sulzer, 2005). Amperometric measurements enable the reconstruction of the vesicle opening kinetics as well as the determination of the aperture angle, α_{op}, inevitable for defining the location of the moving boundary, which defines the diffusion of transmitters (Figure 8.5a) (Amatore et al., 2010). Current spike characteristics such as shape and area below the curve provide the information related to the diffusion coefficient of the neurotransmitters and intravesicular concentration, respectively. Therefore, a mathematical analysis can be carried out based on consideration that the current of the amperometric spike, $i(\tau)$, is convoluted from the opening sigmoid function, $S(\tau)$, and purely a diffusional function, $\psi(\tau)$ (Equation 1, Figure 8.5b). In this manner, mathematical models have been employed in order to extract opening function with a high precision, coupling the membrane dynamics to the exocytotic flux of neurotransmitters. Accordingly, $\psi(\tau)$ is obtained *via* Brownian fits to the descending part of the current trace (Figure 8.5b) and $S(\tau)$ is related to the fluctuation of the uncovered surface of each vesicle as a function of time (Equation 2, Figure 8.5a) (Amatore et al., 2008).

$$i(\tau) = 2F \cdot \int_0^\tau \psi(\tau - u) \cdot \frac{dS(\tau)}{d\tau} \cdot du \tag{1}$$

$$S(t) = \frac{1 - \cos\alpha_{op}(t)}{2} \tag{2}$$

These models allow full Brownian motion fitting of current spikes and thus make it possible to determine the opening sigmoid function S(τ) for each exocytotic event without involving any prior assumptions. The significance of the mathematical models is that they allow clarification of the exocytotic process from energetic aspects as well as prediction of the fusion pore and release dynamics (Amatore et al., 2008).

8.3.2 MODELING SUPPORTS THE HYPOTHESIS THAT VESICLE PORE OPENING DOES NOT NEED TO BE FULL

Amperometric data enables reconstruction of the kinetics of the pore opening. In the case of a full release event, the maximal fusion angle α_{op}^{max} is equal to π (α_{op} α_{op}^{max} in Figure 8.5a). However, bilipidic membranes, studied in biophysics, tend to fuse completely to the membrane due to spontaneous dissipation of the curvature energy. From the aspect of exocytosis and its complexity, vesicle membranes could be restricted by local constraints and as a result partial release might happen (Amatore et al., 2010). The early stage of the exocytosis process involves the migration of vesicles across the cellular matrix. Due to the vesicle size and its cargo/membrane costs, directionally designated physical movements define vesicular transport toward the plasma membrane. In this manner, vesicular migration time as well as the number of vesicles needed for neuronal signaling is reduced. One of the transport mechanisms is based on movements of the vesicles along the actin network of the cell cytoskeleton *via* molecular machines toward specifically partitioned areas of plasma membrane. The membrane partitioning is considered to be a result of its interactions with the cell cytoskeleton. Since the cytoskeleton in the vicinity of the cell membrane is connected by a large number of molecules dispersed in the membrane, it has been termed as "membrane cytoskeleton" and differs from the cytoskeleton in the bulk of the cell (Amatore et al., 2009b). There are many constrains that might control the fusion pore formation and its expansion, such as the proteins incorporated in vesicle and cell membranes, membrane cytoskeleton, lipidic composition of the membranes etc. If this is the case, the maximal fusion angle α_{op}^{max} will be less than π (Figure 8.6) (Amatore et al., 2009a, 2010).

It is possible to evaluate the most probable α_{op}^{max} reached by the vesicle by taking into consideration the concentration distributions of the chemical messengers inside each vesicle obtained by the charge of the current spike. Mathematical models provide evidence that the maximum aperture angle values fall between 3° and 20°, and generally α_{op}^{max} is restricted to relatively small values (α_{op}^{max} <45°). This discovery implies that the vesicle matrix is still covered by its membrane at the end of the neurotransmitter expulsion event, thus showing partial release as a mechanism of exocytosis (Amatore et al., 2010).

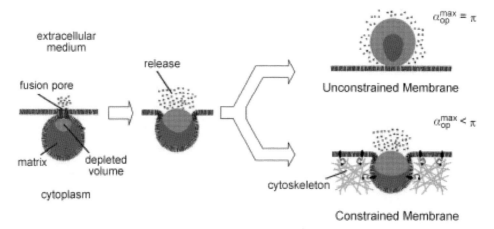

Figure 8.6 Schematic representation of the structural association between vesicle and membrane cytoskeleton network and possible restriction of the α_{op}^{max} due to this relation. Adapted with permission from Amatore et al., 2010.

8.3.3 IMPLICATION THAT LIPIDS CONTROL THE PORE ONCE IT IS OPEN

To date, it is well known that the initial pore formation during vesicle fusion requires involvement of organized biostructures (SNARE proteins, molecular motors, membrane cytoskeleton). However, the nature and structure of the fusion pore after it has been formed is a hotly debated subject. Part of this argument revolves around whether physicochemical factors or biological factors control the pore. Thus, it is essential to investigate the influence of different factors such as membrane proteins and/or lipidic composition on the fusion pore. The patch-clamp technique has been employed to obtain the distribution of pore radii and later the relation to the potential energy that governs the thermodynamics of the pore. The wide distribution of initial fusion pore radii obtained by patch-clamp suggests that the nature of the pore control is not completely protein in nature (Figure 8.7) (Oleinick et al., 2013).

8.4 REGULATION OF OPEN-AND-CLOSED EXOCYTOSIS

8.4.1 EVIDENCE THAT DYNAMIN REGULATES THE OPENING OF THE FUSION PORE

If exocytosis is open and closed, then what regulates the process beyond formation of the fusion pore? Dynamin is a candidate for part of this process. Dynamin is a GTPase mechanochemical enzyme involved in the late steps of endocytosis, where it separates the endocytotic vesicle from the cell membrane. However, several

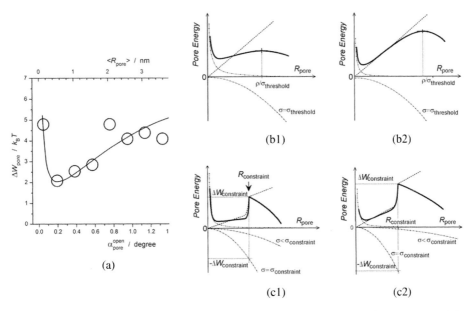

Figure 8.7 Fusion pore potential energy comparison based on experimental (a) and theoretical data (b1, b2, c1 and c2) supports the premise that initial fusion pore have lipidic structure. (a). Experimental trend as a function of the pore opening angle (α_{pore}^{open}) with assumption that pore size distribution is Boltzman controlled. (b1 and b2). Theoretical prediction for the pore's purely lipidic behavior with low or high tension relative to the pore edge, respectively. (c1 and c2). Theoretical prediction when the pore is constrained by scaffolding or biological structures with low or high tension relative to the pore edge, respectively. The overall energy is shown by the solid curve. For more details see Oleinick et al., 2013. Adapted with permission from Oleinick et al., 2013.

recent reports have emphasized its role in exocytosis. In this case, dynamin appears to contribute to the control of the exocytotic pore, and possibly influencing a direct control on the efflux of neurotransmitters.

Dynasore is a selective inhibitor of the GTPase activity of dynamin and has been used to investigate the role of dynamin in exocytosis with amperometric detection (Trouillon and Ewing, 2013). These experiments (Figure 8.8) revealed that dynasore inhibits exocytosis in a dose-dependent manner. Analysis of the exocytotic peaks showed that the inhibition of the GTPase activity of dynamin led to shorter, smaller events. This observation is in good agreement with results published recently from near field microscopy observation of pore dynamics, where the GTPase activity of dynamin was inhibited by dynasore or mutations (Anantharam et al., 2012; Trouillon and Ewing, 2013). The duration of the feet is also reduced, and the constraining effect of the pore is larger on the decaying part of the peak. In terms of pore dynamics, this appears to indicate that the fusion pore opens and closes faster, and that its maximum opening radius is decreased, after dynasore. Thus it has been proposed that dynamin is involved in regulating

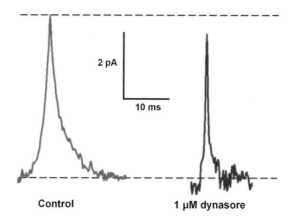

Figure 8.8 Average amperometric peaks obtained from the control and 1 μM dynasore treatments. Amperometry of exocytotic events from single PC12 cells. The working electrode was held at +700 mV *versus* an Ag/AgCl reference electrode using an Axon 200B potentiostat. The output was filtered at 2.1 kHz using a Bessel filter and digitized at 5 kHz. Adapted with permission from Trouillon and Ewing, 2013.

the duration and kinetics of the open and closed version of exocytotic release. The role of dynamin in endocytosis is thought to be tube closure. In exocytosis, dynamin appears to contribute to the dilation and stability of the pore.

In the case of open and closed exocytosis, it appears that the dynamin complex actually contributes to supporting, or framing, the nanotube formed between the vesicle and the membrane. In this case, the pore does not follow its intrinsic kinetics, but is actually controlled by the dynamin complex. The longer t_{foot} in presence of the fully active dynamin might indicate that the foot feature corresponds to the formation of the dynamin assembly and the initiation of the GTPase activity, followed by the pore expansion. This concept, supported by several studies (Trouillon and Ewing, 2013, 2014) suggest that the mechanochemical activity of dynamin is involved in exocytosis, and that this activity relies on its GTPase properties.

8.4.2 EVIDENCE THAT ACTIN REGULATES THE CLOSING OF THE FUSION PORE

A key question in open and closed exocytosis, is what closes the pore? The effect of latrunculin A, an inhibitor of actin cross-linking, on exocytosis in PC12 cells has been investigated, again with single cell amperometry to examine if the actin cytoskeleton might be involved in regulating exocytosis, especially by mediating the constriction of the pore (Trouillon and Ewing, 2014). In an open-and-closed release mode, actin could actually control the fraction of neurotransmitter molecules released by the vesicle.

Following application of latrunculin A to PC12 cells, a 40% increase of the median of the amount released is observed (51% increase of the mean release). The probability to obtain large events (>200,000 molecules released) is also increased. One of the main variations in the peak characteristics after the latrunculin A treatment is the observed increase in peak fall time. The decaying part of the peak is expected to contain some information about the closing of the fusion pore in open-and-closed exocytosis (Mellander et al., 2012). In this case, the increased fall time is evidence that the inhibition of actin polymerization increases the lifetime of the pore, and slows down the collapse of the pore. Thus, it is argued that actin is an important factor in the closing of the lipid pore. These findings strongly support the theory of partial release at PC12 cells, and that actin is involved in regulating the amount of vesicular content released during open-and-closed exocytosis.

Interestingly, actin is known to interact with several membrane-shaping chemical motors, such as dynamin (Pelkmans et al., 2002; Orth and McNiven, 2003; Gu et al., 2010). Together the two proteins appear to be part of the regulation of the fusion pore dynamics in a competitive process between different mechano-chemical transducers, as summarized in Figure 8.9. In this case, dynamin, at least in part, promotes the opening of the pore, and actin its closure. It appears likely that actin and dynamin complement each other in balancing and controlling release during an open-and-closed exocytosis process. The composition of the lipid membrane also plays a critical part. However, this list is not exhaustive, as other polymers, proteins or lipid structures might clearly contribute to this highly dynamic and competitive process.

8.5 POST-SPIKE FOOT IS OBSERVED IN AMPEROMETRY

8.5.1 OBSERVATION OF POST-SPIKE FEET

Mechanistic studies of exocytosis are essential to understand neurotransmission. In the past two decades bioanalytical electrochemical techniques have been developed in order to detect, quantify and characterize the release of neurotransmitters during individual exocytotic events at single cells. As discussed above, amperometry with carbon-fiber electrodes provides high spatio-temporal resolution and has been used to study the fast exocytotic release of signaling molecules. Amperometric data provides quantitative, kinetic as well as mechanistic information related to vesicle fusion events.

A fraction of amperometric transients from exocytosis are often preceded by what is called the "pre-spike foot," a pedestal like feature of the spike, showing a slight increase in current before the leading current peak (Cans and Ewing, 2011). The presence of the pre-spike foot in amperometric traces indicates the leakage of

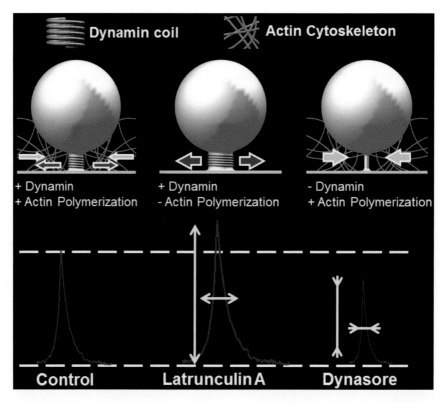

Figure 8.9 Proposed scheme for the contribution of dynamin and actin to exocytosis. Actin appears to force the closing of the pore, whereas dynamin promotes its opening. The resulting typical amperometric spikes are presented in the bottom portion of the figure (not drawn to scale). Exposure to latrunculin A induces higher, wider peaks, and inhibition of the GTPase activity of dynamin with the inhibitor dynasore induces shorter, narrower peaks. Adapted with permission from Trouillon and Ewing, 2014.

neurotransmitter through a fusion pore during the early stage of vesicle fusion with the plasma membrane — again discussed above (Chow et al., 1992). The amperometric spike provides a lot of quantitative information regarding the exocytotic event. The shape of the current peak is related to kinetic information of the release, half-width of the peak suggesting the duration of the event, the amplitude depends on the maximum flux of the molecules released, the rise time shows the time it takes for the fusion pore to open and fall time is defined as the time from 75% to the 25% of the peak (Mellander et al., 2010).

From a mechanistic aspect, traditionally exocytosis has been thought of as an all-or-nothing process (full distension), where vesicles completely collapse into cell membrane and release the entire content into the synapse. However, transient vesicle fusion with partial release of transmitter content has been seen for small synaptic vesicles and large dense core vesicles (Mellander et al., 2010). One can

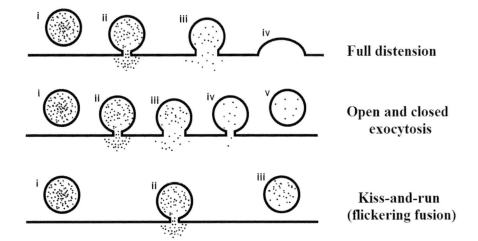

Figure 8.10 Scheme of the three proposed mechanisms for exocytosis: top, full distension of the vesicle into the membrane (i- docking of the vesicle, ii- formation of the fusion pore giving rise to a pre-spike foot, iii- opening of the pore leading to the recording of the 'body' of the peak and iv- full integration of the vesicle into the membrane); middle, open-and-closed (i- docking of the vesicle, ii- formation of the fusion pore giving rise to a pre-spike foot, iii- opening of the pore leading to the recording of the 'body' of the peak, iv- closing of the pore and formation of a fission pore giving rise to a post-spike foot current and v- complete fission of the cell from the membrane); bottom, flickering fusion (i-docking of the vesicle, ii- formation of the fusion pore giving rise to a small, short peak and iii- complete fission of the cell from the membrane). Reproduced with permission from Mellander et al., 2012.

find few possible scenarios for transient vesicular fusion. Sulzer and co-workers reported complex kiss-and-run exocytosis or flickering fusion where a vesicle fuses repeatedly, releasing a fraction of its content during the each fusion event (Staal et al., 2004; Wightman and Haynes, 2004). For open-and-closed exocytosis, the vesicle closes up again following the larger "full" release observed in the amperometric current spike. As discussed above, this is supported by mathematical models of the release event and by the measurement of only 40% release for full release in PC12 cells. Figure 8.10 shows possible mechanistic concepts of exocytosis (Mellander et al., 2012). The importance of fusion pore flickering and open-and-closed exocytosis can be specifically related to natural prevention of the vesicle loss and slow process of endocytosis and vesicle recycling (Staal et al., 2004).

The pre-spike foot is thought to be the fusion pore sticking in its partially open and very small state, allowing transmitter to diffuse out. If open-and-closed exocytosis is the normal mode of release, then it would stand to reason that one would also observe "post-spike feet" in the descending part of the current peak. Indeed, post-spike feet have been reported and an example is shown in Figure 8.11a. This is essentially a direct observation of the closing of the fusion pore after the "normal" release. In order to accurately observe post-spike feet, they are defined as stable current transients three times higher than root mean square (RMS) noise and lasting for at least two milliseconds during the descending region of the peak and followed by the fast decline to the baseline. As the plot is descending, only plateau shapes are included in the analysis. Pre-spike feet were detected for 5.1% of the peaks while post-spike feet were detected for 2.3% of amperometric peaks for control cells. As the number of pre-spike feet events is similar to that for post-spike feet it appears that nearly all transmitter release takes place *via* open-and-closed exocytosis, at least in PC12 cells.

The foot is traditionally described by three features: the lifetime (t_{foot}), the current (I_{foot}), and the charge (Q_{foot}). The value t_{foot} indicates the stability of the fusion pore, I_{foot} depends on the flux of neurotransmitter through the pore, and Q_{foot} represents the charge released through the pore during its lifetime. Mellander et al. have observed differences in the pre- and post-spike foot current in amperometric measurements of open-and-closed exocytosis showing that the amount of neurotransmitter is lost during the full release event, but not all. As expected, the concentration of neurotransmitters in the vesicle interior is lower during closing of the vesicle than during fusion pore opening. This is confirmed by an observed decrease (Figures 8.11a and 8.11b2) in the current between pre- and post-spike foot (Mellander et al., 2012).

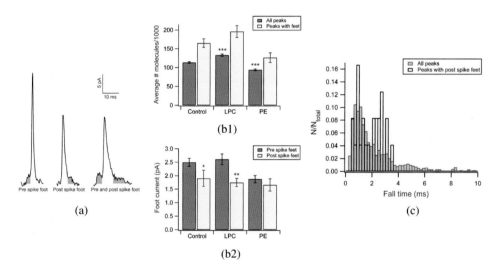

Figure 8.11 (a). Three example peaks from control cells. The first peak displays the traditional plateau foot, here termed pre-spike foot. The second peak shows what is referred to as a post-spike foot, while the third peak displays both types of feet. (b1). Average number of molecules released during single exocytotic events for control cells, cells incubated with lysophosphatidylcholine (LPC), and cells incubated with phosphatidylethanolamine (PE). (b2). Pre- and post-spike foot currents for control cells, cells incubated with LPC and cells incubated with PE. (c). Histogram of fall time for all peaks and peaks with post-spike feet. Reproduced with permission from Mellander et al., 2012.

8.5.2 VARIATION OF LIPIDS CONFIRMS VALIDITY OF POST-SPIKE FEET

The effect of short-term incubation with lipids on pre- and post-spike feet was also investigated. Lipids have been shown to affect exocytosis as the lipid curvature affects the size of the fusion pore and might have some mechanistic bearing on short-term memory (Amatore et al., 2006; Uchiyama et al., 2007). The amount of catecholamine released during exocytosis was found to decrease following incubation with PE and increase following incubation with LPC. This was interpreted as changing the fraction of transmitter released. Consistent with this interpretation, the ratio of the post-spike foot current to the pre-spike foot current was found to increase following incubation with PE and decrease following incubation with LPC compared to control cells (Figures 8.11b1 and 8.11b2). The calculated foot current ratios were then correlated with the amount of messenger released to estimate the vesicular content prior to release. The result was used to calculate the fraction of transmitter released as 39%, a value in apparent agreement with the results from the previous cytometry experiments (Omiatek et al., 2010). These data suggest a model for exocytosis where the vesicles normally open and close and the transmitter released during individual events can be regulated.

8.5.3 SOME EXOCYTOSIS EVENTS APPEAR TO OPEN ALL THE WAY

Amperometric spikes with post-spike foot features have slightly faster dynamics which might indicate that a small fraction of exocytotic events are undergoing full release and opening. The events with the longest fall time do not have post-spike feet. The distribution of fall time for all spikes and spikes with post-spike feet is shown in Figure 8.11c. The fraction of all spikes with slower fall times than the slowest post-spike foot peak was found to be 17%. Some of these slow events might represent vesicle fusion where the fast endocytosis machinery, for some reason, is not employed, but instead the vesicle collapses completely into the cell membrane releasing its entire contents. One can speculate that this happens when a vesicle has lost a key component of its protein machinery and needs to be degraded or recycled. Hence, spike trains were specifically examined for slow as well as for large events and it was found that approximately 2% of all events released more than 200,000 molecules in addition to having base width longer than 50 milliseconds. Thus, it can be argued that most events are open-and-closed exocytosis, and only a small fraction of exocytosis events involve full distension of the vesicle and complete vesicle collapse into cell membrane (Mellander et al., 2012).

8.6 FUTURE PERSPECTIVES

To date, the use of electrochemical cytometry has been aimed at the evaluation of PC12 cells and mouse synaptic vesicles. It is important to examine the content of vesicles in different cell systems to verify the levels of transmitter relative to release. The mechanism of the lysing process in the cytometry method and the coulometric efficiency need to be understood better. Additionally, there is a need to understand and refine the vesicle preparation to be sure that only one type of vesicle is isolated.

Applications of the methods described here will likely include investigations of disease states like Parkinson's disease and perhaps more. It will also be useful to determine the effect of drugs on vesicular content and therefore provide novel targets for pharmaceuticals related to vesicles and disease.

Investigation of open-and-closed exocytosis as the normal mode of release, promises to change the way we look at learning and memory, signal regulation and pharmaceuticals. The fact that only partial release takes place means that release events can be regulated by cells on the level of single vesicles. This is an exciting new angle of research in this area.

ACKNOWLEDGMENTS

We acknowledge the many co-workers who have come before and whose works we cite in this chapter. We acknowledge the support from the European Research Council (ERC Advanced Grant), the Knut and Alice Wallenberg Foundation in Sweden, the Swedish Research Council (VR), and the USA National Institutes of Health (NIH).

REFERENCES

Amatore C, Arbault S, Bonifas I, Guille M (2009a) Quantitative investigations of amperometric spike feet suggest different controlling factors of the fusion pore in exocytosis at chromaffin cells. Biophys Chem 143:124–131.

Amatore C, Arbault S, Bonifas I, Lemaitre F, Verchier Y (2007) Vesicular exocytosis under hypotonic conditions shows two distinct populations of dense core vesicles in bovine chromaffin cells. Chem Phys Chem 8:578–585.

Amatore C, Arbault S, Bouret Y, Guille M, Lemaitre F, Verchier Y (2006) Regulation of exocytosis in chromaffin cells by trans-insertion of lysophosphatidylcholine and arachidonic acid into the outer leaflet of the cell membrane. Chem Bio Chem 7:1998–2003.

Amatore C, Arbault S, Guille M, Lemaitre F (2008) Electrochemical monitoring of single cell secretion: vesicular exocytosis and oxidative stress. Chem Rev 108:2585–2621.

Amatore C, Oleinick AI, Klymenko OV, Svir I (2009b) Theory of long-range diffusion of proteins on a spherical biological membrane: application to protein cluster formation and actin-comet tail growth. Chem Phys Chem 10:1586–1592.

Amatore C, Oleinick AI, Svir I (2010) Reconstruction of aperture functions during full fusion in vesicular exocytosis of neurotransmitters. Chem Phys Chem 11:159–174.

Anantharam A, Axelrod D, Holz RW (2012) Real-time imaging of plasma membrane deformations reveals pre-fusion membrane curvature changes and a role for dynamin in the regulation of fusion pore expansion. J Neurochem 122:661–671.

Anderson BB, Chen GY, Gutman DA, Ewing AG (1999) Demonstration of two distributions of vesicle radius in the dopamine neuron of *Planorbis corneus* from electrochemical data. J Neurosci Methods 88:153–161.

Cans A-S, Ewing AG (2011) Highlights of 20 years of electrochemical measurements of exocytosis at cells and artificial cells. J Solid State Electrochem 15:1437–1450.

Chen GY, Gavin PF, Luo GA, Ewing AG (1995) Observation and quantitation of exocytosis from the cell body of a fully-developed neuron in *Planorbiscorneus*. J Neurosci 15:7747–7755.

Chow RH, Vonruden L, Neher E (1992) Delay in vesicle fusion revealed by electrochemical monitoring of single secretory events in adrenal chromaffin cells. Nature 356:60–63.

Colliver TL, Pyott SJ, Achalabun M, Ewing AG (2000) VMAT-mediated changes in quantal size and vesicular volume. J Neurosci 20:5276–5282.

Dernick G, Gong LW, Tabares L, de Toledo GA, Lindau M (2005) Patch amperometry: high-resolution measurements of single-vesicle fusion and release. Nat Methods 2:699–708.

Detoledo GA, Fernandezchacon R, Fernandez JM (1993) Release of secretory products during transient vesicle fusion. Nature 363:554–558.

Finnegan JM, Pihel K, Cahill PS, Huang L, Zerby SE, Ewing AG, Kennedy RT, Wightman RM (1996) Vesicular quantal size measured by amperometry at chromaffin, mast, pheochromocytoma, and pancreatic beta-cells. J Neurochem 66:1914–1923.

Gu C, Yaddanapudi S, Weins A, Osborn T, Reiser J, Pollak M, Hartwig J, Sever S (2010) Direct dynamin-actin interactions regulate the actin cytoskeleton. EMBO J 29:3593–3606.

Leszczyszyn DJ, Jankowski JA, Viveros OH, Diliberto EJ, Near JA, Wightman RM (1990) Nicotinic receptor-mediated catecholamine secretion from individual chromaffin cells — chemical evidence for exocytosis. J Biol Chem 265:14736–14737.

Matthews G (1996) Synaptic exocytosis and endocytosis: capacitance measurements. Curr Opin Neurobiol 6:358–364.

Mellander L, Cans AS, Ewing AG (2010) Electrochemical probes for detection and analysis of exocytosis and vesicles. Chem Phys Chem 11:2756–2763.

Mellander LJ, Trouillon R, Svensson MI, Ewing AG (2012) Amperometric post spike feet reveal most exocytosis is *via* extended kiss-and-run fusion. Sci Rep 2:907.

Mosharov EV, Sulzer D (2005) Analysis of exocytotic events recorded by amperometry. Nat Methods 2:651–658.

Oleinick A, Lemaitre F, Collignon MG, Svir I, Amatore C (2013) Vesicular release of neurotransmitters: converting amperometric measurements into size, dynamics and energetics of initial fusion pores. Faraday Discuss 164:33–55.

Omiatek DM, Bressler AJ, Cans AS, Andrews AM, Heien ML, Ewing AG (2013) The real catecholamine content of secretory vesicles in the CNS revealed by electrochemical cytometry. Sci Rep 3:1447.

Omiatek DM, Dong Y, Heien ML, Ewing AG (2010) Only a fraction of quantal content is released during exocytosis as revealed by electrochemical cytometry of secretory vesicles. ACS Chem Neurosci 1:234–245.

Omiatek DM, Santillo MF, Heien ML, Ewing AG (2009) Hybrid capillary-microfluidic device for the separation, lysis, and electrochemical detection of vesicles. Anal Chem 81:2294–2302.

Orth JD, McNiven MA (2003) Dynamin at the actin-membrane interface. Curr Opin Cell Biol 15:31–39.

Pelkmans L, Puntener D, Helenius A (2002) Local actin polymerization and dynamin recruitment in SV40-induced internalization of caveolae. Science 296:535–539.

Sombers LA, Maxson MM, Ewing AG (2005) Loaded dopamine is preferentially stored in the halo portion of PC12 cell dense core vesicles. J Neurochem 93:1122–1131.

Staal RGW, Mosharov EV, Sulzer D (2004) Dopamine neurons release transmitter *via* a flickering fusion pore. Nat Neurosci 7:341–346.

Trouillon R, Ewing AG (2013) Amperometric measurements at cells support a role for dynamin in the dilation of the fusion pore during exocytosis. Chem Phys Chem 14:2295–2301.

Trouillon R, Ewing AG (2014) Actin controls the vesicular fraction of dopamine released during extended kiss and run exocytosis. ACS Chem Biol 9:812–820.

Uchiyama Y, Maxson MM, Sawada T, Nakano A, Ewing AG (2007) Phospholipid mediated plasticity in exocytosis observed in PC12 cells. Brain Res 1151:46–54.

Westerink RHS, de Groot A, Vijverberg HPM (2000) Heterogeneity of catecholamine-containing vesicles in PC12 cells. Biochem Biophys Res Commun 270:625–630.

Wightman RM, Haynes CL (2004) Synaptic vesicles really do kiss and run. Nat Neurosci 7:321–322.

Wightman RM, Jankowski JA, Kennedy RT, Kawagoe KT, Schroeder TJ, Leszczyszyn DJ, Near JA, Diliberto EJ, Viveros OH (1991) Temporally resolved catecholamine spikes correspond to single vesicle release from individual chromaffin cells. Proc Natl Acad Sci USA 88:10754–10758.

Wightman RM, Schroeder TJ, Finnegan JM, Ciolkowski EL, Pihel K (1995) Time-course of release of catecholamines from individual vesicles during exocytosis at adrenal-medullary cells. Biophys J 68:383–390.

Woods LA, Gandhi PU, Ewing AG (2005) Electrically assisted sampling across membranes with electrophoresis in nanometer inner diameter capillaries. Anal Chem 77:1819–1823.

CHAPTER 9

COUPLING VOLTAMMETRY WITH OPTOGENETICS TO REVEAL AXONAL CONTROL OF DOPAMINE TRANSMISSION BY STRIATAL ACETYLCHOLINE

Polina Kosillo, Katherine R. Brimblecombe,
Sarah Threlfell and Stephanie J. Cragg

Department of Physiology, Anatomy and Genetics,
and Oxford Parkinson's Disease Centre, Sherrington Building,
University of Oxford, OX1 3PT, UK

9.1 INTRODUCTION

Dopamine (DA) is one of the key neurotransmitters in the function of the basal ganglia, the network of interconnected nuclei that is critically involved in motor control and planning, context-dependent action selection, motivation, addiction, cognition and habit learning (Graybiel et al., 1994; Albin et al., 1995; Knowlton et al., 1996; Middleton and Strick, 2000; Haber et al., 2000; Gerdeman et al., 2003; Schultz, 2006; Cisek and Kalaska, 2010). Dopaminergic innervation of the primary input nucleus of the basal ganglia, the striatum, originates from midbrain DA neurons residing in the substantia nigra pars compacta (SNc) and ventral tegmental area (VTA) (areas A9–A10), that project in a topographic manner *via* the medial forebrain bundle to innervate dorsal and ventral striatal territories through extensively arborizing axons (Bjorklund and Lindvall, 1984; Gerfen et al., 1987; Haber et al., 2000). Phasic activity of DA neurons is hypothesized to encode prediction-related information about rewards (Schultz, 1998; Matsumoto and

Hikosaka, 2009), whilst DA release at the level of the striatum regulates neuronal output and plasticity induction rules (Gerfen and Surmeier, 2011).

The relationship between midbrain DA neuron activity and axonal DA release is incompletely resolved. It is evident that DA axons provide a strategic site for several mechanisms to locally regulate DA release. Neuromodulators acting on dopaminergic axons can gate release events and control the gain, or contrast, of DA release triggered by activity in DA neurons (reviewed in Cragg, 2006; Rice et al., 2011). Furthermore, our recent data reveal that axo-axonic control mechanisms bypass parent soma activity and drive local striatal DA release directly (Threlfell et al., 2012). In this chapter, we review these new insights into axonal mechanisms that govern DA release in the striatum, which we have gained from experimental approaches that couple optogenetics with fast-scan cyclic voltammetry (FCV).

A key input to DA axons which locally regulates DA release arises from striatal cholinergic interneurons (ChIs). Striatal acetylcholine (ACh) gates DA release probability and its short-term plasticity in a frequency-dependent manner *via* nicotinic receptors (nAChRs) on DA axon terminals and muscarinic receptors (mAChRs) on ChIs (Rice and Cragg, 2004; Zhang and Sulzer, 2004; Exley et al., 2008; Threlfell et al., 2010; Cohen et al., 2012). Furthermore, recent evidence shows that synchronous activation of a small network of ChIs directly drives striatal DA release *via* activation of nAChRs (Threlfell et al., 2012; Cachope et al., 2012). Methods for the detection of striatal DA release in real-time are well-established, but previous stimulation approaches have made it difficult to study local mechanisms that regulate release. However, the advent of optogenetic technologies has now enabled the role of defined circuits to be studied with unprecedented specificity, and the role of ACh inputs to DA axons is beginning to be elucidated through this combination of powerful technologies.

9.1.1 DA DETECTION WITH FCV

FCV at carbon-fiber microelectrodes (CFMs) is the current state-of-the-art technique for monitoring identifiable electroactive molecules on a sub-second timescale and with a spatial resolution of microns. FCV at CFMs is ideally suited for exploring the control of DA transmission, especially in the striatum where innervation density of DA axons is high but density of interfering electroactive neurotransmitters, such as 5-hydroxytryptamine (5-HT) and norepinephrine, is low. FCV at CFMs has been used successfully and extensively alongside electrical stimulation at small local bipolar electrodes, to identify resident mechanisms in the striatum that can powerfully regulate how activity in midbrain DA neurons might be translated into release events from striatal axons (e.g., Cragg, 2003; Schmitz et al., 2003; Zhang and Sulzer, 2003; Rice and Cragg, 2004; Zhang and Sulzer, 2004; Exley et al., 2008; Threlfell et al., 2010; Rice et al., 2011).

9.1.2 DRIVING NEURAL ACTIVITY IN LOCAL STRIATAL CIRCUITS: ELECTRICAL STIMULATION *VERSUS* OPTOGENETICS

The use of conventional stimulation methods, such as electrical stimulation, does however have shortcomings for investigating how specific circuits act. Although electrical stimulation is temporally precise, it is non-specific in the cell types that it depolarizes within the vicinity of the stimulating electrodes, and therefore cannot be used to activate a specific pathway or cell type. Investigations into the roles of specific circuits therefore require the combined use of cocktails of receptor antagonists for known interfering transmitters, while other unknown substances that might also be released in response to stimulation cannot be controlled for. Thus, insights from electrical stimulation approaches can be limited.

The recent introduction of optogenetic techniques can overcome this major shortcoming. Optogenetics allows study of the functions of defined circuits and cells with a specificity that has previously been unattainable. This technique has been subject of many recent reviews (Chen et al., 2012; Han, 2012; Packer et al., 2013; Muller and Weber, 2013; Stuber and Mason, 2013). In overview, optogenetics is an optical stimulation technique whereby light-sensitive ion channels are expressed in a population of genetically identifiable cells to allow their activity to be controlled with light. This method achieves sophisticated control over the activity of specific neural circuits. To date, this has been achieved primarily through the use of transgenic rodent driver lines expressing Cre-recombinase under the control of a neuron-type-specific promoter, combined with injection of floxed viral constructs, so that light-sensitive proteins are expressed in a neuron-type-specific manner.

9.1.3 ELUCIDATING THE ROLE OF CHIS IN STRIATAL DA TRANSMISSION

Striatal ChIs are large aspiny neurons, are tonically active, and provide one of the highest levels of cholinergic innervation in the brain (Hoover et al., 1978; Wilson et al., 1990; Kawaguchi, 1992, 1993; Bolam et al., 2000). ACh, like DA, might act at least in part by activation of extrasynaptic receptors *via* volume transmission (Descarries and Mechawar, 2000; Cragg and Rice, 2004). ACh regulates DA transmission *via* nAChRs on DA axons and *via* mAChRs on ChIs in a frequency-dependent manner and through different subtypes of nAChRs and mAChRs in dorsal *versus* ventral striatum (Threlfell and Cragg, 2011). ChIs operate a gain control mechanism on striatal DA release by increasing the initial release probability for single pulse stimulation, whilst triggering subsequent short-term depression during subsequent depolarizations (Rice and Cragg, 2004; Exley et al., 2008). Thus, stereotyped pauses in ChI firing, which are characteristic of ChI activity in reward-related learning (Morris et al., 2004), may result in inactive nAChRs and consequently increase the gain for phasic DA neuron activity that signals reward predictions (Cragg, 2006).

However, pauses are not the only stereotyped activity in ChIs: pauses can be flanked by a preceding and/or rebound brief burst of high frequency firing. Furthermore, ChIs function as a network: electrophysiology cross-correlogram data suggest that pairs of neurons are coupled in their characteristic burst-pause-burst discharge (Morris et al., 2004; Graybiel, 2008). Furthermore, inputs that are known to drive ChIs, such as thalamic projections, provide dense innervations to the striatum (Lapper and Bolam, 1992; Sadikot et al., 1992) and are likely to drive a population response. This synchronous discharge of a ChI network would provide a massive boost to local ACh levels activating nAChRs on striatal DA axon terminals. Our group, therefore, has recently explored whether this local activation of ChIs could also drive striatal DA transmission events. The combined use of optogenetics and FCV at CFMs is a timely experimental strategy to address whether defined activity in ChIs influences striatal DA release.

Answering this question provides major advances in understanding brain function. A direct drive of DA release by ACh would indicate that local cholinergic control of striatal DA transmission goes well beyond simple dynamic gating of DA release in response to activity in midbrain DA neurons. Instead, striatal DA signalling can be driven directly at the axonal level by a different neural circuit that bypasses DA neurons. In turn, the information encoded by ChIs and their long-range afferent inputs that synchronize their activity, might also be translated into a striatal DA signal. Under these circumstances, striatal DA transmission would reflect not only parameters encoding reward-prediction error or related states conveyed by the phasic burst firing of midbrain DA neurons, but also incorporate signalling from other systems, such as thalamic projections, which can provide attentional and contextual information.

In this chapter, we will briefly review our recent findings of how striatal ChIs can directly drive local DA transmission (Threlfell et al., 2012). We will also discuss key methodological issues that must be considered when coupling optogenetic approaches with FCV and pharmacological experiments. We will describe the issues that we have encountered that will be of general relevance whatever the circuit being explored.

9.2 OPTOGENETIC ACTIVATION OF STRIATAL CHOLINERGIC INTERNEURONS DRIVES DOPAMINE TRANSMISSION

Combined use of optogenetics and FCV at CFMs allowed us to explore whether synchronous activation of a small population of striatal ChIs can directly drive local DA transmission. This experimental question would not be possible to address using electrical stimulation that would co-activate the DA axis directly. We stereotaxically injected a floxed viral construct coding for Channel Rhodospin2 (ChR2) fused to enhanced yellow fluorescent protein (eYFP) into the striatum of a transgenic mouse line expressing Cre-recombinase in choline acetyltransferase positive (ChAT+) neurons (ChAT-Cre mice). Using blue light to activate ChIs, recording neuronal activity in ChIs using patch-clamp electrophysiology, and detecting DA with FCV at CFMs, we identified that single action potentials in a small population of ChIs can directly drive DA release by activating nAChRs on striatal DA axon terminals, bypassing the need for activation of midbrain DA neuron cell bodies residing in theSNc (Threlfell et al., 2012).

9.2.1 METHODOLOGY FOR COMBINED FCV AND OPTOGENETICS

We injected Cre-inducible recombinant adeno-associated virus (AAV) vector containing the ChR2 gene sequence fused with eYFP in heterozygous ChAT-Cre transgenic mice. We also performed some experiments exploring inputs to the ChIs originating from thalamic nuclei; for this purpose we targeted our injections to the parafascicular nucleus (Pf) of the thalamus in a CamK2a-Cre mouse line (CamK2a = calcium/calmodulin dependent protein kinase II a). Briefly, mice were anesthetized with isoflurane, placed in a stereotaxic frame and small craniotomies were made bilaterally over the nucleus of interest. With a Hamilton syringe micro-injector, 1 μl of viral construct was injected in the striatum or 300 nl were injected in the thalamus in each hemisphere (Threlfell et al., 2012). After allowing sufficient time for transgene expression, normally between 2–10 weeks, brains were removed and immersed in ice-cold oxygenated artificial cerebrospinal fluid, and acute slices were prepared for voltammetry as previously described (Rice and Cragg, 2004; Threlfell et al., 2010).

ChR2-expressing striatal ChIs or thalamic terminals were visualized by eYFP expression in brain slices of caudate-putamen (CPu) and nucleus accumbens (NAc). ChR2 was activated using a 473 nm diode laser (DL-473, Rapp Optoelectronic) coupled to the immersion objective microscope *via* a fiber-optic cable. We used a laser for spot illumination (15–60 μm, depending on the objective), but see Section 9.3.5 for a discussion of the choice of laser *versus* LED. TTL-driven light pulses supplied to the laser system were used to trigger depolarization in ChIs and were generated out-of-phase with voltammetric scans, in order to prevent possible interference of the photoelectric currents with recorded FCV signals (see 9.3.7.1). Stimulation protocols were designed to reflect physiological firing frequencies of ChIs.

DA release in response to light-induced stimulation of ChR2-expressing neurons was monitored every 2.5 minutes with FCV at CFMs using a Millar

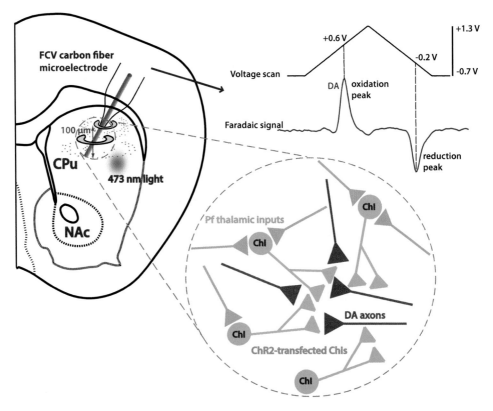

Figure 9.1 Schematic representation of experiments with FCV at CFMs and optogenetic activation of DA release. CFM is inserted into striatal slice to a depth of 100 μm. Triangular voltage scan reveals DA release is evoked by a local 473 nm blue light flash to either ChR2-transfected ChIs or Pf thalamic inputs to ChIs.

voltammeter (Julian Millar, Barts, and the London School of Medicine and Dentistry), as previously described (Rice and Cragg, 2004; Exley et al., 2008; Threlfell et al., 2010). Briefly, the scanning voltage was a triangular waveform applied to carbon-fiber from −0.7 V to +1.3 V and back to −0.7 V against Ag/AgCl reference electrode with a scan rate of 800 V/s (Figure 9.1). Recorded voltammetry sweeps were digitized at 50 kHz using a Digidata 1440A acquisition board and stored for off-line analysis. All the data were acquired using Axoscope 10.2 (Molecular Devices) and analyzed using locally written programs.

9.2.2 KEY FINDINGS

Our experiments showed that light-induced depolarization of striatal ChIs, which recruits a small population of ChIs, can trigger DA release *via* activation of nAChRs on striatal DA axon terminals (Figure 9.1) (Threlfell et al., 2012). Thus, synchrony of firing in the cholinergic network overrides the need for the activation of midbrain DA neurons to trigger striatal DA release, as striatal release events are initiated directly *via* activation of nAChRs present on DA axons. The requirement for a synchronous activation in a group of ChIs was confirmed with manipulation of the firing behavior of individual ChIs. Current injection *via* the patch pipette into individual electrophysiologically confirmed ChIs, elicited action potential firing in the patched cell, while concurrent monitoring of DA with CFM in close proximity to the spiking cell detected no release events. This manipulation showed that activation of individual ChIs was insufficient to trigger detectable levels of DA release, whilst light stimulation of ChR2-transfected ChIs population resulted in DA transients of amplitude comparable to electrically evoked events (Threlfell et al., 2012).

To probe further the physiological relevance of this newly discovered locally-driven release mechanism and to determine whether it could be routinely recruited by endogenous systems during normal brain function, we activated thalamic inputs to ChIs, which are known to synchronize firing of the striatal cholinergic system *in vivo*. These experiments were performed in CamK2a-Cre transgenic line where ChR2-expressing thalamostriatal fibers, originating from the Pf, were light-activated whilst striatal DA release was recorded. Activation of this projection pathway activated ChIs and evoked striatal DA release (Figure 9.1). All the published experiments were performed with FCV recordings in the dorsal striatum (Threlfell et al., 2012), but we also observed ChI-evoked DA release events in the NAc. Our observations were corroborated by another research group who showed that simultaneous recruitment of a population of ChIs in NAc can drive local DA release, both *in vitro* and *in vivo* (Cachope et al., 2012).

Furthermore, regardless of the rate of stimulation of ChIs (which could follow pulse trains up to 40 Hz as evident from electrophysiology data), the amplitude of the recorded DA transients showed no sensitivity to frequency, suggesting that frequency information is not conveyed by individual ChIs (Threlfell et al., 2012). This is consistent with observations in electrical stimulation experiments where frequency sensitivity is absent when nAChRs are activated (Rice and Cragg, 2004). We also performed experiments where we activated striatal DA axon terminals directly following ChR2 injections targeted to SNc in a dopamine transporter promoter (DAT)-Cre mouse line to evoke release of striatal DA by directly activating dopaminergic axis, but this is not the subject of the current chapter.

9.3 CRITICAL METHODOLOGICAL ISSUES

Coupling of FCV and optogenetics offers a powerful approach to probe circuit level interactions between defined cell populations. However, there are a number of methodological issues which we encountered, and these should be given due consideration when planning such experiments.

9.3.1 CHR2 EXPRESSION METHOD: LOCAL TARGETING *VERSUS* TRANSGENIC BREEDING APPROACHES

The Cre-loxP optogenetic approach enables cell type-specific expression of ChR2 using viral vector delivery of double-floxed inverted open reading frame (DIO) constructs carrying the gene sequence from algae *Chlamydomonasreinhardtii* that encodes the light-activated cation channel (Boyden et al., 2005; Zhang et al., 2006; Tsai et al., 2009; Cardin et al., 2009, 2010). Such constructs use a strong general promoter (e.g., EF-1α) to drive high ChR2 expression levels to offset low single channel conductance of ChR2 and are packaged in replication-deficient viral vectors, for example those of AAV, which offer extensive spatial spread and high expression levels (Taymans et al., 2007; Cardin et al., 2009). The Cre-loxP system is ideal for selective optogenetic control of genetically-defined neuronal networks owing to the ready availability of transgenic rodent lines expressing Cre-recombinase under numerous cell-type specific promoters using knock-in or bacterial artificial chromosome (BAC) technology (e.g., Gong et al., 2007; Tsai et al., 2009; Madisen et al., 2010; Taniguchi et al., 2011; Zhao et al., 2011; Weber et al., 2011; Witten et al., 2011; Schonig et al., 2012). Thus, it is possible to virally deliver opsins to a specific brain region using stereotaxic surgery, whereby only Cre-positive cells in the target area will activate ChR2 expression by selective recombination of the loxP sites in the construct resulting in inversion of the transcription frame and subsequent expression of ChR2, driven by a strong ubiquitous promoter.

Notably, for species where transgenesis is more difficult to achieve or the choice of Cre-expressing driver lines is restricted (e.g., primates, rats), approaches have been adopted that use constructions not conditionally targeted, i.e., do not use the Cre-loxP method. For example, opsin expression can be driven directly by a promoter sequence incorporated into a viral construct, which can be either strong general promoters which indiscriminately transfect all cell types in the vicinity of the injection site or cell type-specific promoters which predominantly target a particular neuronal population. These approaches frequently use lentiviral capsids for opsin delivery (Tsubota et al., 2011; Diester et al., 2011; Galvan et al., 2012; Gerits et al., 2012; Han et al., 2009, 2011; Ruiz et al., 2013; Yaguchi et al., 2013), although some rabies virus-based and AAV-based constructs also exist (e.g., Löw et al., 2013; Wickersham et al., 2013). However, there are several considerable shortcomings in employing non-Cre-loxP-targeted approaches compared with Cre-loxP-targeted approaches. For example, with lentiviral vectors where the specificity of ChR2 incorporation depends on cell-type-specific promoters, expression tends to be less selective due to leaky vector expression, and shows greater preference for transduction of excitatory neurons but lower levels of transfection spread and protein expression (Boyden et al., 2005; Han et al., 2009; Huber et al., 2008, Gradinaru et al., 2009; Nathanson et al., 2009a,b). These shortcomings could therefore introduce a confounding influence from activation of a transfected cell population which is not the primary target of investigation, or generally render optogenetic experiments less successful due to low expression levels/transfection spread which could make it difficult to achieve activation levels sufficient to modulate a particular circuit. Low expression is thought to arise because cell type-specific promoters used to drive ChR2 expression are less effective than stronger ubiquitous promoters like EF-1α, which are inserted into Cre-loxP-targeted constructs. These shortcomings of lentiviral vectors could also make it difficult to study function of inhibitory interneurons, for example. However, Cre-recombinase-expressing rat lines are emerging (Weber et al., 2011; Witten et al., 2011; Schonig et al., 2012), and can be used in the future for injections of Cre-loxP-dependent viral vectors in rat studies.

An important consideration for mouse optogenetic experiments is whether to choose transgenic animal lines that do not require local viral injections because they have been engineered (e.g., using BAC technology) to constitutively express ChR2 in a specific neuronal population, and can therefore simply be bred for use without surgical administration of ChR2 vectors. For example, there are transgenic lines currently available which express ChR2 in gamma-aminobutyric acid (GABA) ergic, cholinergic, serotonergic and parvalbumin-expressing neurons (Zhao et al., 2011). There are also ChR2 "reporter" lines that can be crossed with any existing Cre-line to incorporate ChR2 into specific cell types (e.g., Madisen et al., 2010). Both of these approaches can present an attractive alternative to the need for stereotaxic surgery for injection of viral vectors into the brain. Surgery can be time-consuming, and can result in experimental error due to injection mistargeting and/or variability of ChR2 expression levels.

However, there are disadvantages to transgenic lines that constitutively express ChR2. Expression of ChR2 in a transgenic model usually means that ChR2 is incorporated globally into the same neuron type throughout the brain, whereas, by contrast, surgical targeting enables regional selectivity. Also, there may be developmental compensations caused by the constitutive presence of ChR2 which could alter the function of the circuit of interest. In addition, in some specific BAC transgenic lines, the BAC construct may affect expression levels of the endogenous neurotransmitter of interest or its regulatory enzymes/elements. Specifically, incorporation of flanking gene elements in a BAC vector could result in an unexpected change in expression of a gene involved in the pathway under study. A specific example is the ChAT–ChR2–eYFP transgenic line (Ren et al., 2011, Zhao et al., 2011), which, in addition to expressing ChR2 in ChIs, also shows functional overexpression of the vesicular acetylcholine transporter (VAChT), since the incorporated ChR2 BAC construct carries copies of the gene encoding VAChT (Kolisnyk et al., 2013). Consequently, this mouse line has abnormally increased cholinergic tone and therefore any experimental paradigms investigating cholinergic system function using these transgenic mice should be interpreted with caution.

9.3.2 CHOICE OF AAV

When using the Cre-loxP system in transgenic mouse lines for viral delivery of ChR2 by means of stereotaxic surgery, it is important to understand the basic working principles of the viral vectors employed, although the rationale for vector choice is rarely given in the published literature. Here, we discuss the vectors used in our own experiments, AAV-based ChR2 constructs (Cardin et al., 2009; Tsai et al., 2009; Kravitz et al., 2010; Lin, 2011).

Currently available data on AAV-packaged vectors suggests that different serotypes may have differential 'preferences' for a cell type that they predominantly transfect (e.g., Rabinowitz et al., 2002; Broekman et al., 2006; Sanchez et al., 2011; Korecka et al., 2011; Hutson et al., 2012). This factor is important when choosing a serotype, given that preferential transfection of a particular cell type may mean that only low levels of expression in the cell population of interest are present if the wrong type is used. Furthermore, preferential transfection with higher expression levels in a related pathway may lead to confounding data that could mask the signals from the main pathway under investigation.

Another critical issue that we have encountered in our own work is the direction of trafficking of particular ChR2 vectors from the injection site. Although very few articles published to date explicitly distinguish between retrograde and anterograde trafficking of the injectable ChR2 constructs, there is now sufficient evidence to suggest that different serotypes used for AAV-packaging differentially employ axonal transport machinery, in rodents and primates (Masamizu et al., 2011; Ciesielska et al., 2011). For example, AAV serotype 2 is transported only anterogradely, while AAV serotype 6 has the capacity for preferential retrograde transport (Salegio et al., 2013). In our own unpublished work, we have encountered both retrograde and anterograde axonal transport for ChR2, packaged into AAV serotype 5 but not serotype 2, which is consistent with observations made by other groups (e.g., Paterna et al., 2004). The complication introduced by bidirectional axonal transport of ChR2 need not be problematic if local, intrinsic circuits are those transfected, or if the cell type being studied is the only population expressing Cre-recombinase in the injected region. However, where two or more pathways concurrently express Cre-recombinase and are known to send reciprocal collateral projections to each other, careful consideration should be given to the use of anterograde-only vectors, so that functions of each specific pathway can be determined independently of the other.

9.3.2.1 ChR2 expression levels

In our work, we have found that different AAV serotypes injected in the same transgenic lines and brain regions with the same virus volumes/titres lead to different ChR2 expression levels. Specifically, AAV5-based constructs consistently result in higher eYFP levels, presumably reflecting greater expression of ChR2, compared to AAV2-based constructs. This does not appear to be a function of a slower transfection/expression, as extended post-transfection times (eight weeks) for AAV2 do not lead to reach levels of ChR2 seen at much shorter times with an AAV5-based construct. This is true for local expression levels in soma near the site of injection as well as remote axons to where ChR2 must be trafficked, in our hands for two different genotypes: ChR2-injected ChAT-Cre and CamK2a-Cre mice. This observation suggests that effective transfection with ChR2 varies substantially as a function of serotype choice and should be taken into account. A safe strategy to adopt would be to ensure that the virus used is of the same serotype as has been previously reported in the literature for successful experiments with the Cre-driver line and cell type of interest, where this is possible.

9.3.2.2 Axonal expression of ChR2 and post-injection delays

Due consideration should also be given to the times allowed post-surgery to generate sufficient opsin expression at the site of interest. Sufficiently high and stable levels of ChR2 are reached at earliest timepoints at sites most proximal to the injection site. For example, sufficient ChR2 expression is attained in local ChAT-Cre cell bodies following striatal injection, well before that in the striatal axons of long projecting inputs from CamK2a-Cre intralaminar thalamus following thalamic injections. Axonal trafficking of ChR2 from the transfected cell body at the injection site to remote axonal terminals takes time, and therefore animals should be kept for longer prior to being used for experimentation, where terminals of interest are to be light-controlled. For example, with AAV5-packaged ChR2 injections in the striatum of ChAT-Cre mice, successful experiments can be performed as early as two weeks post-surgery, whilst with thalamic injections in CamK2a-Cre mice with the same AAV5-based construct, ChR2 expression at striatal terminals reaches sufficiently high levels at least four weeks post-injection and is completely absent at two weeks.

AAV serotype is also a factor. Our experiments highlighted that different expression times were required for different serotypes to generate sufficient ChR2 expression levels at axonal terminals of long projection pathways. Specifically, following injections of AAV5-packaged ChR2 in the intralaminar thalamus, eYFP fluorescence could be seen in CPu after four weeks, whereas AAV2-based constructs were trafficked to the thalamostriatal terminals at a much slower rate, requiring approximately eight weeks or longer for detection of eYFP-positive fibers in CPu.

9.3.3 CHOICE OF FLUORESCENT TAG

The ability to visualize ChR2 expression is a valuable part of experiments, both during live imaging and for subsequent verification of cell-specific expression. Commercially available opsin constructs normally incorporate a fluorescent tag which allows live visualization of the opsin expression, location and intensity without additional tissue processing. It is important to use a construct in that ChR2 (or other opsin) is fused to a fluorescent protein that is excited by a different wavelength of light to that required to activate the opsin itself. Otherwise, when locating regions of opsin expression by visualising the fluorescent tag, the opsin itself becomes activated prematurely/inappropriately. This will become particularly important for dual opsin experiments where visualization-activation spectra for each opsin and respective fused fluorescent protein have greater capacity for overlap. A combination of very low light intensity and/or infra-red differential interference contrast (IR-DIC) optics to place electrodes can circumvent this issue if necessary, reserving brief visualization only for the initial localization of ChR2-expressing cells/fibers.

When using optogenetic approaches, especially in a newly generated animal line or using a new viral construct, it is essential to verify that the opsin is expressed in the expected region and/or cell type of interest, by double-labeling for expression of the ChR2 construct and a cell-specific marker, typically using immunofluorescence studies. These studies can exploit the fluorescent tag for the opsin (e.g., eYFP) alongside another appropriately selected fluorophore that operates at a different wavelength, for immunofluorescence to indicate a cell-specific marker. This step identifies whether the desired cell population is targeted and therefore will highlight the presence of any possible off-target effects if ChR2 is found to be expressed inappropriately.

9.3.4 LASER OR LED

The choice between a laser and an LED-based system for optical activation depends on the type of stimulation required for the experiment, and has significant budget implications. Use of a laser system offers focused spot illumination achievable by directing the light through the microscope optics *via* a fiber-optic cable. Depending on microscope objective magnification and the size of fiber-optic cable, a spot size as small as a few micrometres can be achieved. However, the brain tissue might affect the incident beam intensity *via* absorption, diffraction, or scattering of the light. Laser systems are generally more expensive to buy and maintain than LEDs, especially for complex set ups. If full-field illumination is satisfactory for the experimental question, then a low cost approach can be achieved by using something as simple as a manipulator-mounted LED. Focused spot illumination can also be achieved to some degree with an LED using a pin-hole system, but for that purpose LED systems need to be coupled to the microscope optics.

The amount of depolarization generated in a given cell by optical stimulation reflects the summation of photocurrents from all activated opsin channels, and is proportional to the light intensity at the target and the illuminated surface area. On a per cell basis, diffused light from an LED system having lower light intensity, but able to cover larger illumination area, could have the same effect as high-intensity illumination from a highly focused laser beam on a smaller surface area. However, in practical terms, experiments running on LED and laser systems can be expected to depend on the geometry of the system under study, the expression pattern of opsin, the question addressed and the measured outcome. For example, when population activation with little spatial specificity is required, a wide field LED approach would be suitable, whereas to study effects of activation of a single synapse/cell, a spot laser might be required. Work performed in our lab shows that both small spot laser illumination and wide field LED illumination can be used interchangeably when studying the population response such as locally evoked DA release detected at a CFM. However, by using a focused laser beam of only 15 μm in diameter to activate ChR2-expression ChIs, we have been able to establish that action potentials synchronized in only a few ChIs are required to directly drive widespread DA release. By contrast, when using ChR2s to activate thalamostriatal terminals to depolarise ChIs to then evoke DA release, the determining factor in the choice of LED *versus* laser was the expression and innervation profile of ChR2-positive thalamic fibers. Thalamic innervation of ChIs is likely to be variable and patchy and therefore the likelihood of driving ChI-evoked DA release by activation of thalamic afferents is lower than when directly driving a population of cholinergic interneurons. Thus, focused spot-illumination of tissue with a laser is less successful than full-field illumination with a LED in experiments that involve activating thalamic projections. When expression levels are high, then both laser and LED activation of ChR2-expressing axon terminals are possible. Thus, where ChR2 expression is low or very diffused/sparse and a release site proves difficult to find, full-field illumination with an LED might be the best approach to ensure that any ChR2-expressing fibers present in the field of view can be activated.

9.3.5 LASER POWER AND PULSE DURATION

In the same way that varying the amplitude or pulse width of electrical stimuli will vary the evoked outcome, the events generated in optogenetic experiments will vary with the power of laser or LED illumination. It is very important to standardize the relative efficacy of illumination on the experimental measures, e.g., DA release, on an experiment-to-experiment basis, since each injected animal will have slightly different ChR2 expression levels following stereotaxic surgery and may require greater or lower illumination power for experiments to be comparable. It is important also to ensure that a chosen stimulation power is not supra-maximal, especially when performing pharmacological manipulations which are expected to alter DA release levels, but also because laser/LED can generate local heat which compromises tissue quality and could erroneously contribute to the experimental results. It is advisable to be able to identify actual power of light output from at the microscope objective/optical fiber with a light power meter in each experimental set-up, to standardize and troubleshoot experiments over time.

9.3.6 ELECTROCHEMICAL ANALYSIS

Successful use of FCV for DA detection requires careful data analysis to ensure that observed currents at the DA oxidation and reduction potentials are attributable to DA, rather than to interfering factors. Several parameters, such as changes in local pH and extracellular Ca^{2+} concentration due to electrical stimulation or other activity, or electroactive drugs, can contribute to background shifts that superpose with DA peaks. When combining optogenetic stimulation with FCV, there are fewer artifacts associated with whole field stimulation, but there are other factors to consider.

9.3.6.1 Photoelectric effect of CFM photo-activation

One particularly important consideration is that direct illumination of the CFM can generate photoelectric currents. These may interfere with data acquisition by introducing artefacts in the cyclic voltammogram, similar to artefacts reported in combined electrophysiology/optogenetics recordings with silicone probes, tungsten electrodes, etc. (Han et al., 2009), and thus could either obscure identification of the electroactive compound detected or distort quantification of the final concentration. It is possible to avoid direct illumination of the CFM with a focussed spot laser system but this is more problematic with full-field illumination, e.g., with an LED. For this latter application, light pulses can be time locked electronically to be delivered out-of-phase with the intermittent voltammetric waveform scans, which prevents such interference.

9.3.6.2 Drugs as photo- and electroactive compounds

Drugs are often used to probe the roles of specific channels and receptors in the DA release. When using FCV, it is important to be aware of the potential for drugs to be electroactive or to quench DA signals; drugs with reactive groups can oxidize/reduce by the applied potentials giving rise to competing currents or can alter the sensitivity of the CFM to DA. This problem remains when coupling FCV with optogenetics with the added potential problem for drugs to be photoactive. Drugs that are unstable under optogenetic light stimuli can have breakdown products that alter the sensitivity of the CFM to DA or have additional off-target effects and, depending on the reaction kinetics, may substantially change the concentration of the applied drug. An example of a photoactive drug is nifedipine. Nifedipine is a commonly used L-type calcium channel blocker, which under light activation was found to modify DA voltammograms in size, shape and potential (Figure 9.2).

9.3.7 PHYSIOLOGICAL RELEVANCE

It is also important to consider whether the findings from optogenetic experiments indicate mechanisms that operate under physiological conditions; *in vivo*,

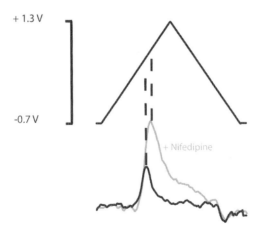

Figure 9.2 Nifedipine has photoactive effects on DA detection. DA can be identified by oxidation and reduction peak potentials of ~+0.6 V and −0.2 V. Following a 473 nm light pulse, nifedipine (10 μM) modifies the DA voltammogram in a manner that does not occur with electrical stimulation (not shown).

and these can be supported by parallel observations from alternative or complementary methods.

9.3.7.1 Physiological firing frequencies

Like all ion channels, opsins have specific open/closing kinetics that limit their frequency of reactivation. Most commonly used ChR2 variants show relatively slow recovery from channel inactivation, requiring around 25 milliseconds, which means they are restricted to frequencies of activation of ~40 Hz (Boyden et al., 2005; Ishizuka et al., 2006; Lin et al., 2009; Gunaydin et al., 2010). This feature can be a limitation for optogenetic studies in which the neural system of interest exhibits higher firing frequencies than 40 Hz or where it would otherwise be desirable to drive the system at higher frequencies. In general, it is important to check using electrophysiological recording whether the neurons of interest follow the optical stimulation with spiking. Some versions of ChR2, such as ChETA, exhibit faster channel kinetics than ChR2 and have been reported to drive activation at frequencies of up to 100 Hz. However, this opsin has lower single channel conductance and higher desensitization rates than ChR2 (Gunaydin et al., 2010; Lin, 2011) and therefore in practice it can be difficult to generate sufficiently high expression levels of ChETA to drive sufficient activity of a system of interest. In our hands, experiments with ChETA were unsuccessful at driving DA release (unpublished observations).

9.3.7.2 Synchrony and selective activation of projection pathways

Caution is required when interpreting the results emerging from optogenetic experiments. For example, recent findings, such as glutamate/DA co-release or GABA/DA co-release have been observed employing optogenetic approaches (Stuber et al., 2010; Tecuapetla et al., 2010; Tritsch et al., 2012) and do not yet clearly tally with anatomical data. It is important to ensure that any discovered mechanism brought into light by optogenetic techniques, correlates with data from other approaches, and has physiological relevance. It will continue to be important to have supporting evidence from other approaches, such as anatomical or ultra-structural data, to corroborate, for example, that the co-released neurotransmitters have the appropriate machinery for co-packaging, and are present and localized to the same synaptic terminals. Furthermore, optical activation, while specific for selective neurons, may produce an artificial state of macro-synchronization of a particular network, which may have limited correlation with normal brain function. Hence, in our own experiments exploring the effects of activating ChIs, we chose also to activate inputs to ChIs, to explore whether ACh-driven DA release could be recruited by endogenous inputs to ChIs that normally synchronize their activity (Threlfell et al., 2012).

9.4 SUMMARY

By coupling optogenetic activation with DA detection using FCV at CFMs, we have shown that ACh circuits intrinsic to the striatum, can act locally and directly on DA axons, bypassing activity in midbrain DA neurons, to drive DA release. This insight shows that local influences at the strategic end point of DA neurons, not only gate DA transmission in response to DA neuron activity, as previously thought, but moreover, play an independent role in generating DA signals, and presumably function, in the striatum. There are unparalleled strengths of optogenetic techniques that have allowed this mechanism to be revealed for DA, but this method is not without limitations, and requires careful considerations. Nonetheless, the data we review here suggest not only that axonic actions of ACh on DA transmission systems might be more important than previously thought, but generally suggest that axo-axonic interactions between neurotransmitter actions can be powerful and strategic sites for governing neuronal signal integration.

ACKNOWLEDGMENT

The authors are grateful to their funding bodies: Parkinson's UK, the Medical Research Council (UK) and the Clarendon Fund (University of Oxford).

REFERENCES

Albin RL, Young AB, Penney JB (1995) The functional anatomy of disorders of the basal ganglia. Trends Neurosci 18:63–64.

Bjorklund A, Lindvall O (1984) Dopamine-containing systems in the CNS. In: Handbook of chemical neuroanatomy (Hokfelt T, Bjorklund A, eds), pp 55–122. New York: Elsevier.

Bolam JP, Hanley JJ, Booth PA, Bevan MD (2000) Synaptic organisation of the basal ganglia. J Anat 196:527–542.

Boyden ES, Zhang F, Bamberg E, Nagel G, Deisseroth K (2005) Millisecond-timescale, genetically targeted optical control of neural activity. Nat Neurosci 8:1263–1268.

Broekman ML, Comer LA, Hyman BT, Sena-Esteves M (2006) Adeno-associated virus vectors serotyped with AAV8 capsid are more efficient than AAV-1 or -2 serotypes for widespread gene delivery to the neonatal mouse brain. Neuroscience 138:501–510.

Cachope R, Mateo Y, Mathur BN, Irving J, Wang HL, Morales M, Lovinger DM, Cheer JF. (2012). Selective activation of cholinergic interneurons enhances accumbal phasic dopamine release: setting the tone for reward processing. Cell Rep 2:33–41.

Cardin JA, Carlén M, Meletis K, Knoblich U, Zhang F, Deisseroth K, Tsai LH, Moore CI. (2009) Driving fast-spiking cells induces gamma rhythm and controls sensory responses. Nature 459:663–667.

Cardin JA, Carlén M, Meletis K, Knoblich U, Zhang F, Deisseroth K, Tsai LH, Moore CI. (2010) Targeted optogenetic stimulation and recording of neurons *in vivo* using cell-type-specific expression of channelrhodopsin-2. Nat Protoc 5:247–254.

Chen Q, Zeng Z, Hu Z (2012) Optogenetics in neuroscience: what we gain from studies in mammals. Neurosci Bull 28:423–434.

Ciesielska A, Mittermeyer G, Hadaczek P, Kells AP, Forsayeth J, Bankiewicz KS (2011) Anterograde axonal transport of AAV2-GDNF in rat basal ganglia. Mol Ther 19:922–927.

Cisek P, Kalaska JF (2010) Neural mechanisms for interacting with a world full of action choices. Annu Rev Neurosci 33:269–298.

Cohen BN, Mackey ED, Grady SR, McKinney S, Patzlaff NE, Wageman CR, McIntosh JM, Marks MJ, Lester HA, Drenan RM (2012) Nicotinic cholinergic mechanisms causing elevated dopamine release and abnormal locomotor behavior. Neuroscience 200:31–41.

Cragg SJ (2003) Variable dopamine release probability and short-term plasticity between functional domains of the primate striatum. J Neurosci 23:4378–4385.

Cragg SJ (2006) Meaningful silences: how dopamine listens to the ACh pause. Trends Neurosci 29:125–131.

Cragg SJ, Rice ME (2004) DAncing past the DAT at a DA synapse. Trends Neurosci 27:270–277.

Descarries L, Mechawar N (2000) Ultrastructural evidence for diffuse transmission by monoamine and acetylcholine neurons of the central nervous system. Prog Brain Res 125:27–47.

Diester I, Kaufman MT, Mogri M, Pashaie R, Goo W, Yizhar O, Ramakrishnan C, Deisseroth K, Shenoy KV (2011) An optogenetic toolbox designed for primates. Nat Neurosci 14:387–397.

Ding JB, Guzman JN, Peterson JD, Goldberg JA, Surmeier DJ (2010) Thalamic gating of corticostriatal signaling by cholinergic interneurons. Neuron 67:294–307.

Exley R, Clements MA, Hartung H, McIntosh JM, Cragg SJ (2008) Alpha6-containing nicotinic acetylcholine receptors dominate the nicotine control of dopamine neurotransmission in nucleus accumbens. Neuropsychopharmacology 33:2158–2166.

Galvan A, Hu X, Smith Y, Wichmann T (2012) *In vivo* optogenetic control of striatal and thalamic neurons in non-human primates. PLoS One 7:e50808.

Gerdeman GL, Partridge JG, Lupica CR, Lovinger DM (2003) It could be habit forming: drugs of abuse and striatal synaptic plasticity. Trends Neurosci 26:184–192.

Gerfen CR, Herkenham M, Thibault J (1987) The neostriatal mosaic: II. patch- and matrix-directed mesostriatal dopaminergic and non-dopaminergic systems. J Neurosci 7:3915–3934.

Gerfen CR, Surmeier DJ (2011) Modulation of striatal projection systems by dopamine. Annu Rev Neurosci 34:441–466.

Gerits A, Farivar R, Rosen BR, Wald LL, Boyden ES, Vanduffel W (2012) Optogenetically induced behavioral and functional network changes in primates. Curr Biol 22:1722–1726.

Gong S, Doughty M, Harbaugh CR, Cummins A, Hatten ME, Heintz N, Gerfen CR (2007) Targeting Cre-recombinase to specific neuron populations with bacterial artificial chromosome constructs. J Neurosci 27:9817–9823.

Gradinaru V, Mogri M, Thompson KR, Henderson JM, Deisseroth K (2009) Optical deconstruction of parkinsonian neural circuitry. Science 324:354–359.

Graybiel AM (2008) Habits, rituals, and the evaluative brain. Annu Rev Neurosci 31:359–387.

Graybiel AM, Aosaki T, Flaherty AW, Kimura M (1994) The basal ganglia and adaptive motor control. Science 265:1826–1831.

Gunaydin LA, Yizhar O, Berndt A, Sohal VS, Deisserith K, Hegemann P (2010) Ultrafast optogenetic control. Nat Neurosci 13:387–392.

Haber SN, Fudge JL, McFarland NR (2000) Striatonigrostriatal pathways in primates form an ascending spiral from the shell to the dorsolateral striatum. J Neurosci 20:2369–2382.

Han X (2012) *In vivo* application of optogenetics for neural circuit analysis. ACS Chem Neurosci 3:577–584.

Han X, Chow BY, Zhou H, Klapoetke NC, Chuong A, Rajimehr R, Yang A, Baratta MV, Winkle J, Desimone R, Boyden ES (2011) A high-light sensitivity optical neural silencer: development and application to optogenetic control of non-human primate cortex. Front Syst Neurosci 5:18.

Han X, Qian X, Bernstein JG, Zhou H, Franzesi GT, Stern P, Bronson RT, Graybiel AM, Desimone R, Boyden ES (2009) Millisecond-timescale optical control of neural dynamics in the non-human primate brain. Neuron 62:191–198.

Hoover DB, Muth EA, Jacobowitz DM (1978) A mapping of the distribution of acetylcholine, choline acetyltransferase and acetylcholinesterase in discrete areas of rat brain. Brain Res 153:295–306.

Huber D, Petreanu L, Ghitani N, Ranade S, Hromadka T, Mainen Z, Svoboda K (2008) Sparse optical microstimulation in barrel cortex drives learned behavior in freely moving mice. Nature 451:61–64.

Hutson TH, Verhaagen J, Yáñez-Muñoz RJ, Moon LD (2012) Corticospinal tract transduction: a comparison of seven adeno-associated viral vector serotypes and a non-integrating lentiviral vector. Gene Ther 19:49–60.

Ishizuka T, Kakuda M, Araki R, Yawo H (2006) Kinetic evaluation of photosensitivity in genetically engineered neurons expressing green algae light-gated channels. Neurosci Res 54:85–94.

Kawaguchi Y (1992) Large aspiny cells in the matrix of the rat neostriatum *in vitro*: physiological identification, relation to the compartments and excitatory postsynaptic currents. J Neurophysiol 67:1669–1682.

Kawaguchi Y (1993) Physiological, morphological, and histochemical characterization of three classes of interneurons in rat neostriatum. J Neurosci 13:4908–4923.

Knowlton BJ, Mangels JA, Squire LR (1996) A neostriatal habit learning system in humans. Science 273:1399–1402.

Kolisnyk B, Guzman MS, Raulic S, Fan J, Magalhaes AC, Feng G, Prado MA (2013) ChAT-ChR2-EYFP mice have enhanced motor endurance but show deficits in attention and several additional cognitive domains. J Neurosci 33:10427–10438.

Korecka JA, Schouten M, Eggers R, Ulusoy A, Bossers K, Verhaagen J (2011) Comparison of AAV serotypes for gene delivery to dopaminergic neurons in the substantianigra. InTech: Viral gene therapy (Xu K, ed), pp 205–224.

Kravitz AV, Freeze BS, Parker PR, Kay K, Thwin MT, Deisseroth K, Kreitzer AC (2010) Regulation of parkinsonian motor behaviors by optogenetic control of basal ganglia circuitry. Nature 466:622–626.

Lapper SR, Bolam JP (1992) Input from the frontal cortex and the parafascicular nucleus to cholinergic interneurons in the forsal striatum of the rat. Neuroscience 51:533–545.

Lin JY (2011) A user's guide to channelrhodopsin variants: features, limitations and future developments. Exp Physiol 96:19–25.

Lin JY, Lin MZ, Steinbach P, Tsien RY (2009) Characterization of engineered channelrhodopsin variants with improved properties and kinetics. Biophys J 96:1803–1814.

Löw K, Aebischer P, Schneider BL (2013) Direct and retrograde transduction of nigral neurons with AAV 6, 8, and 9 and intraneuronal persistence of viral particles. Hum Gene Ther 24:613–629.

Madisen L, Zwingman TA, Sunkin SM, Oh SW, Zariwala HA, Gu H, Ng LL, Palmiter RD, Hawrylycz MJ, Jones AR, Lein ES, Zeng H (2010) A robust and high-throughput Cre reporting and characterization system for the whole mouse brain. Nat Neurosci 13:133–140.

Masamizu Y, Okada T, Kawasaki K, Ishibashi H, Yuasa S, Takeda S, Hasegawa I, Nakahara K (2011) Local and retrograde gene transfer into primate neuronal pathways *via* adeno-associated virus serotypes 8 and 9. Neuroscience 193:249–258.

Matsumoto M, Hikosaka O (2009) Two types of dopamine neuron distinctly convey positive and negative motivational signals. Nature 459:837–841.

Middleton FA, Strick PL (2000) Basal ganglia and cerebellar loops: motor and cognitive circuits. Brain Res Rev 31:236–250.

Morris G, Arkadir D, Nevet A, Vaadia E, Bergman H (2004) Coincident but distinct messages of midbrain dopamine and striatal tonically active neurons. Neuron 43:133–143.

Müller K, Weber W (2013) Optogenetic tools for mammalian systems. Mol Biosyst 9:596–608.

Nathanson JL, Jappelli R, Scheeff ED, Manning G, Obata K, Brenner S, Callaway EM (2009a) Short promoters in viral vectors drive selective expression in mammalian inhibitory neurons, but do not restrict activity to specific inhibitory cell-types. Front Neural Circuits 3:19.

Nathanson JL, Yanagawa Y, Obata K, Callaway EM (2009b) Preferential labeling of inhibitory and excitatory cortical neurons by endogenous tropism of adeno-associated virus and lentivirus vectors. Neuroscience 161:441–450.

Packer AM, Roska B, Häusser M (2013) Targeting neurons and photons for optogenetics. Nat Neurosci 16:805–815.

Paterna JC, Feldon J, Büeler H (2004) Transduction profiles of recombinant adeno-associated virus vectors derived from serotypes 2 and 5 in the nigrostriatal system of rats. J Virol 78:6808–6817.

Rabinowitz JE, Rolling F, Li C, Conrath H, Xiao W, Xiao X, Samulski RJ (2002) Cross-packaging of a single adeno-associated virus (AAV) type 2 vector genome into multiple AAV serotypes enables transduction with broad specificity. J Virol 76:791–801.

Ren J, Qin C, Hu F, Tan J, Qiu L, Zhao S, Feng G, Luo M (2011) Habenula "cholinergic" neurons co-release glutamate and acetylcholine and activate postsynaptic neurons *via* distinct transmission modes. Neuron 69:445–452.

Rice ME, Cragg SJ (2004) Nicotine amplifies reward-related dopamine signals in striatum. Nat Neurosci 7:583–584.

Rice ME, Patel JC, Cragg SJ (2011) Dopamine release in the basal ganglia. Neuroscience 198:112–137.

Ruiz O, Lustig BR, Nassi JJ, Cetin AH, Reynolds JH, Albright TD, Callaway EM, Stoner GR, Roe AW (2013) Optogenetics through windows on the brain in the non-human primate. J Neurophysiol 110:1455–1467.

Sadikot AF, Parent A, Smith Y, Bolam J P (1992) Efferent connections of the centromedian and parafascicular thalamic nuclei in the squirrel monkey: a light and electron microscopic study of the thalamostriatal projection in relation to striatal heterogeneity. J Comp Neurol 320:228–242.

Salegio EA, Samaranch L, Kells AP, Mittermeyer G, San Sebastian W, Zhou S, Beyer J, Forsayeth J, Bankiewicz KS (2013) Axonal transport of adeno-associated viral vectors is serotype-dependent. Gene Ther 20:348–352.

Sanchez CE, Tierney TS, Gale JT, Alavian KN, Sahin A, Lee JS, Mulligan RC, Carter BS (2011) Recombinant adeno-associated virus type 2 pseudotypes: comparing safety, specificity, and transduction efficiency in the primate striatum. Laboratory investigation. J Neurosurg 114:672–680.

Schmitz Y, Benoit-Marand M, Gonon F, Sulzer D (2003) Presynaptic regulation of dopaminergic neurotransmission. J Neurochem 87:273–289.

Schönig K, Weber T, Frömmig A, Wendler L, Pesold B, Djandji D, Bujard H, Bartsch D (2012) Conditional gene expression systems in the transgenic rat brain. BMC Biol 10:77.

Schultz W (1998) Predictive reward signal of dopamine neurons. J Neurophysio 180:1–27.

Schultz W (2006) Behavioral theories and the neurophysiology of reward. Annu Rev Psychol 57:87–115.

Stuber GD, Hnasko TS, Britt JP, Edwards RH, Bonci A (2010) Dopaminergic terminals in the nucleus accumbens but not the dorsal striatum corelease glutamate. J Neurosci 30:8229–8233.

Stuber GD, Mason AO (2013) Integrating optogenetic and pharmacological approaches to study neural circuit function: current applications and future directions. Pharmacol Rev 65:156–170.

Taniguchi H, He M, Wu P, Kim S, Paik R, Sugino K, Kvitsiani D, Fu Y, Lu J, Lin Y, Miyoshi G, Shima Y, Fishell G, Nelson SB, Huang ZJ (2011) A resource of Cre driver lines for genetic targeting of GABAergic neurons in cerebral cortex. Neuron 71:995–1013.

Taymans JM, Vandenberghe LH, Haute CV, Thiry I, Deroose CM, Mortelmans L, Wilson JM, Debyser Z, Baekelandt V (2007) Comparative analysis of adeno-associated viral vector serotypes 1, 2, 5, 7, and 8 in mouse brain. Hum Gene Ther 18:195–206.

Tecuapetla F, Patel JC, Xenias H, English D, Tadros I, Shah F, Koos T (2010) Glutamatergic signaling by mesolimbic dopamine neurons in the nucleus accumbens. J Neurosci 30:7105–7110.

Threlfell S, Clements MA, Khodai T, Pienaar IS, Exley R, Wess J, Cragg SJ (2010) Striatal muscarinic receptors promote activity dependence of dopamine transmission *via* distinct receptor subtypes on cholinergic interneurons in ventral *versus* dorsal striatum. J Neurosci 30:3398–3408.

Threlfell S, Cragg SJ (2011) Dopamine signaling in dorsal *versus* ventral striatum: the dynamic role of cholinergic interneurons. Front Syst Neurosci 5:11.

Threlfell S, Lalic T, Platt NJ, Jennings KA, Deisseroth K, Cragg SJ (2012) Striatal dopamine release is triggered by synchronized activity in cholinergic interneurons. Neuron 75:58–64.

Tritsch NX, Ding JB, Sabatini BL (2012) Dopaminergic neurons inhibit striatal output through non-canonical release of GABA. Nature 490:262–266.

Tsai HC, Zhang F, Adamantidis A, Stuber GD, Bonci A, de Lecea L, Deisseroth K (2009) Phasic firing in dopaminergic neurons is sufficient for behavioral conditioning. Science 324:1080–1084.

Tsubota T, Ohashi Y, Tamura K, Sato A, Miyashita Y (2011) Optogenetic manipulation of cerebellar Purkinje cell activity *in vivo*. PLoS One 6:e22400.

Weber T, Schönig K, Tews B, Bartsch D (2011) Inducible gene manipulations in brain serotonergic neurons of transgenic rats. PLoS One 6:e28283.

Wickersham IR, Sullivan HA, Seung HS (2013) Axonal and subcellular labeling using modified rabies viral vectors. Nat Commun 4:2332.

Wilson CJ, Chang HT, Kitai ST (1990) Firing patterns and synaptic potentials of identified giant aspiny interneurons in the rat neostriatum. J Neurosci 10:508–519.

Witten IB, Steinberg EE, Lee SY, Davidson TJ, Zalocusky KA, Brodsky M, Yizhar O, Cho SL, Gong S, Ramakrishnan C, Stuber GD, Tye KM, Janak PH, Deisseroth K (2011) Recombinase-driver rat lines: tools, techniques, and optogenetic application to dopamine-mediated reinforcement. Neuron 72:721–733.

Yaguchi M, Ohashi Y, Tsubota T, Sato A, Koyano KW, Wang N, Miyashita Y (2013) Characterization of the properties of seven promoters in the motor cortex of rats and monkeys after lentiviral vector-mediated gene transfer. Hum Gene Ther Methods 24:333–344.

Zhang F, Wang LP, Boyden ES, Deisseroth K (2006) Channelrhodopsin-2 and optical control of excitable cells. Nat Methods 3:785–792.

Zhang H, Sulzer D (2003) Glutamate spillover in the striatum depresses dopaminergic transmission by activating group I metabotropic glutamate receptors. J Neurosci 23:10585–10592.

Zhang H, Sulzer D (2004) Frequency-dependent modulation of dopamine release by nicotine. Nat Neurosci 7:581–582.

Zhao S, Ting JT, Atallah HE, Qiu L, Tan J, Gloss B, Augustine GJ, Deisseroth K, Luo M, Graybiel AM, Feng G (2011) Cell type-specific channelrhodopsin-2 transgenic mice for optogenetic dissection of neural circuitry function. Nat Methods 8:745–752.

CHAPTER 10

ELECTROCHEMICAL RECORDINGS DURING DEEP BRAIN STIMULATION IN ANIMALS AND HUMANS: WINCS, MINCS, AND CLOSED-LOOP DBS

Charles D. Blaha[*], Su-Youne Chang[†,‡], Kevin E. Bennet[†,§] and Kendall H. Lee[†,‡]

[*]Department of Psychology, University of Memphis, Memphis, TN 38152, [†]Departments of Neurologic Surgery and [‡]Physiology and Biomedical Engineering, and [§]Division of Engineering, Mayo Clinic, Rochester, MN 55905

10.1 INTRODUCTION

The Mayo Neural Engineering Laboratory and associated colleagues have been studying the mechanisms of Deep Brain Stimulation (DBS) action using the *in vivo* electrochemical monitoring techniques; fast-scan cyclic voltammetry (FSCV) and constant or fixed potential amperometry (FPA). Previously, we developed a neurochemical monitoring device called (Wireless Instantaneous Neurochemical Concentration Sensing system (WINCS), and confirmed its functionality for FSCV monitoring during animal and human DBS neurosurgery. WINCS was successfully employed in wireless, near real-time monitoring of dopamine (DA) and adenosine (ADO) release during DBS neurosurgery in small (rat) and large animal (pig) models (Bledsoe et al., 2009; Agnesi et al., 2009; Shon et al., 2010 a, b). To expand the utility of the original device, we developed a novel wirelessly controlled stimulation device called Mayo Investigational Neuromodulation

Control System (MINCS). In animal (rat) testing, MINCS evoked DA release detectable in the striatum, as measured by simultaneous and interleaved FSCV. Importantly, the controlled release of DA was detected without stimulation artifact during the application of variable and wirelessly controlled high-frequency stimulation (Chang et al., 2013). We have further developed carbon-based microsensors compatible with WINCS recording protocols that are suitable for intraoperative human neurochemical recordings (Chang et al., 2012a) in neurological patients undergoing DBS surgery. In this chapter, we review hypotheses on mechanisms of DBS action, the development of WINCS and MINCS, and animal and human recordings that serve as the first steps toward the potential future development and applications of closed-loop DBS systems based on *in vivo* electrochemical recordings.

10.1.1 DBS IS A STATE-OF-THE-ART NEUROSURGICAL THERAPY

More than 100,000 people have been successfully implanted with DBS stimulation electrodes worldwide, and this number is expected to grow exponentially. DBS is now Food and Drug Administration (FDA) approved and is in routine clinical use for treatment of essential tremor (Lind et al., 2008; Papavassiliou et al., 2008; Hariz et al., 2008), Parkinson's disease (PD) (Hou et al., 2010; Follet and Torres-Russotto, 2012; Benabid et al., 2009), dystonia (Vidailhet et al. 2005; Speelman et al., 2010; Ostrem et al., 2011), and obsessive-compulsive disorder (Komotar et al., 2009; Denys et al., 2010; Burdick and Foote, 2011). Furthermore, disorders such as depression (Mayberg et al., 2005; Puigdemont et al., 2011), epilepsy (Pereira et al., 2012; Lee et al., 2006; Boon et al., 2007), Tourette syndrome (Porta et al., 2009; Hariz and Robertson, 2010; Savica et al., 2012; Cavanna et al., 2011), and chronic pain (Thomas et al., 2009; Sillay et al., 2010) are also being investigated to determine if DBS is a viable treatment option. The general procedure for DBS involves calculating target stereotactic coordinates through software that merges magnetic resonance imaging (MRI) of the patient brain with an atlas. Microelectrode unit recording is then used to verify a trajectory based on MRI data using region-specific neural activity as functional landmarks, although a correlation between these techniques can lead to discrepancies in targeting (Schlaier et al., 2011). If suitable results are not obtained, the electrode is removed and the recording procedure is repeated. Once a trajectory is verified, the microelectrode is withdrawn and the stimulating electrode implanted, secured to the skull and connected to a battery powered pulse generator that is implanted subcutaneously in the chest area (similar to a cardiac pacemaker). After surgical recovery, DBS is initiated and optimal stimulation parameters (amplitude, frequency, pulse width, and active contact) are empirically determined for therapeutic benefit. While DBS has been proven effective, we submit that this still-emerging neurosurgical technique can be substantially improved by instrumentation and bioprobes supporting intraoperative and potentially long-term neurochemical monitoring for both electrochemical recording and more efficient electrical stimulation.

Because the therapeutic effects of DBS are similar to those of a lesion, DBS has been thought to silence neurons at the site of stimulation (Bergman et al., 1990; Patel et al., 2003). Other hypotheses concerning the mechanism of action of DBS have suggested that (1) stimulated neurons are held in a depolarized state (depolarization blockade) in which they are unable to generate action potentials, (2) the neural network is disrupted by the additional impulses generated by DBS, and (3) DBS produces net inhibition in the network either by preferential activation of gamma-aminobutyric acid (GABA) ergic inhibitory neurons or disruption of the properties of the network itself when driven at high rates (Benazzouz et al., 1995; Lozano et al., 2002; Magarinos-Ascone et al., 2002; Garcia et al., 2003, 2005).

Clinically, bilateral stimulation of the subthalamic nucleus (STN) reverses the three cardinal motor symptoms in PD patients: akinesia, rigidity, and tremor (Limousin et al., 1998; Volkmann, 2004) and decreases or eliminates the need for L-3,4-dihydroxyphenylalanine (L-DOPA) (Moro et al., 1999; Molinuevo et al., 2000). DBS of the STN is most effective in PD patients who respond well to L-DOPA (Breit et al., 2004), suggesting that effective DBS requires endogenous DA production. DBS may even elicit dyskinesias that resemble those seen when excess L-DOPA is given (Limousin et al., 1998). Taken together, these clinical observations suggest that an increase in DA within the basal ganglia may, in part, account for the therapeutic efficacy of STN DBS in PD patients.

In contrast, several positron emission tomography (PET) studies using [^{11}C]-raclopride binding to measure DA release have failed to demonstrate significant raclopride displacement despite significant improvements in motor performance following STN DBS (Abosch et al., 2003; Hilker et al., 2003; Thobois et al., 2003; Strafella et al., 2003). These *in vivo* findings suggest that STN stimulation does not mediate its anti-Parkinsonian effects *via* terminal release of DA. However, PET scanning with raclopride has relatively poor temporal resolution and requires an increase of greater than 90% of baseline measures in order to detect a change in DA efflux (Hilker et al., 2003; Volkow et al., 1993). As well, adaptive changes in DA receptor populations, such as D2 receptor internalization and/or recycling occurring over long-term STN stimulation, have been suggested to interfere with PET quantification of DA release in these patients (Laruelle, 2000). Thus, whether STN DBS improves PD symptoms *via* the release of DA remains an important question which is of paramount importance.

Although DBS use has increased dramatically, with clinical indications expanding to both neurologic and psychiatric diseases, there is a great need to improve this functional neurosurgical technique. DBS for movement disorders typically involves placing a permanently implanted 1.27 mm diameter macroelectrode (with four contacts each 1.5 mm in width and 1.5 mm apart) in a specific region of the brain using high-resolution MRI or CT for localization. To treat the full spectrum of PD (tremor, rigidity, bradykinesia), the STN (an almond-shaped region roughly 10 mm in its largest dimension) is the highly effective stimulation site. A major challenge is an incomplete understanding of DBS mechanisms, especially why stimulation of specific targets is effective. Furthermore, DBS technology and procedures have remained largely unchanged for the past 20 years. Indeed, the DBS procedure is quite lengthy, requiring the awake patient to undergo difficult hours of electrophysiological recording required to confirm the anatomy scanned by MRI, with each pass of the microelectrode increasing the risk of intracranial hemorrhage (Park et al., 2011). Even with this extensive implantation procedure, stimulating electrodes are misplaced, or, due to the large size of the currently available DBS electrodes, precise localized stimulation is not achieved and there is a risk of unwanted stimulation side effects. These effects include: (1) unwanted motor contractions due to stimulation of the internal capsule immediately lateral to the STN, (2) paresthesias (sensory symptoms) due to stimulation slightly posterior to the STN, (3) eye signs suggestive of third cranial nerve pathway stimulation if the electrode is slightly medial to the STN, (4) depression, and even suicidal ideation (Deuschl et al., 2013). Another important consideration is that once a trajectory is chosen by anatomical planning and microelectrode recording, the stimulating electrode is "blindly" implanted without direct guidance from any additional parameter that might allow the neurosurgeon to know with confidence that the particular implant will be therapeutically effective. The one exception to this "blindness" is tremor, where immediate relief is seen during the awake neurosurgical procedure. Thus, patients and attending physicians may not know for days to months whether DBS surgery has provided any benefit. Stimulation parameters are adjusted during post-operative programming to optimize clinical benefit, but this trial-and-error process requires many hours of neurologist and patient time, and is ultimately limited by the location of the stimulation electrode, which is fixed at this critical therapeutic juncture. While we thus firmly submit that neurotransmitter measurements during human DBS surgery will address many of these critical issues by supporting guided stimulating electrode placement, suitable instrumentation and microsensors are not currently available to realize these important goals.

10.1.2 DBS MEDIATED DA RELEASE IN THE BASAL GANGLIA: PRE-CLINICAL STUDIES

In agreement with the hypothesis that STN DBS improves motor symptoms of PD by striatal DA release, there are several animal studies that show that STN DBS increases striatal DA levels. For example, *in vivo* microdialysis studies have shown that STN DBS increases the striatal DA metabolites 3,4-dihydroxyphenylacetic acid (DOPAC), homovanillic acid (HVA), and tyrosine hydroxylase activity in normal and 6-hydroxydopamine (6-OHDA) lesioned rats (Meissner et al., 2001, 2002, 2003; Paul et al., 2000). With one exception (Bruet et al., 2001), DBS-evoked increases in striatal DA dialysate could not be detected without first inhibiting DA re-uptake with nomifensine and stimulating for prolonged durations (20 minutes) (Meissner et al., 2003). *In vivo* monitoring of slow (minutes–hours) changes in DA release is easily accomplished using these conventional microdialysis techniques. However, analysis of more rapid changes in DA release in the absence of DA re-uptake inhibition that may result from STN DBS requires an equally rapid 'real-time' detection and monitoring system. For detection, sensitive carbon-fiber microelectrodes (CFM) and enzymatic sensors used with these methods permit submicromolar monitoring of central DA and glutamate release. As such, to establish the functional characteristics of the dorsal striatal complex circuitry, we have utilized electrochemical recording procedures established for reliable monitoring of DA and glutamate release and re-uptake in dopaminergic and glutamatergic terminal sites in the brain *in vivo* (Blaha and Phillips, 1996; Michael and Wightman, 1999; Suaud-Chagny, 2004; Wilson and Gifford, 2005).

10.2 PRELIMINARY NEUROCHEMICAL STUDIES

The use of these *in vivo* electrochemical methods to ascertain the neural circuitry underlying DBS makes it possible to circumvent the assumptions that are necessary with the use of radiolabeled compounds or *in vitro* methods. As such, they provide a much more realistic measure of the factors and neurotransmitter systems that mediate DA and glutamate neurotransmission. The electrochemical procedure FSCV and FPA offer the best temporal resolution of all *in vivo* electrochemical methods to date (5–10 samples/second for FSCV; 10K samples/second for FPA) (Kimble et al., 2009; Venton et al., 2002).

In Figure 10.1a, FPA in combination with CFMs permitted quantitative detection of striatal DA overflow (efflux) evoked by electrical stimulation of excitatory inputs to DA cells in the substantia nigracompacta (SNc), such as those originating in the hindbrain pedunculopontine tegmental nucleus (PPT) (Forster and

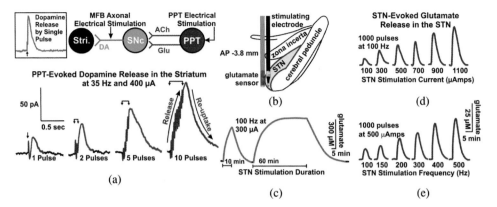

Figure 10.1 (a) DA release in the striatum of urethane anesthetized rats evoked by 1 to 10 pulses of electrical stimulation of PPT glutamatergic and cholinergic projections to SNc DA cells. Note that the increase in DA release is time locked to the stimulation (pulse artifacts are superimposed on the rising portion of the signal) and the recovery to baseline is a reflection of clearance of DA mainly *via* pre-synaptic re-uptake. INSET: rapid response of DA to stimulation of DA axons in the medial forebrain bundle (MFB) illustrating that the relatively slower PPT evoked response is trans-synaptically mediated. Glutamate release in the STN evoked by electrical stimulation of the STN at various current (c) durations, (d) intensities, and (e) frequencies. (b) Positioning of a glutamate sensor adjacent to a bipolar stimulating electrode in the STN (see Lee et al., 2007).

Blaha, 2003). Overflow or efflux of synaptic transmitter is referred to as release throughout this chapter. Recently, Dommett et al. (2005) have shown that FPA can be used to monitor striatal DA release in response to natural stimuli, such as light pulses. Visual stimuli evoked increases in striatal DA release *via* a direct input to the superior colliculus from the retina that, in turn, activated midbrain dopaminergic cells at short latencies. Collectively, these studies have confirmed the utility of fast electrochemical recording procedures to measure DA transmission driven by poly-synaptic pathways, such as those we have proposed to mediate DBS-evoked DA neurotransmission (Lee et al., 2008, 2009; Shah et al., 2010).

With respect to quantifying glutamate release using FPA, recent development of enzyme-coated platinum microelectrodes based on work by Hu et al. (1994) have shown a high degree of reliability as a selective, sensitive and rapidly responding glutamate sensor *in vivo* (Wilson and Gifford, 2005). This glutamate sensor system provides an additional quantitative measure of potential glutamatergic transmission in the dorsal striatal complex circuitry interfacing with SNc DA cells. Our preliminary studies have shown that electrical stimulation can evoke frequency and intensity dependent increases in glutamate release recorded locally at the site of stimulation using these procedures (Figures 10.1b–10.1e).

10.2.1 STN DBS AFFECTS STRIATAL DA RELEASE IN THE ANESTHETIZED RAT

Our neurochemical studies have focused on determining the functional consequences of STN DBS in terms of rapid changes in striatal DA release elicited by relatively brief and prolonged electrical stimulation of the STN in the urethane anesthetized rat. Brief STN stimulation (15 pulses at 300 μA and 50 Hz) resulted in a stimulus time-locked increase in striatal DA release as measured by FPA in combination with CFMs (Figure 10.2). The selectivity of the recording microelectrode to STN stimulation-evoked DA release was confirmed by systemic injection of the DA re-uptake inhibitor nomifensine, which resulted in a significant increase in DA oxidation current, compared to serotonin (fluoxetine) and noradrenaline (desipramine) re-uptake inhibitors, which did not alter the STN stimulation-evoked striatal response (Lee et al., 2006). These results support the hypothesis

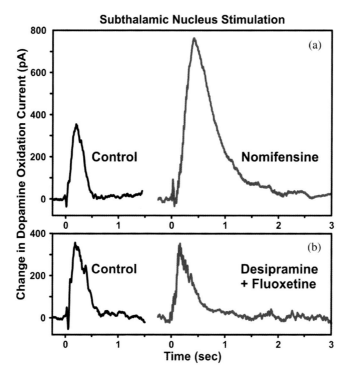

Figure 10.2 Striatal DA is increased with brief STN stimulation. The selectivity of the response was confirmed by (a) systemic injection of the DA re-uptake inhibitor nomifensine (red line), which resulted in a significant increase in DA oxidation current, compared to (b) serotonin (fluoxetine) and norepinephrine (desipramine) re-uptake inhibitors (blue line) which did not significantly increase the STN stimulation-evoked response.

that STN DBS results in quantifiable striatal DA release. However, as these studies were performed in rats with intact SNc dopaminergic neurons, it is crucial to perform systematic measurements in an animal model of PD, such as 6-OHDA lesions, where the SNc neurons are selectively and partially destroyed.

With regards to an animal model of PD, our preliminary results have shown that, in combination with L-DOPA, repetitive stimulations in 6-OHDA lesioned rats are capable of facilitating DA release to levels comparable to that seen with stimulation in intact rats and are consistent with recent findings that high frequency stimulations modulate the action of L-DOPA (Oueslati, et al., 2007). Microinfusion of the neurotoxin 6-OHDA onto DA cell bodies in the SNc results in the selective degeneration of dopaminergic cells in that region. As shown in Figure 10.3, compared to intact rats, 6-OHDA lesions resulted in rats exhibiting a marked attenuation in MFB stimulation-evoked DA release in the striatum (16.4±8.3% of 100% intact responses) as monitored using FPA in combination with CFMs. Most significantly, compared to intact animals receiving a systemic injection of saline, systemic administration of a relatively low dose of L-DOPA to lesioned rats resulted in a near complete recovery in the magnitude of the evoked DA responses (84.87 ± 16.22% of 100% intact responses at 30 minutes post-injection). Since L-DOPA increased the DA response in the lesioned animals to the statistical equivalent of the non-lesioned (saline-treated) animals, it can be inferred that on-line FPA or FSCV would be capable of (1) detecting depleted extracellular levels of DA in the striatum and (2) enhancements in these levels in response to L-DOPA treatment across various dose ranges in Parkinson's patients (Blaha et al., 2008). These data highlight the relevance of monitoring dopaminergic transmission during STN DBS and provide a framework for the development of a closed-loop neuroprosthesis with chemical sensing feedback and neuromodulation to maintain neurotransmitter levels consistent with optimal therapeutic efficacy.

10.2.2 EFFECTS OF VARIOUS STN DBS PARAMETERS ON STRIATAL DA RELEASE IN THE ANESTHETIZED RAT

Stimulation frequencies in the range of 37 to 75 Hz and current intensities in the range of 200–600 μA evoked maximal DA responses in the striatum (Figures 10.4a and 10.4b) (see Lee et al., 2006). More prolonged STN stimulation evoked a DA response that peaked within 15–20 applied pulses and fell off to ~30% of pre-stimulus baseline levels despite continuous stimulation. Stimulation dorsomedial to STN, corresponding to a portion of the MFB containing ascending nigrostriatal dopaminergic axons, resulted in an increase in striatal DA release that plateaued five seconds into stimulation (Figure 10.4c).

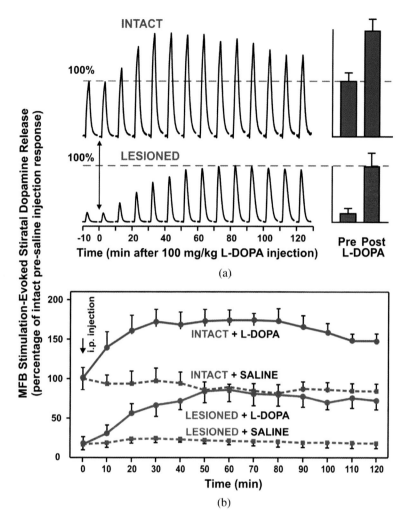

Figure 10.3 (a) Representative example of MFB stimulation-evoked (20 pulses at 100 Hz applied every 10 minutes) striatal DA release in a urethane anesthetized intact rat and a rat sustaining neurotoxic 6-OHDA lesioning of the nigrostriatal dopaminergic pathway before and following systemic injection of L-DOPA (100 mg/kg i.p.). Note the recovery of stimulated DA release to 100% baseline levels of evoked striatal DA release following administration of L-DOPA in lesioned animals. (b) Mean ± SEM time courses of effects of L-DOPA injections, compared to saline administration, on striatal DA release in intact and lesioned rats before and following systemic injection of L-DOPA. Solid and dashed blue lines: intact mice that received L-DOPA and saline, respectively. Percentage changes are with respect to intact pre-saline treated mice.

These results suggest that DBS in PD patients typically applied to the STN and adjacent regions increases brain extracellular DA levels, but also indicate that the pattern and magnitude of DA release varies significantly depending on the site and nature of stimulation. This latter finding has therapeutic implications

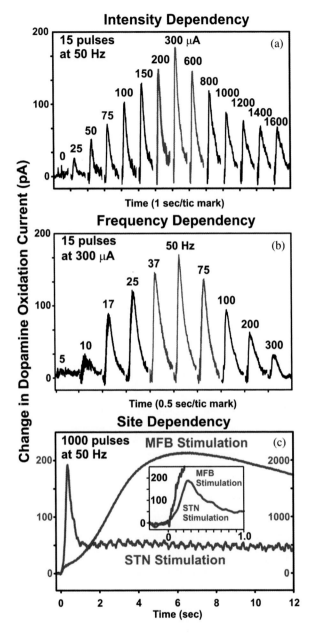

Figure 10.4 Intensity, frequency, and site dependency of STN stimulation evoked striatal DA release of the urethane anesthetized rat. (a) The optimal current intensity for STN stimulation was 300 μA, with the stimulation-evoked DA response falling off at 600 μA and greater. (b) The optimal frequency for STN stimulation was 50 Hz, with the stimulation-evoked DA response falling off at 75 Hz and greater. (c) Prolonged STN stimulation evoked a transient response that peaked within 20 applied pulses (red line), as compared to a 10-fold greater and more sustained DA response to stimulation of the MFB (blue line). INSET: DA release evoked by stimulation of the MFB (blue line) was significantly faster in onset compared to STN stimulation when viewed on the same scale as the STN response (red line).

for the site and pattern of stimulation in PD patients, which so far have not been fully explored.

10.3 NEUROTRANSMITTER MONITORING SYSTEMS

10.3.1 WINCS

To examine the functional anatomical and neurochemical effects of DBS, the Mayo Clinic has developed a novel WINCS device specifically designed to monitor neurochemical release during both experimental and clinical DBS surgical procedures (Figure 10.5). As such, research subject safety, signal fidelity, and integration with existing DBS surgical procedures, and MRI pre-, intra-, and postoperative analysis have been key priorities during the development of WINCS. This device, designed in compliance with FDA-recognized consensus standards for medical electrical device safety, consists of a relatively small, wireless, sterilizable, battery-powered unit that can interface with CFM or enzyme-based microsensors for real-time monitoring of neurotransmitter release in the brain (Bledsoe et al., 2009; Agnesi et al., 2009; Shon et al., 2010a,b).

For our studies, the WINCS device has significant advantages over other commercially available wireless recording systems as it offers (1) an advanced microprocessor with superior analog to digital conversion, greater internal memory, and faster clock speed, (2) wirelessly programmable waveform parameters (scan bias, range and rate) using an advanced Bluetooth module for wireless communication, (3) a higher precision voltage reference for the micro-processor, (4) a low-power mode to preserve battery life, voltage sensing, and low-power alert, and most

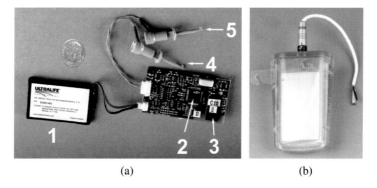

Figure 10.5 Photograph showing the (a) lithium-polymer battery (1), micro-processor (2), Bluetooth transmission chip (3), electrochemical recording electrode lead (4), and reference electrode lead (5) of the WINCS device. Photograph (b) showing the WINCS device encased in its sterilizable polycarbonate case.

importantly (5) proven safety and feasibility in human subjects undergoing DBS neurosurgery.

We have demonstrated our ability to measure *in vivo* real-time oxygen, DA, ADO, histamine, and serotonin release with CFMs and glutamate release with an enzyme-linked biosensor during DBS (Lee et al., 2007; Agnesi et al., 2009; Bledsoe et al., 2009; Chang et al., 2012b; Shon et al., 2010a,b). We have also shown that we can use WINCS to co-detect *in vivo* changes in ADO and DA concentrations in small and large animal models of DBS (Shon et al., 2010a). More importantly, our results (Shon et al., 2010b) have shown that STN DBS elicits DA and ADO release in the caudate nucleus of isoflurane anesthetized pigs. Additionally, we examined striatal DA release evoked by STN DBS in awake monkeys. We identified a site-dependency of DA release by stimulating at multiple points along a trajectory passing through the thalamus and STN. Greater DA release was observed when the stimulating electrode was within the dorsal STN. Our results showed that the amount of DA release depends critically on the location of the stimulating electrode (Gale et al., 2013).

10.3.2 MINCS

In order to provide brain stimulation using a wide range of stimulation parameters we developed the MINCS (Figure 10.6). MINCS is capable of providing wirelessly-controlled neurostimulation (Chang et al., 2013). The neurochemical recording device, WINCS, and the synchronized stimulator, MINCS, are linked *via* an optical connection. Both units connect wirelessly *via* Bluetooth technology

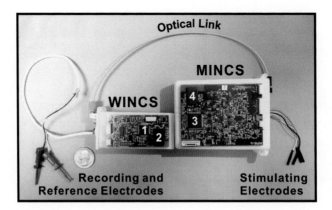

Figure 10.6 MINCS–WINCS hardware. Photograph of the MINCS–WINCS hardware showing relative size, optical connection, and recording and stimulating electrode leads. WINCS (1-microcontroller, 2-Bluetooth module); MINCS (3-microcontroller, 4-Bluetooth module).

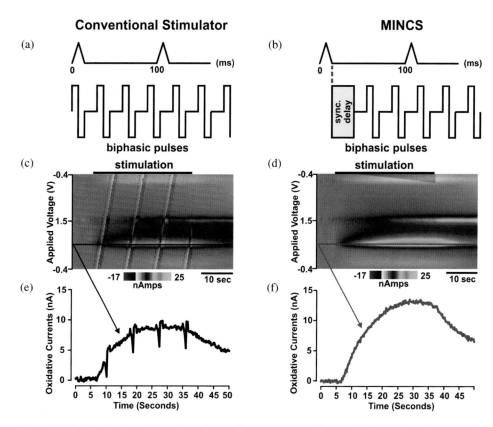

Figure 10.7 Elimination of stimulus pulse artifacts using optically synced MINCS stimulations with WINCS neurochemical recordings. Comparison of the stimulus pulse sequences provided by a conventional stimulator (a) and MINCS (b) in relation to FSCV scans (triangular waveforms). (c and d) Comparison of color plots of striatal DA release acquired *in vivo* from MFB stimulations (30 seconds stimulation at 60 Hz, 200 µA, two milliseconds biphasic pulse duration) in the anesthetized rat with (right) and without (left) stimulus pulse synchronization. Stimulus pulse artifacts are readily apparent as repeating diagonal bands in the color plot (left). (e and f) Time courses of stimulation-evoked changes in DA oxidation current extracted from each respective color plot record at the applied voltage of +0.6 V (right, solid red arrow; left, solid black arrow). Stimulus pulse artifacts (negative-going spikes in current) were readily apparent using a conventional stimulator, while entirely absent using MINCS.

to a base station, which commands both systems *via* WincsWare software. Leads from each device provide connections to the recording and reference electrodes and four stimulating electrodes. Using MINCS and WINCS, we stimulated the MFB of anesthetized rats and showed that we can evoke striatal DA release. Similarly, we have shown that STN DBS elicits DA and ADO release distally in the striatum of isoflurane anesthetized pigs. In these experiments, we were able to evoke, identify, and record DA release without interference from the electrical stimulation pulses (Figure 10.7).

10.4 RECORDING ELECTRODE DEVELOPMENT

10.4.1 MICROELECTRODE FOR HUMAN NEUROCHEMICAL RECORDING

To examine the functional anatomical and neurochemical effects of DBS, our team has recently obtained an institutional review board (IRB) approval for use of WINCS in 15 human patients to test the safety and feasibility of obtaining neurochemical monitoring during clinical DBS surgical procedures. As such, research subject safety, signal fidelity and integration of WINCS with existing DBS surgical procedures have now been successfully tested. Figure 10.8 demonstrates the successful development of a 1st generation human WINCS neurochemical recording electrode (WINCStrode, Figure 10.8a), where the proven human tungsten electrophysiology electrode was modified at the tip with a 30 μm diameter carbon-fiber that was insulated with polyimide. The WINCStrode successfully recorded DA (not shown) and ADO (Figures 10.8c–10.8e) in a flow cell analysis system. Further, the WINCS and WINCStrode were successfully implemented in measuring ADO during DBS lead implantation (see Figure 10.13).

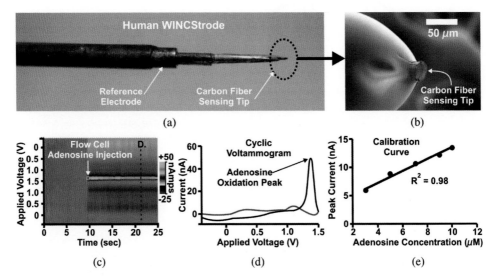

Figure 10.8 WINCS electrode for human application. (a) Photograph of the human WINCS electrode (WINCStrode) where the human electrophysiology electrode was modified at the tip, incorporating a 30 μm carbon-fiber insulated with silicone. (b) Scanning Electron Micrograph of the tip of WINC Strode showing the carbon-fiber sensing tip. (c) Pseudo color plot of calibration data using WINCS and WINCStrode in a flow cell demonstrating ADO oxidation current. The x-axis is time, y-axis is the voltage scan from −0.4 V to +1.5 V and back to −0.4 V at 400 V/second, and color signifying current measurement (nanoAmp). (d) Cyclic voltammogram obtained following injection of ADO as marked in (c) (D.-dashed line). (e) Calibration curve obtained for ADO (3 to 10 μM) demonstrating correlation coefficient of 0.98.

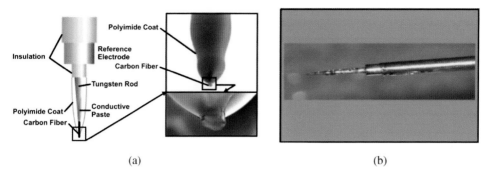

Figure 10.9 (a) Schematic representation and scanning electron microscopy of human FSCV electrode. (b) Photograph of the electrode tip.

It is worth noting that neurochemical recording electrodes employed in basic research using *in vivo* electrochemical techniques are typically glass-insulated 5–10 μm diameter carbon-fiber electrodes, as opposed to the 30 μm diameter carbon-fiber comprising our 1st generation WINCStrode described above. This allows for higher spatial resolution and minimization of tissue damage. However, these electrodes are fragile and not suitable for intraoperative or chronic human recordings. For this reason, we have further developed a safer and more durable electrode for human use (Figure 10.9). This electrode is made of carbon-fiber insulated with polyimide-fused silica. The electrode sensing tip consists of a 7 μm diameter and a 100–110 μm shaft. We have recently employed this electrode for human recordings as part of an IRB approved study (unpublished data). In addition, we are currently developing a flexible electrode for chronic implantation.

10.5 NEUROCHEMICAL RECORDINGS IN LARGE ANIMALS AND NEUROLOGICAL PATIENTS

10.5.1 DA AND ADO RECORDING IN THE PIG

Before conducting *in vivo* neurochemical recordings in human neurological patients, it is necessary to demonstrate STN DBS in a large mammalian preparation as a model to examine the functional anatomical and neurochemical effects of this interventive neurosurgical treatment. Easy to obtain and maintain, the white domestic pig is proving to be an economically viable alternative to the expensive non-human primate, and thus is becoming an increasingly popular animal for neuroscience research. Several characteristics make the pig an excellent model for mock human DBS surgery: (1) the adult pig brain (~160 g) is comparable in size to that of the rhesus monkey (~100 g) and baboon (~140 g); (2) a pig brain atlas is available, and it demonstrates significant similarities with non-human primate

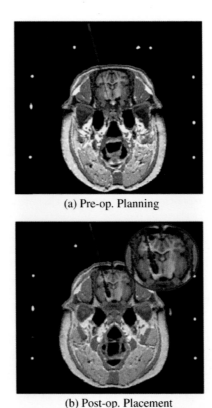

(a) Pre-op. Planning

(b) Post-op. Placement

Figure 10.10 A coronal MRI and corresponding fiducials (white circles) on the MRI-compatible head frame for pre-operative planning and calculation of the trajectory coordinates (red line) for implantation of a DBS electrode in the pig STN (a). A coronal MRI showing the post-operative confirmation of the DBS electrode placement in the pig STN (b). Note that the MRI shown in this example was for a relatively large Medtronic 3389 human DBS electrode in the pig STN.

and human (Felix et al., 1999); (3) pig brain development is complete by ~five months (Dobbing, 1964), which enables the use of younger pigs (20–50 kg) with adult-sized brains for easier mobilization and dissection during cranial surgery; (4) 1-methyl-4-phenyl-1,2,3,6-tetrahydropyridine(MPTP) produces an effective pig PD model (Danielsen, 2000; Dall et al., 2002; Cumming et al., 2003).

In the animal operating room, the pig is initially sedated with Telazol (5 mg/kg i.m.) and xylazine (2 mg/kg i.m.), then intubated with endo-tracheal tube and ventilated with an artificial ventilator. General anesthesia is then maintained with isoflurane (1%) for the remaining experimental procedure. In the prone position, the pig is initially placed in the MRI-compatible stereotactic head frame. With the pig under general anesthesia, the MRI-compatible stereotactic head frame and localizer box create nine fiducials to enable localization of MR images in stereotactic space (Figure 10.10).

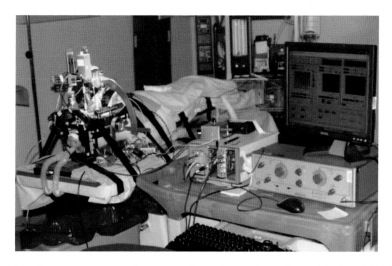

Figure 10.11 Experimental setup for *in vivo* electrochemical recordings of DA and ADO release evoked by site-specific electrical stimulation in the brain of the large DBS animal (pig) model.

The pig is then transported to the MRI scanner for imaging with a General Electric Signa 3.0 T MRI research system with EchoSpeed LX Version 9.1 software. This scanner is identical to that used with the human DBS protocol consisting of magnetization-prepared rapid acquisition with gradient echo (MP-RAGE) sequences, 1.5 mm slice thickness, and 24 cm field of view. Using the COMPASS navigational software, MRI data is then merged with the pig atlas (Felix et al., 1999) and stereotactic coordinates for the DBS electrode into the STN is defined.

Subsequently, the pig is returned to an animal operating room, where a large midline incision of the skin is made to expose the cranial landmarks of bregma and lambda. This is followed by a 14 mm burr hole made on the skull using a high-speed drill in line with our electrode trajectory coordinates. An alpha-omega computer-controlled microdrive that is attached to the Leksell stereotactic arc is then placed onto the head frame (Figure 10.11). In turn, a single CFM mounted onto the microdrive is then lowered to the recording target (in the case of STN DBS, the placement of the tip of the CFM is the head of the caudate nucleus). In addition, a 14 mm burr hole is drilled on the contralateral side of the skull to allow the placement of an Ag/AgCl reference electrode in contact with surface cortical tissue to complete the circuit necessary for the FSCV of DBS-evoked release of neurotransmitters.

Our pig experiments have demonstrated that STN DBS elicits DA and ADO release distally in the caudate nucleus, shown in Figure 10.12 (see Shon et al., 2010b). For these experiments, we implanted a single CFM into the caudate nucleus of isoflurane (1%) anesthetized pigs and FSCV recordings were taken during brief

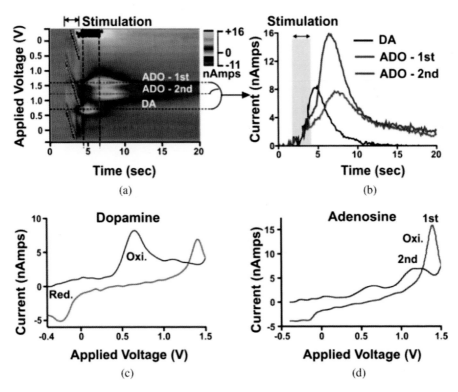

Figure 10.12 *In vivo* DA and ADO release measured with WINCS-based FSCV at CFMs in the CN of the isoflurane anesthetized pig. (a) Electrical stimulation (140 Hz, 0.5 milliseconds pulse width, for two seconds) of the STN evoked both DA and ADO release in the CN. The color plot shows the appearance of DA release immediately during and after stimulation, while the peak corresponding to ADO release was delayed. (b) Current *versus* time plot at +1.5V (blue), +1.0V (red), and +0.6V (black line) showing ADO first and second oxidation (ADO- 1st, blue and ADO- 2nd, red) and DA (DA, black) release following electrical stimulation (yellow box). (c) and (d) Background subtracted voltammograms for DA and ADO, respectively, demonstrate simultaneous measurement of DA and ADO release (black and blue vertical dashed line in a).

(two seconds) electrical stimulation of the human Medtronic 3389 DBS electrode implanted in the ipsilateral STN. With high frequency stimulation (140 Hz), ADO and DA were clearly released (Figure 10.12a). The temporal patterns and magnitude of DA (black line) and ADO release (1st peak blue line and oxidation product 2nd peak red line) evoked by STN DBS are shown in Figure 10.12b. The voltammograms obtained with WINCS in the pig revealed one peak at +0.6 V for DA oxidation and two oxidation peaks for ADO (1st peak near +1.5 V and 2nd peak near +1.0 V), as shown in Figures 10.12c and 10.12d, respectively. Our now established pig model has permitted testing of our human neurochemical recording electrodes, as well as study of distal neurotransmitter release by STN DBS.

10.5.2 HUMAN NEUROCHEMICAL RECORDINGS

From the knowledge and experienced gained from tests conducted in our large animal (pig) experiments, we have successfully used WINCS to monitor DA and ADO release in the hippocampus of human patients undergoing resective surgery for medically intractable epilepsy (unpublished data). In addition, we have now successfully tested the WINCS and WINCStrode system in human patients with PD and Essential Tremor. As shown in Figure 10.13a, WINCS is attached to the stereotaxic Leksell stereotactic headframe. As shown in Figure 10.13b, we placed a WINCStrode 2 mm anterior to the intended DBS target within the ventral intermediate (VIM) thalamus in an essential tremor patient followed by the placement of Medtronic 3387 electrode, to study the effect of microthalamotomy, where tremor is abolished with lead insertion. Figure 10.13c demonstrates that in this patient, lead insertion resulted in tremor arrest as seen by before and after handwriting sample. Tremor arrest corresponded to local ADO release (Figures 10.13d–10.13f) as measured by FSCV obtained by WINCS and the WINCStrode

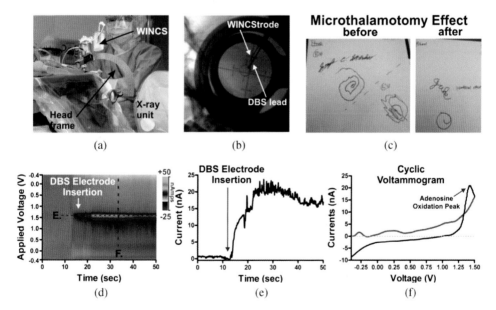

Figure 10.13 Neurochemical changes evoked by DBS in the VIM of the thalamus in patients with essential tremor. (a) Surgical set-up. (b) X-ray of the positioning of the WINCStrode and DBS lead in patient's thalamus. (c) Handwriting sample before and after DBS electrode implantation (L, left hand; R, right hand). (d) Pseudo-color plot showing an instantaneous increase at around $+1.45 \pm 0.03$ V (n = 7 patients) upon electrode insertion. (e) Current *versus* time plot as depicted at letter E. in (d). (f) Background subtracted cyclic voltammogram at F. in (d); the solid black line indicates the current detected during the ascending portion of the applied voltage waveform; the red line, by reverse-going voltage protocol.

during insertion of the DBS electrode into the VIM thalamic target site. These results demonstrate the first application of WINCS during DBS neurosurgery in human patients (see Chang et al., 2012a; Kasasbeh et al., 2013).

10.6 CONCLUSION AND FUTURE DIRECTIONS

The findings of this chapter provide information concerning the world's first wireless human electrochemical recording system. WINCS was specifically designed for human use and we have successfully tested WINCS in human patients undergoing DBS neurosurgery. It is conceivable that these initial intraoperative recordings in human patients during DBS may lay the foundation for an implantable closed-loop "smart" device incorporating microsensor, feedback control, and neuromodulation to optimize neurotransmitter levels continuously for improved clinical efficacy.

ACKNOWLEDGMENT

This work was supported by National Institutes of Health (K08 NS 52232, R01 NS 75013 and R01 NS 70872).

REFERENCES

Abosch A, Kapur S, Lang AE, Hussey D, Sime E, Miyasaki J, Houle S, Lozano AM (2003) Stimulation of the subthalamic nucleus in Parkinson's disease does not produce striatal dopamine release. Neurosurgery 53:1095–1102.

Agnesi F, Tye SJ, Bledsoe J, Griessenauer C, Kimble CJ, Sieck GC, Bennet KE, Garris PA, Blaha CD, Lee KH (2009) Wireless instantaneous neurotransmitter concentration system-based amperometric detection of dopamine, adenosine, and glutamate for intraoperative neurochemical monitoring. J Neurosurg 111:701–711.

Benabid AL, Chabardes S, Mitrofanis J, Pollak P (2009) Deep brain stimulation of the subthalamicnucleus for the treatment of Parkinson'sdisease. Lancet Neurol 8:67–81.

Benazzouz A, Piallat B, Pollak P, Benabid AL (1995) Responses of substantianigra pars reticulata and globuspallidus complex to high frequency stimulation of the subthalamic nucleus in rats: electrophysiological data. Neurosci Lett 189:77–80.

Bergman H, Wichmann T, DeLong MR (1990) Reversal of experimental parkinsonism by lesions of the subthalamic nucleus. Science 249:1436–1438.

Blaha CD, Phillips AG (1996) A critical assessment of electrochemical procedures applied to the measurement of dopamine and its metabolites during drug-induced and species-typical behaviors. BehavPharmacol 7:675–708.

Blaha CD, Lester DB, Ramsson ES, Lee KH, Garris PA (2008) Striatal dopamine release evoked by subthalamic stimulation in intact and 6-OHDA-lesioned rats: Relevance to deep brain stimulation in Parkinson's disease. In: Monitoring molecules in neuroscience (Phillips PEM, Sandberg SG, Ahn S, Phillips AG, eds), pp395–397. Vancouver, Canada: University of British Columbia.

Bledsoe JM, Kimble C, Covey DP, Blaha CD, Agnesi F, Mohseni P, Whitlock S, Johnson DM, Horne A, Bennet K, Lee KH, Garris PA (2009) Development of the wireless instantaneous neurotransmitter concentration system for intraoperative neurochemical monitoring using fast-scan cyclic voltammetry. J Neurosurg 111:712–723.

Boon P, Vonck K, De Herdt V, Van Dycke A, Goethals M, Goossens L, Van Zandijcke M, De Smedt T, Dewaele I, Achten R, Wadman W, Dewaele F, Caemaert J, Van Roost D (2007) Deep brain stimulation in patients with refractorytemporal lobe epilepsy. Epilepsia 48:1551–1560.

Breit S, Schulz JB, Benabid AL (2004) Deep brain stimulation. Cell Tissue Res 318:275–288.

Bruet N, Windels F, Bertrand A, Feuerstein C, Poupard A, Savasta M (2001) High frequency stimulation of the subthalamic nucleus increases the extracellular contents of striatal dopamine in normal and partially dopaminergic denervated rats. J Neuropathol Exp Neurol 60:15–24.

Burdick AP, Foote KD (2011) Advancing deepbrain stimulation for obsessive-compulsive-disorder. Expert Rev Neurother 11:341–344.

Cavanna AE, Eddy CM, Mitchell R, Pall H, Mitchell I, Zrinzo L, Foltynie T, Jahanshahi M, Limousin P, Hariz MI, Rickards H (2011) An approach to deep brain stimulation for severe treatment-refractory Tourette syndrome: The UK perspective. Br J Neurosurg 25:38–44.

Chang SY, Jay T, Muñoz J, Kim I, Lee KH (2012b) Wireless fast-scan cyclic voltammetry measurement of histamine using WINCS— a proof-of-principle study. Analyst 137:2158–2165.

Chang SY, Kim I, Marsh MP, Jang DP, Hwang SC, Van Gompel JJ, Goerss SJ, Kimble CJ, Bennet KE, Garris PA, Blaha CD, Lee KH (2012a) Wireless fast-scan cyclic voltammetry to monitor adenosine in patients with essential tremor during deep brain stimulation. Mayo ClinProc 87:760–765.

Chang SY, Kimble CJ, Kim I, Knight EJ, Kasasbeh A, Paek SB, Kressin KR, Boesche JB, Whitlock SV, Eaker DR, Horne AE, Min HP, Marsh MP, Blaha CD, Bennet KE, Lee KH (2013) Development of the mayo investigational neuromodulation control system: toward a closed-loop electrochemical-feedback system for deep brain stimulation. J Neurosurg 119:1556–1565.

Cumming P, Gillings NM, Jensen SB, Bjarkam C, Gjedde A (2003) Kinetics of the uptake and distribution of the dopamine D(2,3) agonist (R)-N-1-[^{11}C]n-propylnorapomorphine in brain of healthy and MPTP-treated Gottingen miniature pigs. Nucl Med Biol 30:547–553.

Dall AM, Danielsen EH, Sorensen JC, Andersen F, Moller A, Zimmer J, Gjedde AH, Cumming P (2002) Quantitative [18F] fluorodopa/PET and histology of fetal mesencephalic dopaminergic grafts to the striatum of MPTP-poisoned minipigs. Cell Transplant 11:733–746.

Danielsen EH (2000) The DaNeX study of embryonic mesencephalic, dopaminergic tissue grafted to a minipig model of Parkinson's disease: preliminary findings of effect of MPTP poisoning on striatal dopaminergic markers. Cell Transplant 9:247–259.

Denys D, Mantione M, Figee M, van denMunckhof P, Koerselman F, Westenberg H, Bosch A, Schuurman R (2010) Deep brain stimulationof the nucleus accumbens for treatment refractory obsessive-compulsive disorder. Arch Gen Psychiatry 67:1061–1068.

Deuschl G, Paschen S, Witt K (2013) Clinical outcome of deep brain stimulation for Parkinson's disease. Handb Clin Neurol 116:107–128.

Dobbing J (1964) The influence of early nutrition on the development and myelination of the brain./Proc Royal Society London Series B, Containing Papers of a Biological Character/ 159:503–509.

Dommett E, Coizet V, Blaha CD, Martindale J, Lefebvre V, Walton N, Mayhew JEW, Overton PG, Redgrave P (2005) How visual stimuli activate dopaminergic neurons at short-latency. Science 307:1476–1479.

Felix B, Leger ME, Albe-Fessard D, Marcilloux JC, Rampin O, Laplace JP (1999) Stereotaxic atlas of the pig brain. Brain Res Bull 49:1–137.

Follett KA, Torres-Russotto D (2012) Deep brain stimulation of globuspallidus interna, subthalamic nucleus, and pedunculopontine nucleus for Parkinson's disease: which target? Parkinsonism Relat Disord 18:S165–S167.

Forster GL, Blaha CD (2003) Pedunculopontine tegmental stimulation evokes striatal dopamine efflux by activation of acetylcholine and glutamate receptors in the midbrain and pons of the rat. Eur J Neurosci 17:751–762.

Gale JT, Lee KH, Amirnovin R, Roberts DW, Williams Z, Blaha CD, Eskandar EN (2013) Local stimulation-evoked dopamine release in the primate caudate putamen: evidence for the role of dopamine in plasticity. Stereotact Funct Neurosurg 91:355–363.

Garcia L, Audin J, D'Alessandro G, Bioulac B, Hammond C (2003) Dual effect of high-frequency stimulation on subthalamic neuron activity. J Neurosci 23:8743–8751.

Garcia L, D'Alessandro G, Bioulac B, Hammond C (2005) High-frequency stimulation in Parkinson's disease: more or less? Trends Neurosci 28:209–216.

Hariz GM, Blomstedt P, Koskinen LO (2008) Longterm effect of deep brain stimulation for essential tremor on activities of daily living and health-related quality of life. Acta Neurol Scand 118:387–394.

Hariz MI, Robertson MM (2010) Gilles de la Tourette syndrome and deep brain stimulation. Eur J Neurosci 32:1128–1134.

Hilker R, Voges J, Ghaemi M, Lehrke R, Rudolf J, Koulousakis A, Herholz K, Wienhard K, Sturm V, Heiss WD (2003) Deep brain stimulation of the subthalamic nucleus does not increase the striatal dopamine concentration in parkinsonian humans. Mov Disord 18:41–48.

Hou B, Jiang T, Liu R (2010) Deep-brain stimulation for Parkinson's disease. N Engl J Med 363:987–988.

Hu Y, Mitchell KM, Albahadily FN, Michaelis EK, Wilson GS (1994) Direct measurement of glutamate release in the brain using a dual enzyme-based electrochemical sensor. Brain Res 659:117–125.

Kasasbeh A, Lee K, Bieber A, Bennet K, Chang SY (2013) Wireless neurochemical monitoring in humans. Stereotact Funct Neurosurg. 91:141–147.

Kimble CJ, Johnson DM, Winter BA, Whitlock SV, Kressin KR, Horne AE, Robinson JC, Bledsoe JM, Tye SJ, Chang SY, Agnesi F, Griessenauer CJ, Covey D, Shon YM, Bennet KE, Garris PA, Lee KH. (2009) Wireless instantaneous neurotransmitter concentration sensing system (WINCS) for intraoperative neurochemical monitoring. ConfProc IEEE Eng Med Biol Soc 2009:4856–4859.

Komotar RJ, Hanft SJ, Connolly ES Jr (2009) Deep brain stimulation for obsessive compulsive disorder. Neurosurgery 64:N13.

Laruelle M (2000) Imaging synaptic neurotransmission with *in vivo* binding competition techniques: a critical review. J Cereb Blood Flow Metab 20:423–451.

Lee KH, Blaha CD, Garris PA, Mohseni P, Horne AE, Bennet KE, Agnesi F, Bledsoe JM, Lester DB, Kimble C, Min H-K, Kim Y-B, Cho Z-H (2009) Evolution of deep brain stimulation: Human electrometer and smart devices supporting the next generation of therapy. Neuromodulation 12:85–103.

Lee KH, Blaha CD, Harris BT, Cooper S, Hitti FL, Leiter JC, Roberts DW, Kim U (2006) Dopamine efflux in the rat striatum evoked by electrical stimulation of the subthalamic nucleus: potential mechanism of action in Parkinson's disease. Eur J Neurosci 23:1005–1014.

Lee KH, Blaha CD, Bledsoe JM (2008) Mechanisms of action of deep brain stimulation: a review. In: A textbook of neuromodulation (Krames E, Hunter PP, Rezai A, eds), pp54–79. Amsterdam, The Netherlands: Elsevier.

Lee KH, Kristicm K, van Hoff R, Hitti FL, Blaha CD, Harris B, Roberts DW, Leiter JC (2007) High frequency stimulation of the subthalamic nucleus increases glutamate in the subthalamic nucleus of rats as demonstrated by *in vivo* enzyme-linked glutamate sensor. Brain Res 1162:121–129.

Lee KJ, Jang KS, Shon YM (2006) Chronic deep brain stimulation of subthalamic and anterior thalamic nuclei for controlling refractorypartial epilepsy. Acta Neurochir Suppl 99:87–91.

Limousin P, Krack P, Pollak P, Benazzouz A, Ardouin C, Hoffmann D, Benabid AL (1998) Electrical stimulation of the subthalamic nucleus in advanced Parkinson's disease. N Engl J Med 339:1105–1111.

Lind G, Schechtmann G, Lind C, Winter J, Meyerson BA, Linderoth B (2008) Subthalamic stimulation for essential tremor. Short- and long-term results and critical target area.Stereotact Funct Neurosurg 86:253–258.

Lozano AM, Dostrovsky J, Chen R, Ashby P (2002) Deep brain stimulation for Parkinson's disease: disrupting the disruption. Lancet Neurol 1:225–231.

Magarinos-Ascone C, Pazo JH, Macadar O and Buno W (2002) High-frequency stimulation of the subthalamic nucleus silences subthalamic neurons: A possible cellular mechanism in Parkinson's disease. Neuroscience 115:1109–1117.

Mayberg HS, Lozano AM, Voon V, McNeely HE, Seminowicz D, Hamani C, Schwalb JM, Kennedy SH (2005) Deep brain stimulation for treatment-resistant depression. Neuron 45:651–660.

Meissner W, Harnack D, Paul G, Reum T, Sohr R, Morgenstern R, Kupsch A (2002) Deep brain stimulation of subthalamic neurons increases striatal dopamine metabolism and induces contralateral circling in freely moving 6-hydroxydopamine-lesioned rats. Neurosci Lett 328:105–108.

Meissner W, Harnack D, Reese R, Paul G, Reum T, Ansorge M, Kusserow H, Winter C, Morgenstern R, Kupsch A (2003) High-frequency stimulation of the subthalamic nucleus enhances striatal dopamine release and metabolism in rats. J Neurochem 85:601–609.

Meissner W, Reum T, Paul G, Harnack D, Sohr R, Morgenstern R, Kupsch A (2001) Striatal dopaminergic metabolism is increased by deep brain stimulation of the subthalamic nucleus in 6-hydroxydopamine lesioned rats. Neurosci Lett 303:165–168.

Michael DJ, Wightman RM (1999) Electrochemical monitoring of biogenic amine neurotransmission in real time. J Pharm Biomed Anal 19:33–46.

Molinuevo JL, Valldeoriola F, Tolosa E, Rumia J, Valls-Sole J, Roldan H, Ferrer E (2000) Levodopa withdrawal after bilateral subthalamic nucleus stimulation in advanced Parkinson's disease. Arch Neurol 57:983–988.

Moro E, Scerrati M, Romito LM, Roselli R, Tonali P, Albanese A (1999) Chronic subthalamic nucleus stimulation reduces medication requirements in Parkinson's disease. Neurology 53:85–90.

Ostrem JL, Racine CA, Glass GA, Grace JK, Volz MM, Heath SL, Starr PA (2011) Subthalamic nucleus deep brain stimulation in primary cervical dystonia. Neurology 76:870–878.

Oueslati A, Sgambato-Faure V, Melon C, Kachidian P, Gubellini P, Amri M, Kerkerian-Le Goff L, Salin P (2007) High-frequency stimulation of the subthalamic nucleus potentiates L-DOPA-induced neurochemical changes in the striatum in a rat model of Parkinson's disease. J Neurosci 27:2377–2386.

Papavassiliou E, Rau G, Heath S, Abosch A, Barbaro NM, Larson PS, Lamborn K, Starr PA (2008) Thalamic deep brain stimulation for essential tremor: Relation of lead location to outcome. Neurosurgery 62:884–894.

Park JH, Chung SJ, Lee CS, Jeon SR (2011) Analysis of hemorrhagic risk factors during deep brain stimulation surgery for movement disorders: Comparison of the circumferential paired and multiple electrode insertion methods. Acta Neurochir (Wien) 153:1573–1578.

Patel NK, Heywood P, O'Sullivan K, McCarter R, Love S, Gill SS (2003) Unilateral subthalamotomy in the treatment of Parkinson's disease. Brain 126:1136–1145.

Paul G, Reum T, Meissner W, Marburger A, Sohr R, Morgenstern R, Kupsch A (2000) High frequency stimulation of the subthalamic nucleus influences striatal dopaminergic metabolism in the naive rat. Neuroreport 11:441–444.

Pereira EA, Green AL, Stacey RJ, Aziz TZ (2012) Refractory epilepsy and deep brain stimulation. J Clin Neurosci 19:27–33.

Porta M, Brambilla A, Cavanna AE, Servello D, Sassi M, Rickards H, Robertson MM (2009) Thalamic deep brain stimulation for treatment refractory Tourette syndrome: Two-year outcome. Neurology 73:1375–1380.

Puigdemont D, Perez-Egea R, Portella MJ, Molet J, de Diego-Adelino J, Gironell A, Radua J, Gomez-Anson B, Rodriguez R, SerraM, de QC, Artigas F, Alvarez E, Perez V (2011) Deep brain stimulation of the subcallosalcingulate gyrus: further evidence in treatment-resistant major depression. Int J Neuro psycho pharmacol 22:1–13.

Savica R, Stead M, Mack KJ, Lee KH, Klassen BT (2012) Deep brain stimulation in Tourette syndrome: A description of 3 patients with excellent outcome. Mayo Clin Proc 87:59–62.

Schlaier JR, Habermeyer C, Warnat J, Lange M, Janzen A, Hochreiter A, Proescholdt M, Brawanski A, Fellner C (2011) Discrepancies between the MRI- and the electrophysiologically defined subthalamic nucleus. Acta Neurochir (Wien) 153:2307–2318.

Shah RS, Chang SY, Min HK, Cho ZH, Blaha CD, Lee KH (2010) Deep brain stimulation: Technology at the cutting edge. J Clin Neurol 6:167–182.

Shon YM, Lee KH, Goerss SJ, Kim IY, Kimble C, Van Gompel JJ, Bennet K, Blaha CD, Chang SY (2010b) High frequency stimulation of the subthalamic nucleus evokes striatal dopamine release in a large animal model of human DBS neurosurgery. Neurosci Lett 475:136–140.

Shon YM, Chang SY, Tye SJ, Kimble CJ, Bennet KE, Blaha CD, Lee KH (2010a) Co-monitoring of adenosine and dopamine using the wireless instantaneous neuro transmitter concentration system: Proof of principle. J Neurosurg 112:539–548.

Sillay KA, Sani S, Starr PA (2010) Deep brain stimulation for medically intractable cluster headache. Neurobiol Dis 38:361–368.

Speelman JD, Contarino MF, Schuurman PR, Tijssen MA, de Bie RM (2010) Deep brain stimulation for dystonia: Patient selection andoutcomes. Eur J Neurol 17:102–106.

Strafella AP, Sadikot AF, Dagher A (2003) Subthalamic deep brain stimulation does not induce striatal dopamine release in Parkinson's disease. Neuroreport 14:1287–1289.

Suaud-Chagny MF (2004) *In vivo* monitoring of dopamine overflow in the central nervous system by amperometric techniques combined with carbon-fiber electrodes. Methods 33:322–329.

Thobois S, Fraix V, Savasta M, Costes N, Pollak P, Mertens P, Koudsie A, Le Bars D, Benabid AL, Broussolle E (2003) Chronic subthalamic nucleus stimulation and striatal D2 dopamine receptors in Parkinson's disease — a [(11)C]-raclopride PET study. J Neurol 250:1219–1223.

Thomas L, Bledsoe JM, Stead M, Sandroni P, Gorman D, Lee KH (2009) Motor cortex and deepbrain stimulation for the treatment of intractableneuropathic face pain. Curr Neurol Neurosci Rep 9:120–126.

Venton BJ, Troyer KP, Wightman RM (2002) Response times of carbon-fiber microelectrodes to dynamic changes in catecholamine concentration. Anal Chem 74:539–546.

Vidailhet M, Vercueil L, Houeto JL, Krystkowiak P, Benabid AL, Cornu P, Lagrange C, Tezenas du MS, Dormont D, Grand S, Blond S, Detante O, Pillon B, Ardouin C, Agid Y, Destee A, Pollak P (2005) Bilateral deep-brain stimulation of the globus pallidus in primary generalized dystonia. N Engl J Med 352:459–467.

Volkmann J (2004) Most effective stimulation site in subthalamic deep brain stimulation for Parkinson's disease. Mov Disord 19:1050–1054.

Volkow ND, Fowler JS, Wang GJ, Dewey SL, Schlyer D, MacGregor R, Logan J, Alexoff D, Shea C, Hitzemann R, Angrist B, Wolf AP (1993) Reproducibility of repeated measures of carbon–11–raclopride binding in the human brain. J Nucl Med 34:609–613.

Wilson GS, Gifford R (2005) Biosensors for real-time *in vivo* measurements. Biosens Bioelectron 20:2388–2403.

CHAPTER 11

SWEET LEAF: NEUROCHEMICAL ADVANCES REVEAL HOW CANNABINOIDS AFFECT BRAIN DOPAMINE CONCENTRATIONS

Erik B. Oleson*,† and Joseph F. Cheer*

*University of Colorado Denver
†University of Maryland Baltimord

11.1 INTRODUCTION

Cannabinoids are pharmacologically defined as a class of chemical compounds — comprising phytocannabinoids, chemically synthesized cannabinoids and endocannabinoids — that bind to the cannabinoid CB1 and/or CB2 receptors. Although the precise historical date varies (Small and Cronquist, 1976; Russo, 2007), it is well accepted that cannabinoids were one of the earliest phytochemicals used by humanity — possibly even shaping the evolution of our nervous system (McPartland and Guy, 2004; McPartland et al., 2006). Cannabis is the most commonly abused illicit drug in the world (Degenhardt et al., 2008). In addition to traditional preparation of cannabis, synthetic cannabinoids are beginning to emerge as a commonly abused illicit drug (Wiley et al., 2011; Loewinger et al., 2013). Indeed, a recent report indicates that synthetic cannabinoids are the second most used illicit drug amongst adolescents in the United States, trailing only to cannabis itself (Johnston et al., 2013). Despite the widespread use of cannabinoids, until recently, we knew very little regarding the neurochemical mechanisms through which they produce their rewarding effects due to technical limitations. Through cutting-edge scientific advances, it is now well accepted that exogenous (phyto-, synthetic) and endogenous cannabinoids increase tonic and phasic dopamine concentrations in the nucleus accumbens, similarly to all known drugs of abuse (Di Chiara and Imperato, 1988; Cheer et al., 2007). Tonic and phasic dopamine concentrations are thought to arise from two distinct patterns of dopaminergic neural activity occurring in the freely-moving animal. A tonic dopamine concentration theoretically originates from dopamine neurons firing in a regular pacemaker pattern of low frequency events (1–5 Hz) (Grace, 1991; Floresco et al., 2003), and is detectable using *in vivo* techniques that sample neurochemicals on a time-scale of minutes (e.g., *in vivo* microdialysis). Phasic release events theoretically originate from dopamine neurons firing in high frequency bursts (≥20 Hz) (Grace, 1991; Floresco et al., 2003; Sombers et al., 2009), and are detectable using techniques that sample neurochemicals on a time-scale of seconds (e.g., fast-scan cyclic voltammetry (FSCV)). The purpose of this chapter is to describe how cannabinoids increase accumbal dopamine concentrations within the context of technological advances that facilitated their scientific discoveries. While we will first provide an overview of the contributions provided by *in vivo* microdialysis, an emphasis will be placed on discoveries obtained using FSCV performed in freely-moving rats (Robinson et al., 2003; Heien and Wightman, 2006). As the current manuscript will focus exclusively on neurochemical studies, we would like to refer the reader to other reviews describing the invaluable contributions made using electrophysiological approaches (Lupica and Riegel, 2005; Melis et al., 2012; Melis and Pistis, 2012; Oleson and Cheer, 2012).

11.2 REAL-TIME MONITORING OF CANNABINOID ACTIONS ON DOPAMINE

11.2.1 CANNABINOIDS, MICRODIALYSIS, AND TONIC DOPAMINE CONCENTRATIONS

Using microdialysis, the Gardner lab provided the first *in vivo* neurochemical evidence that cannabinoids increase tonic dopamine concentrations in the nucleus accumbens (Ton et al., 1988; Chen et al., 1990; Chen et al., 1991; Chen et al., 1993). Improvements in neurochemical detection and cannabinoid pharmacology allowed experimenters using microdialysis techniques to further elucidate that CB1 receptors contribute to these cannabinoid-induced increases in dopamine tone, as pretreating animals with CB1 receptor antagonists prevented cannabinoid-induced increases in accumbal dopamine concentration (Tanda et al., 1997). Interestingly, cannabinoid CB2 receptors may also modulate accumbal dopamine tone, albeit in an opposite manner from CB1 receptors. One recent microdialysis study (Xi et al., 2011) reported that intranasal administration of the cannabinoid CB2 agonist

JWH133 decreased, whereas intra-accumbal administration of the CB2 agonist AM630 increased, the concentration of dopamine detected in the nucleus accumbens. The use of microdialysis techniques also led to the demonstration that various synthetic cannabinoids increase accumbal dopamine concentrations, in spite of exhibiting divergent structural profiles (Tanda et al., 1997; Lecca et al., 2006).

11.2.2 NOVEL INSIGHTS FROM FSCV

Despite the seminal insights *in vivo* microdialysis studies provided regarding the neural mechanisms through which cannabinoids produce rewarding effects, several questions remained. Due to the temporal limitations of microdialysis, it remained unknown whether cannabinoids increase phasic dopamine concentrations in the nucleus accumbens. Furthermore, speculation remained whether cannabinoids increase dopamine concentrations directly in the nucleus accumbens, presumably by altering dopamine uptake (Hershkowitz and Szechtman, 1979; Poddar and Dewey, 1980; Chen et al., 1993), in addition to activating cannabinoid CB1 receptors in the midbrain (French et al., 1997; Gessa et al., 1998; Melis et al., 2000). To address these questions, Cheer and colleagues (2004) measured the effect of a single intravenous injection (i.v.) of the synthetic cannabinoid WIN55,212-2 on phasic dopamine release events in the nucleus accumbens shell of freely-moving rats using FSCV. WIN55,212-2 rapidly increased both the frequency and amplitude of phasic dopamine release events (Figures 11.1a and 11.1b), an effect that was observed over a six-minute period. This time-course was unique to that previously reported using microdialysis. For example, Tanda et al., 1997 showed that a tonic dopamine concentration reached maximal levels 20–30 minutes following an i.v. injection of WIN55,212-2, and then declined over a 50-minute period. Next, Cheer and colleagues (2004) addressed whether changes in dopamine uptake contribute to the observed increases in accumbal dopamine concentration. To investigate whether cannabinoids alter dopamine uptake, they stimulated dopamine release from the medial forebrain bundle repeatedly, before and after injecting WIN55,212-2. Electrically evoked dopamine concentrations are sufficient in concentration to allow for the changes in dopamine transporter occupancy produced by various drugs to be discerned (Wu et al., 2001; Oleson et al., 2008; Oleson et al., 2009; Covey et al., 2013). Using this approach, WIN55,212-2 failed to change the clearance rate of stimulated dopamine release (Figure 11.1a), demonstrating that cannabinoid-induced increases in accumbal dopamine concentration do not involve alterations in dopamine uptake (Cheer et al., 2004).

To assess whether cannabinoid-induced increases in accumbal dopamine concentration require CB1 receptor activation, they then challenged WIN55,212-2

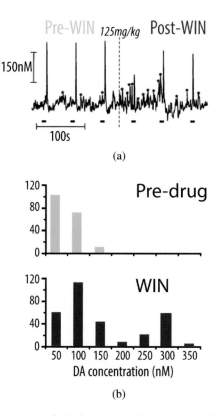

Figure 11.1 Cannabinoids increase phasic dopamine release events in the nucleus accumbens shell. (a) Representative dopamine concentration traces show phasic dopamine release events before (left) and after (right) administration of the synthetic cannabinoid WIN55,212-2 (125 μg/kg i.v.). Electrically evoked dopamine release was also evoked every minute by medial forebrain bundle stimulation (biphasic, 0.4 second, 60 Hz, 120 μA; black bars). (b) WIN55,212-2 (125 μg/kg i.v.) increases the frequency (ordinate) and amplitude (abscissa) of phasic dopamine release events. Reproduced with permission from Cheer et al., 2004.

treated rats with the cannabinoid CB1 receptor antagonist rimonabant (SR141716). Rimonabant abolished the cannabinoid-induced increases in phasic dopamine release (Figure 11.2a) at a dose that failed to independently alter release (Figure 11.2b). Furthermore, additional injections of WIN55,212-2 failed to increase release after CB1 receptors were occupied by rimonabant (Figures 11.2a and 11.2b).

11.2.3 A CURRENT THEORY ON THE PHARMACOLOGICAL MECHANISM OF CANNABINOID ACTION

The FSCV results showing that exogenous cannabinoids fail to alter dopamine uptake but require CB1 receptor activation to increase dopamine release, combined

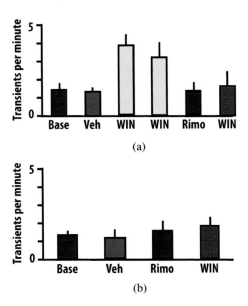

Figure 11.2 Cannabinoids increase phasic dopamine release events through a CB1 receptor dependent mechanism. (a) WIN55,212-2 (125 μg/kg i.v., left WIN; 250 μg/kg i.v., right WIN) increased the frequency of release events *versus* vehicle (Veh) and baseline (Base). These increases were blocked and prevented by rimonabant (Rimo). (b) Rimonabant (300 μg/kg i.v.), when administered alone, failed to alter the frequency of dopamine release events and prevented WIN55,212-2 from increasing the frequency of release events in a cannabinoid agonist naive rat. Reproduced with permission from Cheer et al., 2004.

with anatomical studies demonstrating that midbrain dopamine neurons lack CB1 receptors (Julian et al., 2003), offered support to a pharmacological mechanism of cannabinoid action previously proposed by Lupica and Riegel (2005). These investigators theorized that, similar to opiates, cannabinoids increase dopamine release by indirectly disinhibiting dopamine neurons. According to their conceptualization, cannabinoid activation of G protein (Gi/o)-coupled CB1 receptors located on gamma-aminobutyric acid (GABA) terminals within the midbrain decreases GABA release onto ventral tegmental area (VTA) dopamine neurons. The reduced GABA tone theoretically would decrease activation of GABA receptors on VTA dopamine neurons, thus resulting in a disinhibition of dopaminergic neural activity and a consequent increase in dopamine release events.

11.3 THE ENDOCANNABINOID SYSTEM

11.3.1 INTRODUCTION

All previously described studies involved the investigation of exogenous cannabinoids (i.e., phyto- or synthetic); however, the recently discovered endogenous cannabinoid, or endocannabinoid system, influences dopamine release as well. While we are still in a discovery phase regarding the intricacies of this complex system, it is currently accepted that the brain's endocannabinoid system is composed of lipid signaling molecules (2-arachidonoylglycerol and anandamide were the first and remain the best characterized endocannabinoids), their G protein-coupled receptor targets (CB1 and CB2) (Matsuda et al., 1990; Gong et al., 2006), synthesizing enzymes (diacylglycerol lipase) (Bisogno et al., 2003), degradative enzymes (monoacylglycerol lipase and fatty acid amide hydrolase (FAAH)) (Dinh et al., 2002), and a putative transport system (Di Marzo, 2009; Fu et al., 2011). The recognition that additional uncharacterized endocannabinoids, binding sites, binding dynamics and downstream effects exist (Solinas et al., 2008; Console-Bram et al., 2012), however, deserves acknowledgment when considering how endocannabinoids affect dopamine dynamics in the behaving animal.

11.3.2 ENDOCANNABINOIDS, MICRODIALYSIS AND TONIC DOPAMINE CONCENTRATIONS

A relatively limited number of microdialysis studies attempted to describe how endocannabinoids influence dopamine transmission within the nucleus accumbens. Intravenously administered anandamide was first reported to increase dopamine tone in the nucleus accumbens, an effect that was augmented by inhibiting its degradative enzyme, FAAH (Solinas et al., 2006; Solinas et al., 2007). Paradoxically, however, inhibiting anandamide's degradative enzymatic system was also reported to decrease nicotine-induced increases in accumbal dopamine concentration (Scherma et al., 2008). Additional insights were provided by microdialysis studies assessing the effects of endocannabinoid uptake inhibitors on dopamine concentration. In these studies, it was found that the endocannabinoid uptake inhibitors (9Z)-N-[1-((R)-4-hydroxybenzyl)-2-hydroxyethyl]-9-octadecenamide (OMDM-2), N-(4-hydroxy-2-methylphenyl)-5Z,8Z,11Z,14Z-eicosatetraenamide (VDM11) and N-arachidonoylaminophenol (AM404) decrease accumbal dopamine concentrations (Murillo-Rodríguez et al., 2012) — in addition to those evoked by nicotine (Scherma et al., 2012). It should be noted that many of the aforementioned endocannabinoid uptake inhibitors also exhibit an affinity for anandamide's degradative enzyme, FAAH (López-Rodríguez et al., 2003; Fowler et al., 2004; Vandevoorde and Fowler, 2005), and increase anandamide to a greater extent than 2-arachidonoylglycerol (Van der Stelt et al., 2006). Due largely to a lack of pharmacological compounds capable of selectively increasing 2-arachidonoylglycerol, to date, the effects of 2-arachidonoylglycerol on dopamine tone remain unclear.

11.3.3 NOVEL INSIGHTS FROM FSCV

By taking advantage of the high temporal resolution provided by FSCV, investigators determined that the endocannabinoid system is necessary to detect drug-induced increases in phasic dopamine concentration within the nucleus accumbens shell. Cheer and colleagues (2007) first demonstrated that the drugs nicotine, ethanol and cocaine all increase phasic dopamine release events in the nucleus accumbens shell, albeit in different ways. Despite exhibiting distinct pharmacological mechanisms of action, treating rats with the cannabinoid CB1 receptor antagonist rimonabant abolished the ability of each drug to increase phasic dopamine release events. The finding that CB1 receptor activation is necessary for abused drugs from various pharmacological classes to increase phasic dopamine concentrations implied that endocannabinoid inhibition of GABA release onto dopamine neurons is necessary for drugs of abuse to increase dopamine release (Cheer et al., 2007). This finding is likely due to a unique feature of endocannabinoids; i.e., unlike classical neurotransmitters, endocannabinoids are not stored and released from vesicles, but rather are synthesized and released from vesicles, but rather are synthesized and released 'on demand' during specific neural events (Freund et al., 2003). Thus, it is likely that phasic bursts of dopaminergic neural activity result in the synthesis and release of endocannabinoids (Melis et al., 2004). These observations led the Cheer lab to speculate that the critical interaction between endocannabinoids and phasic dopamine release events may not only involve the actions of drugs of abuse, but may pertain to appetitive behavior in general.

It is well accepted that phasic dopamine release events encode the earliest predictor of reward in appetitive tasks (Schultz et al., 1997; Wassum et al., 2012), and theorized that the resulting increases in accumbal dopamine concentration are required to promote reward directed behavior (Nicola, 2010). If endocannabinoids are necessary for the dopaminergic encoding of reward predictive cues; then CB1 receptor antagonism should uniformly decrease the concentration of cue-evoked dopamine release and reward seeking behavior. To investigate whether endocannabinoids modulate dopaminergic encoding of cues during reward seeking, Oleson and colleagues (2012) measured release in the nucleus accumbens shell during brain stimulation reward and maintained behavior before and after administering the CB1 receptor antagonist rimonabant. Specifically, animals responded on a lever to receive electrical currents delivered to the VTA. Each lever press resulted in the delivery of the electrical currents (i.e., brain stimulation reward) and the simultaneous retraction of the lever. After a fixed time-out period, a cue light placed above the lever illuminated prior to lever extension, thereby serving as the earliest predictor of reward availability. CB1 receptor antagonism concurrently decreased reward seeking (Figure 11.3a) and cue-evoked dopamine concentrations

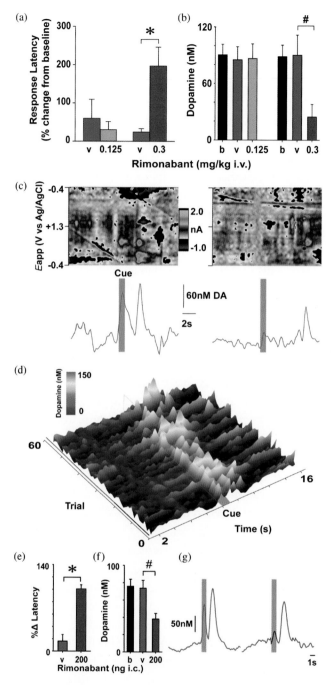

Figure 11.3 a) Response latency (a metric of reward seeking) for brain stimulation reward maintained in the brain stimulation reward task. A high (0.3 mg/kg i.v.; red bar) but not low (0.125 mg/kg i.v.; orange bar) dose of rimonabant increased the latency to respond for brain stimulation reward in comparison to vehicle (v, blue bar). b) Mean dopamine concentration observed during the first second of cue-presentation under baseline (b), vehicle (v), and drug conditions.

Figure 11.3 (*Continued*) Rimonabant at a high (0.3 mg/kg i.v.; red bar) but not low (0.125 mg/kg i.v.; orange bar) dose decreased the concentration of cue-evoked dopamine in comparison to vehicle. c) Representative color plots (top) and dopamine concentration traces (bottom) show the effects of rimonabant on cue-evoked dopamine events in individual trials. Top: representative color plots topographically depict the voltammetric data with time on the x-axis, applied scan potential (Eapp) on the y-axis and background-subtracted faradaic current shown on the z-axis in pseudocolor. Dopamine can be identified by an oxidation peak (green) at +0.6 V and a smaller reduction peak (yellow) at −0.2 V. Bottom: corresponding traces show the concentration of dopamine (nM) detected at the time of cue-presentation (gray bar) following vehicle (left; blue trace) and rimonabant (right; red trace) administration. d) A representative surface-plot shows changes in dopamine concentration (z-axis) across trials (y-axis) during baseline (black line), vehicle (blue line), and rimonabant (red line) conditions. Data are centered around lever presentation on the x-axis. e) Disrupting endocannabinoid signaling within the VTA is sufficient to decrease reward seeking. Intra-tegmental rimonabant (200 ng i.c.; red bar) significantly increased response latency in comparison to vehicle (v, blue bar). f) Mean dopamine concentrations observed during first second of cue-presentation under baseline (b), vehicle (v), and drug conditions. Intrategmental rimonabant (200 ng i.c.; red bar) significantly decreased the concentration of cue-evoked dopamine in comparison to vehicle. g) Representative dopamine concentration traces from individual trials after vehicle (left; blue trace) and rimonabant (200 ng i.c.; right; red trace) treatment. Reproduced with permission from Oleson et al., 2012.

←

(Figure 11.3b). Representative color plots and corresponding dopamine concentration traces show the effect of CB1 receptor antagonism on release in single trials (Figure 11.3c), whereas the representative surface plot shows the effect of CB1 receptor antagonism on release across trials (Figure 11.3d). To assess whether the observed decreases in cue-evoked dopamine concentration were primarily due to CB1 receptor antagonism within the VTA, they assessed the effects of intra-tegmental rimonabant on release and reward seeking. Similar to systemically administered rimonabant, CB1 antagonism in the VTA alone was sufficient to decrease reward seeking (Figure 11.3f) and cue-evoked dopamine release (Figure 11.3h) (Oleson et al., 2012). Together, these data suggested that endocannabinoid signaling within the VTA is necessary to sculpt ethologically relevant patterns of dopamine signaling during appetitive behavior.

The question remained; however, which endocannabinoid is responsible for modulating phasic dopamine release within the VTA, 2-arachidonoylglycerol or anandamide? To assess this, Oleson and colleagues (2012) first treated animals with newly developed pharmacological compounds that selectively target each molecule's degradatory enzyme. 2-arachidonoylglycerol is hydrolyzed by monoacylglycerol lipase (MAGL), and can be increased by treating animals with the MAGL inhibitor JZL184; whereas, anandamide is hydrolyzed by FAAH and can be increased by treating animals with the FAAH inhibitor URB597 (Cravatt et al., 1996; Fegley et al., 2005; Long et al., 2009; Pan et al., 2009). They found that systemically treating rodents with JZL184, but not URB597, increased various metrics of reward seeking behavior (Oleson et al., 2012). These increases in reward

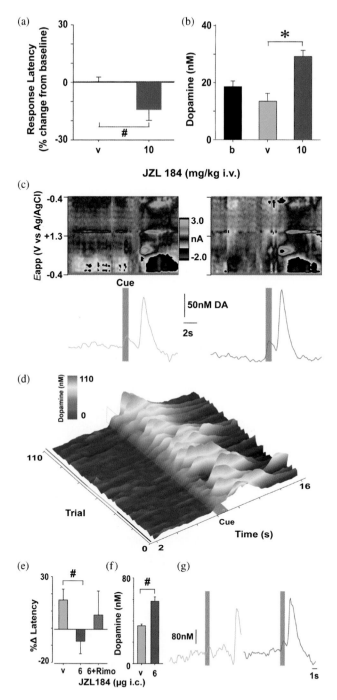

Figure 11.4 (a) Augmenting 2-arachidonoylglycerol levels facilitated reward seeking in a brain stimulation reward task. JZL184 (10 mg/kg i.v., purple bar) decreased response latency in comparison to vehicle (v, blue bar). (b) Facilitated reward seeking was accompanied by an increase in cue-evoked dopamine concentration. (c) Representative color plots (top) and dopamine concentration

Figure 11.4 (*Continued*) traces (bottom) show the effects of JZL184 (right, purple trace) in comparison to vehicle (left, green trace) during individual trials. (d) A representative surface plot illustrates changes in dopamine concentration across trials (y-axis) under baseline (black line), vehicle (green line), and rimonabant (purple line) conditions. (e) Augmenting 2-arachidonoylglycerol in the VTA is sufficient to facilitate reward seeking. JZL184 (6 mg, ipsilateral, purple bar) decreased response latency in comparison to dimethyl sulfoxide (DMSO) (green bar). Post-treatment with a sub-threshold dose of rimonabant (1.25 mg/kg i.v.) reversed the JZL184-induced decrease in reward latency. (f) Facilitated reward seeking occurred simultaneously with an increase in cue-evoked dopamine concentration in comparison to vehicle. (g) Representative traces show the effects of intra-tegmental vehicle (left, green trace) and JZL184 (right, purple trace) on cue-evoked dopamine concentration in individual trials. Reproduced with permission from Oleson et al., 2012.

←

seeking were accompanied by augmented cue-evoked dopamine concentrations (Figures 11.4a and 11.4b). Figures 11.4c and 11.4d illustrate the effects of JZL184 on cue-evoked dopamine concentrations during individual trials and across trials, respectively. To investigate the specific role of intra-tegmental 2-arachidonoylglycerol in regulating the neural mechanism of reward seeking, they then infused JZL184 into the VTA during cue-motivated behavior. As occurred following systemic JZL184 administration, intra-tegmental JZL184 facilitated reward seeking and cue-evoked dopamine concentrations (Figures 11.4e–11.4g). Together, these observations suggest that 2-arachidonoylglycerol levels in the VTA facilitate mesolimbic dopaminergic regulation of appetitive behavior.

The observation that disrupting VTA, endocannabinoid signaling decreased, while augmenting 2-arachidonoylglycerol facilitated cue-evoked dopamine signaling and appetitive behavior, lent credence to the previously proposed theoretical construct by which cannabinoids modulate brain reward function (Lupica and Riegel, 2005; Cheer et al., 2007). In addition, these FSCV data identified 2-arachidonoylglycerol as the primary endocannabinoid involved — a finding that is consistent with the notion that 2-arachidonoylglycerol is the principle endocannabinoid for multiple forms of synaptic plasticity across different brain regions (Melis et al., 2004; Tanimura et al., 2010). Synthesizing the neurochemical data with previous work from other branches of neuroscience led us conclude that 2-arachidonoylglycerol modulates the neural mechanisms of appetitive behavior by disinhibiting dopaminergic neural activity, similarly to exogenous cannabinoids. A depiction of our complete conceptualization is presented as a cartoon in Figure 11.5.

11.4 FUTURE PROSPECTS

While tremendous advances in our understanding of the brain's endocannabinoid system occurred over the last two decades, we still stand today on the edge of a new frontier. We remain unclear on exactly how many lipid molecules exist with

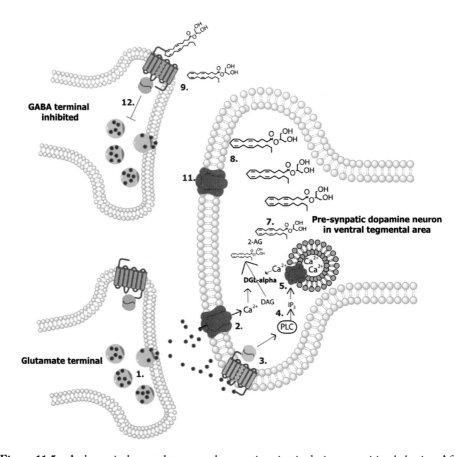

Figure 11.5 A theoretical ventral tegmental area microcircuit during appetitive behavior. After encountering a cue predicting a large reward, conditioned glutamate release occurs in the VTA (1), thus resulting in Ca^{2+} influx into the dopamine neuron (2) and activation of G protein-coupled receptors (e.g., mGluR1/5) (3). G protein-coupled receptor stimulation activates phospholipase C (PLC), which ultimately leads to the formation of inositol trisphosphate (IP3) and diacylglycerol (DAG) (4). IP3 binds to IP3 receptors, resulting in the mobilization of intracellular Ca^{2+} stores (5). Elevated intracellular Ca^{2+} activates the enzyme diacylglycerol lipase-alpha (DGL-alpha) (6), which hydrolyzes DAG to form 2-arachidonoylglycerol (7). 2-arachidonoylglycerol traverses the plasma membrane into the extra-synaptic space (8), where it retrogradely activates cannabinoid CB1 receptors on pre-synaptic GABA terminals (9). Activation of the Gi/o subunit of CB1 receptor suppresses GABA release (10). Decreased GABA activation of $GABA_A$ receptors (11) on dopamine neurons disinhibits the dopaminergic neural activity, thus facilitating cue-evoked dopamine signaling during reward-directed behavior. Reproduced with permission from Lee et al., 2012.

the capacity to target traditional cannabinoid receptors (e.g., CB1), and what other receptor targets may also produce downstream effects. Undoubtedly, enhancing our ability to pharmacologically harness the endocannabinoid system will provide new avenues into the treatment of disease. The immediate questions we are focusing on involve the role of the endocannabinoid system in regulating mesolimbic dopamine function during negative affect, which might lead to the therapeutic advances in drug addiction, post-traumatic stress disorder and depression.

ACKNOWLEDGMENT

Funding for this work was provided by NIH grants R01DA022340 (J.F.C.) and F32DA032266 (E.B.O.).

REFERENCES

Bisogno T, Howell F, Williams G, Minassi A, Cascio MG, Ligresti A, Matias I, Schiano-Moriello A, Paul P, Williams EJ (2003) Cloning of the first sn1-DAG lipases points to the spatial and temporal regulation of endocannabinoid signaling in the brain. J Cell Biol 163:463–468.

Cheer JF, Wassum KM, Heien ML, Phillips PE, Wightman RM (2004) Cannabinoids enhance subsecond dopamine release in the nucleus accumbens of awake rats. J Neurosci 24:4393–4400.

Cheer JF, Wassum KM, Sombers LA, Heien ML, Ariansen JL, Aragona BJ, Phillips PE, Wightman RM (2007) Phasic dopamine release evoked by abused substances requires cannabinoid receptor activation. J Neurosci 27:791–795.

Chen J, Paredes W, Lowinson JH, Gardner EL (1991) Strain-specific facilitation of dopamine efflux by Δ^9-tetrahydrocannabinol in the nucleus accumbens of rat: an *in vivo* microdialysis study. Neurosci Lett 129:136–140.

Chen J, Marmur R, Pulles A, Paredes W, Gardner EL (1993) Ventral tegmental microinjection of Δ^9-tetrahydrocannabinol enhances ventral tegmental somatodendritic dopamine levels but not forebrain dopamine levels: evidence for local neural action by marijuana's psychoactive ingredient. Brain Res 621:65–70.

Chen J, Paredes W, Li J, Smith D, Lowinson J, Gardner EL (1990) Δ^9-tetrahydrocannabinol produces naloxone-blockable enhancement of presynaptic basal dopamine efflux in nucleus accumbens of conscious, freely-moving rats as measured by intracerebral microdialysis. Psychopharmacology 102:156–162.

Console-Bram L, Marcu J, Abood ME (2012) Cannabinoid receptors: nomenclature and pharmacological principles. Prog Neuropsychopharmacol Biol Psychiatry 38:4–15.

Covey DP, Juliano SA, Garris PA (2013) Amphetamine elicits opposing actions on readily releasable and reserve pools for dopamine. PloS One 8:e60763.

Cravatt BF, Giang DK, Mayfield SP, Boger DL, Lerner RA, Gilula NB (1996) Molecular characterization of an enzyme that degrades neuromodulatory fatty-acid amides. Nature 384:83–87.

Degenhardt L, Chiu WT, Sampson N, Kessler RC, Anthony JC, Angermeyer M, Bruffaerts R, de Girolamo G, Gureje O, Huang Y (2008) Toward a global view of alcohol, tobacco, cannabis, and cocaine use: findings from the WHO world mental health surveys. PLoS Med 5:e141.

Di Chiara G, Imperato A (1988) Drugs abused by humans preferentially increase synaptic dopamine concentrations in the mesolimbic system of freely moving rats. Proc Natl Acad Sci USA 85:5274–5278.

Di Marzo V (2009) The endocannabinoid system: its general strategy of action, tools for its pharmacological manipulation and potential therapeutic exploitation. Pharmacol Res 60:77–84.

Dinh T, Carpenter D, Leslie F, Freund T, Katona I, Sensi S, Kathuria S, Piomelli D (2002) Brain monoglyceride lipase participating in endocannabinoid inactivation. Proc Natl Acad Sci USA 99:10819–10824.

Fegley D, Gaetani S, Duranti A, Tontini A, Mor M, Tarzia G, Piomelli D (2005) Characterization of the fatty acid amide hydrolase inhibitor cyclohexyl carbamic acid 3'-carbamoyl-biphenyl-3-yl ester (URB597): effects on anandamide and oleoylethanolamide deactivation. J Pharmacol Exp Ther 313:352–358.

Floresco SB, West AR, Ash B, Moore H, Grace AA (2003) Afferent modulation of dopamine neuron firing differentially regulates tonic and phasic dopamine transmission. Nat Neurosci 6:968–973.

Fowler CJ, Tiger G, Ligresti A, López-Rodríguez ML (2004) Selective inhibition of anandamide cellular uptake *versus* enzymatic hydrolysis — a difficult issue to handle. Eur J Pharmacol 492:1–11.

French ED, Dillon K, Wu X (1997) Cannabinoids excite dopamine neurons in the ventral tegmentum and substantia nigra. Neuroreport 8:649–652.

Freund TF, Katona I, Piomelli D (2003) Role of endogenous cannabinoids in synaptic signaling. Physiol Rev 83:1017–1066.

Fu J, Bottegoni G, Sasso O, Bertorelli R, Rocchia W, Masetti M, Guijarro A, Lodola A, Armirotti A, Garau G (2011) A catalytically silent FAAH-1 variant drives anandamide transport in neurons. Nat Neurosci 15:64–69.

Gessa G, Melis M, Muntoni A, Diana M (1998) Cannabinoids activate mesolimbic dopamine neurons by an action on cannabinoid CB1 receptors. Eur J Pharmacol 341:39–44.

Gong J-P, Onaivi ES, Ishiguro H, Liu Q-R, Tagliaferro PA, Brusco A, Uhl GR (2006) Cannabinoid CB2 receptors: immunohistochemical localization in rat brain. Brain Res 1071:10–23.

Grace A (1991) Phasic *versus* tonic dopamine release and the modulation of dopamine system responsivity: a hypothesis for the etiology of schizophrenia. Neuroscience 41:1–24.

Heien ML, Wightman RM (2006) Phasic dopamine signaling during behavior, reward, and disease states. CNS Neurol Disord Drug Targets 5:99–108.

Hershkowitz M, Szechtman H (1979) Pretreatment with δ^1-tetrahydrocannabinol and psychoactive drugs: effects on uptake of biogenic amines and on behavior. Eur J Pharmacol 59:267–276.

Johnston LD, O'Malley PM, Bachman JG, Schulenberg JE (2013) Monitoring the future, national results on drug use: 2012 overview, key findings on adolescent drug use, pp14–16. Ann Arbor: Institute for Social Research, The University of Michigan.

Julian M, Martin A, Cuellar B, Rodriguez De Fonseca F, Navarro M, Moratalla R, Garcia-Segura LM (2003) Neuroanatomical relationship between type 1 cannabinoid receptors and dopaminergic systems in the rat basal ganglia. Neuroscience 119:309–318.

Lecca D, Cacciapaglia F, Valentini V, Gronli J, Spiga S, Di Chiara G (2006) Preferential increase of extracellular dopamine in the rat nucleus accumbens shell as compared to that in the core during acquisition and maintenance of intravenous nicotine self-administration. Psychopharmacology 184:435–446.

Lee AM, Oleson EB, Diergaarde L, Cheer JF, Pattij T (2012) Cannabinoids and value-based decision making: implications for neurodegenerative disorders. Basal Ganglia 2:131–138.

Loewinger G, Oleson E, Cheer J (2013) Using dopamine research to generate rational cannabinoid drug policy. Drug Test Anal 5:22–26.

Long JZ, Nomura DK, Vann RE, Walentiny DM, Booker L, Jin X, Burston JJ, Sim-Selley LJ, Lichtman AH, Wiley JL (2009) Dual blockade of FAAH and MAGL identifies behavioral processes regulated by endocannabinoid crosstalk *in vivo*. Proc Natl Acad Sci USA 106:20270–20275.

López-Rodríguez ML, Viso A, Ortega-Gutiérrez S, Fowler CJ, Tiger G, de Lago E, Fernández-Ruiz J, Ramos JA (2003) Design, synthesis, and biological evaluation of new inhibitors of the endocannabinoid uptake: comparison with effects on fatty acid amidohydrolase. J Med Chem 46:1512–1522.

Lupica CR, Riegel AC (2005) Endocannabinoid release from midbrain dopamine neurons: a potential substrate for cannabinoid receptor antagonist treatment of addiction. Neuropharmacology 48:1105–1116.

Matsuda LA, Lolait SJ, Brownstein MJ, Young AC, Bonner TI (1990) Structure of a cannabinoid receptor and functional expression of the cloned cDNA. Nature 346:561–564.

McPartland JM, Guy GW (2004) The evolution of cannabis and coevolution with the cannabinoid receptor — a hypothesis. In: The medicinal uses of cannabis and cannabinoids, pp71–101. London: Pharmaceutical Press.

McPartland JM, Matias I, Di Marzo V, Glass M (2006) Evolutionary origins of the endocannabinoid system. Gene 370:64–74.

Melis M, Pistis M (2012) Hub and switches: endocannabinoid signalling in midbrain dopamine neurons. Philos Trans R Soc Lond B Biol Sci 367:3276–3285.

Melis M, Gessa GL, Diana M (2000) Different mechanisms for dopaminergic excitation induced by opiates and cannabinoids in the rat midbrain. Prog Neuropsychopharmacol Biol Psychiatry 24:993–1006.

Melis M, Muntoni AL, Pistis M (2012) Endocannabinoids and the processing of value-related signals. Front Pharmacol 3:7.

Melis M, Perra S, Muntoni AL, Pillolla G, Lutz B, Marsicano G, Di Marzo V, Gessa GL, Pistis M (2004) Prefrontal cortex stimulation induces 2-arachidonoyl-glycerol-mediated suppression of excitation in dopamine neurons. J Neurosci 24:10707–10715.

Murillo-Rodríguez E, Palomero-Rivero M, Millán-Aldaco D, Di Marzo V, Drucker-Colín R (2012) The administration of endocannabinoid uptake inhibitors OMDM-2 or VDM-11 promotes sleep and decreases extracellular levels of dopamine in rats. Physiol Behav 109:88–95.

Nicola SM (2010) The flexible approach hypothesis: unification of effort and cue-responding hypotheses for the role of nucleus accumbens dopamine in the activation of reward-seeking behavior. J Neurosci 30:16585–16600.

Oleson EB, Cheer JF (2012) A brain on cannabinoids: The role of dopamine release in reward seeking. Cold Spring Harb Perspect Med 2:8.

Oleson EB, Salek J, Bonin KD, Jones SR, Budygin EA (2009) Real-time voltammetric detection of cocaine-induced dopamine changes in the striatum of freely moving mice. Neurosci Lett 467:144–146.

Oleson EB, Talluri S, Childers SR, Smith JE, Roberts DC, Bonin KD, Budygin EA (2008) Dopamine uptake changes associated with cocaine self-administration. Neuropsychopharmacology 34:1174–1184.

Oleson EB, Beckert MV, Morra JT, Lansink CS, Cachope R, Abdullah RA, Loriaux AL, Schetters D, Pattij T, Roitman MF (2012) Endocannabinoids shape accumbal encoding of cue-motivated behavior *via* CB1 receptor activation in the ventral tegmentum. Neuron 73:360–373.

Pan B, Wang W, Long JZ, Sun D, Hillard CJ, Cravatt BF, Liu QS (2009) Blockade of 2-arachidonoylglycerol hydrolysis by selective monoacylglycerol lipase inhibitor 4-nitrophenyl 4-(dibenzo [d][1, 3] dioxol-5-yl (hydroxy) methyl) piperidine-1-carboxylate (JZL184) enhances retrograde endocannabinoid signaling. J Pharm Exp Ther 331:591–597.

Poddar M, Dewey W (1980) Effects of cannabinoids on catecholamine uptake and release in hypothalamic and striatal synaptosomes. J Pharm Exp Ther 214:6–67.

Robinson DL, Venton BJ, Heien ML, Wightman RM (2003) Detecting subsecond dopamine release with fast-scan cyclic voltammetry *in vivo*. Clin Chem 49:1763–1773.

Russo EB (2007) History of cannabis and its preparations in saga, science, and sobriquet. Chem Biodivers 4:1614–1648.

Scherma M, Panlilio LV, Fadda P, Fattore L, Gamaleddin I, Le Foll B, Justinova Z, Mikics E, Haller J, Medalie J (2008) Inhibition of anandamide hydrolysis by cyclohexyl carbamic acid 3′carbamoyl-3-yl ester (URB597) reverses abuse-related behavioral and neurochemical effects of nicotine in rats. J Pharm Exp Ther 327:482–490.

Scherma M, Justinová Z, Zanettini C, Panlilio LV, Mascia P, Fadda P, Fratta W, Makriyannis A, Vadivel SK, Gamaleddin I (2012) The anandamide transport inhibitor AM404 reduces the rewarding effects of nicotine and nicotine-induced dopamine elevations in the nucleus accumbens shell in rats. Br J Pharmacol 165:2539–2548.

Schultz W, Dayan P, Montague PR (1997) A neural substrate of prediction and reward. Science 275:1593–1599.

Small E, Cronquist A (1976) A practical and natural taxonomy for cannabis. Taxon 25:405–435.

Solinas M, Goldberg SR, Piomelli D (2008) The endocannabinoid system in brain reward processes. Br J Pharmacol 154:369–383.

Solinas M, Justinova Z, Goldberg SR, Tanda G (2006) Anandamide administration alone and after inhibition of fatty acid amide hydrolase (FAAH) increases dopamine levels in the nucleus accumbens shell in rats. J Neurochem 98:408–419.

Solinas M, Tanda G, Justinova Z, Wertheim CE, Yasar S, Piomelli D, Vadivel SK, Makriyannis A, Goldberg SR (2007) The endogenous cannabinoid anandamide produces δ^9-tetrahydrocannabinol-like discriminative and neurochemical effects that are enhanced by inhibition of fatty acid amide hydrolase but not by inhibition of anandamide transport. J Pharm Exp Ther 321:370–380.

Sombers LA, Beyene M, Carelli RM, Wightman RM (2009) Synaptic overflow of dopamine in the nucleus accumbens arises from neuronal activity in the ventral tegmental area. J Neurosci 29:1735–1742.

Tanda G, Pontieri FE, Di Chiara G (1997) Cannabinoid and heroin activation of mesolimbic dopamine transmission by a common μ1 opioid receptor mechanism. Science 276:2048–2050.

Tanimura A, Yamazaki M, Hashimotodani Y, Uchigashima M, Kawata S, Abe M, Kita Y, Hashimoto K, Shimizu T, Watanabe M (2010) The endocannabinoid 2-arachidonoylglycerol produced by diacylglycerol lipase α mediates retrograde suppression of synaptic transmission. Neuron 65:320–327.

Ton JMNC, Gerhardt GA, Friedemann M, Etgen AM, Rose GM, Sharpless NS, Gardner EL (1988) The effects of Δ^9-tetrahydrocannabinol on potassium-evoked release of dopamine in the rat caudate nucleus: an *in vivo* electrochemical and *in vivo* microdialysis study. Brain Res 451:59–68.

Van der Stelt M, Mazzola C, Esposito G, Matias I, Petrosino S, De Filippis D, Micale V, Steardo L, Drago F, Iuvone T (2006) Endocannabinoids and β-amyloid-induced neurotoxicity *in vivo*: effect of pharmacological elevation of endocannabinoid levels. Cell Mol Life Sci 63:1410–1424.

Vandevoorde S, Fowler CJ (2005) Inhibition of fatty acid amide hydrolase and monoacylglycerol lipase by the anandamide uptake inhibitor VDM11: evidence that VDM11 acts as an FAAH substrate. Br J Pharmacol 145:885–893.

Wassum KM, Ostlund SB, Maidment NT (2012) Phasic mesolimbic dopamine signaling precedes and predicts performance of a self-initiated action sequence task. Biol Psychiatry 71:846–854.

Wiley JL, Marusich JA, Huffman JW, Balster RL, Thomas BF (2011) Hijacking of basic research: the case of synthetic cannabinoids. Methods Rep RTI Press 2011.

Wu Q, Reith ME, Wightman RM, Kawagoe KT, Garris PA (2001) Determination of release and uptake parameters from electrically evoked dopamine dynamics measured by real-time voltammetry. J Neurosci Methods 112:119–133.

Xi ZX, Peng XQ, Li X, Song R, Zhang HY, Liu QR, Yang HJ, Bi GH, Li J, Gardner EL (2011) Brain cannabinoid CB2 receptors modulate cocaine's actions in mice. Nat Neurosci 14:1160–1166.

CHAPTER 12

PROBING SEROTONIN NEUROTRANSMISSION: IMPLICATIONS FOR NEUROPSYCHIATRIC DISORDERS

Kevin M. Wood, David Cepeda and Parastoo Hashemi

Wayne State University

12.1 INTRODUCTION

Serotonin is a unique neurotransmitter. About 98% of the body's serotonin is located outside of the brain (Cooper et al., 2003), however, as a neuromodulator it plays important, expansive roles. Serotonin is thought to control many essential processes including appetite, sleep, mood, memory, cognition, movement and reward. Imbalances in the serotonergic system are associated with many neuropsychiatric disorders.

Autism (Chugani et al., 1997; Sutcliffe et al., 2005), post-traumatic stress disorder (PTSD) (Lee et al., 2005), anxiety (Lesch et al., 1996; Sen et al., 2004; Ramboz et al., 1998), schizophrenia (Inayama et al., 1996; Laruelle et al., 1993), bipolar disorder (Furlong et al., 1998; Young et al., 1994), and addiction (Muller et al. 2007; Sellers et al., 1992) all display distinct deregulations of the serotonergic system.

The most notorious of serotonin's associations, however, is depression. Depression debilitates millions of Americans every year; it is a mental illness that can destroy feelings of joy gained from previous exciting activities. Fatigue, hopelessness, irritability, and excessive sadness can plague the sufferer. In 2011, antidepressants and antipsychotics dominated the pharmaceutical market (Lindsley, 2012).

Modern antidepressants typically exaggerate serotonin's effects in the synapse by slowing down its reuptake. Despite their mainstream usage, antidepressants generate controversy because of variable clinical efficacy (Smith et al., 2002; Cipriani et al., 2009) and systemic side effects (Ferguson, 2001; Masand and Gupta, 2002). Moreover, it is common for patients to take antidepressants for 3–4 weeks without experiencing therapeutic benefit (Gelenberg and Chesen, 2000; Onder and Tural, 2003). During this period patients can experience heightened depression and suicidal tendencies (Stone et al., 2009).

Serotonin's role in depression is poorly understood. In recent times there is a growing belief that elevated serotonin levels after antidepressant treatment modifies other neuronal systems (D'Aquila et al., 2000; Lacasse and Leo, 2005). These modifications may be responsible for clinical mood elevation. In agreement with this notion, some antidepressants combine serotonin reuptake inhibition with dopaminergic and noradrenergic uptake inhibition (Richelson and Pfenning, 1984). To decipher more clearly serotonin's roles in the brain, real-time chemical analysis is useful. Neurotransmission is a rapid process (<seconds). The ability to localize a chemical serotonin signal in real-time can allow for direct correlation of serotonin signaling with pharmacological and behavioral events.

Chemical serotonin analysis in real-time has been challenging for analytical chemists for two main reasons. First, endogenous serotonin levels in the brain are low, which creates challenges for collection-based methods. Second, as an electroactive molecule, serotonin is highly detrimental to electrode surfaces due to its surface fouling effects (Jackson et al., 1995; Hashemi et al., 2009). New technological breakthroughs are pushing the sensitivity limits of analytical methods to measure serotonin signaling in various models. In this chapter we review the most current advances in real-time serotonin monitoring. We outline the utility of such measurements for studying serotonin function in the brain and ultimately for improving future therapies to treat serotonin related brain disorders.

12.2 REAL-TIME SEROTONIN MONITORING

12.2.1 MICRODIALYSIS

Since microdialysis was pioneered in the 70s and 80s for collecting monoamines (Ungerstedt and Pycock, 1974; Zetterstrom et al., 1983; Sharp et al., 1986), the technique has been used in over 2000 studies exploring serotonin. These studies have broken ground and have provided seminal information on absolute level serotonin changes that are governed physiologically, pharmacologically and behaviorally.

A semi-permeable membrane bound probe is implanted into a brain region of interest. A solution resembling the brain's fluids is perfused through the probe; analytes pass through the membrane *via* concentration gradients and can be delivered or collected as such. Typically offline analysis characterizes and quantifies the dialysate chemistry. While providing multi-analyte analysis at a single location, it is difficult to use traditional microdialysis for studying neurotransmission. First, the large probe size (125 µm radius) cannot sample from small, localized regions. Second, offline analysis limits the temporal resolution of this measurement to around 10 minutes.

Lately, researchers have challenged the temporal limitations of microdialysis. Using segmented flow and mass spectroscopic analysis, Song et al. increased their temporal resolution to five seconds (Song et al., 2012). Additionally, Scott et al. wirelessly integrated a microdialysis probe to a microfluidic device with amperometric detection to analyze dialysate at one minute intervals (Scott et al., 2013). Neither of these methods has yet been applied to serotonin monitoring. In 2013 Yang et al. increased their microdialysis sampling frequency and reported serotonin levels every three minutes (Yang et al., 2013). The group analyzed serotonin in mice during a circadian cycle. More recently, Weber and colleagues used high pressure and temperature microdialysis to measure serotonin in a freely-moving rodent striatum at one-minute intervals (Zhang et al., 2013).

While these emerging technologies for microdialysis measurements have unearthed important information on fast neurotransmitter dynamics, in depth physiological studies on serotonin are still pending. More commonly employed detection methods for serotonin are the electrochemical methods.

12.2.2 CONSTANT POTENTIAL AMPEROMETRY

The fastest serotonin measurements are aided by the inherent electroactivity of the serotonin molecule. Serotonin easily lends itself to electrochemistry, allowing fast, local measurements at microelectrodes. Carbon-fiber microelectrodes, typically of 5–7 µm diameter, offer high sensitivity and rapid surface kinetics. Constant potential amperometry, abbreviated simply to amperometry, provides µs temporal resolution. Here, carbon-fiber microelectrodes are held at a constant potential, sufficient to oxidize or reduce the analyte of interest. A faradaic current quantifies changes in analyte concentration. Amperometry has provided valuable information on exocytosis and quantal release of serotonin (Ge et al., 2011; Bruns et al., 2000; Alvarez de Toledo et al., 1993).

Because several biologically important molecules such as dopamine, ascorbic acid, 3,4-Dihydroxyphenylacetic acid (DOPAC) and histamine oxidize at similar electrochemical potentials, it is very difficult to use amperometry to study serotonin in complex *in vivo* matrices. Amperometry is suited to well-controlled, single-cell studies, typically in non-neuronal mast cells. In this way, Haynes and colleagues investigated the effect of foreign bodies on cell function and corresponding serotonin release in an elegant group of studies. Certain metal nanoparticles are found in modern consumer goods and carry unknown toxicities. Haynes and colleagues found that serotonin transmission was modified in mast cells exposed to Au, Ag and immunotoxicants (Marquis et al., 2010; Marquis and Haynes, 2010). These studies signal the utility and need for real-time measurements of nanoparticles on celular function.

Few studies have examined serotonin in neuronal cells (Bruns et al., 2000; Bruns and Jahn, 1995). These examined neurons of the leech.

Physiological characterizations in model cells provide important information for serotonin exocytosis events. For a more general picture of the role that serotonin plays in its native neuronal environment, it is most useful to measure it *in vivo*.

12.2.3 HIGH SPEED CHRONOAMPEROMETRY

In contrast to constant potential amperometry, high-speed chronoamperometry adds a good degree of chemical selectivity to the measurement. In this method a square wave potential step is applied to the carbon-fiber microelectrode at oxidation and reduction potentials sufficient for serotonin's reversible electrochemical couple. After the step application, an exponentially decaying current at the electrode surface is measured. The slope of decay is used to characterize serotonin. Ratios of the reductive to oxidative current decay slopes vary between serotonin and other analytes, thereby providing selectivity (Schenk et al., 1983). Although this method is theoretically sound, experimental *in vivo* analysis still demands a higher level of selectivity. In this regard, scientists inject a bolus of serotonin close to the electrode surface and measure its clearance. High-speed chronoamperometry is particularly well suited to measuring serotonin clearance *via* the serotonin transporters (SERTs). SERT dynamics (or kinetics) and pharmacodynamics have been studied in depth with chronoamperometry (Daws et al., 1997, 2005; Perez and Andrews, 2005; Benmansour et al., 2002, 1999; Montanez et al., 2003) as well as disease studies of the SERT in autism (Veenstra-VanderWeele et al., 2012) and drugs of abuse (Callaghan et al., 2007; Thompson et al., 2011; Daws et al., 2006).

The most biologically relevant chronoamperometry studies inject serotonin exogenously. To study native serotonin, more physiologically relevant stimuli are necessary.

12.2.4 FAST-SCAN CYCLIC VOLTAMMETRY (FSCV)

FSCV can measure serotonin with a high level of selectivity in *in vivo* models without exogenous injection of serotonin. FSCV also employs carbon-fiber microelectrodes implanted directly into brain tissue. A substrate-specific electrochemical waveform is applied to the electrode to oxidize/reduce a given analyte. This process yields a cyclic voltammogram that simultaneously identifies and quantifies the analyte based on the potential and current positions of the oxidation/reduction peaks. The waveform is applied every 100 milliseconds, each cyclic voltammogram is background-subtracted and selective analyte concentrations can be reported within milliseconds. Traditionally, FSCV has been used to detect dopamine in a range of models spanning from single cells to awake, behaving animals (Cahill et al., 1996; John and Jones, 2007; Phillips et al., 2003; Kozminski et al., 1998; Cacciapaglia et al., 2011). There have been electrochemical and physiological challenges for serotonin FSCV. One major impasse for serotonin FSCV is electrode fouling, a process inherent to serotonin electro-oxidation. Serotonin's oxidation products polymerize and coat the electrode surface, impeding electrode kinetics. Modifications to the electrochemical waveform allow serotonin to be measured *in vitro* (Jackson et al., 1995), in tissue slice preparations (Bunin et al., 1998; Bunin and Wightman, 1998) and in drosophila (Borue et al., 2009). *In vivo*, however, a thin, uniform electro-coated Nafion® layer is necessary to maintain the surface integrity of the electrode in the face of high concentrations of serotonin metabolites (Hashemi et al., 2009).

Serotonin FSCV requires that serotonin be evoked from its cell bodies and detected in terminal regions. A second challenge for serotonin FSCV stems from the physiologically low levels of serotonin in the rodent terminal brain regions and the inaccessible location of the serotonin cell bodies. The serotonin cell bodies are located in the dorsal raphe nucleus (DRN). This small, diffuse structure lies directly under a dense, critical vascular network. Physically targeting this structure is not only spatially challenging but can often be fatal if the vessels are disrupted. A technically simpler and physiologically safer way to elicit serotonin is to stimulate one of its major axonal bundles, the medial forebrain bundle (MFB) (Hashemi et al., 2011). Additionally, Nafion® coatings on the electrode surface increase the sensitivity of carbon-fiber microelectrodes so that low levels of released serotonin can be measured in a serotonin dense terminal region, the substantianigra pars reticulata (SNR) (Hashemi et al., 2012; Lama et al., 2012; Pathirathna et al., 2012; Wood and Hashemi, 2013).

12.3 THE RELEVANCE OF REAL-TIME SEROTONIN MEASUREMENTS TO PHYSIOLOGY AND DISEASE

The bulk of real-time chemical serotonin analysis has focused on serotonin's multiple receptors and the SERTs. In this section we review recent experiments that have unearthed new information on serotonin's native mechanisms in the brain and during pharmacological manipulations.

12.3.1 ENDOGENOUS *IN VIVO* PROCESSES

In neurons serotonin is synthesized from the amino acid tryptophan, which is converted to 5-hydroxytryptophan, and subsequently to serotonin *via* decarboxylation. Serotonin then actively enters synaptic vesicles through the vesicular monoamine transporter (VMAT) *via* a proton–ATPase pump. Serotonin is stored in these vesicles until an action potential causes calcium influx into the terminal, which leads to exocytosis. Released serotonin acts on postsynaptic receptors and presynaptic autoreceptors and is then reuptaken up into the presynaptic terminal by the SERT. Back in the cell, it can be repackaged into vesicles or degraded by monoamine oxidase (MAO). FSCV has nuanced our understanding of these basic steps of serotonin neurotransmission.

Synapsins are phosphoproteins that are thought to be involved in vesicle trafficking and transmitter release. Specifically it is suggested that synapsins bind the vesicles to the cytoskeleton, thereby creating a reserve pool of transmitter and also dock and/or fuse vesicles to the plasma membrane before release (Hilfiker et al., 1999). Caron, Wightman and colleagues used FSCV to highlight the role of synapsins on serotonin transmission in synapsin knockout mice (Kile et al., 2010). Interestingly, serotonin release was unchanged in these transgenic mice; while dopamine release was strongly enhanced. This indicates that synapsins more strongly regulate dopamine release.

Serotonin is under a high degree of regulation. FSCV characterizations of serotonin release in rat SNR tissue slice preparations suggest that electrically released serotonin has a similar kinetic profile to that of dopamine (Bunin et al., 1998; Bunin and Wightman, 1998). *In vivo* studies, however, paint a more complex picture. Serotonin and dopamine can be measured *in vivo* simultaneously in the rat SNR and nucleus accumbens (NAc) respectively (Hashemi et al., 2012). Serotonin release is approximately 300 times less than dopamine release despite a common electrical stimulation of the MFB and high tissue content levels (Palkovits et al, 1974; Kile et al., 2010). This shows that intact, physiological circuitry strongly controls serotonin release.

Pharmacologically inhibiting serotonin synthesis and vesicular packaging have very little effect on electrically stimulated serotonin release (Hashemi et al., 2012). This is consistent with high intracellular levels due to a tight degree of release regulation. In the face of high intracellular serotonin levels, electrically stimulated serotonin release is insensitive to disruptions in synthesis and packaging. In contrast, serotonin is extremely sensitive to re-uptake and metabolism inhibition. This is an interesting finding because two major antidepressant classes target the re-uptake and metabolism systems. Pharmacologically inhibiting both reputake and metabolism can lead to the potentially fatal serotonin syndrome, which has also been characterized with FSCV (Hashemi et al., 2012).

While the mechanisms of *in vivo* serotonin regulation remain to be established, it is apparent that such a tight level of control is physiologically important. Pharmacologically inhibiting both re-uptake and metabolism can lead to the potentially fatal serotonin syndrome, which has also been characterized with FSCV (Hashemi et al., 2012).

12.3.2 SEROTONIN MODULATION

The brain had a large, compact structure where activity in one location influences activity in another. Neuroreceptors are the biological site of chemical communication and serotonin is a particularly complex transmitter because it acts on at least 14 known serotonin receptor subtypes (Barnes and Sharp, 1999). Furthermore, serotonin activity can influence other neurotransmitters, and other neurotransmitters can influence serotonin. This is termed modulation.

FSCV has uncovered a tight degree of modulation between serotonin and histamine. Histamine is a signaling molecule generally known for its involvement in the immune response, but it is also a neurotransmitter involved in many systemic and neuronal functions (Haas and Panula, 2003; Haas et al, 2008). Histamine receptors were found to modulate serotonin dynamics. In the rodent brain SNR, histamine H_2 receptors (H_2R) located on serotonin neurons are constitutively active (or always "on" in the absence of a bound ligand) to suppress serotonin exocytosis (Threlfell et al., 2008) and histamine H_3 receptors (H_3Rs) function similarly (Threlfell et al., 2004). Voltammetric analysis has shown that this latter receptor may weakly inhibit serotonin release in the substantianigra (Threlfell et al., 2010). Furthermore in tissue slice preparations and *in vivo*, activation of histamine receptors has been found to decrease serotonin activity (Threlfell et al., 2008; Hashemi et al., 2011).

Serotonin has multiple autoreceptors (Barnes and Sharp, 1999). Classically autoreceptors are thought to control cell excitability and modulate cell firing. However, Daws and colleagues showed a contrasting role for the 5-HT_{1B} receptor. Using chronoamperometry, the group found that 5-HT_{1B} receptors modulate serotonin uptake *via* the SERT (Daws et al., 2000). Similarly, adenosine and norepinephrine have been found to modulate serotonin uptake *via* A3 adenosine receptors (Zhu et al., 2007)and *via* α_2-adrenoceptors (Ansah et al., 2003).

These studies highlight that modulation of serotonin is important in regulating its transmission. This agrees with the notion that too much serotonin activity in the brain can be highly physiologically detrimental.

12.3.3 SEROTONIN CLEARANCE

After serotonin is released it can be reuptaken back into the presynaptic neuron *via* SERTs; the site of action of most antidepressants. SERTs are coupled receptors (GPCR) whose intracellular signaling cascades are provoked by the binding of serotonin molecules. The most popular antidepressants are the selective serotonin reuptake inhibitors (SSRIs) (Lindsley, 2012). The neurochemical effects of SSRIs on serotonin have been studied electrochemically (Hashemi et al., 2012; Wood and Hashemi, 2013; Bunin et al., 1998; Daws et al., 1997).

SSRIs decrease the clearance of serotonin exogenously injected or electrically stimulated. *In vivo* SSRI effects on serotonin in mice are not static (Wood and Hashemi, 2013). Serotonin clearance after an acute SSRI dose continued to increase two hours after intraperitoneal administration, when the drug should no longer be present in the system (Kreilgaard et al., 2008).

Michaelis–Menten models have examined SERT kinetics (Bunin et al., 1998; Bunin and Wightman, 1998; Best et al., 2010). The Michaelis–Menten constant, K_m, is an index of transporter affinity, whereas V_{max}, maximum rate of uptake, is an index of transporter efficiency. Bunin et. al. reported V_{max} for serotonin in the DRN and SNR of the rat brain from the descending slope, after pausing stimulation resulted in [5-HT] $>> K_m$ (Bunin et al., 1998). They reported V_{max} in the DRN as 1300 ± 20 nM s^{-1} and in the SNR as 570 ± 70 nM s^{-1}. Similarly, John et al. determined these kinetic parameters in the ventral tegmental area (VTA) and substantia nigra of mice before and after transporter inhibition and with dopamine transporter (DAT) and SERT knockout mice (John et al., 2006).

Daws and colleagues have provided interesting evidence for non-SERT serotonin reuptake mechanisms. They advanced two uptake mechanisms for serotonin: uptake 1 and uptake 2 (Daws et al., 2013), first proposed more than 40 years ago (Shaskan and Snyder, 1970; Burgen and Iversen, 1965). Specifically, uptake 1 is high affinity, low efficiency uptake by SERTs on serotonin neurons, and uptake 2 is low affinity, high efficiency uptake by organic cation transporters (OCTs), plasma membrane monoamine transporters (PMATs), norepinephrine transporters (NETs), and DATs of other neurons and glial cells (Daws, 2009). Inhibiting OCTs and PMATs with decynium-22 enhanced the neurochemical and behavioral antidepressant effects in SERT deficient mice (Baganz et al., 2008; Horton et al., 2013). The group also demonstrated serotonin uptake by NETs (Daws et al., 1998). They hypothesize that when, in some patients, SSRIs are not therapeutically beneficial, it is due to the activity of uptake 2 transporters which are not targeted by SSRIs (Daws et al., 2013; Schildkraut and Mooney, 2004).

Our most recent work also suggest that serotonin is cleared *via* two mechanisms (Wood et al., 2014). Unpublished data from Hashemi and colleagues suggests

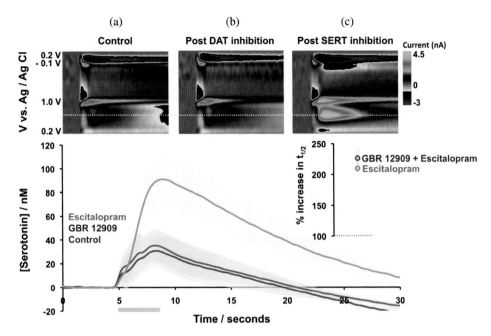

Figure 12.1 Top panel- color plots displaying potential *versus* time with current in false color. (a) Serotonin in the SNR. (b) Serotonin 30 minutes after GBR 12909. (c) Serotonin 60 minutes after subsequent escitalopram. Bottom panel- average [serotonin] *versus* time with no drug, 30 minutes after GBR 12909 and 60 minutes after subsequent escitalopram. Stimulation of MFB is denoted with purple line under the plot. Inset- % increase in $t_{1/2}$ of serotonin clearance in mice treated with escitalopram only and in mice treated with GBR 12909 and escitalopram.

that uptake 2 has less contribution from DAT than from the other proposed uptake 2 mechanisms. Figure 12.1 shows a serotonin signal isolated in mouse SNR upon medial forebrain bundle (MFB) stimulation. In the top panel, (a) is a control signal with no administration of drug, (b) is taken 30 minutes post-DAT inhibition with GBR 12909 (10 mg kg^{-1}) and (c) is taken 60 minutes after subsequent SERT inhibition with escitalopram (10 mg kg^{-1}). In the bottom panel, [serotonin] *versus* time is presented for this experiment in five mice (± SEM). Inset is a chart showing % increase in $t_{1/2}$ of serotonin clearance from control levels in mice treated with escitalopram only or with GBR 12909 and escitalopram. It is seen that DAT inhibition does not, by itself, increase the $t_{1/2}$ of serotonin clearance. Moreover, when compared to mice who were treated only with escitalopram, GBR + escitalopram treatment did not significantly alter serotonin clearance (Figure 12.1 inset). It is therefore likely that other transporters, such as the OCTs, play the most important roles for uptake 2 mediated serotonin clearance.

It is again evident that serotonin is subjected to a high degree of regulation, *via* multiple, intricate uptake mechanisms. All seem geared toward removing serotonin from the synapse as efficiently as possible.

12.3.4 SEROTONIN'S INVOLVEMENT IN DISEASE

Serotonin has many regulatory roles in the central nervous system including circadian rhythm, hunger, mood, sexual drive, memory, stress response, anger, motor control, thermoregulation, and others (Berger et al., 2009; Muller and Jacobs, 2009). Dysregulations in the serotonergic system have been linked to addiction and countless neuropsychiatric disorders. Real-time electrochemical monitoring has given new insights into the serotonin neurochemistry underlying some of these.

The serotonergic system has been implicated in addiction. Cocaine targets both SERTs and the DATs. In one study, when DAT function was impaired by either using a DAT knockout mouse model or *via* DAT inhibition, subsequent SERT inhibition increased the levels of dopamine (Mateo et al., 2004). In alcohol studies, ethanol administration was found to decrease serotonin clearance through a non-SERT mediated mechanism (Daws et al., 2006). Additionally, long-term p-methoxyamphetamine (PMA) treatment also decreased serotonin clearance (Callaghan et al., 2007). This may be due to neurodegenerative effects of this substance over time.

The majority of neuropsychiatric serotonin targeting pharmacologically are antidepressants. Nearly 50% of patients are unresponsive to antidepressant drug treatment (Baghai et al., 2006). There are different hypotheses for this effect and a drive to create more effective agents. Currently, screening tests for new antidepressants do not adequately reflect clinical effectiveness. The most common screening test for antidepressant efficacy is a behavioral test in rodents, the forced swim test (FST) (Petit-Demouliere et al., 2005). The animals undergo a behavioral paradigm of learned helplessness, reflected by the amount of time they swim in a water-filled vessel. A positive screen for effective antidepressants occurs when animals swim for a longer period of time after antidepressant treatment. While many agents cause this effect and pass the screen, they are later found to have little or no clinical effectiveness. There is a demand for a more specific screening tool.

Unpublished research by Hashemi and colleagues suggests that FSCV can provide a neurochemical index of antidepressant clinical efficacy. A 2009 meta-analysis by Cipriani et al. ranked the 12 most common antidepressants in order of clinical efficacy (Cipriani et al., 2009). When a clinically effective antidepressant (escitalopram) is compared to a less effective one (fluoxetine) with FSCV in mice *in vivo*, immediate differences can be detected. Figure 12.2 shows a comparison of the serotonin response in the mouse SNR for two hours after acute escitalopram (Figure 12.2a) and fluoxetine (Figure 12.2b) administration. The effects of escitalopram on the serotonin signal are not only fast, but they are exaggerated in comparison to fluoxetine. This is promising evidence that FSCV can be used to neurochemically test antidepressants.

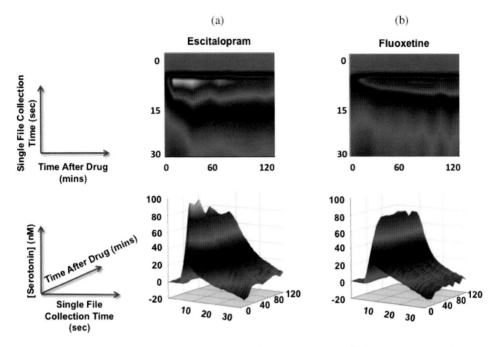

Figure 12.2 Comparison of acute doses of escitalopram (10 mg kg^{-1}) (a) to fluoexetine (20 mg kg^{-1}) (b). Both panels show stimulated [serotonin] during a single file collection with time after drug administration. The color scale in the bottom panel reflects the current changes in the top.

12.4 CONCLUSIONS

Real-time serotonin monitoring can be achieved in a number of ways. Fast microdialysis provides chemical information on multiple analytes and their interactions with serotonin. Electrochemical methods yield information on serotonin dynamics, modulation and transport in the brain. This can open the doors for studying the intricacies of serotonin related diseases. Finally, FSCV is emerging as a potential tool to neurochemically screen effective antidepressants. Such a tool could transform the efficiency with which new therapies are developed, improving clinical care for depressed patients.

REFERENCES

Alvarez de Toledo G, Fernandez-Chacon R, Fernandez JM (1993) Release of secretory products during transient vesicle fusion. Nature 363:554–558.

Ansah TA, Ramamoorthy S, Montanez S, Daws LC, Blakely RD (2003) Calcium-dependent inhibition of synaptosomal serotonin transport by the alpha 2-adrenoceptor agonist 5-bromo-N-[4,5-dihydro-1H-imidazol-2-yl]-6-quinoxalinamine (UK14304). J Pharmacol Exp Ther 305:956–965.

Baganz NL, Horton RE, Calderon AS, Owens WA, Munn JL, Watts LT, Koldzic-Zivanovic N, Jeske NA, Koek W, Toney GM, Daws LC (2008) Organic cation transporter 3: keeping the brake on extracellular serotonin in serotonin-transporter-deficient mice. Proc Natl Acad Sci USA 105:18976–18981.

Baghai TC, Moller HJ, Rupprecht R (2006) Recent progress in pharmacological and non-pharmacological treatment options of major depression. Curr Pharm Des 12:503–515.

Barnes NM, Sharp T (1999) A review of central 5-HT receptors and their function. Neuropharmacology 38:1083–1152.

Benmansour S, Cecchi M, Morilak DA, Gerhardt GA, Javors MA, Gould GG, Frazer A (1999) Effects of chronic antidepressant treatments on serotonin transporter function, density, and mRNA level. J Neurosci 19:10494–10501.

Benmansour S, Owens WA, Cecchi M, Morilak DA, Frazer A (2002) Serotonin clearance *in vivo* is altered to a greater extent by antidepressant-induced downregulation of the serotonin transporter than by acute blockade of this transporter. J Neurosci 22:6766–6772.

Berger M, Gray JA, Roth BL (2009) The expanded biology of serotonin. Annu Rev Med 60:355–366.

Best J, Nijhout HF, Reed M (2010) Serotonin synthesis, release and reuptake in terminals: a mathematical model. Theor Biol Med Model 7:34.

Borue X, Cooper S, Hirsh J, Condron B, Venton BJ (2009) Quantitative evaluation of serotonin release and clearance in Drosophila. J Neurosci Methods 179:300–308.

Bruns D, Jahn R (1995) Real-time measurement of transmitter release from single synaptic vesicles. Nature 377:62–65.

Bruns D, Riedel D, Klingauf J, Jahn R (2000) Quantal release of serotonin. Neuron 28:205–220.

Bunin MA, Prioleau C, Mailman RB, Wightman RM (1998) Release and uptake rates of 5-hydroxytryptamine in the dorsal raphe and substantia nigra reticulata of the rat brain. J Neurochem 70:1077–1087.

Bunin MA, Wightman RM (1998) Quantitative evaluation of 5-hydroxytryptamine (serotonin) neuronal release and uptake: an investigation of extrasynaptic transmission. J Neurosci 18:4854–4860.

Burgen AS, Iversen LL (1965) The inhibition of noradrenaline uptake by sympathomimetic amines in the rat isolated heart. Br J Pharmacol Chemother 25:34–49.

Cacciapaglia F, Wightman RM, Carelli RM (2011) Rapid dopamine signaling differentially modulates distinct microcircuits within the nucleus accumbens during sucrose-directed behavior. J Neurosci 31:13860–13869.

Cahill PS, Walker QD, Finnegan JM, Mickelson GE, Travis ER, Wightman RM (1996) Microelectrodes for the measurement of catecholamines in biological systems. Anal Chem 68:3180–3186.

Callaghan PD, Owens WA, Javors MA, Sanchez TA, Jones DJ, Irvine RJ, Daws LC (2007) *In vivo* analysis of serotonin clearance in rat hippocampus reveals that repeated administration of p-methoxyamphetamine (PMA), but not 3,4-methylenedioxymethamphetamine (MDMA), leads to long-lasting deficits in serotonin transporter function. J Neurochem 100:617–627.

Chugani DC, Muzik O, Rothermel R, Behen M, Chakraborty P, Mangner T, da Silva EA, Chugani HT (1997) Altered serotonin synthesis in the dentatothalamocortical pathway in autistic boys. Ann Neurol 42:666–669.

Cipriani A, Furukawa TA, Salanti G, Geddes JR, Higgins JP, Churchill R, Watanabe N, Nakagawa A, Omori IM, McGuire H, Tansella M, Barbui C (2009) Comparative efficacy and acceptability of 12 new-generation antidepressants: a multiple-treatments meta-analysis. Lancet 373:746–758.

Cooper JR, Bloom FE, Roth RH (2003) The biochemical basis of neuropharmacology. New York: Oxford University Press.

D'Aquila PS, Collu M, Gessa GL, Serra G (2000) The role of dopamine in the mechanism of action of antidepressant drugs. Eur J Pharmacol 405:365–373.

Daws LC (2009) Unfaithful neurotransmitter transporters: focus on serotonin uptake and implications for antidepressant efficacy. Pharmacol Ther 121:89–99.

Daws LC, Gould GG, Teicher SD, Gerhardt GA, Frazer A (2000) 5-HT(1B) receptor-mediated regulation of serotonin clearance in rat hippocampus *in vivo*. J Neurochem 75:2113–2122.

Daws LC, Koek W, Mitchell NC (2013) Revisiting serotonin reuptake inhibitors and the therapeutic potential of "uptake-2" in psychiatric disorders. ACS Chem Neurosci 4:16–21.

Daws LC, Montanez S, Munn JL, Owens WA, Baganz NL, Boyce-Rustay JM, Millstein RA, Wiedholz LM, Murphy DL, Holmes A (2006) Ethanol inhibits clearance of brain serotonin by a serotonin transporter-independent mechanism. J Neurosci 26: 6431–6438.

Daws LC, Montanez S, Owens WA, Gould GG, Frazer A, Toney GM, Gerhardt GA (2005) Transport mechanisms governing serotonin clearance *in vivo* revealed by high-speed chronoamperometry. J Neurosci Methods 143:49–62.

Daws LC, Toney GM, Davis DJ, Gerhardt GA, Frazer A (1997) *In vivo* chronoamperometric measurements of the clearance of exogenously applied serotonin in the rat dentate gyrus. J Neurosci Methods 78:139–150.

Daws LC, Toney GM, Gerhardt GA, Frazer A (1998) *In vivo* chronoamperometric measures of extracellular serotonin clearance in rat dorsal hippocampus: contribution of serotonin and norepinephrine transporters. J Pharmacol Exp Ther 286:967–976.

Ferguson JM (2001) SSRI antidepressant medications: adverse effects and tolerability. Prim Care Companion J Clin Psychiatry 3:22–27.

Furlong RA, Ho L, Walsh C, Rubinsztein JS, Jain S, Paykel ES, Easton DF, Rubinsztein DC (1998) Analysis and meta-analysis of two serotonin transporter gene polymorphisms in bipolar and unipolar affective disorders. Am J Med Genet 81:58–63.

Ge S, Woo E, Haynes CL (2011) Quantal regulation and exocytosis of platelet dense-body granules. Biophys J 101:2351–2359.

Gelenberg AJ, Chesen CL (2000) How fast are antidepressants? J Clin Psychiatry 61:712–721.

Haas H, Panula P (2003) The role of histamine and the tuberomamillary nucleus in the nervous system. Nat Rev Neurosci 4:121–130.

Haas HL, Sergeeva OA, Selbach O (2008) Histamine in the nervous system. Physiol Rev 88:1183–1241.

Hashemi P, Dankoski EC, Lama R, Wood KM, Takmakov P, Wightman RM (2012) Brain dopamine and serotonin differ in regulation and its consequences. Proc Natl Acad Sci USA 109:11510–11515.

Hashemi P, Dankoski EC, Petrovic J, Keithley RB, Wightman RM (2009) Voltammetric detection of 5-hydroxytryptamine release in the rat brain. Anal Chem 81:9462–9471.

Hashemi P, Dankoski EC, Wood KM, Ambrose RE, Wightman RM (2011) *In vivo* electrochemical evidence for simultaneous 5-HT and histamine release in the rat substantia nigra pars reticulata following medial forebrain bundle stimulation. J Neurochem 118:749–759.

Hilfiker S, Pieribone VA, Czernik AJ, Kao HT, Augustine GJ, Greengard P (1999) Synapsins as regulators of neurotransmitter release. Philos Trans R Soc Lond B Biol Sci 354:269–279.

Horton RE, Apple DM, Owens WA, Baganz NL, Cano S, Mitchell NC, Vitela M, Gould GG, Koek W, Daws LC (2013) Decynium-22 enhances SSRI-induced antidepressant-like effects in mice: uncovering novel targets to treat depression. J Neurosci 33:10534–10543.

Inayama Y, Yoneda H, Sakai T, Ishida T, Nonomura Y, Kono Y, Takahata R, Koh J, Sakai J, Takai A, Inada Y, Asaba H (1996) Positive association between a DNA sequence variant in the serotonin 2A receptor gene and schizophrenia. Am J Med Genet 67:103–105.

Jackson BP, Dietz SM, Wightman RM (1995) Fast-scan cyclic voltammetry of 5-hydroxytryptamine. Anal Chem 67:1115–1120.

John CE, Budygin EA, Mateo Y, Jones SR (2006) Neurochemical characterization of the release and uptake of dopamine in ventral tegmental area and serotonin in substantia nigra of the mouse. J Neurochem 96:267–282.

John CE, Jones SR (2007) Fast scan cyclic voltammetry of dopamine and serotonin in mouse brain slices. In: Electrochemical methods for neuroscience (Michael AC, Borland LM, eds), pp 49–62. Boca Raton (FL): CRC Press.

Kile BM, Guillot TS, Venton BJ, Wetsel WC, Augustine GJ, Wightman RM (2010) Synapsins differentially control dopamine and serotonin release. J Neurosci 30:9762–9770.

Kozminski KD, Gutman DA, Davila V, Sulzer D, Ewing AG (1998) Voltammetric and pharmacological characterization of dopamine release from single exocytotic events at rat pheochromocytoma (PC12) cells. Anal Chem 70:3123–3130.

Kreilgaard M, Smith DG, Brennum LT, Sanchez C (2008) Prediction of clinical response based on pharmacokinetic/pharmacodynamic models of 5-hydroxytryptamine reuptake inhibitors in mice. Br J Pharmacol 155:276–284.

Lacasse JR, Leo J (2005) Serotonin and depression: a disconnect between the advertisements and the scientific literature. PLoS Med 2:e392.

Lama RD, Charlson K, Anantharam A, Hashemi P (2012) Ultrafast detection and quantification of brain signaling molecules with carbon fiber microelectrodes. Anal Chem 84:8096–8101.

Laruelle M, Abi-Dargham A, Casanova MF, Toti R, Weinberger DR, Kleinman JE (1993) Selective abnormalities of prefrontal serotonergic receptors in schizophrenia. A postmortem study. Arch Gen Psychiatry 50:810–818.

Lee HJ, Lee MS, Kang RH, Kim H, Kim SD, Kee BS, Kim YH, Kim YK, Kim JB, Yeon BK, Oh KS, Oh BH, Yoon JS, Lee C, Jung HY, Chee IS, Paik IH (2005) Influence of the serotonin transporter promoter gene polymorphism on susceptibility to post-traumatic stress disorder. Depress Anxiety 21:135–139.

Lesch KP, Bengel D, Heils A, Sabol SZ, Greenberg BD, Petri S, Benjamin J, Muller CR, Hamer DH, Murphy DL (1996) Association of anxiety-related traits with a polymorphism in the serotonin transporter gene regulatory region. Science 274:1527–1531.

Lindsley CW (2012) The top prescription drugs of 2011 in the United States: antipsychotics and antidepressants once again lead CNS therapeutics. ACS Chem Neurosci 3:630–631.

Marquis BJ, Haynes CL (2010) Evaluating the effects of immunotoxicants using carbon fiber microelectrode amperometry. Anal Bioanal Chem 398:2979–2985.

Marquis BJ, Liu Z, Braun KL, Haynes CL (2010) Investigation of noble metal nanoparticle zeta-potential effects on single-cell exocytosis function *in vitro* with carbon-fiber microelectrode amperometry. Analyst 136:3478–3486.

Masand PS, Gupta S (2002) Long-term side effects of newer-generation antidepressants: SSRIS, venlafaxine, nefazodone, bupropion, and mirtazapine. Ann Clin Psychiatry 14:175–182.

Mateo Y, Budygin EA, John CE, Jones SR (2004) Role of serotonin in cocaine effects in mice with reduced dopamine transporter function. Proc Natl Acad Sci USA 101:372–377.

Montanez S, Owens WA, Gould GG, Murphy DL, Daws LC (2003) Exaggerated effect of fluvoxamine in heterozygote serotonin transporter knockout mice. J Neurochem 86:210–219.

Muller CP, Carey RJ, Huston JP, De Souza Silva MA (2007) Serotonin and psychostimulant addiction: focus on 5-HT1A-receptors. Prog Neurobiol 81:133–178.

Muller CP, Jacobs B (2009) Handbook of the behavioral neurobiology of serotonin. London: Academic Press.

Onder E, Tural U (2003) Faster response in depressive patients treated with fluoxetine alone than in combination with buspirone. J Affect Disord 76:223–227.

Palkovits M, Brownstein M, Saavedra JM (1974) Serotonin content of the brain stem nuclei in the rat. Brain Res 80:237–249.

Pathirathna P, Yang Y, Forzley K, McElmurry SP, Hashemi P (2012) Fast-scan deposition-stripping voltammetry at carbon-fiber microelectrodes: real-time, subsecond, mercury free measurements of copper. Anal Chem 84:6298–6302.

Perez XA, Andrews AM (2005) Chronoamperometry to determine differential reductions in uptake in brain synaptosomes from serotonin transporter knockout mice. Anal Chem 77:818–826.

Petit-Demouliere B, Chenu F, Bourin M (2005) Forced swimming test in mice: a review of antidepressant activity. Psychopharmacology (Berl) 177:245–255.

Phillips PE, Stuber GD, Heien ML, Wightman RM, Carelli RM (2003) Subsecond dopamine release promotes cocaine seeking. Nature 422:614–618.

Ramboz S, Oosting R, Amara DA, Kung HF, Blier P, Mendelsohn M, Mann JJ, Brunner D, Hen R (1998) Serotonin receptor 1A knockout: an animal model of anxiety-related disorder. Proc Natl Acad Sci USA 95:14476–14481.

Richelson E, Pfenning M (1984) Blockade by antidepressants and related compounds of biogenic amine uptake into rat brain synaptosomes: most antidepressants selectively block norepinephrine uptake. Eur J Pharmacol 104:277–286.

Schenk JO, Miller E, Rice ME, Adams RN (1983) Chronoamperometry in brain slices: quantitative evaluations of *in vivo* electrochemistry. Brain Res 277:1–8.

Schildkraut JJ, Mooney JJ (2004) Toward a rapidly acting antidepressant: the normetanephrine and extraneuronal monoamine transporter (uptake 2) hypothesis. Am J Psychiatry 161:909–911.

Scott DE, Grigsby RJ, Lunte SM (2013) Microdialysis sampling coupled to microchip electrophoresis with integrated amperometric detection on an all-glass substrate. Chemphyschem 14:2288–2294.

Sellers EM, Higgins GA, Sobell MB (1992) 5-HT and alcohol abuse. Trends Pharmacol Sci 13:69–75.

Sen S, Burmeister M, Ghosh D (2004) Meta-analysis of the association between a serotonin transporter promoter polymorphism (5-HTTLPR) and anxiety-related personality traits. Am J Med Genet B Neuropsychiatr Genet 127B:85–89.

Sharp T, Zetterstrom T, Christmanson L, Ungerstedt U (1986) p-Chloroamphetamine releases both serotonin and dopamine into rat brain dialysates *in vivo*. Neurosci Lett 72:320–324.

Shaskan EG, Snyder SH (1970) Kinetics of serotonin accumulation into slices from rat brain: relationship to catecholamine uptake. J Pharmacol Exp Ther 175:404–418.

Smith D, Dempster C, Glanville J, Freemantle N, Anderson I (2002) Efficacy and tolerability of venlafaxine compared with selective serotonin reuptake inhibitors and other antidepressants: a meta-analysis. Br J Psychiatry 180:396–404.

Song P, Hershey ND, Mabrouk OS, Slaney TR, Kennedy RT (2012) Mass spectrometry "sensor" for *in vivo* acetylcholine monitoring. Anal Chem 84:4659–4664.

Stone M, Laughren T, Jones ML, Levenson M, Holland PC, Hughes A, Hammad TA, Temple R, Rochester G (2009) Risk of suicidality in clinical trials of antidepressants in adults: analysis of proprietary data submitted to US Food and Drug Administration. BMJ 339:b2880.

Sutcliffe JS., Delahanty RJ, Prasad HC, McCauley JL, Han Q, Jiang L, Li C, Folstein SE, Blakely RD (2005) Allelic heterogeneity at the serotonin transporter locus (SLC6A4) confers susceptibility to autism and rigid-compulsive behaviors. Am J Hum Genet 77:265–279.

Thompson BJ, Jessen T, Henry LK, Field JR, Gamble KL, Gresch PJ, Carneiro AM, Horton RE, Chisnell PJ, Belova Y, McMahon DG, Daws LC, Blakely RD (2011) Transgenic elimination of high-affinity antidepressant and cocaine sensitivity in the presynaptic serotonin transporter. Proc Natl Acad Sci USA 108:3785–3790.

Threlfell S, Cragg SJ, Kallo I, Turi GF, Coen CW, Greenfield SA (2004) Histamine H3 receptors inhibit serotonin release in substantia nigra pars reticulata. J Neurosci 24:8704–8710.

Threlfell S, Exley R, Cragg SJ, Greenfield SA (2008) Constitutive histamine H2 receptor activity regulates serotonin release in the substantia nigra. J Neurochem 107:745–755.

Threlfell S, Greenfield SA, Cragg SJ (2010) 5-HT(1B) receptor regulation of serotonin (5-HT) release by endogenous 5-HT in the substantia nigra. Neuroscience 165:212–220.

Ungerstedt U, Pycock C (1974) Functional correlates of dopamine neurotransmission. Bull Schweiz Akad Med Wiss 30:44–55.

Veenstra-VanderWeele J, Muller CL, Iwamoto H, Sauer JE, Owens WA, Shah CR, Cohen J, Mannangatti P, Jessen T, Thompson BJ, Ye R, Kerr TM, Carneiro AM, Crawley JN, Sanders-Bush E, McMahon DG, Ramamoorthy S, Daws LC, Sutcliffe JS, Blakely RD (2012) Autism gene variant causes hyperserotonemia, serotonin receptor hypersensitivity, social impairment and repetitive behavior. Proc Natl Acad Sci USA 109:5469–5474.

Wood KM, Hashemi P (2013) Fast-scan cyclic voltammetry analysis of dynamic serotonin reponses to acute escitalopram. ACS Chem Neurosci 4:715–720.

Wood KM, Zeqia A, Nijhout HF, Reed MC, Best J, Hashemi P (2014) Voltammetric and mathematical evidence for dual transport of serotonin clearance *in vivo*. J. Neurochem

Yang H, Thompson AB, McIntosh BJ, Altieri SC, Andrews AM (2013) Physiologically relevant changes in serotonin resolved by fast microdialysis. ACS Chem Neurosci 4:790–798.

Young LT, Warsh JJ, Kish SJ, Shannak K, Hornykeiwicz O (1994) Reduced brain 5-HT and elevated NE turnover and metabolites in bipolar affective disorder. Biol Psychiatry 35:121–127.

Zetterstrom T, Sharp T, Marsden CA, Ungerstedt U (1983) *In vivo* measurement of dopamine and its metabolites by intracerebral dialysis: changes after d-amphetamine. J Neurochem 41:1769–1773.

Zhang J, Jaquins-Gerstl A, Nesbitt KM, Rutan SC, Michael AC, Weber SG (2013) *In Vivo* monitoring of serotonin in the striatum of freely moving rats with one minute temporal resolution by online microdialysis-capillary high-performance liquid chromatography at elevated temperature and pressure. Anal Chem 85:9889–9897.

Zhu CB, Steiner JA, Munn JL, Daws LC, Hewlett WA, Blakely RD (2007) Rapid stimulation of presynaptic serotonin transport by A(3) adenosine receptors. J Pharmacol Exp Ther 322:332–340.

CHAPTER 13

VOLTAMMETRIC ANALYSIS OF LOSS AND GAIN OF DOPAMINE FUNCTION

Paul A. Garris*, Kristen A. Keefe[†]

*Illinois State University, [†]University of Utah

13.1 INTRODUCTION

The objective of this chapter is to highlight contributions of the real-time monitoring technique of fast-scan cyclic voltammetry (FSCV) at a carbon-fiber microelectrode (CFM) for understanding the neurobiological underpinnings of pathological states characterized by either a loss or gain of dopamine function. Loss of dopamine function was investigated in the context of the dopamine-depleted conditions of Parkinson's disease (PD) and methamphetamine (METH)-induced neurotoxicity, whereas gain of dopamine function was investigated in the context of acute effects of amphetamine (AMPH), a highly addictive psychostimulant. Our neurochemical investigations utilizing FSCV at a CFM have not only focused on establishing the status of the two modes of dopamine signaling, tonic and phasic (Schultz, 2007b), in these pathological states, but also on identifying alterations in the key presynaptic mechanisms of dopamine release and uptake (Venton et al., 2003) that drive deviations in signaling modes from normal and are potential targets for compensatory responses.

Preservation of motor function despite the progressive loss of dopamine neurons during the extended pre-clinical phase of PD represents a truly remarkable example of brain plasticity that was first recognized nearly a half century ago (Hornykiewicz, 1966). In this chapter, we focus on compensatory adaptations by surviving dopamine neurons that underlie this phenomenon. While potent, these compensatory adaptations are not able to maintain dopamine signaling fully. We thus also focus on deficits in dopamine signaling and their potential relevance to cognitive dysfunction in PD and in METH-induced neurotoxicity (Marshall and O'Dell, 2012;Arnaldi et al., 2012). Finally, similar to other abused substances, AMPH generates a hyperdopamine state that plays an important role in drug reinforcement and addiction (Di Chiara and Imperato, 1988). However, this psychostimulant is thought to elevate brain dopamine levels by targeting unique actions on dopamine neurons (Sulzer, 2011). The third focus of this chapter is therefore AMPH activation of dopamine signaling.

Analytical characteristics of FSCV at a CFM (Robinson et al., 2008) are well suited for these investigations of lost and gained dopamine function. Foremost is the capability to provide sensitive and temporally, spatially, and chemically resolved measurements of dopamine. As such, dynamic changes in brain extracellular dopamine, whether artificially evoked by physiological stimulation or occurring naturally and associated with behavior, can be captured with high fidelity. Accompanying analyses that, for example, enhance chemical resolution, quantify phasic dopamine signals, and resolve presynaptic mechanisms of dopamine release and uptake, have further extended the utility of this real-time monitoring technique. Collectively, these powerful analytical features of FSCV at a CFM have provided unique neurobiological insight into dopamine function in PD and METH-induced neurotoxicity and after acute administration of AMPH. As we discuss herein, this insight has allowed us to challenge established models of compensatory adaptation by dopamine neurons and AMPH action.

13.2 MODES OF DOPAMINE SIGNALING: OVERVIEW

To set the stage for discussing pathological states, it is important to first provide an overview of normal dopamine signaling. Indeed, the loss and gain of dopamine function is fundamentally mediated at the level of brain circuits by altered dopamine control over target cells. Dopamine neurons signal in two modes, tonic and phasic (Carelli and Wightman, 2004; Grace et al., 2007; Schultz, 2007b). During tonic dopamine signaling, the combination of low-frequency firing by dopamine neurons and heteroreceptor activation of dopamine terminals generates a low, ambient level of extracellular dopamine bathing target cells called dopamine tone. In contrast, during phasic dopamine signaling, high-frequency burst firing of dopamine neurons produces short-lived spikes in extracellular dopamine called dopamine transients that temporally modulate target cells. Dopamine tone is considered to result from the balance of dopamine release and uptake and activate high-affinity D2 dopamine receptors, whereas dopamine transients ride on top of this steady-state level, are dominated by dopamine release, and activate low-affinity D1 dopamine receptors (Venton et al., 2003; Dreyer et al., 2010).

Elucidating the function of tonic and phasic dopamine signaling is an ongoing pursuit. As discussed next in Section 13.3.1, PD reveals a critical role of dopamine tone in motoric behavior. In general, this signaling mode is thought to enable the dopamine-related functions of movement, cognition, and motivation (Schultz, 2007a). However, difficulty in quantifying dopamine tone has prevented establishing a more definitive role. For example, as a monitoring technique that measures the relative difference between analyte concentrations, FSCV lacks the ability to determine absolute concentration. On the other hand, microdialysis, which exhibits the requisite sensitivity and selectivity to quantify dopamine tone absolutely, suffers from complications associated with implantation damage due to large probe size (Watson et al., 2006).

Phasic dopamine signaling is activated by rewards and their predictive cues, which is supportive of proposed roles in reward learning (Schultz, 2007a; Glimcher, 2011) and incentive salience (i.e., reward "wanting") (Berridge, 2007; Flagel et al., 2011). FSCV at a CFM is well suited for monitoring dopamine transients. This analytical feature has been recently coupled with transgenic and optogenetic approaches to demonstrate that phasic dopamine signals are necessary and sufficient for cue-directed reward seeking in appetitive behavior (Zweifel et al., 2009; Tsai et al., 2009; Steinberg et al., 2013).

13.3 DOPAMINE SIGNALING IN THE DOPAMINE-DEPLETED BRAIN

The overarching question of this section is how modes of dopamine signaling respond to the partial dopamine depletions observed in the striatum during the pre-clinical phase of PD and in METH-induced neurotoxicity. As we discuss next, the short answer appears to be that they respond differently, with a compensatory normalization of tonic dopamine signaling but degradation of phasic dopamine signaling. While the former is consistent with preserved motor function and the later with cognitive dysfunction, these differential responses are not consistent with the established model of compensatory adaptation.

13.3.1 PASSIVE STABILIZATION OF TONIC DOPAMINE SIGNALING

Although PD is associated with the loss of nigrostriatal dopamine neurons, the relationship between cardinal symptoms and neurodegeneration is not straightforward. Indeed, resting tremor, rigidity, and akinesia emerge only when striatal dopamine depletion becomes severe (>~80%) (Hornykiewicz and Kish, 1987). This complex relationship suggests that potent neuroadaptations preserve motor function during the pre-clinical phase of PD. Studies using microdialysis in the

6-hydroxydopamine (6-OHDA)-lesioned rat model of PD suggest that normalization of dopamine tone is the primary compensatory mechanism in pre-clinical Parkinsonism (Abercrombie et al., 1990). Zigmond and co-workers proposed a model to account for the maintenance of dopamine tone (Zigmond et al., 1990). In this model, up-regulation of dopamine release by surviving dopamine neurons drives the diffusion of dopamine from innervated to denervated regions, thus maintaining dopamine tone through the striatum. Moreover, up-regulated release is maintained by up-regulated dopamine synthesis, and the loss of dopamine transporters accompanying the loss of dopamine neurons passively enhances dopamine diffusion into denervated regions.

While up-regulated dopamine synthesis (Wolf et al., 1989) and enhanced dopamine diffusion (van Horne et al., 1992) are well established compensatory mechanisms, only indirect measurements of up-regulated dopamine release were available at the time the Zigmond model was proposed. For these measurements, dopamine released from a striatal slice by electrical stimulation was collected *via* superfusion (Stachowiak et al., 1987; Snyder et al., 1990). Although a competitive dopamine uptake inhibitor was present in the superfusate, dopamine uptake is only slowed under these conditions and not blocked. Thus, evoked measurements in this preparation reflected the interplay of dopamine release, uptake and diffusion.

FSCV at a CFM was used to provide a direct measurement of dopamine release in 6-OHDA-lesioned rats in order to test directly the hypothesis that this presynaptic mechanism is up-regulated with dopamine depletion (Figure 13.1) (Bergstrom and Garris, 2003). This real-time monitoring technique confers several advantages to the study of compensatory adaptation by dopamine neurons. Critical is the collection of a voltammogram or "chemical signature" to identify the evoked signal as dopamine (Figure 13.1c, INSET). Just as important is recording temporally resolved changes in dopamine. This capability permits assessing effects of dopamine depletion on electrically evoked signals with either tonic- or phasic-like dynamics. Indeed, notice in Figure 13.1c that dopamine depletion does not alter the amplitude of steady-state signals evoked by a 20-Hz train, consistent with normalized tonic dopamine signaling, but decreases the amplitude of peak-shaped signals evoked by a 60-Hz train, suggesting degradation of phasic dopamine signaling. The significance of this latter result is discussed in more detail in Section 13.3.3. Moreover, the temporally resolved evoked dopamine signals can be analyzed kinetically to quantify dopamine release and uptake mechanisms (Wightman et al., 1988). Indeed, kinetic analysis demonstrated that dopamine release and uptake decreased in proportion to the degree of dopamine depletion, consistent with no active compensation of either presynaptic mechanism.

The "Passive Stabilization" model was proposed to account for the maintenance of dopamine tone in the partially dopamine-depleted striatum without a

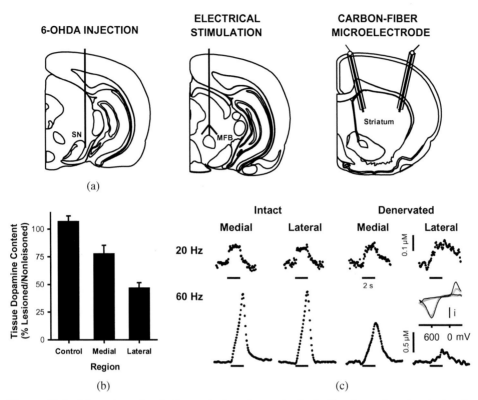

Figure 13.1 Measuring electrically evoked tonic- and phasic-like dopamine signals in the 6-OHDA-lesioned rat. (a). Brain placements: Left- Cannula in the substantia nigra (SN) for injecting 6-OHDA. Middle- Stimulating electrode in the medial forebrain bundle (MFB). Right- CFMs in medial and lateral dorsal striatum. (b). Tissue dopamine content. (c). Dopamine levels electrically evoked by trains delivered at 20-Hz (tonic) and 60-Hz (phasic). Solid line under trace demarcates stimulation. INSET: Background subtracted cyclic voltammogram for dopamine. Recordings were collected under urethane anesthesia. Reproduced from Bergstrom and Garris, 2003.

compensatory up-regulation of dopamine release (Bergstrom and Garris, 2003). In this model, dopamine tone is considered a steady-state concentration determined by the balance of dopamine release, which increases tone, and dopamine uptake, which decreases tone. Because sites for dopamine release and uptake are found on dopamine neurons and are thus lost proportionately with denervation, their effects offset and dopamine tone is passively maintained. "Passive Stabilization" also explains the degradation of evoked phasic-like dopamine signals. Because the amplitude of these peak-shaped signals is dominated by dopamine release, the former tracks the latter and decreases in proportion to dopamine depletion. Taken together, "Passive Stabilization" reconciles differential effects of dopamine depletion on dopamine signaling: normalized tonic and degraded phasic dopamine signaling.

13.3.2 COMPENSATORY UP-REGULATION OF EXOCYTOTIC DOPAMINE RELEASE

It could be argued that FSCV at a CFM advanced the investigation of compensatory adaptation by dopamine neurons by providing a direct measurement of dopamine release. Documented differences in the status of this presynaptic mechanism with dopamine depletion may thus be due to measurement technique. However, subsequent work using FSCV at a CFM in monkey striatal slices depleted of dopamine by 1-methyl-4-phenyl-1,2,3,6-tetrahydropyridine (MPTP) suggests discrepancies originate elsewhere. Indeed, dopamine release measured voltammetrically in this preparation was shown to be up-regulated with partial dopamine depletion and consistent with indirect indices of nicotine- and K^+- evoked ^{3}H-dopamine release from perfused synaptosomes (McCallum et al., 2006; Perez et al., 2008). Discrepancies between determinations of dopamine release by FSCV at a CFM may therefore have an alternative origin, such as differences between rat and monkey or *in vivo* and *in vitro* preparations.

Dopamine release is not static, but depends upon release history (Montague et al., 2004). Perhaps due to direct stimulation of dopamine terminals, brain slices exhibit novel time-dependent mechanisms of dopamine release (Phillips et al., 2002; Cragg, 2003). These mechanisms may contribute to the discrepancies regarding the status of dopamine release in the setting of dopamine depletion as measured with FSCV at a CFM *in vivo* in anesthetized rats *versus in vitro* in the slice preparation. Thus, FSCV studies in the 6-OHDA-lesioned rat were performed to extend these measurements to the time-dependent realm (Figure 13.2) (Bergstrom et al., 2011). In this design, pulse trains were delivered every five minutes to fatigue dopamine neurons. Signals evoked by the first train represent time-independent dopamine release and are equivalent to previous measurements in the 6-OHDA-lesioned striatum that exhibited no up-regulation of dopamine release (Figure 13.1). However, whereas the amplitude of evoked signals degrades in intact rats during the fatiguing protocol, the amplitude in lesioned rats does not (Figure 13.2a), suggestive of a compensatory up-regulation of time-dependent dopamine release.

To further explore this compensatory adaptation, a model of time-dependent dopamine release (Montague et al., 2004) was used to simulate experimental results. This model contains terms for short- and long-term depression and short-term facilitation, processes that alter instantaneous dopamine release based on release history. Interestingly, the model describes evoked responses in both intact and lesioned animals with high fidelity if the term for long-term depression was removed from the latter (Figure 13.2b). This result suggests that dopamine depletion may enhance mobilization of the reserve pool to maintain dopamine release

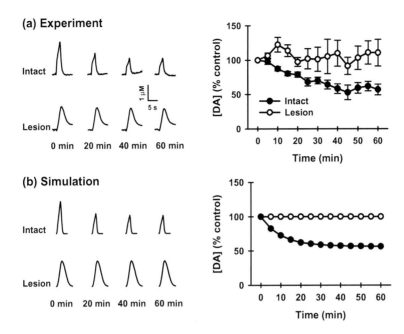

Figure 13.2 Compensation of time-dependent dopamine release with dopamine depletion. (a). Measurements of electrically evoked dopamine levels in intact and 6-OHDA-lesioned rats using FSCV at a CFM. Recordings were collected under urethane anesthesia. (b). Evoked responses calculated from the Montague model (Montague et al., 2004). Modified from Bergstrom et al., 2011.

during the fatiguing protocol. How this novel compensatory adaption fits with "Passive Stabilization" is not known. One possibility is that dopamine tone can be maintained passively, but that dopamine transients rely on up-regulated dopamine release during intense activity so that phasic signaling is not further degraded. Nonetheless, these results when taken together suggest that dopamine neurons exploit both active and passive mechanisms when mounting a compensatory response to dopamine depletion.

13.3.3 DEFICITS IN PHASIC DOPAMINE SIGNALING

Studies in anesthetized rodents demonstrate that electrically evoked phasic-like dopamine signals recorded in the striatum degrade with dopamine depletions mimicking PD (Figure 13.1) (Garris et al., 1997; Bezard et al., 2000; Bergstrom and Garris, 2003; Bergstrom et al., 2011). These deficits in phasic dopamine signaling could underlie cognitive dysfunction observed in PD (Arnaldi et al., 2012), as is predicted by neurocomputational modeling (Wiecki and Frank, 2010). Functional magnetic resonance imaging (MRI) in PD patients has also detected reduced striatal signals representing "reward prediction error", which are thought to reflect diminished dopamine transients (Schonberg et al., 2010). While a link between diminished phasic dopamine signaling and cognition in PD is thus provocative, this hypothesis is not based on direct evidence supporting deficits in naturally occurring dopamine transients.

Cognitive dysfunction is also observed with chronic METH use in addicts (Scott et al., 2007), which appears to be associated with decreases in markers for striatal dopamine terminals (Volkow et al., 2001). Binge dosing regimens of METH administered to rats reliably mimic striatal dopamine loss and other long-term neurotoxic effects of METH use in humans (Marshall and O'Dell, 2012). Moreover, partial dopamine depletions observed with METH-induced neurotoxicity in rats are associated with long-term cognitive and behavioral sequelae (Chapman et al., 2001; Son et al., 2011). METH-induced neurotoxicity in rats could therefore be a viable model to investigate the link between deficits in phasic dopamine signaling and cognition. Indeed, a binge dosing regimen of METH produces similar effects on dopamine signaling, i.e., normalized dopamine tone, diminished electrically evoked phasic-like dopamine signals, and un-compensated dopamine release (Cass and Manning, 1999; Howard et al., 2011; Loewinger et al., 2012), as reported for the 6-OHDA-lesioned rat model of PD (see Section 13.3.1 above).

The effects of METH-induced neurotoxicity have been more recently investigated on so-called spontaneous dopamine transients using FSCV at a CFM in awake, freely behaving rats (Figure 13.3) (Howard et al., 2013a). These naturally occurring dopamine transients are not ostensibly associated with external stimuli, but may be important for reward learning in social interactions (Robinson et al., 2002) and drug abuse (Willuhn et al., 2010). This application of FSCV at a CFM highlights several of its analytical features, including measurements of behaviorally relevant signals in un-anesthetized animals, quantifying descriptive characteristics (i.e., frequency, amplitude, and duration) of dopamine transients, three-dimensional plotting of voltammograms for analyte identification, and principal component regression (PCR), a chemometrics approach for resolving dopamine from a complex, multi-analyte signal (Robinson et al., 2008).

As shown in Figure 13.3a, transients are observed in raw FSCV recordings collected in both saline- and METH-pretreated rats and identified as dopamine by the voltammogram plotted below in pseudo-color. Moreover, PCR further confirms these transients as dopamine. Most importantly, METH-induced neurotoxicity decreases the amplitude of dopamine transients (Figures 13.3b and 13.3c). Interestingly, the decrease in amplitude is less than that observed for striatal dopamine content, electrically evoked phasic-like dopamine signals, and dopamine release (15 *versus* ~60%, respectively), suggesting some compensation of spontaneous dopamine transients (Howard et al., 2013a). Nevertheless, this is the

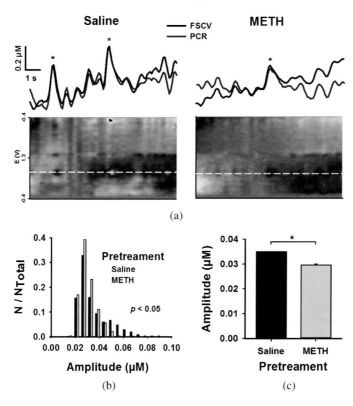

Figure 13.3 METH-induced neurotoxicity decreases the amplitude of naturally occurring dopamine transients. (a). Dopamine transients (*) in recordings determined by FSCV and PCR. Pseudocolor plots under traces sequentially display voltammograms in time. (b). Histograms comparing transient amplitude after saline- and METH-pretreatment. A significant effect of pretreatment was determined by Kolmogorov–Smirnov two-sample test. (c). Comparison of saline- and METH-pretreatment on transient amplitude. A significant effect of pretreatment was determined by Wilcoxon's ranked sum test of the median. Modified from Howard et al., 2013a.

first demonstration of a deficit in naturally occurring dopamine transients for any dopamine-depleting condition and supports the hypothesis linking dysfunctional phasic dopamine signaling and cognition.

13.3.4 DEFICITS IN POSTSYNAPTIC GENE EXPRESSION

Although it is now clear that partial striatal dopamine loss, such as that occurring in the pre-clinical phases of PD or in the context of METH-induced neurotoxicity, is associated with impairments in phasic dopamine signaling and cognition, the molecular substrates linking impaired phasic dopamine signaling to impaired cognitive function remain to be fully determined. However, data suggest that deficits in post-synaptic gene expression, particularly in striatonigral efferent neurons, may arise as a consequence of altered phasic dopamine signaling and, therefore, may underlie cognitive deficits associated with partial striatal dopamine loss. First, partial loss of striatal dopamine, whether induced by administration of 6-OHDA or METH, is associated with impaired basal expression of preprotachykinin (*ppt*) mRNA expression in dopamine D1 receptor-expressing striatonigral efferent neurons, but not preproenkephalin (*ppe*) mRNA expression, in dopamine D2 receptor-expressing striatopallidal neurons (Nisenbaum et al., 1996; Chapman et al., 2001; Johnson-Davis et al., 2002). Second, METH-induced partial striatal dopamine loss is associated with disrupted regulation of *Arc* (activity regulated, cytoskeletal-associated) in dopamine D1 receptor-expressing striatonigral neurons (Daberkow et al., 2008; Barker-Haliski et al., 2012). *Arc* is an effector immediate early gene (IEG) critical for long-term consolidation of learning and memory, including striatally based learning (Hearing et al., 2011; Pastuzyn et al., 2012). Thus, these findings suggest that impaired regulation of striatal efferent neuron gene expression, particularly the regulation of plasticity-related IEGs in striatonigral efferent neurons, may link impaired phasic dopamine signaling to cognitive deficits associated with partial striatal dopamine loss.

In light of the important role of phasic dopamine signaling in regulating striatal neuron function and the disruption of such signaling and striatal function in the context of partial striatal dopamine loss, it follows that restoration of phasic signaling should restore striatal function. In fact, prior electrophysiological data suggest that striatonigral and striatopallidal efferent neurons are particularly sensitive to phasic *versus* tonic DA signaling, respectively (Onn et al., 2000). Further, recent modeling data suggest that phasic dopamine signals selectively increase dopamine D1 receptor occupancy (Dreyer et al., 2010). Consistent with this model, stimulation of the MFB in a bursting pattern, rather than in single pulses, potently excites striatal neurons in a dopamine D1 receptor-dependent manner (Gonon, 1997) and results in increased mRNA expression of the IEG *zif 268* selectively in striatonigral efferent neurons (Chergui et al., 1997). Likewise, we have recently reported that electrical stimulation of the MFB in a phasic-like, but not a tonic-like, manner preferentially increases *Arc* mRNA expression, and this effect predominates in striatonigral neurons (Figure 13.4) (Howard et al., 2013b). Importantly, the effect is apparent in both normal rats and those with METH-induced disruption of phasic dopamine signaling. These findings suggest that it should be possible to restore phasic signaling and, thus, striatal efferent neuron function, under conditions in which there is partial loss of striatal dopamine innervation. Additionally, these findings suggest that restoration of phasic dopamine signaling may be efficacious in reversing cognitive deficits associated with partial striatal dopamine loss by restoring the regulation of critical plasticity-related molecules.

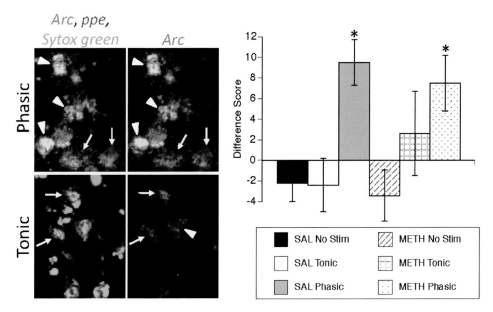

Figure 13.4 Expression of *Arc* mRNA in striatal efferent neurons in response to phasic- or tonic-like stimulation of the MFB. Left panel: representative fluorescent *in situ* hybridization images of *Arc* mRNA expression in presumed striatonigral neurons (*Arc*-positive/preproenkephalin *(ppe)*-negative; arrowheads) and striatopallidal neurons (*Arc*-positive/*ppe*-positive; arrows) in rats that received phasic-like (top) or tonic-like (bottom) stimulation. *Arc* is orange, *ppe* is blue/violet, SYTOX® nuclear stain is green. Images in the left-most column show all three channels, whereas images in the right-most column show only the *Arc* channel. Right panel: graph showing the 'Difference Score', defined as the difference between the average gray value of *Arc*-positive/*ppe*-negative (i.e., striatonigral) and *Arc*-positive/*ppe*-positive (i.e., striatopallidal) neurons. *Significantly different from the respective Tonic and No Stimulation (Stim) groups, $p<0.05$. Modified from Howard et al., 2013b.

13.4 MECHANISM OF AMPH ACTION

The overarching question of this section is how AMPH acts on dopamine neurons to generate a hyperdopamine state? As we discuss next, the short answer is very different from the historic view, which posits that this highly addictive psychostimulant activates tonic dopamine signaling *via* an action potential-independent mechanism. Rather, recent work using FSCV at a CFM in awake, freely behaving rats suggests that the predominant AMPH action is an activation of phasic dopamine signaling.

13.4.1 ACTIVATION, NOT DISRUPTION, OF PHASIC DOPAMINE SIGNALING

It is well established that all abused drugs act directly or indirectly on dopamine neurons to generate a hyperdopamine state (DiChiara and Imperato, 1988).

A large group of abused drugs spanning several classes, including ethanol, nicotine, cocaine, and cannabinoids, have now been shown by FSCV at a CFM to activate dopamine transients (Cheer et al., 2004, 2007). This effect is consistent with an emerging hypothesis in addiction research that abused drugs hijack brain reward circuits by hyperactivating dopamine transients, leading to aberrant reward learning and overvaluation of cues predicting drug availability (Hyman et al., 2006; Willuhn et al., 2010). However, AMPH and its structurally related analog, METH, have long been considered to act distinctly from other abused drugs to elevate brain dopamine levels by disrupting normal patterns of action potential-dependent dopamine signaling.

The historic action of AMPH is inducing non-exocytotic dopamine release called efflux by depleting vesicular dopamine stores and reversing the direction of the dopamine transporter (Sulzer, 2011). FSCV at a CFM concurrently demonstrates these key actions of AMPH, along with inhibiting the dopamine transporter by acting as a substrate, in striatal brain slices (Jones et al., 1998; Schmitz et al., 2001). However, voltammetric techniques in anesthetized rats have yielded conflicting results, with studies showing increases, decreases, or no change in electrically evoked dopamine signals (Kuhr et al., 1986; May et al., 1988) and no evidence for increased tonic dopamine signaling *via* efflux (Wiedemann et al., 1990). More recent work with FSCV at a CFM has suggested that observed discrepancies are due to differences between *in vitro* and *in vivo* preparations and stimulation parameters used for eliciting exocytotic dopamine release in order to assess vesicular dopamine depletion (Ramsson et al., 2011a). Indeed, phasic-like stimulation in anesthetized rats reveals an activation of action potential-dependent dopamine signaling by AMPH that is mediated by up-regulated dopamine release and inhibited dopamine uptake (Ramsson et al., 2011b). However, AMPH-induced dopamine efflux was still difficult to discern, perhaps due to an anesthesia artifact.

FSCV at a CFM was subsequently used to assess the effects of AMPH on dopamine signaling in awake, freely behaving rats (Figure 13.5) (Daberkow et al., 2013). This application extends analytical features highlighted above for real-time dopamine monitoring by the simultaneous assessment of behavior. Interestingly, AMPH was found to activate, rather than disrupt, phasic dopamine signaling by augmenting the amplitude, frequency, and duration of dopamine transients coincident with up-regulating vesicular dopamine release and inhibiting dopamine uptake. A high, but not low, dose of AMPH elicited a small increase in the baseline dopamine signal. However, this increase may not be solely attributed to efflux, because dopamine transients were recorded riding on top of this envelope. Established motoric effects of AMPH were observed concurrently, indicating that these novel actions occurred in animals behaviorally responding normally to this psychostimulant.

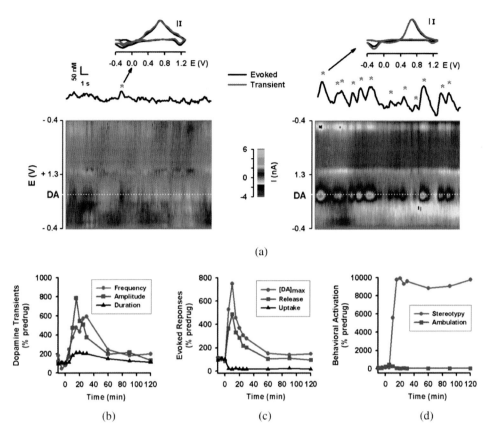

Figure 13.5 AMPH activates phasic dopamine signaling. (a). FSCV recordings of dopamine transients (*) before (left) and after (right) administration of AMPH (10 mg/kg). (b). Characteristics of dopamine transients. (c). Characteristics of electrical evoked phasic-like dopamine signals. (d). Behavioral activation. AMPH was injected at 0 minute. Reproduced from Daberkow et al., 2013.

AMPH effects on phasic dopamine signaling were also investigated in a discriminative stimulus task (Figure 13.6) (Daberkow et al., 2013). In this paradigm, animals were trained to respond to a cue by lever pressing for food reward (DS+), but not to another cue in which food reward was not delivered with lever pressing. Consistent with a role for dopamine in appetitive behavior (Berridge, 2007; Schultz, 2007a), dopamine transients are elicited by the DS+ cue predicting food reward. This phasic dopamine signal is augmented by low-dose AMPH. However, high-dose AMPH robustly activates spontaneous dopamine transients, which interfere with operant responding and DS+-evoked signals. Taken together, these results suggest that AMPH augments, rather than disrupts, phasic dopamine signaling and supports hyperactivation of dopamine transients as a unifying mechanism for abused drugs.

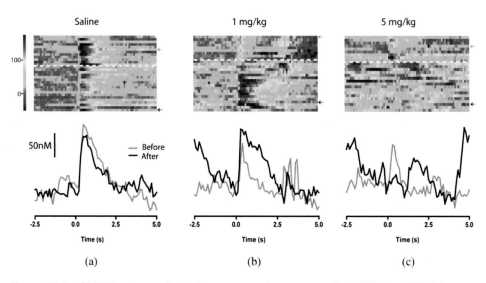

Figure 13.6 AMPH activates phasic dopamine signaling measured by FSCV at a CFM during a discriminative stimulus task. (a). Saline. (b). Low-dose (1 mg/kg) AMPH. (c). High-dose (5 m/kg) AMPH. The DS+ cue was delivered at 0 second. Top: dopamine heat maps during individual trials. Color scale in nM. The white dashed line demarcates drug injection. Bottom: exemplar trials collected before and after drug injection. Reproduced from Daberkow et al., 2013.

13.4.2 AMPH-INDUCED UP-REGULATION OF EXOCYTOTIC DOPAMINE RELEASE

The AMPH-induced up-regulation of vesicular dopamine release observed *in vivo* with FSCV at a CFM (Figure 13.5) (Ramsson et al., 2011b; Daberkow et al., 2013) is unexpected, based on the historic view that this psychostimulant depletes vesicular dopamine stores (Sulzer, 2011). However, other dopamine transporter-inhibitors, such as cocaine and methylphenidate, have been shown to elicit a similar effect on dopamine release, perhaps by mobilizing the reserve pool through actions on synaptic proteins such as synapsin and α-synuclein (Venton et al., 2006; Chadchankar et al., 2012). AMPH could thus be acting in this manner or through a unique mechanism, such as activating dopamine synthesis or inhibiting the dopamine degradative enzyme, monoamine oxidase (Sulzer, 2011). These latter actions would indirectly up-regulate dopamine release by increasing cytosolic dopamine levels, vesicular packaging, and vesicular content.

An experimental design using FSCV at a CFM in anesthetized rats and modified from Venton et al., 2006 was employed to investigate the mechanism by which AMPH up-regulates vesicular dopamine release (Figure 13.7) (Avelar et al., 2013). This design assessed the effects of AMPH on electrically evoked phasic-like dopamine signals in a treatment condition consisting of supraphysiological

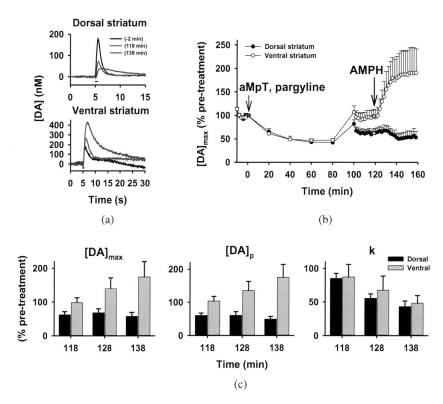

Figure 13.7 Vesicular dopamine release is sensitive to pharmacological inhibition of dopamine synthesis and degradation in the dorsal, but not ventral, striatum. (a). Exemplar recordings of electrically evoked phasic-like dopamine signals. (b). Time course of changes in the maximal concentration of evoked dopamine ($[DA]_{max}$) during the experiment. Treatment with depleting electrical stimulation and drugs was applied at 0 minute. (c). AMPH effects on: evoked phasic-like dopamine signals ($[DA]_{max}$, left); dopamine release (concentration of dopamine elicited per stimulus pulse or $[DA]_p$, middle); dopamine uptake (first-order rate constant or k, right). Data in (b) and (c) are expressed as a percent of pre-treatment (% pre-treatment). Reproduced from Avelar et al., 2013.

("depleting") electrical stimulation to manipulate the readily releasable pool of dopamine, alpha-methyl-para-tyrosine (aMpT; 200 mg/kg) to inhibit dopamine synthesis, and pargyline (75 mg/kg) to inhibit monoamine oxidase. Interestingly, the ability of AMPH to increase evoked, phasic-like dopamine signals was prevented in the dorsal, but not ventral, striatum by aMpT and pargyline (Figures 13.7a and 13.7b). This differential effect of AMPH in striatal sub-regions was due specifically to an action on vesicular dopamine release, not dopamine uptake. Indeed, in the presence of aMpT, AMPH increased dopamine release in the ventral, but not dorsal, striatum, but inhibited dopamine uptake similarly in both striatal sub-regions (Figure 13.7c). A similar effect was observed when aMpT or pargyline was administered separately.

Taken together, these results suggest that AMPH up-regulates vesicular dopamine release by different mechanisms in striatal sub-regions. AMPH appears to target dopamine synthesis and degradation in the dorsal striatum, but targets an alternative mechanism, perhaps mobilization of the reserve dopamine pool, in the ventral striatum. Moreover, several inhibitors of the dopamine transporter, including AMPH, cocaine and methylphenidate, nomifensine, amfonelic acid, and WIN 35428, have now been shown to up-regulate vesicular dopamine release (Ewing et al., 1983; Lee et al., 2001; Venton et al., 2006; Ramsson et al., 2011b; Chadchankar et al., 2012; Daberkow et al., 2013). Thus, up-regulating vesicular dopamine release may be a fundamental presynaptic mechanism targeted by this important class of psychostimulants.

13.4.3 RECONCILING PHASIC ACTIVATION AND DOPAMINE EFFLUX

The historic action of AMPH is to re-distribute dopamine from vesicular to cytosolic pools and drive dopamine through the dopamine transporter in the reverse direction (Sulzer, 2011). Ostensibly, this non-exocytotic, action potential-independent form of dopamine release called efflux is inconsistent with the AMPH-induced activation of phasic dopamine signaling and up-regulation of vesicular dopamine release shown recently using FSCV at a CFM *in vivo* (Figures 13.5 and 13.7) (Ramsson et al., 2011b; Daberkow et al., 2013; Avelar et al., 2013). However, there is also ample evidence using the technique of microdialysis to suggest that AMPH concurrently increases dopamine tone *in vivo* through a Ca^{2+}- and action potential-independent mechanism (Westerink et al., 1989; Hurd and Ungerstedt, 1989; Nomikos et al., 1990;Benwell et al., 1993). It is thus important to reconcile these apparently discrepant effects on tonic and phasic dopamine signaling *in vivo* to elucidate more fully the mechanism of AMPH action on dopamine neurons.

One hypothesis for reconciling the AMPH-induced dual activation of phasic dopamine signaling and dopamine efflux is that AMPH elicits different actions on distinct dopamine storage pools. For example, AMPH effects on electrically evoked dopamine levels are stimulus dependent, with short trains revealing increases and long trains revealing decreases (Ramsson et al., 2011a,b). Because work on model synapses demonstrates that short and long trains interrogate the readily releasable and reserve pools, respectively, it is reasonable to speculate that AMPH increases dopamine signals evoked by short trains by up-regulating the readily releasable dopamine pool, but decreases dopamine signals evoked by long trains by depleting the reserve dopamine pool. This postulate would thus reconcile concurrent activation of phasic dopamine signing, *via* up-regulation of the readily releasable dopamine pool, and dopamine efflux, *via* depletion of the reserve dopamine pool.

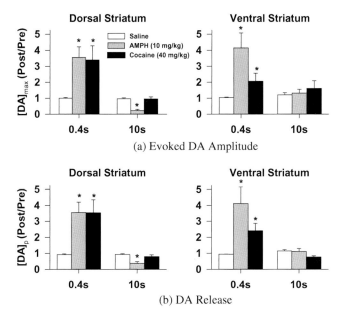

Figure 13.8 Stimulus-dependent effects of AMPH and cocaine on dopamine signaling in the dorsal and ventral striatum. (a). Maximal concentration of electrically evoked dopamine ($[DA]_{max}$). (b). Vesicular dopamine release (concentration of dopamine released per stimulus pulse or $[DA]_p$). All data are expressed as ratio of drug to pre-drug effect (Post/Pre). The duration of the 60-Hz stimulus train is identified in the x-axis. Data were collected under urethane anesthesia. *Significantly different from saline at $p<0.05$. Modified from Covey et al., 2013.

The hypothesis that AMPH acts distinctly on dopamine storage pools was tested using FSCV at a CFM and stimulus trains of different duration (Figure 13.8) (Covey et al., 2013). Trains of 0.4 and 10 seconds were used to interrogate the readily releasable and reserve dopamine pools, respectively. Cocaine, was used as a control, because this psychostimulant up-regulates vesicular dopamine release and competitively inhibits dopamine uptake, but is not a substrate for the dopamine transporter and therefore does not deplete vesicular dopamine stores and elicit dopamine efflux (Venton et al., 2006; John and Jones, 2007). In the dorsal striatum, AMPH increases dopamine signals evoked by short trains, but decreases signals evoked by long trains (Figure 13.8a). Similar effects were observed on vesicular dopamine release, consistent with the hypothesis of AMPH up-regulating the readily releasable pool, but depleting the reserve pool. These differential effects were not observed for AMPH in the ventral striatum and for cocaine in either striatal sub-region (Figures 13.8a and 13.8b). Dopamine tone, quantified by PCR during the first 10 minutes after drug injection but before the first post-drug stimulation, was increased only by AMPH and only in the dorsal striatum, consistent with AMPH-induced depletion of vesicular dopamine stores driving dopamine efflux in this striatal sub-region (Covey et al., 2013). Taken together, these results suggest that differential effects on dopamine storage pools reconciles the dual activation of phasic dopamine signaling and dopamine efflux by AMPH.

13.5 SUMMARY AND CONCLUSION

The studies described herein highlight the use of FSCV at a CFM to investigate dopamine signaling during loss and gain of dopamine function. Using the 6-OHDA-lesioned rat model of PD and METH-induced neurotoxicity, we demonstrated that compensatory adaptation by surviving dopamine neurons maintains tonic dopamine signaling, which is important for preserving motor control, but not phasic dopamine signaling, which may lead to cognitive deficits. This incomplete normalization of dopamine signaling appears to be inevitable and related to how the two signaling modes are distinctly regulated by presynaptic mechanisms of dopamine release and uptake. Indeed, increasing vesicular dopamine release could normalize the amplitude of dopamine transients, because they are release-dominated signals, but would concomitantly hyperelevate dopamine tone, because the amplitude of this signal is determined by the balance of release and uptake. We also demonstrated that AMPH activates, as opposed to disrupts, phasic dopamine signaling. This new view of AMPH action permits proposing hyperactivation of dopamine transients as a unifying mechanism of abused drugs in the addiction process and reconciles the observed dual activation of phasic dopamine signaling and dopamine efflux by this psychostimulant.

ACKNOWLEDGMENT

This work was supported by NIH NS 35298-01 and -02, DA 024036, DA 021770, and NSF DBI 075461.

REFERENCES

Abercrombie ED, Bonatz AE, Zigmond MJ (1990) Effects of L-dopa on extracellular dopamine in striatum of normal and 6-hydroxydopamine-treated rats. Brain Res 525:36–44.

Arnaldi D, Morbelli S, Morrone E, Campus C, Nobili F (2012) Cognitive impairment in degenerative parkinsonisms: role of radionuclide brain imaging. Q J Nucl Med Mol Imaging 56:55–67.

Avelar AJ, Juliano SA, Garris PA (2013) Amphetamine augments vesicular dopamine release in the dorsal and ventral striatum through different mechanisms. J Neurochem 125:373–385.

Barker-Haliski ML, Oldenburger K, Keefe KA (2012) Disruption of subcellular Arc/Arg 3.1 mRNA expression in striatal efferent neurons following partial monoamine loss induced by methamphetamine. J Neurochem 123:845–855.

Benwell ME, Balfour DJ, Lucchi HM (1993) Influence of tetrodotoxin and calcium on changes in extracellular dopamine levels evoked by systemic nicotine. Psychopharmacology (Berl) 112:467–474.

Bergstrom BP, Garris PA (2003) 'Passive stabilization' of striatal extracellular dopamine across the lesion spectrum encompassing the presymptomatic phase of Parkinson's disease: a voltammetric study in the 6-OHDA-lesioned rat. J Neurochem 87:1224–1236.

Bergstrom BP, Sanberg SG, Andersson M, Mithyantha J, Carroll FI, Garris PA (2011) Functional reorganization of the presynaptic dopaminergic terminal in parkinsonism. Neuroscience 193:310–322.

Berridge KC (2007) The debate over dopamine's role in reward: the case for incentive salience. Psychopharmacology (Berl) 191:391–431.

Bezard E, Jaber M, Gonon F, Boireau A, Bloch B, Gross CE (2000) Adaptive changes in the nigrostriatal pathway in response to increased 1-methyl-4-phenyl-1,2,3,6-tetrahydropyridine-induced neurodegeneration in the mouse. Eur J Neurosci 12:2892–2900.

Carelli RM, Wightman RM (2004) Functional microcircuitry in the accumbens underlying drug addiction: insights from real-time signaling during behavior. Curr Opin Neurobiol 14:763–768.

Cass WA, Manning MW (1999) Recovery of presynaptic dopaminergic functioning in rats treated with neurotoxic doses of methamphetamine. J Neurosci 19:7653–7660.

Chadchankar H, Ihalainen J, Tanila H, Yavich L (2012) Methylphenidate modifies overflow and presynaptic compartmentalization of dopamine *via* an alpha-synuclein-dependent mechanism. J Pharmacol Exp Ther 341:484–492.

Chapman DE, Hanson GR, Kesner RP, Keefe KA (2001) Long-term changes in basal ganglia function after a neurotoxic regimen of methamphetamine. J Pharmacol Exp Ther 296:520–527.

Cheer JF, Wassum KM, Heien ML, Phillips PE, Wightman RM (2004) Cannabinoids enhance subsecond dopamine release in the nucleus accumbens of awake rats. J Neurosci 24:4393–4400.

Cheer JF, Wassum KM, Sombers LA, Heien ML, Ariansen JL, Aragona BJ, Phillips PE, Wightman RM (2007) Phasic dopamine release evoked by abused substances requires cannabinoid receptor activation. J Neurosci 27:791–795.

Chergui K, Svenningsson P, Nomikos GG, Gonon F, Fredholm BB, Svensson TH (1997) Increased expression of NGFI-A mRNA in the rat striatum following burst stimulation of the medial forebrain bundle. Eur J Neurosci 9:2370–2382.

Covey DP, Juliano SA, Garris PA (2013) Amphetamine elicits opposing actions on readily releasable and reserve pools for dopamine. PLoS One 8:e60763.

Cragg SJ (2003) Variable dopamine release probability and short-term plasticity between functional domains of the primate striatum. J Neurosci 23:4378–4385.

Daberkow DP, Brown HD, Bunner KD, Kraniotis SA, Doellman MA, Ragozzino ME, Garris PA, Roitman MF (2013) Amphetamine paradoxically augments exocytotic dopamine release and phasic dopamine signals. J Neurosci 33:452–463.

Daberkow DP, Riedy MD, Kesner RP, Keefe KA (2008) Effect of methamphetamine neurotoxicity on learning-induced Arc mRNA expression in identified striatal efferent neurons. Neurotox Res 14:307–315.

Di Chiara G, Imperato A (1988) Drugs abused by humans preferentially increase synaptic dopamine concentrations in the mesolimbic system of freely moving rats. Proc Natl Acad Sci USA 85:5274–5278.

Dreyer JK, Herrik KF, Berg RW, Hounsgaard JD (2010) Influence of phasic and tonic dopamine release on receptor activation. J Neurosci 30:14273–14283.

Ewing AG, Bigelow JC, Wightman RM (1983) Direct *in vivo* monitoring of dopamine released from two striatal compartments in the rat. Science 221:169–171.

Flagel SB, Clark JJ, Robinson TE, Mayo L, Czuj A, Willuhn I, Akers CA, Clinton SM, Phillips PE, Akil H (2011) A selective role for dopamine in stimulus-reward learning. Nature 469:53–57.

Garris PA, Walker QD, Wightman RM (1997) Dopamine release and uptake rates both decrease in the partially denervated striatum in proportion to the loss of dopamine terminals. Brain Res 753:225–234.

Glimcher PW (2011) Understanding dopamine and reinforcement learning: the dopamine reward prediction error hypothesis. Proc Natl Acad Sci USA 108 3:15647–15654.

Gonon F (1997) Prolonged and extrasynaptic excitatory action of dopamine mediated by D1 receptors in the rat striatum *in vivo*. J Neurosci 17:5972–5978.

Grace AA, Floresco SB, Goto Y, Lodge DJ (2007) Regulation of firing of dopaminergic neurons and control of goal-directed behaviors. Trends Neurosci 30:220–227.

Hearing MC, Schwendt M, McGinty JF (2011) Suppression of activity-regulated cytoskeleton-associated gene expression in the dorsal striatum attenuates extinction of cocaine-seeking. Int J Neuropsychopharmacol 14:784–795.

Hornykiewicz O (1966) Dopamine (3-hydroxytyramine) and brain function. Pharmacol Rev 18:925–964.

Hornykiewicz O, Kish SJ (1987) Biochemical pathophysiology of Parkinson's disease. Adv Neurol 45:19–34.

Howard CD, Daberkow DP, Ramsson ES, Keefe KA, Garris PA (2013a) Methamphetamine-induced neurotoxicity disrupts naturally occurring phasic dopamine signaling. Eur J Neurosci 38:2078–2088.

Howard CD, Keefe KA, Garris PA, Daberkow DP (2011) Methamphetamine neurotoxicity decreases phasic, but not tonic, dopaminergic signaling in the rat striatum. J Neurochem 118:668–676.

Howard CD, Pastuzyn ED, Barker-Haliski ML, Garris PA, Keefe KA (2013b) Phasic-like stimulation of the medial forebrain bundle augments striatal gene expression despite methamphetamine-induced partial dopamine denervation. J Neurochem 125:555–565.

Hurd YL, Ungerstedt U (1989) Ca^{2+} dependence of the amphetamine, nomifensine, and Lu 19-005 effect on *in vivo* dopamine transmission. Eur J Pharmacol 166:261–269.

Hyman SE, Malenka RC, Nestler EJ (2006) Neural mechanisms of addiction: the role of reward-related learning and memory. Annu Rev Neurosci 29:565–598.

John CE, Jones SR (2007) Voltammetric characterization of the effect of monoamine uptake inhibitors and releasers on dopamine and serotonin uptake in mouse caudate-putamen and substantia nigra slices. Neuropharmacology 52:1596–1605.

Johnson-Davis KL, Hanson GR, Keefe KA (2002) Long-term post-synaptic consequences of methamphetamine on preprotachykinin mRNA expression. J Neurochem 82:1472–1479.

Jones SR, Gainetdinov RR, Wightman RM, Caron MG (1998) Mechanisms of amphetamine action revealed in mice lacking the dopamine transporter. J Neurosci 18:1979–1986.

Kuhr WG, Bigelow JC, Wightman RM (1986) *In vivo* comparison of the regulation of releasable dopamine in the caudate nucleus and the nucleus accumbens of the rat brain. J Neurosci 6:974–982.

Lee TH, Balu R, Davidson C, Ellinwood EH (2001) Differential time-course profiles of dopamine release and uptake changes induced by three dopamine uptake inhibitors. Synapse 41:301–310.

Loewinger GC, Beckert MV, Tejeda HA, Cheer JF (2012) Methamphetamine-induced dopamine terminal deficits in the nucleus accumbens are exacerbated by reward-associated cues and attenuated by CB1 receptor antagonism. Neuropharmacology 62:2192–2201.

Marshall JF, O'Dell SJ (2012) Methamphetamine influences on brain and behavior: unsafe at any speed? Trends Neurosci 35:536–545.

May LJ, Kuhr WG, Wightman RM (1988) Differentiation of dopamine overflow and uptake processes in the extracellular fluid of the rat caudate nucleus with fast-scan *in vivo* voltammetry. J Neurochem 51:1060–1069.

McCallum SE, Parameswaran N, Perez XA, Bao S, McIntosh JM, Grady SR, Quik M (2006) Compensation in pre-synaptic dopaminergic function following nigrostriatal damage in primates. J Neurochem 96:960–972.

Montague PR, McClure SM, Baldwin PR, Phillips PE, Budygin EA, Stuber GD, Kilpatrick MR, Wightman RM (2004) Dynamic gain control of dopamine delivery in freely moving animals. J Neurosci 24:1754–1759.

Nisenbaum LK, Crowley WR, Kitai ST (1996) Partial striatal dopamine depletion differentially affects striatal substance P and enkephalin messenger RNA expression. Brain Res Mol Brain Res 37:209–216.

Nomikos GG, Damsma G, Wenkstern D, Fibiger HC (1990) *In vivo* characterization of locally applied dopamine uptake inhibitors by striatal microdialysis. Synapse 6:106–112.

Onn SP, West AR, Grace AA (2000) Dopamine-mediated regulation of striatal neuronal and network interactions. Trends Neurosci 23:S48–S56.

Pastuzyn ED, Chapman DE, Wilcox KS, Keefe KA (2012) Altered learning and Arc-regulated consolidation of learning in striatum by methamphetamine-induced neurotoxicity. Neuropsychopharmacology 37:885–895.

Perez XA, Parameswaran N, Huang LZ, O'Leary KT, Quik M (2008) Pre-synaptic dopaminergic compensation after moderate nigrostriatal damage in non-human primates. J Neurochem 105:1861–1872.

Phillips PE, Hancock PJ, Stamford JA (2002) Time window of autoreceptor-mediated inhibition of limbic and striatal dopamine release. Synapse 44:15–22.

Ramsson ES, Covey DP, Daberkow DP, Litherland MT, Juliano SA, Garris PA (2011a) Amphetamine augments action potential-dependent dopaminergic signaling in the striatum *in vivo*. J Neurochem 117:937–948.

Ramsson ES, Howard CD, Covey DP, Garris PA (2011b) High doses of amphetamine augment, rather than disrupt, exocytotic dopamine release in the dorsal and ventral striatum of the anesthetized rat. J Neurochem 119:1162–1172.

Robinson DL, Heien ML, Wightman RM (2002) Frequency of dopamine concentration transients increases in dorsal and ventral striatum of male rats during introduction of conspecifics. J Neurosci 22:10477–10486.

Robinson DL, Hermans A, Seipel AT, Wightman RM (2008) Monitoring rapid chemical communication in the brain. Chem Rev 108:2554–2584.

Schmitz Y, Lee CJ, Schmauss C, Gonon F, Sulzer D (2001) Amphetamine distorts stimulation-dependent dopamine overflow: effects on D2 autoreceptors, transporters, and synaptic vesicle stores. J Neurosci 21:5916–5924.

Schonberg T, O'Doherty JP, Joel D, Inzelberg R, Segev Y, Daw ND (2010) Selective impairment of prediction error signaling in human dorsolateral but not ventral striatum in Parkinson's disease patients: evidence from a model-based fMRI study. Neuroimage 49:772–781.

Schultz W (2007a) Behavioral dopamine signals. Trends Neurosci 30:203–210.

Schultz W (2007b) Multiple dopamine functions at different time courses. Annu Rev Neurosci 30:259–288.

Scott JC, Woods SP, Matt GE, Meyer RA, Heaton RK, Atkinson JH, Grant I (2007) Neurocognitive effects of methamphetamine: a critical review and meta-analysis. Neuropsychol Rev 17:275–297.

Snyder GL, Keller RW Jr, Zigmond MJ (1990) Dopamine efflux from striatal slices after intracerebral 6-hydroxydopamine: evidence for compensatory hyperactivity of residual terminals. J Pharmacol Exp Ther 253:867–876.

Son JH, Latimer C, Keefe KA (2011) Impaired formation of stimulus-response, but not action-outcome, associations in rats with methamphetamine-induced neurotoxicity. Neuropsychopharmacology 36:2441–2451.

Stachowiak MK, Keller RWJ, Stricker EM, Zigmond MJ (1987) Increased dopamine efflux from striatal slices during development and after nigrostriatal bundle damage. J Neurosci 7:1648–1654.

Steinberg EE, Keiflin R, Boivin JR, Witten IB, Deisseroth K, Janak PH (2013) A causal link between prediction errors, dopamine neurons and learning. Nat Neurosci 16:966–973.

Sulzer D (2011) How addictive drugs disrupt presynaptic dopamine neurotransmission. Neuron 69:628–649.

Tsai HC, Zhang F, Adamantidis A, Stuber GD, Bonci A, de Lecea L, Deisseroth K (2009) Phasic firing in dopaminergic neurons is sufficient for behavioral conditioning. Science 324:1080–1084.

van Horne C, Hoffer BJ, Stromberg I, Gerhardt GA (1992) Clearance and diffusion of locally applied dopamine in normal and 6-hydroxydopamine-lesioned rat striatum. J Pharmacol Exp Ther 263:1285–1292.

Venton BJ, Seipel AT, Phillips PE, Wetsel WC, Gitler D, Greengard P, Augustine GJ, Wightman RM (2006) Cocaine increases dopamine release by mobilization of a synapsin-dependent reserve pool. J Neurosci 26:3206–3209.

Venton BJ, Zhang H, Garris PA, Phillips PE, Sulzer D, Wightman RM (2003) Real-time decoding of dopamine concentration changes in the caudate-putamen during tonic and phasic firing. J Neurochem 87:1284–1295.

Volkow ND, Chang L, Wang GJ, Fowler JS, Leonido-Yee M, Franceschi D, Sedler MJ, Gatley SJ, Hitzemann R, Ding YS, Logan J, Wong C, Miller EN (2001) Association of dopamine transporter reduction with psychomotor impairment in methamphetamine abusers. Am J Psychiatry 158:377–382.

Watson CJ, Venton BJ, Kennedy RT (2006) *In vivo* measurements of neurotransmitters by microdialysis sampling. Anal Chem 78:1391–1399.

Westerink BH, Hofsteede RM, Tuntler J, De Vries JB (1989) Use of calcium antagonism for the characterization of drug-evoked dopamine release from the brain of conscious rats determined by microdialysis. J Neurochem 52:722–729.

Wiecki TV, Frank MJ (2010) Neurocomputational models of motor and cognitive deficits in Parkinson's disease. Prog Brain Res 183:275–297.

Wiedemann DJ, Basse-Tomusk A, Wilson RL, Rebec GV, Wightman RM (1990) Interference by DOPAC and ascorbate during attempts to measure drug-induced changes in neostriatal dopamine with Nafion-coated, carbon-fiber electrodes. J Neurosci Methods 35:9–18.

Wightman RM, Amatore C, Engstrom RC, Hale PD, Kristensen EW, Kuhr WG, May LJ (1988) Real-time characterization of dopamine overflow and uptake in the rat striatum. Neuroscience 25:513–523.

Willuhn I, Wanat MJ, Clark JJ, Phillips PE (2010) Dopamine signaling in the nucleus accumbens of animals self-administering drugs of abuse. Curr Top Behav Neurosci 3:29–71.

Wolf ME, Zigmond MJ, Kapatos G (1989) Tyrosine hydroxylase content of residual striatal dopamine nerve terminals following 6-hydroxydopamine administration: a flow cytometric study. J Neurochem 53:879–885.

Zigmond MJ, Abercrombie ED, Berger TW, Grace AA, Stricker EM (1990) Compensations after lesions of central dopaminergic neurons: some clinical and basic implications. Trends Neurosci 13:290–296.

Zweifel LS, Parker JG, Lobb CJ, Rainwater A, Wall VZ, Fadok JP, Darvas M, Kim MJ, Mizumori SJ, Paladini CA, Phillips PE, Palmiter RD (2009) Disruption of NMDAR-dependent burst firing by dopamine neurons provides selective assessment of phasic dopamine-dependent behavior. Proc Natl Acad Sci USA 106:7281–7288.

CHAPTER 14

MEASUREMENTS OF DOPAMINE RELEASE AND UPTAKE IN HUNTINGTON'S DISEASE MODEL RODENTS

Sam V. Kaplan[*], Stephen C. Fowler[†], and Michael A. Johnson[*]

[*]Department of Chemistry and R.N. Adams Institute for Bioanalytical Chemistry and [†]Department of Pharmacology and Toxicology, University of Kansas, Lawrence

14.1 INTRODUCTION

Huntington's disease (HD) is an autosomal dominant, neurodegenerative, movement disorder that is characterized by motor dysfunction, altered behavior, and cognitive impairment. HD is caused by a cytosine-adenine-guanine (CAG) repeat expansion on chromosome 4 of the *IT15* gene, which encodes for the huntingtin protein (Martin and Gusella, 1986). An average person normally has 16 to 20 CAG repeats on the huntingtin gene (*htt*) while a person with HD will have 40 or more (Snell et al., 1993). This protein is ubiquitously expressed and is highly concentrated in the central nervous system (CNS) (MacDonald et al., 1993). The mutation results in a polyglutamine (polyQ) expansion near the N-terminus of *htt*. HD affects 4 to 10 out of 100,000 people of western European descent and 0.4 out of 100,000 people of Asian descent (Pringsheim et al., 2012). Symptom onset for HD typically occurs between 35 and 50 years of age and the age of onset is directly correlated to the length of the expanded CAG repeat (Andrew et al., 1993; Duyao et al., 1993; MacMillan et al., 1993; Snell et al., 1993). With an extremely long CAG repeat typically of 52 or more, symptomatic onset may begin in juveniles (Andrew et al., 1993; Duyao et al., 1993). This early onset most often also includes a more rapid progression of the disease and increased severity (Snell et al., 1993).

Documentation of the existence of HD can be found as early as the mid-14th century. Although little was known about the cause of the disease, it was described as an epidemic of what was called "a dancing mania" (Zuccato et al., 2010). It wasn't until later that the term "chorea" was used by the famed student of all sciences, Paracelsus (more specifically, Philippus Aureolus Theophrastus Bombastus von Hohenheim) in the late 15th and early 16th centuries (Zuccato et al., 2010). This term, which is derived from the Greek word for "dance," was used to describe the involuntary muscle movements that result from the disorder. HD eventually spread to the new world in the early 17th century. In the American colonies, HD was coined as "San Vitus" or "That disorder" (Zuccato et al., 2010). Due to the uncontrollable, unusually rhythmic movements associated with HD, people with the disorder were often observed by the top physicians and were misdiagnosed with demonic possession. This diagnosis was tragic as the appropriate cure at the time was public execution; it is assumed that at least one of the victims of the Salem witch trials was burned at the stake because she suffered from HD (Zuccato et al., 2010). Despite a few previous attempts, an accurate medical description of HD was unavailable until 1872 when Dr. George Huntington published his findings in the Medical and Surgical Reporter of Philadelphia (Huntington, 2004). Hence, the disease was named after the young doctor. It is interesting to note that this was one of only two publications he would ever write.

The progression of HD typically occurs over the course of 15 to 20 years from the time of initial clinical diagnosis to death (Martin and Gusella, 1986). During the early stages, a slow deterioration in intellectual capabilities and minor emotional changes occur (Kremer, 2002). Other symptoms may include weight loss, interruptions in the patient's sleep cycle, or changes in sexual behavior, and may be caused by hypothalamic dysfunction (Politis et al., 2008). Furthermore, patients may begin to exhibit minor motor abnormalities such as restlessness, abnormal eye movements, and unusual hand movements (Penney et al., 1990). As the syndrome associated with HD progresses, motor function deficits gradually worsen. The clinical diagnosis of HD is most commonly based on the development of chorea, which is exhibited by 90% of patients, and is characterized by excessive, non-rhythmic, abrupt, unintentional movements (Barbeau et al., 1981; Martin and Gusella, 1986). The severity of this phenotypical response to HD ranges from excessive fidgeting and exaggerated gesture to severely violent, disabling movements (Barbeau et al., 1981; Kremer, 2002). In the later stages of HD, chorea is often replaced with other, more severe movement abnormalities such as bradykinesia, dystonia, rigidity, and complete loss of coordination (Hayden, 1981; Penney et al., 1990). Eventually, patients with HD lose their ability to support themselves. Reasons for this loss of independence include the complete loss of mobility and ability to communicate. Due to its progressive nature, the effects of late stage HD are ultimately fatal. Death can be caused by physical injuries (falling), malnutrition, choking, and, most commonly, pneumonia (Lanska et al., 1988a). Along with the physical abnormalities and cognitive impairments discussed, HD is also characterized by psychiatric manifestations that may include aggression, anxiety, depression, and the initiation or progression of addictive behaviors. The progression of such psychiatric changes can often lead to necessary institutionalization and/or an increase in suicidal thoughts and action (Lanska et al., 1988b; Kremer, 2002). The observance of motor function impairments in conjunction with these psychiatric and behavior abnormalities can be used to diagnose HD. Additionally, due to its autosomal dominant pattern of inheritance, information regarding familial history has also been used to diagnose HD. Nevertheless, definitive HD diagnosis can only be confirmed by genetic testing. The development of a reliable, DNA-based test has made it possible to accurately diagnose HD without obtaining standard clinical information (Kremer, 2002). Despite the available information regarding pathology, the underlying mechanism leading to neuronal degeneration is not well understood. HD, therefore, remains incurable and much effort has been and is currently focused on understanding the mechanisms of neuronal degradation.

In this chapter we highlight the work carried out by our group to identify altered mechanisms of neurotransmitter release in HD. We specifically focused on the measurement of striatal dopamine (DA) with fast-scan cyclic voltammetry (FSCV) at carbon-fiber electrodes. This technique allows the measurement of the release and uptake of DA, as well as other electroactive neurotransmitters and neuromodulators, with sub-second temporal resolution. Our work has revealed alterations in DA release properties in multiple species and strains of rodents that model HD. Moreover, these release impairments result in motor behavioral alterations. In order to place these results in proper context we first discuss important background information, provided in the subsequent three sections. First, mechanisms of neuronal dysfunction, with emphasis on signaling pathways that control movement, are discussed. Second, we provide a brief overview of genetically altered rodent models used in our studies and other studies mentioned in this chapter. Next, we summarize microdialysis findings obtained using HD model rodents. In the final section, we emphasize our voltammetry results obtained *in vivo* and *ex vivo*.

14.2 POTENTIAL MECHANISMS OF NEURONAL DYSFUNCTION AND DEGRADATION IN HD

The striatum, a brain region richly innervated with DA-releasing terminals, undergoes extensive atrophy throughout the course of HD. About 90 to 95% of the neuronal cell bodies in the striatum are gamma-aminobutyric acid (GABA) type medium spiny neurons, which receive input from dopaminergic neurons that project from the substantia nigra pars compacta (SN_c). This projection, known as the nigrostriatal pathway, provides dopaminergic input to D1 and D2 receptors located on different populations of medium spiny neurons. The medium spiny neurons project to the pallidum through two distinct pathways: the direct pathway, which provides inhibitory tone to the globus pallidus internal segment (GPi), and the indirect pathway, which provides inhibitory tone to the globus pallidus external segment (GPe). Activation of the direct pathway facilitates movement, while activation of the indirect pathway inhibits movement. Previous evidence has indicated that the indirect pathway neurons are most vulnerable to degeneration as HD progresses; thus, the loss of this pathway results in the expression of excess movement in the form of chorea, the hallmark motor feature of HD (Kandel et al., 2000).

The mechanisms that underlie the degeneration of these striatal medium spiny neurons have been extensively investigated. It has been shown that mutant *htt* forms insoluble aggregates and nuclear inclusions in medium spiny neurons of patients with HD (DiFiglia et al., 1995; Davies et al., 1997; Scherzinger et al., 1997). The role of aggregates in the pathology of HD is considered to be of key importance as their presence suggests that mutant *htt* is processed differently in those afflicted with HD. Additionally, mutant *htt* acts to impede the function of normal huntingtin protein, which is essential for the viability of the neuronal populations that degenerate in HD (Cattaneo et al., 2005). However, it is thought that the toxicity of mutant *htt* itself, a direct result of the polyQ extension, contributes to the majority of the disease pathogenesis (Ikeda et al., 1996; Mangiarini et al., 1996). This toxicity appears to be enhanced by a mechanism that involves the activation of D2 family DA receptors and the presence of reactive oxygen species (Charvin et al., 2005).

14.3 GENETICALLY ENGINEERED HD MODEL RODENTS

Prior to the discovery of the HD mutation and the subsequent development of genetically-engineered HD model rodents, chemical induction of an HD-like syndrome in otherwise normal rats had been among the most popular methods of modeling HD. One such treatment involved injecting rats with 3-nitropropionic acid (3-NP), an irreversible mitochondrial complex II inhibitor (Alston et al., 1977; Coles et al., 1979). The toxic effects of 3-NP were discovered in the 1980s after individuals developed HD like symptoms following the consumption of moldy sugar cane (Hu et al., 1986; He, 1987). It was later found that 3-NP produced striatal lesions that mimic those found in HD and induced cell death in the caudate putamen that led to dystonia and athetosis.

Since the discovery of the causative mutation of HD, numerous genetically-engineered animal models of HD have been developed. The R6 line of HD model mice was created using a 1.9 kb gene fragment containing promoter sequences and exon 1 of the mutant human HD gene (Mangiarini et al., 1996). Among the most commonly used models are R6/1 and R6/2 mice, which possess CAG repeat lengths of about 115 and 144, respectively. Both models only express exon 1 of the 67 exon gene. This exon encodes 3% of the N-terminal region of the protein, suggesting that the expanded polyQ tract is responsible for the disease (Li et al., 2005). As would be expected, the longer repeat length is associated with the expression of a more quickly progressing motor phenotype in mice: R6/1 mice express overt motor symptoms at 15–21 weeks of age while R6/2 mice express symptoms at 9–11 weeks.

There have since been other genetic HD mouse lines that include larger portions of the gene. For example, yeast artificial chromosome (YAC) lines that contain between 18 to 128 CAG repeats on the whole gene have been reported (Hodgson et al., 1999). Of these lines, YAC 128 mice have been used most extensively because they experience progressive neurodegeneration and motor impairments (Van Raamsdonk, 2005a,b; Zuccato et al., 2010). Unlike R6/2 mice, YAC mice tend to live normal lifespans. The phenotype includes hyperkinesis (excessive locomotor movement) at three months of age, progressive motor impairment at six months of age, neurodegeneration at nine months of age, and hypokinesis at 12 months of age (Slow, 2003; Slow et al., 2005; Van Raamsdonk, 2005a,b). Moreover, the number of striatal neurons decreases by 15% by 12 months of age. Additionally, mutant mice exhibit nuclear huntingtin aggregates by 18 months that are visible at the light microscopy level (Slow, 2003). For a good review on the development of genetic HD model mice, see Levine et al., 2004.

In an attempt to accommodate the need for more complex experiments while recapitulating the neurological HD syndrome, von Hörsten et al. (2003) developed a transgenic HD (HDtg) model rat line. Like the R6/2 mouse, the HDtg rat possesses a truncated portion of the gene encoding huntingtin, except that the CAG repeat length is significantly shorter: 51 in the rat *versus* 144 in the R6/2 mouse. Therefore, it is not surprising that the overt motor behavioral phenotype is expressed more gradually by HDtg rats compared to R6/2 mice. In HDtg rats there are no detectable behavioral abnormalities at five months of age. Spatial and working memory abnormalities appear at 10 months and coordination loss becomes relevant between 10–15 months of age. Noticeable striatal atrophy occurs at 12 months of age.

This chapter primarily addresses work that has been accomplished using R6/1 and R6/2 mice, HDtg rats, and rats treated with 3-NP. We have conducted force-plate behavioral analyses and fast-scan voltammetric (FSV) neurochemical analyses to study the effects of the HD mutation on the behavioral and neurological phenotype. However, in order to provide additional context for our studies, we discuss first other important neurochemical findings obtained using microdialysis.

14.4 MICRODIALYSIS STUDIES IN HD MODEL RODENTS

Microdialysis studies aimed at understanding neurotransmitter alterations in HD model mice have been conducted by other groups. Several features make microdialysis an attractive method for measuring neurotransmitter levels in the brain. First, it allows the sampling of virtually any chemical species, including small molecule neurotransmitters and neuropeptides, which will diffuse into the probe. Furthermore, microdialysis allows the real-time, continuous sampling of endogenous biological compounds *in vivo*. Moreover, the concentrations of a specific analyte or set of analytes can be measured over long periods of time. Sampling times can range from minutes to days and perhaps even weeks depending on the probe and its surrounding environment. This capability is quite attractive for studying analyte concentration changes that may occur over long periods of time. However, most commercially available microdialysis probes are large (about 200 μm) compared to carbon-fiber microelectrodes used in our work (7 μm), resulting in the potential for additional tissue damage upon implantation (Khan and Michael, 2003). This damage may influence detected neurotransmitter release (Borland et al., 2005). Moreover, sampling intervals for many molecules tend to be on the order of minutes or hours (Westerink, 1995; Nandi and Lunte, 2009), limiting temporal resolution.

In one of the first studies aimed at quantifying DA levels in HDtg model mice, Petersén et al. (2002) found that extracellular DA levels were diminished in 16-week old R6/1 mice in comparison to age-matched wild type (WT) control mice (Petersén et al., 2002). Moreover, even though these extracellular DA levels were decreased, post mortem tissue content studies showed no significant difference in total DA levels between R6/1 and WT controls (Petersén et al., 2002). Further studies, in which malonate was locally applied *in vivo* by infusion through microdialysis probes, were also conducted. Malonate is a mitochondrial complex II inhibitor that induces DA release and ultimately neuronal cell death (Petersén et al., 2002; Fernandez-Gomez et al., 2005). Malonate-induced increases of extracellular DA levels were significantly less in R6/1 mice compared to WT control mice. Conversely, increases in extracellular glutamate, aspartate, and GABA were not substantially different. More recent *in vivo* microdialysis studies carried out in R6/2 and YAC128 mice have reaffirmed and expanded on the results obtained by Petersén et al. (2002) in R6/1 mice, demonstrating that extracellular DA concentrations were diminished in R6/2 mice and YAC128 mice (Callahan and Abercrombie, 2011). Taken together, these microdialysis studies revealed that decreases of extracellular DA are not simply due to the loss of neuronal populations or the inability to synthesize DA, but rather the ability of neurons to release DA is impaired (Hickey et al., 2002; Petersén et al., 2002). In order to probe the underlying mechanisms responsible for decreased striatal DA release, we applied FSCV at carbon-fiber microelectrodes (CFMs) to measure sub-second DA release events both *in vivo* and in brain slices, as described in subsequent sections of this chapter.

14.5 VOLTAMMETRIC MEASUREMENTS OF DA RELEASE AND UPTAKE IN HD MODEL RODENTS

14.5.1 FSCV

FSCV at CFMs is an electrochemical method that offers good specificity and temporal resolution for the measurement of electroactive neurotransmitters (Baur et al., 1988; Robinson et al., 2003). Commonly, when using FSCV to quantify DA release in brain tissue, a triangular waveform is applied at the CFM. A typical waveform used for this purpose is −0.4 to +1.0 to −0.4 V (*versus* Ag/AgCl reference electrode) applied linearly at a scan rate of 300 V/s (Robinson et al., 2003). Our electrodes are 5 to 7 μm in diameter and are made from carbon-fiber purchased from Goodfellow Cambridge Ltd. (Huntingdon, England), but the type of carbon-fiber used may vary in size and shape based on the goal of the individual study. As the potential at the working electrode becomes more positive, the surface becomes increasingly electron deficient. Upon reaching a voltage threshold, typically around +0.6 V, DA is oxidized to dopamine-ortho-quinone. This two-electron oxidation produces a faradaic current that is proportional to the amount of analyte present at the electrode surface. Similarly, a current is formed on the backsweep by the reduction of the this quinone. These currents are masked by a large non-faradaic current caused by formation of the electric double layer. Fortunately, this charging current is quite stable between scans and, therefore, can be removed by subtracting a background trace from the trace in which the faradaic current is present (Howell et al., 1986). This subtraction process produces a cyclic voltammogram (CV) which serves as an electrochemical signature for DA.

14.5.2 DA RELEASE AND UPTAKE MEASUREMENTS IN R6/2 AND R6/1 HD MODEL MICE

Previous behavioral studies have demonstrated that, compared to WT controls, R6/2 mice have a blunted locomotor response to cocaine and methamphetamine, inhibitors of DA uptake, suggesting an impairment of DA release (Hickey et al., 2002; Johnson et al., 2006). These behavioral studies, taken with the microdialysis results of Petersen et al. (2002), prompted us to investigate how mechanisms of DA release and uptake are impacted in HD model rodents. To measure these parameters, FSCV was applied to striatal brain slices acutely harvested from R6/2 mice and age-matched WT control mice. Exocytotic release was induced by the application of a single, biphasic electrical stimulus pulse (350 μA current, four milliseconds duration) (Johnson et al., 2006). Upon analyzing brain slices from subjects at ages ranging from 6 to 14 weeks, a dramatic attenuation of DA release was found in R6/2 mice compared to WT mice. This attenuation progressively worsened with age, in agreement with diminished locomotor activity (Johnson et al., 2006). Experiments in which exogenous DA iontophoretically was injected revealed no difference in the extracellular space due to neuronal atrophy in R6/2 mice in comparison to WT controls (Johnson et al., 2007). Thus, the attenuation in DA release is not simply a result of neurotransmitter dilution, but rather a diminished capacity for DA release in the striatum.

The plots of stimulated release were also modeled using simplex-based modeling software developed by Prof. Paul Garris (Illinois State University, Normal, IL). Values of V_{max}, the maximum rate of uptake by the DA transporter, were extracted from these modeled curves. Importantly, V_{max} was unchanged between genotype and age group, indicating that uptake was not impaired in R6/2 mice. Moreover, the uptake inhibition constants (K_i) for cocaine and methamphetamine, which indicate how strongly the drugs bind to the DA transporter protein molecules, did not significantly differ between R6/2 and WT control mice. Collectively, these results suggest that impairments in DA release, rather than uptake, are responsible for locomotor abnormalities observed in R6/2 mice, either in the presence or absence of psycho-stimulants (Johnson et al., 2006). Similar FSCV experiments have revealed that DA release is also attenuated in R6/1 HD model mice (Ortiz et al., 2011). These findings are consistent with those found in R6/2 mice in that the DA release attenuation progressed with age and the onset of the overt phenotype. Interestingly, DA uptake, again measured as V_{max}, was impaired in R6/1 HD model mice. Although the specific mechanisms underlying this impairment are not clear, the ages of the mice may also play a role, given that R6/1 mice develop significant deficits in rotorod performance at 13 to 20 weeks of age, while R6/2 mice become severely impaired at 8 to 12 weeks of age (Carter et al., 1999; Lüesse et al., 2001).

14.5.3 MECHANISMS OF NEUROTRANSMITTER RELEASE IMPAIRMENT

Although it had been confirmed up to this point that striatal DA release impairment is a component of neuronal dysfunction associated with HD, the underlying cellular mechanisms were not yet known. Several factors could decrease the ability of neurons to release DA, including impaired entry of extracellular calcium through voltage-gated calcium channels (VGCCs), diminished intracellular DA content, decreased loading of vesicles with DA, and decreased number of DA-containing vesicles available for release.

The DA release impairments that we noted were likely not caused by the diminished ability of VGCCs to allow extracellular calcium passage into the cell: when extracellular concentrations of calcium present in the perfusion buffer were increased, equal changes in DA release were found in R6/2 and WT brain slices (Johnson et al., 2006). Moreover, DA content was roughly the same in young R6/2 and WT mice (Hickey et al., 2002) even though release was sharply decreased (Johnson et al., 2006), indicating that DA content was not the only cause of the release impairments we found.

It was clear, therefore, that alterations of intracellular mechanisms play a role in this DA release impairment; thus, subsequent studies focused on identifying these potential mechanisms. One set of studies involved the measurement of quantal release of catecholamine measured by constant potential amperometry from adrenal chromaffin cells. These cells arise from the neural crest during development of the mammalian nervous system and are, therefore, thought to be similar to neurons (Pancrazio et al., 1994). Cells were harvested and cultured from R6/2 mice and age-matched WT control mice. Initial studies revealed that less catecholamine was released per vesicle in R6/2 cells compared to WT cells, suggesting incomplete vesicle loading. Furthermore, impairment of adenosine triphosphate (ATP) production by exposure to 3-NP decreased release from WT cells compared to that measured from R6/2 cells, suggesting an ATP-dependent mechanism (Johnson et al., 2006). The neuronal vesicular monoamine transporter (VMAT2) is driven by a pH gradient supplied by an ATP-dependent proton pump (Schuldiner, 1994). Therefore, one interpretation of these findings is that R6/2 mice suffer from a chronic depletion of ATP that can be induced in their WT counterparts.

Striatal degeneration is a hallmark feature of HD (Kowall et al., 1987; Brouillet et al., 2005); therefore, we focused our studies on the characterization of DA release in brain striatum in order to investigate in more depth alterations of intracellular mechanisms that may contribute to the observed DA release impairments. When applying a stimulation pulse train of 120 pulses to the brain slice, we noticed that DA release increased with the increasing frequency of application in slices from WT mice, but not in slices from R6/2 mice. Neurotransmitter reserve pools are present in neurons to supply additional vesicles of neurotransmitter during periods of increased synaptic activity; therefore, we hypothesized here that the DA reserve pool was depleted in R6/2 mice. To identify alterations in how DA is packaged within the reserve pool, FSCV was used in combination with selected pharmacological agents in striatal brain slices.

The initial experimental paradigm is shown in Figure 14.1. The brain slice tissue surrounding the carbon-fiber working electrode was stimulated with a single

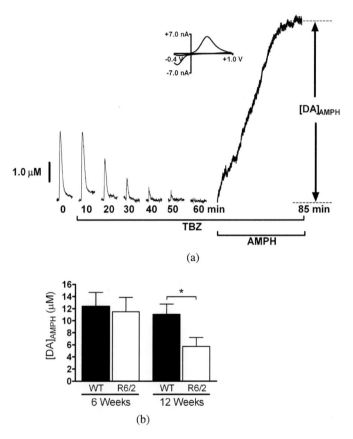

Figure 14.1 Measurement of AMPH-induced DA efflux after TBZ treatment. The representative data shown in (a) were collected from a brain slice harvested from a 6-week old WT mouse. While stimulating the slice periodically with a four milliseconds duration electrical pulse every five minutes, the slice was treated with TBZ. The DA release peak disappeared because the VMAT was inhibited, thereby depleting vesicles of DA. Subsequent treatment of the slice with 20 μM AMPH caused a surge in DA release from vesicular reserve pool and free cytosolic DA stores. The provided CV confirms the presence of DA at the peak release time. (b) The AMPH-induced efflux of DA is significantly less in 12-week old R6/2 mice compared to WT mice (*$p<0.05$; $n = 7$ R6/2 mice and 8 WT mice), but no difference was found at 6-weeks of age ($p>0.05$; $n = 6$ R6/2 and 7 WT mice). Reproduced with permission from Ortiz et al., 2010. Copyright Wiley.

electrical pulse (350 μA current, four milliseconds duration) once every five minutes and evoked DA release was measured. After the peak current stabilized between successive measurements, tetrabenazine (TBZ), which blocks DA from entering vesicles, was added to the artificial cerebral spinal fluid (aCSF) perfusate. Throughout this time, brain slices were stimulated every five minutes, as before. After nearly an hour, the DA release peak disappeared because the releasable vesicles were depleted of DA. At this point, DA was present only in reserve pool vesicles or as free DA in the cytosol. The brain slice was then treated with d-amphetamine (AMPH) sulfate, introduced into the gravity feed line along with the TBZ that was already present. AMPH causes the reverse transport of DA, transporting it from the intracellular space to the extracellular space. Additionally, it empties intracellular vesicles of DA. Thus, treatment with AMPH after the TBZ treatment results in the efflux of cytosolic plus reserve pool DA into the extracellular space. A representative sample of the data obtained from this procedure is shown in Figure 14.1. We observed an age-dependent effect in which intracellular stores were depleted in 12-week old mice but not 6-week old mice.

The TBZ data provided information on the quantity of free cytosolic DA plus reserve pool DA packaged in vesicles. To quantitate reserve pool DA alone, we repeated the sample experimental paradigm described in the previous paragraph, except that the methyl ester form of α-methyl-p-tyrosine (αMPT) was used instead of TBZ (Ortiz et al., 2010; Figure 14.2). This procedure, followed again by treatment with AMPH, resulted in the selective measurement of reserve pool DA within presynaptic terminals. Similar to the results obtained using TBZ, we found that the total amount of DA released by reverse transport from R6/2 slices was the same as WT slices at six weeks of age, but less than WT at 12 weeks of age. These findings are especially interesting given that R6/2 mice develop the overt motor phenotype at 9 to 11 weeks of age (Mangiarini et al., 1996), consistent with the idea that reserve pool depletion contributes to the HD motor syndrome. More studies are required to confirm this relationship.

We next examined in more detail the reserve pool depletion observed in R6/2 mice (Ortiz et al., 2010). The goal here was to determine if reserve pool depletion was a result of fewer vesicles or incomplete loading of vesicles. Again, stimulated DA release was measured before treatment with αMPT. Once DA release disappeared, however, the brain slice was treated with cocaine and application of the electrical stimulus pulses every five minutes were continued. Although the principal action of cocaine is inhibition of the DA transporter, cocaine also causes reserve pool vesicles to mobilize (Venton, 2006). As expected, the stimulated DA release signal re-appeared in both R6/2 and WT brain slices (Figure 14.3); however, the duration of this signal was shorter-lived in R6/2 slices (about 30 minutes) than in WT slices (about 90 minutes). Moreover, peak DA release from R6/2 and

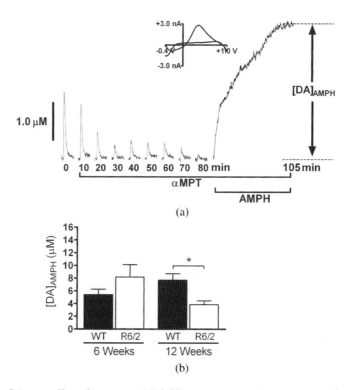

Figure 14.2 Selective efflux of reserve pool DA. The representative data shown in (a) were collected from a brain slice harvested from a 12-week old R6/2 mouse. The procedure for this experiment was identical to that in Figure 14.3, except that αMPT was used to deplete all intracellular DA other than that present in the reserve pool. The provided CV confirms the presence of DA at the peak release time. (b) The AMPH-induced efflux of DA is significantly less in 12-week old R6/2 mice compared to age-matched WT mice (*$p<0.05$; $n = 5$ R6/2 mice and 4 WT mice), but no difference was found at six weeks of age ($p>0.05$; $n = 5$ R6/2 and 3 WT mice). Reproduced with permission from Ortiz et al., 2010. Copyright Wiley.

WT slices was the same, suggesting that, although fewer reserve pool vesicles are available for mobilization in R6/2 mice, they are still fully-loaded with DA. Interestingly, this finding contradicts the data previously obtained from adrenal chromaffin cells, which indicated that the vesicular uptake of catecholamine was disrupted (Johnson et al., 2007). The mechanism underlying this decrease in available reserve pool DA vesicles is currently not known.

14.5.4 DA RELEASE AND UPTAKE MEASUREMENTS IN HD MODEL RATS

14.5.4.1 Treatment with 3-NP

We used FSCV to study the effect of treatment with 3-NP, which is toxic to neurons, on DA release and uptake measured *in vivo* (Kraft et al., 2009). In order to

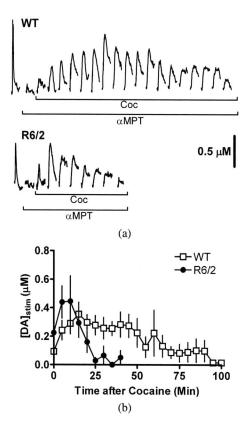

Figure 14.3 The number of reserve pool DA vesicles is diminished in 12-week old R6/2 mice. (a) Striatal brain slices from WT and R6/2 mice were stimulated every five minutes, and evoked DA was measured. Treatment of the slices with αMPT resulted in diminishment of the DA release. Subsequent introduction of cocaine (Coc) caused the DA release peak to come back due to the mobilization of reserve pool vesicles. (b) Peak DA release obtained after cocaine treatment was the same between WT and R6/2 mice, but the duration of measurable release was greater in WT than R6/2 mice. Reproduced with permission from Ortiz et al., 2010. Copyright Wiley.

induce striatal atrophy that resembled the HD phenotype, Lewis rats were treated continuously with 3-NP for five days using subcutaneously implanted osmotic pumps. After treatment, striatal DA release was induced by remote electrical stimulation of the medial forebrain bundle and measured in the dorsal striatum with FSCV. Consistent with findings obtained in transgenic mice, DA release was sharply diminished even though DA content was unchanged. The data were then modeled using the same simplex-based software that we used to analyze the R6/1 data (Figure 14.4), and we found that V_{max} was decreased in 3-NP-treated rats, similar to our observations in R6/1 mice. It is important to note that DA release and uptake were impaired similarly at multiple measurement depths; thus, the toxicity due to 3-NP administration appears to have similar impact at different locations within the striatum.

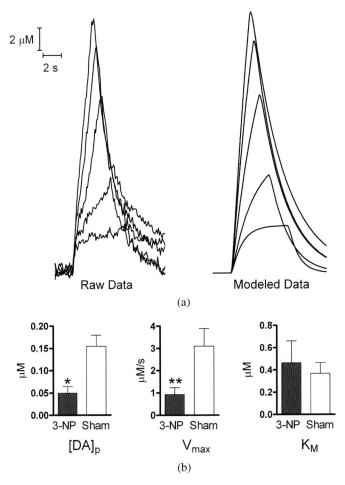

Figure 14.4 DA release and uptake are impaired in rats treated with 3-NP. (a) Shown are representative stimulated release plots obtained from urethane anesthetized rats. The medial forebrain bundle was stimulated with 120 electrical stimulus pulses applied at 20, 30, 40, 50, and 60 Hz. Evoked DA release was measured in the striatum. Each increase in frequency produced a greater release of DA. (b) Pooled measurements of DA release concentration per stimulus pulse ($[DA]_p$), V_{max}, and K_m in multiple 3-NP-treated and sham control rats are shown. Treatment with 3-NP resulted in decreased $[DA]_p$ and V_{max} (*$p<0.05$, $n = 5$ sham and $n = 4$ 3NP-treated; **$p<0.001$, $n = 5$ sham and $n = 4$ 3NP-treated); however, K_m was unaffected. Reproduced with permission from Kraft et al., 2009. Copyright Elsevier.

14.5.4.2 HDtg rats

Our aim in studying HDtg rats was to compare DA release properties to locomotor abnormalities observed in this HD model. To measure behavior, we used a force-plate actometer and a newly-developed force-sensing runway (Ortiz et al., 2012). This work was performed in collaboration with Stephen Fowler, a coauthor of this chapter, who developed both instruments and analyzed the behavioral data. Initial behavioral studies involved measuring focused stereotypy on the force-plate actometer. Briefly, the force-plate actometer is a device consisting of a light-weight, stiff honeycomb surface (carbon-fiber or Nomex® composite material) supported at the corners by four force transducers. These transducers measure forces used to calculate the position of the rat's center of gravity with millimeter precision. Data are collected from the transducers at high rates (adjustable, but 100 points/second, in our case); therefore, the real-time z-axis forces can be measured and processed to identify characteristic frequencies of movement.

We exploited this technique to measure the effect of AMPH injection on behavior in HDtg and WT rats (Ortiz et al., 2012). AMPH, when administered at a proper dose, induces a condition of low mobility and in-place stereotypic head movements in rats. This condition, known as focused stereotypy, can be quantitatively measured and analyzed using the force-plate actometer. Figure 14.5 shows

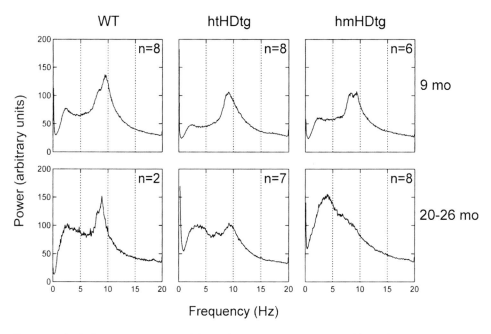

Figure 14.5 HDtg rats express altered focused stereotypy compared to WT rats. Shown are the group average power spectra for WT (−/−), heterozygous HDtg (+/−), and homozygous HDtg (+/+) rats. The force variation was expressed as a percent of body weight. The behavioral data shown were collected 30 minutes after the 4th time the rats were injected with 2.5 mg/kg d-AMPH sulfate for a period of one hour. Two-way ANOVA revealed a significant age effect (F [1,33] = 21.696, $p<0.001$), genotype effect (F [2, 33] = 3.581, $p<0.05$), and age-by-genotype interaction (F [2, 33] = 3.312, $p<0.05$). Values of n per group are indicated in the figure. Reproduced with permission from Ortiz et al., 2012. Copyright Elsevier.

the effect of AMPH injection in HDtg and WT rats, young (nine months) and aged (20–26 months). These data, which are power spectra of the z-axis forces, illustrate the progression of the phenotype in that the young rats express a greater degree of stereotypy at 10 Hz than the older rats. As homozygous HDtg rats (two copies of the mutant *htt* gene) age, the power spectrum shows a frequency shift from 10 Hz (predominately in-place rhythmic head movements) to about 4 Hz (mostly locomotion). Moreover, it is apparent that there is a gene-dosing effect since this shift to lower frequency occurs only partially in hemizygous HDtg rats (one copy of the mutant *htt* gene).

Strong evidence indicates that enhanced extracellular levels of DA in the dorsal striatum are involved in the induction of focused stereotypy (Creese and Iversen, 1974; Kelly et al., 1975; Kelly, 1977; Kuczenski and Segal, 1989); therefore, one potential interpretation of these behavioral findings is that not enough DA is present in reserve for HDtg rats to fully express focused stereotypy at 10 Hz. This interpretation would agree with our previous findings of decreased reserve pool vesicle mobilization in R6/2 mice (Ortiz et al., 2010). DA release, measured using FSCV, was impaired in brain slices harvested from homozygous HDtg rats (Ortiz et al., 2012), especially when a multiple pulse stimulation regimen was applied (Figure 14.6). However, it must be noted that DA content measured in the striatum of homozygous HDtg rats did not differ from that measured in WT rats. This finding runs against the idea of a diminished reserve pool in HDtg rats because the reserve pool makes up most of the DA present. Clearly, a more in-depth analysis of the DA reserve pool in these rats, including experiments similar to those carried out on R6/2 mice, should be conducted to confirm or refute this idea.

We also used the force-sensing runway to analyze alterations in gait of HDtg rats (Ortiz et al., 2012; Figure 14.7). Inspection of raw force data traces reveal a progressively worsening gait dysrhythmia in homozygous HDtg rats compared to their WT controls. Moreover, this dysrhythmia, as measured by force amplitude at the mid-run force point, was found in homozygous, but not hemizygous HDtg rats. Therefore, we conclude that two copies of the mutant HD gene are required for the expression of gait disturbances in this HD model. Evidence from studies carried out on HD model mice indicates that dysfunction of nigrostriatal DA release is an underlying cause of motor control impairment in HD model mice. These results, from the runway, suggest that this is also the case in homozygous HDtg rats.

In summary, we draw several conclusions from the studies performed on HDtg rats. First, these rats produce a behavioral phenotype, which is measurable with the force-plate actometer, with or without pharmacological intervention.

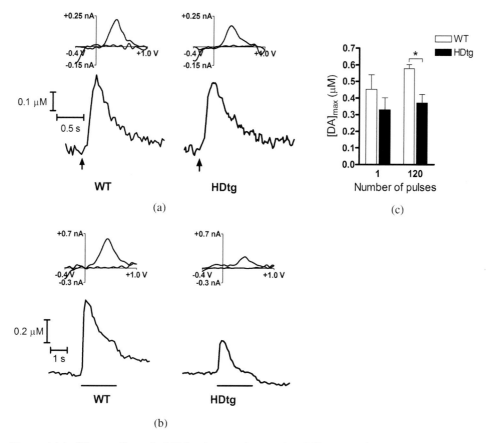

Figure 14.6 Electrically-evoked DA release is decreased in HDtg rats. A single stimulus pulse (a) and a series of 120 stimulus pulses (b) were used to evoke DA release in striatal brain slices from HDtg and WT rats. CVs, located above each plot, indicate the presence of DA. (c) When values obtained from multiple animals were averaged, DA release was found to be diminished in slices from HDtg rats compared to WT controls. A significant effect of genotype on DA release (F [1, 12] = 5.40, $p<0.05$, $n = 4$ WT and 5 HDtg rats) was indicated by two-way ANOVA. Reproduced with permission from Ortiz et al., 2012. Copyright Elsevier.

Moreover, the behavioral phenotype is strongly influenced by the number of mutant gene copies: homozygous HDtg rats express a more robust phenotype than hemizygous rats. Finally, like the other HD models tested, homozygous HDtg rats exhibit a decrease in DA release compared to their respective, age-matched WT controls. However, this decrease in release does not appear to be as dramatic as that observed in R6/2 mice or 3-NP-treated rats. Thus, HDtg rats appear to have a less pronounced phenotype than the other HD models we analyzed.

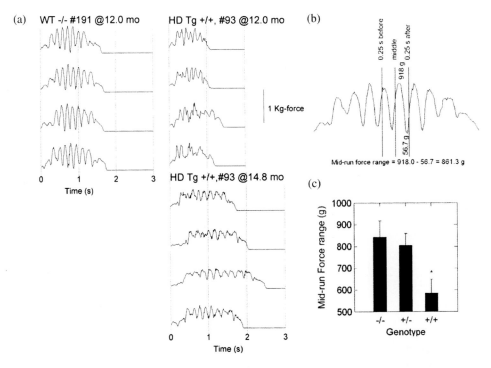

Figure 14.7 Progressive gait disturbances were measured in Homozygous HDtg rats. (a) Raw data traces show the forces produced while rats are ambulating along the force-sensing runway. Gait dysrhythmia, occurring progressively with age, was indicated by decreased force production by a homozygous HDtg (+/+) rat at 12 and 14.8 months of age. Also included for comparison are traces from WT (−/−) rats. (b) The "mid-run force" range, a parameter that was used to quantify the amplitude of force caused by locomotion, was calculated by subtracting the minimum force produced from the maximum force, produced within a half second range in the middle of the run. An example is shown. (c) The mid-run force amplitude was significantly decreased in homozygous HDtg rats (+/+), but not heterozygous (+/−) HDtg rats (*$p<0.05$, t-test; $n = 5$, WT; 6, heterozygous HDtg; 5, homozygous HDtg). Graph is plotted as mean ± SEM. Reproduced with permission from Ortiz et al., 2012. Copyright Elsevier.

14.6 CONCLUSION

Collectively, our work here has established that multiple mouse and rat models of HD have a diminished capacity to release DA, and, in some cases, take up DA. Moreover, as shown by behavioral experiments, performed by our group and others, a close association between this decreased capacity to release DA and behavioral impairments exists. It is interesting to note that the release of other biogenic molecules, including glutamate (Li, 2003), acetylcholine (Vetter et al., 2003), and insulin (Bjorkqvist, 2005) is also diminished in HD model mice. These results, along with our measurements of impaired catecholamine release from adrenal chromaffin cells, suggest that neurotransmitter release as a general phenomenon is diminished in animals that model HD, and may also occur in HD patients.

ACKNOWLEDGMENTS

The University of Kansas and The R. N. Adams Institute for Bioanalytical Chemistry, Hereditary Disease Foundation, Huntington's Disease Society of America, and The Lifespan Institute and NICHD Grant P30 HD002528.

REFERENCES

Alston TA, Mela L, Bright HJ (1977) 3-Nitropropionate, the toxic substance of Indigofera, is a suicide inactivator of succinate dehydrogenase. Proc Natl Acad Sci USA 74:3767–3771.

Andrew SE, Goldberg YP, Kremer B, Telenius H, Theilmann J, Adam S, Starr E, Squitieri F, Lin B, Kalchman MA (1993) The relationship between trinucleotide (CAG) repeat length and clinical features of Huntington's disease. Nat Genet 4:398–403.

Barbeau A, Duvoisin RC, Gerstenbrand F, Lakke JP, Marsden CD, Stern G (1981) Classification of extrapyramidal disorders. Proposal for an international classification and glossary of terms. J Neurol Sci 51:311–327.

Baur JE, Kristensen EW, May LJ, Wiedemann DJ, Wightman RM (1988) Fast-scan voltammetry of biogenic amines. Anal Chem 60:1268–1272.

Bjorkqvist M (2005) The R6/2 transgenic mouse model of Huntington's disease develops diabetes due to deficient beta-cell mass and exocytosis. Hum Mol Genet 14:565–574.

Borland LM, Shi G, Yang H, Michael AC (2005) Voltammetric study of extracellular dopamine near microdialysis probes acutely implanted in the striatum of the anesthetized rat. J Neurosci Methods 146:149–158.

Brouillet E, Jacquard C, Bizat N, Blum D (2005) 3-Nitropropionic acid: a mitochondrial toxin to uncover physiopathological mechanisms underlying striatal degeneration in Huntington's disease. J Neurochem 95:1521–1540.

Callahan JW, Abercrombie ED (2011) *In vivo* dopamine efflux is decreased in striatum of both fragment (R6/2) and full-length (YAC128) transgenic mouse models of Huntington's disease. Front Syst Neurosci 5:61.

Carter RJ, Lione LA, Humby T, Mangiarini L, Mahal A, Bates GP, Dunnett SB, Morton AJ (1999) Characterization of progressive motor deficits in mice transgenic for the human Huntington's disease mutation. J Neurosci 19:3248–3257.

Cattaneo E, Zuccato C, Tartari M (2005) Normal huntingtin function: an alternative approach to Huntington's disease. Nat Rev Neurosci 6:919–930.

Charvin D, Vanhoutte P, Pagès C, Borelli E, Caboche J (2005) Unraveling a role for dopamine in Huntington's disease: the dual role of reactive oxygen species and D2 receptor stimulation. Proc Natl Acad Sci USA 102:12218–12223.

Coles CJ, Edmondson DE, Singer TP (1979) Inactivation of succinate dehydrogenase by 3-nitropropionate. J Biol Chem 254:5161–5167.

Creese I, Iversen SD (1974) The role of forebrain dopamine systems in amphetamine induced stereotyped behavior in the rat. Psychopharmacologia 39:345–357.

Davies SW, Turmaine M, Cozens BA, DiFiglia M, Sharp AH, Ross CA, Scherzinger E, Wanker EE, Mangiarini L, Bates GP (1997) Formation of neuronal intranuclear inclusions underlies the neurological dysfunction in mice transgenic for the HD mutation. Cell 90:537–548.

DiFiglia M, Sapp E, Chase K, Schwarz C, Meloni A, Young C, Martin E, Vonsattel JP, Carraway R, Reeves SA (1995) Huntingtin is a cytoplasmic protein associated with vesicles in human and rat brain neurons. Neuron 14:1075–1081.

Duyao M, Ambrose C, Myers R, Novelletto A, Persichetti F, Frontali M, Folstein S, Ross C, Franz M, Abbott M (1993) Trinucleotide repeat length instability and age of onset in Huntington's disease. Nat Genet 4:387–392.

Fernandez-Gomez FJ, Galindo MF, Gómez-Lázaro M, Yuste VJ, Comella JX, Aguirre N, Jordán J (2005) Malonate induces cell death *via* mitochondrial potential collapse and delayed swelling through an ROS-dependent pathway. Br J Pharmacol 144:528–537.

Hayden MR (1981) Huntington's chorea. Berlin: Springer-Verlag.

He FS (1987) Extrapyramidal lesions caused by mildewed cane poisoning (with a report of 3 cases). Zhonghua Yi Xue Za Zhi 67:395–396.

Hickey MA, Reynolds GP, Morton AJ (2002) The role of dopamine in motor symptoms in the R6/2 transgenic mouse model of Huntington's disease. J Neurochem 81:46–59.

Hodgson JG, Agopyan N, Gutekunst CA, Leavitt BR, LePiane F, Singaraja R, Smith DJ, Bissada N, McCutcheon K, Nasir J (1999) A YAC mouse model for Huntington's disease with full-length mutant huntingtin, cytoplasmic toxicity, and selective striatal neurodegeneration. Neuron 23:181–192.

Howell JO, Kuhr WG, Ensman RE, Mark Wightman R (1986) Background subtraction for rapid scan voltammetry. J Electroanal Chem Interfacial Electrochem 209:77–90.

Hu W, Liang XT, Cheng XM, Wang YP, Liu XJ, Luo XY, Li YW (1986) The isolation and structure identification of a toxic substance, 3-nitropropionic acid, produced by Arthrinium from mildewed sugar cane. Chin J Prev Med 20:321–323.

Huntington G (2004) On chorea. Landmarks Med Genet Class Pap Comment 51:4.

Ikeda H, Yamaguchi M, Sugai S, Aze Y, Narumiya S, Kakizuka A (1996) Expanded polyglutamine in the Machado–Joseph disease protein induces cell death *in vitro* and *in vivo*. Nat Genet 13:196–202.

Johnson MA, Rajan V, Miller CE, Wightman RM (2006) Dopamine release is severely compromised in the R6/2 mouse model of Huntington's disease: attenuated dopamine release in R6/2 mice. J Neurochem 97:737–746.

Johnson MA, Villanueva M, Haynes CL, Seipel AT, Buhler LA, Wightman RM (2007) Catecholamine exocytosis is diminished in R6/2 Huntington's disease model mice. J Neurochem 103:2102–2110.

Kandel ER, Schwartz JH, Jessell TM (2000) Principles of neural science. New York: McGraw-Hill.

Kelly PH (1977) Drug-induced motor behavior. Handb Psychopharmacol 8:295–331.

Kelly PH, Seviour PW, Iversen SD (1975) Amphetamine and apomorphine responses in the rat following 6-OHDA lesions of the nucleus accumbens septi and corpus striatum. Brain Res 94:507–522.

Khan AS, Michael AC (2003) Invasive consequences of using micro-electrodes and microdialysis probes in the brain. TrAC Trends Anal Chem 22:503–508.

Kowall NW, Ferrante RJ, Martin JB (1987) Patterns of cell loss in Huntington's disease. Trends Neurosci 10:24–29.

Kraft JC, Osterhaus GL, Ortiz AN, Garris PA, Johnson MA (2009) *In vivo* dopamine release and uptake impairments in rats treated with 3-nitropropionic acid. Neuroscience 161:940–949.

Kremer B (2002) Oxford monographs on medical genetics. In: Huntington's disease, pp 28–61. New York: Oxford UP.

Kuczenski R, Segal D (1989) Concomitant characterization of behavioral and striatal neurotransmitter response to amphetamine using *in vivo* microdialysis. J Neurosci 9:2051–2065.

Lanska DJ, Lanska MJ, Lavine L, Schoenberg BS (1988a) Conditions associated with Huntington's disease at death: a case-control study. Arch Neurol 45:878.

Lanska DJ, Lavine L, Lanska MJ, Schoenberg BS (1988b) Huntington's disease mortality in the United States. Neurology 38:769.

Levine MS, Cepeda C, Hickey MA, Fleming SM, Chesselet MF (2004) Genetic mouse models of Huntington's and Parkinson's diseases: illuminating but imperfect. Trends Neurosci 27:691–697.

Li H (2003) Abnormal association of mutant huntingtin with synaptic vesicles inhibits glutamate release. Hum Mol Genet 12:2021–2030.

Li JY, Popovic N, Brundin P (2005) The use of the R6 transgenic mouse models of Huntington's disease in attempts to develop novel therapeutic strategies. NeuroRx 2:447–464.

Lüesse H-G, Schiefer J, Spruenken A, Puls C, Block F, Kosinski CM (2001) Evaluation of R6/2 HD transgenic mice for therapeutic studies in Huntington's disease: behavioral testing and impact of diabetes mellitus. Behav Brain Res 126:185–195.

MacDonald ME, Ambrose CM, Duyao MP, Myers RH, Lin C, Srinidhi L, Barnes G, Taylor SA, James M, Groot N (1993) A novel gene containing a trinucleotide repeat that is expanded and unstable on Huntington's disease chromosomes. Cell 72:971–983.

MacMillan JC, Snell RG, Tyler A, Houlihan GD, Fenton I, Cheadle JP, Lazarou LP, Shaw JD, Harper PS (1993) Molecular analysis and clinical correlations of the Huntington's disease mutation. Lancet 342:954–958.

Mangiarini L, Sathasivam K, Seller M, Cozens B, Harper A, Hetherington C, Lawton M, Trottier Y, Lehrach H, Davies SW (1996) Exon 1 of the HD gene with an expanded CAG repeat is sufficient to cause a progressive neurological phenotype in transgenic mice. Cell 87:493–506.

Martin JB, Gusella JF (1986) Huntington's disease. N Engl J Med 315:1267–1276.

Nandi P, Lunte SM (2009) Recent trends in microdialysis sampling integrated with conventional and microanalytical systems for monitoring biological events: a review. Anal Chim Acta 651:1–14.

Ortiz AN, Kurth BJ, Osterhaus GL, Johnson MA (2010) Dysregulation of intracellular dopamine stores revealed in the R6/2 mouse striatum. J Neurochem 112:755–761.

Ortiz AN, Kurth BJ, Osterhaus GL, Johnson MA (2011) Impaired dopamine release and uptake in R6/1 Huntington's disease model mice. Neurosci Lett 492:11–14.

Ortiz AN, Osterhaus GL, Lauderdale K, Mahoney L, Fowler SC, von Hörsten S, Riess O, Johnson MA (2012) Motor function and dopamine release measurements in transgenic Huntington's disease model rats. Brain Res 1450:148–156.

Pancrazio JJ, Johnson PA, Lynch C (1994) A major role for calcium-dependent potassium current in action potential repolarization in adrenal chromaffin cells. Brain Res 668:246–251.

Penney JB, Young AB, Shoulson I, Starosta-Rubenstein S, Snodgrass SR, Sanchez-Ramos J, Ramos-Arroyo M, Gomez F, Penchaszadeh G, Alvir J (1990) Huntington's disease in Venezuela: 7 years of follow-up on symptomatic and asymptomatic individuals. Mov Disord 5:93–99.

Petersén A, Puschban Z, Lotharius J, NicNiocaill B, Wiekop P, O'Connor WT, Brundin P (2002) Evidence for dysfunction of the nigrostriatal pathway in the R6/1 line of transgenic Huntington's disease mice. Neurobiol Dis 11:134–146.

Politis M, Pavese N, Tai YF, Tabrizi SJ, Barker RA, Piccini P (2008) Hypothalamic involvement in Huntington's disease: an *in vivo* PET study. Brain 131:2860–2869.

Pringsheim T, Wiltshire K, Day L, Dykeman J, Steeves T, Jette N (2012) The incidence and prevalence of Huntington's disease: a systematic review and meta-analysis. Mov Disord 27:1083–1091.

Robinson DL, Venton BJ, Heien ML, Wightman RM (2003) Detecting subsecond dopamine release with fast-scan cyclic voltammetry *in vivo*. Clin Chem 49:1763–1773.

Scherzinger E, Lurz R, Turmaine M, Mangiarini L, Hollenbach B, Hasenbank R, Bates GP, Davies SW, Lehrach H, Wanker EE (1997) Huntingtin-encoded polyglutamine expansions form amyloid-like protein aggregates *in vitro* and *in vivo*. Cell 90: 549–558.

Schuldiner S (1994) A molecular glimpse of vesicular monoamine transporters. J Neurochem 62:2067–2078.

Slow EJ (2003) Selective striatal neuronal loss in a YAC128 mouse model of Huntington disease. Hum Mol Genet 12:1555–1567.

Slow EJ, Graham RK, Osmand AP, Devon RS, Lu G, Deng Y, Pearson J, Vaid K, Bissada N, Wetzel R (2005) Absence of behavioral abnormalities and neurodegeneration *in vivo* despite widespread neuronal huntingtin inclusions. Proc Natl Acad Sci USA 102:11402–11407.

Snell RG, MacMillan JC, Cheadle JP, Fenton I, Lazarou LP, Davies P, MacDonald ME, Gusella JF, Harper PS, Shaw DJ (1993) Relationship between trinucleotide repeat expansion and phenotypic variation in Huntington's disease. Nat Genet 4:393–397.

Van Raamsdonk JM (2005a) Selective degeneration and nuclear localization of mutant huntingtin in the YAC128 mouse model of Huntington's disease. Hum Mol Genet 14:3823–3835.

Van Raamsdonk JM (2005b) Cognitive dysfunction precedes neuropathology and motor abnormalities in the YAC128 mouse model of Huntington's disease. J Neurosci 25:4169–4180.

Venton BJ (2006) Cocaine increases dopamine release by mobilization of a synapsin-dependent reserve pool. J Neurosci 26:3206–3209.

Vetter JM, Jehle T, Heinemeyer J, Franz P, Behrens PF, Jackisch R, Landwehrmeyer GB, Feuerstein TJ (2003) Mice transgenic for exon 1 of Huntington's disease: properties of cholinergic and dopaminergic pre-synaptic function in the striatum: ACh and DA in mice transgenic for HD. J Neurochem 85:1054–1063.

Westerink BH (1995) Brain microdialysis and its application for the study of animal behavior. Behav Brain Res 70:103–124.

Zuccato C, Valenza M, Cattaneo E (2010) Molecular mechanisms and potential therapeutical targets in Huntington's disease. Physiol Rev 90:905–981.

CHAPTER 15

CHARACTERIZING NEUROPEPTIDE RELEASE: FROM ISOLATED CELLS TO INTACT ANIMALS

Agatha E. Maki and Jonathan V. Sweedler
University of Illinois at Urbana-Champaign

15.1 INTRODUCTION

Neuropeptides, cell-to-cell signaling molecules that act as neurotransmitters, neuromodulators or hormones, impact a large variety of neuronal processes. The term neuropeptides refers to peptides that are made in neurons and synthesized from larger precursor proteins called prohormones (Strand, 1999). These prohormones are stored in vesicles along with processing enzymes that cleave the prohormone into multiple smaller peptides, some of which eventually become bioactive (Seidah and Chretien, 1997; Steiner, 1998; Hökfelt et al., 2000). Peptides also undergo post-translational modifications such as C-terminal amidation, acetylation, glycosylation, phosphorylation, and sulfation (Hökfelt et al., 2000). The process of neuropeptide synthesis includes: (1) expression of a pre-prohormone gene, where the pre-prohormone is synthesized in the endoplasmic reticulum, (2) signal peptide cleavage (often occurring before the end of expression) at which point the intact pre-prohormone is called a prohormone, (3) processing of the prohormone in the Golgi apparatus, and (4) packaging into dense-core vesicles where further processing occurs. This process results in the production of a single biologically active peptide or a variety of peptides that are cell- or context-specific.

After cleavage, the type of peptides generated varies greatly for different prohormones (Hökfelt et al., 2000). Some prohormones yield a series of related peptides, such as proenkephalin which is processed into multiple peptides with similar biological effects (Reiner, 1987). In contrast, the pro-opiomelanocortin prohormone gives rise to a number of biologically active peptide products, which include adrenocorticotropin, α-melanocyte-stimulating hormone (MSH), β-lipotropin, β-MSH, β-endorphin, and γ-MSH, in addition to some peptides that appear to be biologically inert (de Wied, 1999; Strand, 1999; Hökfelt et al., 2000). These are examples of "classical" neuropeptides that can be identified from a known prohormone. The synthesis of classical neuropeptides takes place within the secretory pathway. Unlike classical transmitters, they are released upon simulation from a range of locations including the soma, terminal or even neuronal processes (Bruns and Jahn, 1995; Fricker and Sweedler, 2010). Once released, biologically active peptides participate in complex signaling pathways involving a variety of cellular receptors, mostly G-protein-coupled receptors. Another category of peptides, "non-classical" neuropeptides, are processed from non-secretory proteins and not prohormones (Fricker, 2010; Gelman and Fricker, 2010). They are more challenging to identify because they are not derived from known precursor molecules. These peptides may be constitutively released and include fragments from nuclear, lysosomal, mitochondrial, and membrane proteins (Fricker, 2010). Due to the complexity of neuropeptide synthesis and processing, sensitive and versatile methods of analysis are needed to characterize neuropeptides from tissues.

Although many peptides can be detected in a tissue homogenate, these samples will include processing intermediates and even degradation products from peptides after release. Given that many peptides interact with their cognate receptor based on only a partial sequence, processing intermediates can have substantial bioactivity, even if they are never released. Thus, while a peptidomics experiment can generate long lists of peptides from a brain region (Svensson et al., 2007; Clynen et al., 2010; Lee et al., 2010), and some may have activity against the appropriate receptor, they still may not be biologically active peptides because they are not released. Therefore, the field of peptidomics is greatly interested in techniques that are capable of characterizing peptides released from specific neuronal tissues, whether a brain region or specific cell.

A variety of methodologies have emerged to aid in analyzing neuropeptide release: sampling approaches such as microdialysis, separation approaches such as capillary electrophoresis (CE) and liquid chromatography (LC), and detection approaches such as laser induced fluorescence (LIF) and mass spectrometry (MS), including electrospray ionization (ESI) and matrix-assisted laser desorption/ionization (MALDI). In this chapter, we discuss the evolution of these analytical approaches and highlight studies that have used these methods to characterize neuropeptide release from samples ranging from cultured individual cells to intact animals.

15.2 HISTORY OF CHARACTERIZING NEUROPEPTIDE RELEASE

Since their discovery and characterization in the late 1960s, neuropeptides have been analyzed using a variety of techniques. Radioimmunoassay and immunohistochemistry (Yalow and Berson, 1960) are among some of the earliest methods of profiling neuropeptides. In particular, immunohistochemistry allows for the localization of a peptide or set of peptides through their reactivity to specific antibodies, which can be visualized by fluorescence microscopy (Falck et al., 1962). This is a powerful technique for determining the anatomical localization and co-localization of peptides. One of the first examples of peptide expression in the brain was the discovery of dynorphin peptide in granule cells and their mossy fibers in the hippocampus (Watson et al., 1982; McGinty et al., 1983). Another early example reported endogenous expression of corticotropin-releasing hormone in olivocerebellar fibers (Palkovits et al., 1987; Sakanaka et al., 1987). Although immunohistochemistry is a sensitive method for detecting peptide expression, it has drawbacks; it requires analyte pre-selection and is limited in its utility for determining dynamic changes.

Over the past two decades, chemically information-rich technologies — CE–ESI–MS, LC–ESI–MS, direct (with no pre-separation) MALDI–MS — have enabled a plethora of novel peptide discovery efforts by allowing scientists to characterize literally hundreds of new neuropeptides in a range of models (Li et al., 2000; Hummon et al., 2006; Li and Sweedler, 2008). In CE, an electric field in a capillary is used to separate peptides based on size (Kennedy et al., 1989; Hogan and Yeung, 1992). Work from the labs of Jorgenson, Ewing, and Yeung pioneered the use of CE-based methods to analyze single cells (Wallingford and Ewing, 1988; Kennedy et al., 1989; Hogan and Yeung, 1992). CE is well-suited for single cell analysis due to its optimization for smaller sample volumes (Lapainis and Sweedler, 2008; Rubakhin et al., 2011, 2013), and when attached to MS, CE has been used for characterizing the metabolome of a range of cell types (Lapainis et al., 2009; Nemes et al., 2012). LC is another commonly used separation approach for peptide characterization when coupled to ESI–MS (Svensson et al., 2007; Clynen et al., 2010; Lee et al., 2010). ESI is a soft ionization technique that is well-suited for investigating neuropeptides and other larger, easily fragmented peptides (Shibdas and Shyamalava, 2012). It allows for the separation of peptides that are structurally quite similar (Smith and Hanley, 1997). It is a high-resolution separation method that can rapidly detect neuropeptide expression and release (Strand, 1999). While analytical scale LC requires large sample volumes, many neurochemical studies use smaller scale LC systems as these are compatible with the smaller volumes inherent to most neurochemical studies (Haskins et al., 2001).

MALDI–MS, a technique first reported in the late 1980s (Karas and Hillenkamp, 1988; Tanaka et al., 1988), is one of the more successful methods for studying peptides. Samples are co-crystallized with an organic matrix which is then excited by a pulsed laser, leading to desorption and ionization of the peptides. The ionized peptides are then analyzed using the mass analyzer, typically time-of-flight (TOF) (Guerrera and Kleiner, 2005). Using direct MALDI with a variety of sampling approaches (Rubakhin and Sweedler, 2007, 2008; Rubakhin et al., 2007; Neupert et al., 2012), a number of peptides have been characterized in a range of models such as the common neuronal model *Aplysia californica* (Li et al., 2001; Sweedler et al., 2002; Jing et al., 2007, 2010; Romanova et al., 2012). Lastly, direct MALDI–MS has evolved as an imaging approach that is capable of providing spatial information on peptides from tissues such as brain slices (Monroe et al., 2008; Chen et al., 2010; Lanni et al., 2012; Sturm et al., 2013).

Separation and direct MS-based characterization approaches have greatly evolved to being quite versatile and sensitive. As highlighted below, they have enabled a wide variety of discovery efforts for *in vitro*, *ex vivo*, and *in vivo* samples.

15.3 CURRENT METHODS OF ANALYZING NEUROPEPTIDE RELEASE

15.3.1 CELLULAR METHODS

Measuring neuropeptide release is more difficult than measuring the contents of a cell or brain region; releasate contains less material and there are sampling constraints during collection (Li and Sweedler, 2008). Progress in peptidomics research has led to the development of a variety of analytical methods integrated with CE, LC, and MALDI–MS for examining neuropeptide release from single cells. Several methods utilizing CE and solid-phase extraction (SPE) coupled with MALDI–TOF–MS have been used to successfully detect neuropeptide secretion from a single neuron (Chen and Lillard, 2001; Rubakhin et al., 2001). The detection of activity-dependent release from individual neurons is a desirable method of characterizing neuropeptides. Another innovative approach has been the use of micrometer-sized SPE beads that can be placed in a specific location of a cell or structure and then analyzed using offline MALDI–MS (Hatcher et al., 2005). Figure 15.1 shows an application of this technique for the collection of neuropeptide release from a single *Aplysia californica* bag cell neuron. Further advances in cellular sampling have led to the development of methods to analyze neuropeptide release in various brain subcellular fractions. For example, Fuller and Arriaga, 2003, developed a method to separate organelles and nuclei to analyze the release

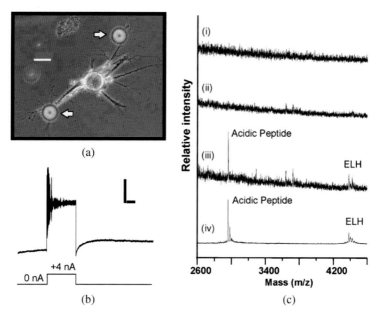

Figure 15.1 Single-bead SPE collection coupled with MALDI–MS to analyze neuropeptide release from a single *Aplysia californica* bag cell neuron. (a) Localization of SPE beads (arrows) on neuronal processes of a cultured bag cell neuron. (b) Action potentials in cultured bag cell neurons obtained with 4 nA of depolarizing current. (c) Mass spectra from single SPE bead releasates (i) before, (ii) during, and (iii) after stimulation. A characteristic mass spectrum of acidic peptide confirms the identity of a (iv) cultured bag cell neuron. Reprinted with permission from Hatcher et al., 2005. Copyright 2005 American Chemical Society.

of their contents using CE. We developed a novel sampling technique coupled to LC–MALDI–MS to analyze neuropeptide release from KCl-stimulated synaptoneurosomal preparations (Figure 15.2a) (Annangudi et al., 2010). More recently, advances in the field of microfluidics has enabled control of the neuronal environment to measure peptide release from single neurons through coupling with MS instrumentation (Croushore and Sweedler, 2013; Guo et al., 2013). Microfluidic devices offer highly regulated conditions, and allow high throughput analyses that are difficult to implement with other methods.

Overall, the aforementioned approaches for analyzing neuropeptide release from neuronal cells and preparations can enable direct characterization of peptides released from a neuron or set of neurons. However, one of the major disadvantages of neuropeptide characterization from individual cells is that they do not allow us to gain insight into the functioning of neuronal networks *in vivo*. Intact tissue sampling of nervous tissue, however, can provide a more global approach to sampling and characterizing neuropeptide release.

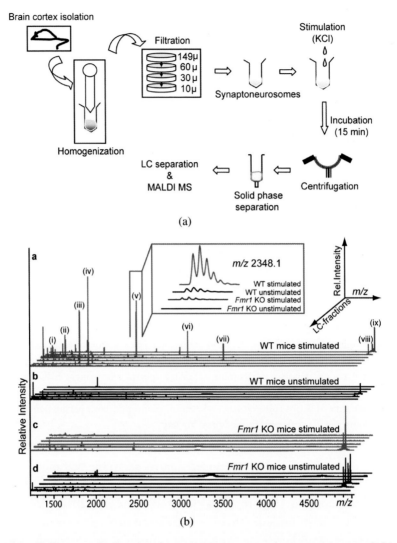

Figure 15.2 Collection of releasates from synaptoneurosomal preparations coupled to LC–MALDI–MS. (a) Schematic work-flow of the preparation of samples for analysis. (b) Mass profiles from releasates of the WT synaptoneurosomes show a number of high intensity peaks compared to the fragile X mental retardation 1(*Fmr*1) knockout (KO) synaptoneurosomes. MS/MS analysis or accurate mass match was used to identify peaks as peptides derived from (i) PEP-19, (ii) cholecystokinin 12, (iii) unknown mass-to-charge ratio (*m/z*) 1776.7, (iv) unknown *m/z* 1923.7, (v) stathmin (vi) orexin B, (vii) unknown *m/z* 3387.0, (viii) thymosin β4, (ix) thymosin β10. Inset shows an expanded mass region from the fraction containing stathmin peptide at *m/z* 2348.1. Adapted and reprinted with permission from Annangudi et al., 2010. Copyright 2010 American Chemical Society.

15.3.2 INTACT TISSUE METHODS

The characterization of neuropeptides from brain tissue provides additional spatial and temporal information that could not otherwise be provided using the aforementioned cellular approaches. In addition, physiological levels of neuropeptide release can be observed. Recently, methods coupled with MS detection have been developed to analyze release from *ex vivo* brain slices. Hatcher et al. (2008) developed an *ex vivo* sampling method integrated with MALDI–TOF–MS to analyze neuropeptide release from the suprachiasmatic nucleus (SCN) (Figure 15.3). A specialized tissue chamber provided the optimal environment for maintaining a living brain slice. C18 ZipTip pipette tips (Millipore) were positioned directly over the SCN and liquid was drawn into the tips, allowing for the collection of extracellular releasate for analysis using MALDI–TOF–MS. This method, which can be used to analyze releasates from localized brain regions, allowed us to demonstrate that the mouse model of Fragile X syndrome (FXS) has reduced neuropeptide release in the cerebral cortex and SCN (Annangudi et al., 2010). Fournier et al. (2003) implemented a similar technique using MALDI–MS to directly collect neuropeptides from the supraoptic nucleus in rat brain slices. In another on-tissue approach, extracellular releasates from *ex vivo* brain slices were collected with a particle-embedded monolithic capillary using a poly (stearyl methacrylate-co-ethylene glycol dimethacrylate) monolith (Fan et al., 2011). Fourier transform MS has also been successfully used in conjunction with an *ex vivo* technique to analyze neuropeptide release directly from neuroendocrine tissue of *Cancer borealis*, or the Jonah crab (Kutz et al., 2004). All in all, analyzing neuropeptide release *ex vivo* from a brain slice bridges single cell and cultured neuron experiments with the *in vivo* analysis of neuropeptides, allowing for real-time sampling in a live animal.

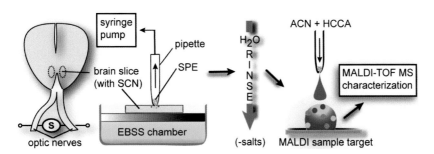

Figure 15.3 Analysis of neuropeptide release from an *ex vivo* brain slice containing the SCN. Experimental set-up includes obtaining a sagittal brain slice, incubating in a brain slice chamber, collecting releasates with a ZipTip pipette tip placed on the SCN, de-salting, and characterizing analytes with MALDI–MS. Reprinted with permission from Hatcher et al., 2008. Copyright 2008 Proc Natl Acad Sci USA.

15.3.3 IN VIVO METHODS

Approaches using *in vivo* sampling techniques coupled to analytical methods are critical in the field of peptidomics. They allow for real-time collection of analytes, while still utilizing well-known CE, LC, and MS approaches for data acquisition. One sampling technique that has been widely used in conjunction with analytical methods is microdialysis, which allows for continuous collection with minimal disturbance of the physiological system (Davies et al., 2000). In microdialysis, a probe containing a dialysis membrane is inserted into the tissue of interest to collect samples, while an isotonic solution is pumped through the probe, allowing for the exchange of analytes with the extracellular fluid. Past innovations using CE for *in vivo* microdialysis coupled with the use of a quadrupole ion trap mass spectrometer have improved the sensitivity of detection of the small analyte concentrations typically acquired from microdialysis (Haskins et al., 2001, 2004; Kennedy et al., 2002). As outlined in other chapters, LC is an effective technique for the online analysis of microdialysis samples. In LC, the temporal resolution of the sample analysis can be manipulated by adjusting the flow rate of separation (Davies et al., 2000), assuming the samples have sufficient analytes to characterize. A drawback of LC can be sample losses, as injection is not continuous. Several groups have reported innovative solutions to overcome this challenge. Steele and Lunte (1995) developed multiple sample loops to allow for the continuous flow of microdialysate, which they used to examine the pharmacokinetics of acetaminophen in rats. Additionally, Leggas et al. (2004) created a method that involved an online injector, used to probe rat cerebrospinal fluid for pharmacokinetic analysis. Recent advances in SPE–LC–MS have resulted in higher analyte concentrations and better throughput, even with smaller sample volumes (Yan et al., 2009; Gode et al., 2012). Furthermore, LC–MS has been documented for the *in vivo* analysis of transmitters such as acetylcholine from rat brain (Keski Rahkonen et al., 2007). Nano ESI–MS has been used for the online analysis of *in vivo* microdialysis samples (Jakubowski et al., 2005), but its widespread application for non-targeted neuropeptide studies has been fairly rare due to the low levels of neuropeptides compared to classical transmitters.

The progress in offline analyses of *in vivo* microdialysis sampling by MS is noteworthy. The offline approaches allow the detection process to be optimized independent of the sampling, but reduce the direct link between sampling and measurement. A thermospray ionization source was one of the first offline MS systems to be used for the analysis of microdialysis samples, specifically, for monitoring the time course of an experimental drug in the nucleus accumbens of rat (Menacherry and Justice, 1990). One of the challenges of using MS in microdialysis is the high salt content in the microdialysate, which can negatively impact the

MS detection process. SPE is a well-utilized method for removing salts from microdialysate for offline analysis, as shown by Prokai et al. (1998). Recently, our group developed a similar approach for collecting and conditioning samples using SPE *via* pre-equilibrated C18 ZipTip pipette tips, which allow for peptide concentration and removal of salts from microdialysate (Figure 15.4a). We have utilized ZipTips for the direct collection of *in vivo* microdialysate samples coupled with MALDI–MS to characterize extracellular peptide release from the ventral hippocampus of rats in response to saline or morphine injection (Figure 15.4b) (Maki et al., 2014). Microdialysis probes are inserted into the hippocampus of

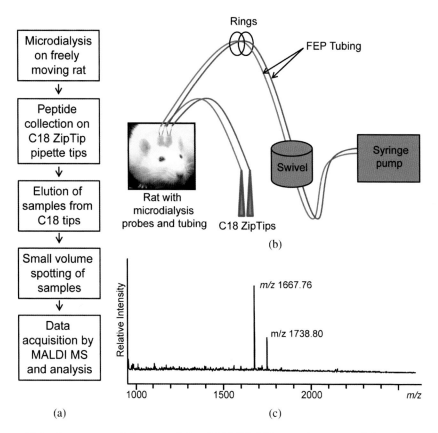

Figure 15.4 *In vivo* microdialysis coupled to MALDI–MS to analyze extracellular releasate from hippocampi of awake and freely-moving rats. (a) Schematic of the work-flow to obtain data from *in vivo* microdialysis. (b) Set-up of fluorinated ethylene propylene (FEP) tubing and C18 ZipTip pipette tips in line with perfusate output for collection of microdialysate from left and right ventral hippocampi. (c) MALDI–MS spectrum showing peaks at *m/z* 1738.80 and *m/z* 1667.76, corresponding to fibrinopeptide A and fibrinopeptide A-derived peptide, respectively, detected at high levels from microdialysis collections.

freely-moving rats, microdialysate is collected directly into ZipTip pipette tips in line with the perfusate output, and then directly spotted onto a MALDI target, after which several released peptides are detected. Although these techniques are a great tool for analyzing samples in a physiological setting, limitations can include difficulty of working with delicate and complex tissues/animals, limited temporal and spatial resolution, and low *in vivo* concentrations of analytes (Perry et al., 2009). Overall, combining the sensitivity of well-known analytical methods with *in vivo* microdialysis sampling provides a powerful tool-set that has great potential for advancing the field of peptidomics.

15.4 CONCLUSIONS

Recent advances in the technologies used to characterize neuropeptides have helped to propel the field of neuropeptide research to new levels with the discovery of greater numbers of peptides from even smaller tissue samples. Single cell analysis offers an innovative means of using *in vitro* sampling to provide localized and neuron-specific peptide information. Over this past decade, the ability to characterize peptides has enabled scientists to identify thousands of brain peptides. However, little information exists on whether these newly discovered peptides function as cell-to-cell signaling molecules.

In addition to the characterization of peptide complements, the development of *ex vivo* methods for analyzing neuropeptide release from slices and cultured cells and tissues has provided important functional information. There has also been a great deal of progress in the development of both online and offline systems for *in vivo* microdialysis methods for characterizing peptide release directly from the brain, offering real-time sampling from live animals. Adding the dimension of measuring peptide release provides a functional context that is critical in determining which of these act as neuropeptides or hormones.

Optimized sampling protocols and newer instrumental enhancements enable peptide release measurements with improved spatial and temporal resolution. These advancements allow analysts to provide a physiological context to the characterization of neuropeptides and perhaps more importantly, enhance our understanding of their functions.

ACKNOWLEDGMENT

Preparation of this manuscript was supported in part by the National Institute on Drug Abuse, Award No. P30DA018310, and the National Institute of Mental Health, Award No. R21 MH100704.

REFERENCES

Annangudi SP, Luszpak AE, Kim SH, Ren S, Hatcher NG, Weiler IJ, Thornley KT, Kile BM, Wightman RM, Greenough WT, Sweedler JV (2010) Neuropeptide release is impaired in a mouse model of Fragile X mental retardation syndrome. ACS Chem Neurosci 1:306–314.

Bruns D, Jahn R (1995) Real-time measurement of transmitter release from single synaptic vesicles. Nature 377:62–65.

Chen R, Jiang X, Conaway MCP, Mohtashemi I, Hui L, Viner R, Li L (2010) Mass spectral analysis of neuropeptide expression and distribution in the nervous system of the lobster *Homarus americanus*. J Proteome Res 9:818–832.

Chen S, Lillard SJ (2001) Continuous cell introduction for the analysis of individual cells by capillary electrophoresis. Anal Chem 73:111–118.

Clynen E, Reumer A, Baggerman G, Mertens I, Schoofs L (2010) Neuropeptide biology in *Drosophila*. Adv Exp Med Biol 692:192–210.

Croushore C, Sweedler J (2013) Microfluidic systems for studying neurotransmitters and neurotransmission. Lab Chip 13:1666–1676.

Davies MI, Cooper JD, Desmond SS, Lunte CE, Lunte SM (2000) Analytical considerations for microdialysis sampling. Adv Drug Del Rev 45:169–188.

de Wied D (1999) Behavioral pharmacology of neuropeptides related to melanocortins and the neurohypophyseal hormones. Eur J Pharmacol 375:1–11.

Falck B, Hillarp NA, Thieme G, Torp A (1962) Fluorescence of catecholamines and related compounds condensed with formaldehyde. J Histochem Cytochem 10:348–354.

Fan Y, Rubakhin S, Sweedler J (2011) Collection of peptides released from single neurons with particle-embedded monolithic capillaries followed by detection with matrix-assisted laser desorption/ionization mass spectrometry. Anal Chem 83:9557–9563.

Fournier I, Day R, Salzet M (2003) Direct analysis of neuropeptides by *in situ* MALDI-TOF mass spectrometry in the rat brain. Neuroendocrinol Lett 24:9–14.

Fricker L (2010) Analysis of mouse brain peptides using mass spectrometry-based peptidomics: Implications for novel functions ranging from non-classical neuropeptides to microproteins. Mol BioSyst 6:1355–1365.

Fricker L, Sweedler J (2010) Fishing for the hidden peptidome in health and disease (drug abuse). AAPS J 12:679–682.

Fuller K, Arriaga E (2003) Analysis of individual acidic organelles by capillary electrophoresis with laser-induced fluorescence detection facilitated by the endocytosis of fluorescently labeled microspheres. Anal Chem 75:2123–2130.

Gelman J, Fricker L (2010) Hemopress in and other bioactive peptides from cytosolic proteins: Are these non-classical neuropeptides? AAPS J 12:279–289.

Gode D, Martin M, Steiner F, Huber C, Volmer D (2012) Rapid narrow band elution for on-line SPE using a novel solvent plug injection technique. Anal Bioanal Chem 404:433–445.

Guerrera I, Kleiner O (2005) Application of mass spectrometry in proteomics. Biosci Rep 25:71–93.

Guo F, French J, Li P, Zhao H, Chan C, Fick J, Benkovic S, Huang T (2013) Probing cell-cell communication with microfluidic devices. Lab Chip 13:3152–3162.

Haskins W, Watson C, Cellar N, Powell D, Kennedy R (2004) Discovery and neurochemical screening of peptides in brain extracellular fluid by chemical analysis of *in vivo* microdialysis samples. Anal Chem 76:5523–5533.

Haskins WE, Wang Z, Watson CJ, Rostand RR, Witowski SR, Powell DH, Kennedy RT (2001) Capillary LC-MS2 at the attomole level for monitoring and discovering endogenous peptides in microdialysis samples collected *in vivo*. Anal Chem 73:5005–5014.

Hatcher NG, Richmond TA, Rubakhin SS, Sweedler JV (2005) Monitoring activity-dependent peptide release from the CNS using single-bead solid-phase extraction and MALDI TOF MS detection. Anal Chem 77:1580–1587.

Hatcher NG, Atkins N, Annangudi SP, Forbes AJ, Kelleher NL, Gillette MU, Sweedler JV (2008) Mass spectrometry-based discovery of circadian peptides. Proc Natl Acad Sci USA 105:12527–12532.

Hogan BL, Yeung ES (1992) Determination of intracellular species at the level of a single erythrocyte *via* capillary electrophoresis with direct and indirect fluorescence detection. Anal Chem 64:2841–2845.

Hökfelt T, Broberger C, Xu ZQ, Sergeyev V, Ubink R, Diez M (2000) Neuropeptides — an overview. Neuropharmacology 39:1337–1356.

Hummon AB, Amare A, Sweedler JV (2006) Discovering new invertebrate neuropeptides using mass spectrometry. Mass Spectrom Rev 25:77–98.

Jakubowski J, Hatcher N, Sweedler J (2005) Online microdialysis-dynamic nanoelectrospray ionization-mass spectrometry for monitoring neuropeptide secretion. J Mass Spectrom 40:924–931.

Jing J, Vilim F, Horn C, Alexeeva V, Hatcher N, Sasaki K, Yashina I, Zhurov Y, Kupfermann I, Sweedler J, Weiss K (2007) From hunger to satiety: Reconfiguration of a feeding network by *Aplysia* neuropeptide Y. J Neurosci 27:3490–3502.

Jing J, Sweedler J, Cropper E, Alexeeva V, Park JH, Romanova E, Xie F, Dembrow N, Ludwar B, Weiss K, Vilim F (2010) Feed forward compensation mediated by the central and peripheral actions of a single neuropeptide discovered using representational difference analysis. J Neurosci 30:16545–16558.

Karas M, Hillenkamp F (1988) Laser desorption ionization of proteins with molecular masses exceeding 10,000 daltons. Anal Chem 60:2299–2301.

Kennedy R, Watson C, Haskins W, Powell D, Strecker R (2002) *In vivo* neurochemical monitoring by microdialysis and capillary separations. Curr Opin Chem Biol 6:659–665.

Kennedy RT, Oates MD, Cooper BR, Nickerson B, Jorgenson JW (1989) Microcolumn separations and the analysis of single cells. Science 246:57–63.

Keski Rahkonen P, Lehtonen M, Ihalainen J, Sarajärvi T, Auriola S (2007) Quantitative determination of acetylcholine in microdialysis samples using liquid chromatography/atmospheric pressure spray ionization mass spectrometry. Rapid Commun Mass Spectrom 21:2933–2943.

Kutz KK, Schmidt JJ, Li L (2004) *In situ* tissue analysis of neuropeptides by MALDI FTMS in-cell accumulation. Anal Chem 76:5630–5640.

Lanni E, Rubakhin S, Sweedler J (2012) Mass spectrometry imaging and profiling of single cells. J Proteomics 75:5036–5051.

Lapainis T, Sweedler J (2008) Contributions of capillary electrophoresis to neuroscience. J Chromatogr 1184:144–158.

Lapainis T, Rubakhin S, Sweedler J (2009) Capillary electrophoresis with electrospray ionization mass spectrometric detection for single-cell metabolomics. Anal Chem 81:5858–5864.

Lee J, Atkins N, Hatcher N, Zamdborg L, Gillette M, Sweedler J, Kelleher N (2010) Endogenous peptide discovery of the rat circadian clock: A focused study of the suprachiasmatic nucleus by ultra high performance tandem mass spectrometry. Mol Cell Proteomics 9:285–297.

Leggas M, Zhuang Y, Welden J, Self Z, Waters C, Stewart C (2004) Microbore HPLC method with online microdialysis for measurement of topotecan lactone and carboxylate in murine CSF. J Pharm Sci 93:2284–2295.

Li L, Sweedler JV (2008) Peptides in the brain: Mass spectrometry-based measurement approaches and challenges. Annu Rev Anal Chem 1:451–483.

Li L, Garden RW, Sweedler JV (2000) Single-cell MALDI: A new tool for direct peptide profiling. Trends Biotechnol 18:151–160.

Li L, Jing J, Floyd PD, Rubakhin SS, Romanova EV, Alexeeva VY, Dembrow NC, Weiss KR, Vilim FS, Sweedler JV (2001) Cerebrin prohormone processing, distribution and action in *Aplysia californica*. J Neurochem 77:1569–1580.

Maki AE, Morris KA, Catherman K, Chen X, Hatcher NG, Gold PE and Sweedler JV (2014) Fibrinogen alpha-chain-derived peptide is upregulated in hippocampus of rats exposed to acute morphine injection and spontaneous alternation testing. Pharmacol Res Perspect 2:e00037.

McGinty JF, Henriksen SJ, Goldstein A, Terenius L, Bloom FE (1983) Dynorphin is contained within hippocampal mossy fibers: Immunochemical alterations after kainic acid administration and colchicine-induced neurotoxicity. Proc Natl Acad Sci USA 80:589–593.

Menacherry SD, Justice JB (1990) *In vivo* microdialysis and thermospray tandem mass spectrometry of the dopamine uptake blocker 1-[2-[bis(4-fluorophenyl)methoxy]ethyl]-4-(3-phenylpropyl)-piper azine (GBR-12909). Anal Chem 62:597–601.

Monroe E, Annangudi S, Hatcher N, Gutstein H, Rubakhin S, Sweedler J (2008) SIMS and MALDI MS imaging of the spinal cord. Proteomics 8:3746–3754.

Nemes P, Knolhoff A, Rubakhin S, Sweedler J (2012) Single-cell metabolomics: Changes in the metabolome of freshly isolated and cultured neurons. ACS Chem Neurosci 3:782–792.

Neupert S, Rubakhin S, Sweedler J (2012) Targeted single-cell microchemical analysis: MS-based peptidomics of individual paraformaldehyde-fixed and immunolabeled neurons. Chem Biol 19:1010–1019.

Palkovits M, Léránth C, Görcs T, Young WS (1987) Corticotropin-releasing factor in the olivocerebellar tract of rats: Demonstration by light- and electron-microscopic immunohistochemistry and *in situ* hybridization histochemistry. Proc Natl Acad Sci USA 84:3911–3915.

Perry M, Li Q, Kennedy R (2009) Review of recent advances in analytical techniques for the determination of neurotransmitters. Anal Chim Acta 653:1–22.

Prokai L, Kim HS, Zharikova A, Roboz J, Ma L, Deng L, Simonsick WJ (1998) Electrospray ionization mass spectrometric and liquid chromatographic-mass spectrometric studies on the metabolism of synthetic dynorphin A peptides in brain tissue *in vitro* and *in vivo*. J Chromatogr 800:59–68.

Reiner A (1987) The distribution of proenkephalin-derived peptides in the central nervous system of turtles. J Comp Neurol 259:65–91.

Romanova E, Sasaki K, Alexeeva V, Vilim F, Jing J, Richmond T, Weiss K, Sweedler J, Kirchmair R (2012) Urotensin II in invertebrates: From structure to function in *Aplysia californica*. PLoS ONE 7:e48764.

Rubakhin S, Sweedler J (2007) Characterizing peptides in individual mammalian cells using mass spectrometry. Nat Protoc 2:1987–1997.

Rubakhin S, Sweedler J (2008) Quantitative measurements of cell-cell signaling peptides with single-cell MALDI MS. Anal Chem 80:7128–7136.

Rubakhin S, Lanni E, Sweedler J (2013) Progress toward single cell metabolomics. Curr Opin Biotechnol 24:95–104.

Rubakhin S, Romanova E, Nemes P, Sweedler J (2011) Profiling metabolites and peptides in single cells. Nat Methods 8:S20–S29.

Rubakhin SS, Page JS, Monroe BR, Sweedler JV (2001) Analysis of cellular release using capillary electrophoresis and matrix assisted laser desorption/ionization-time of flight-mass spectrometry. Electrophoresis 22:3752–3758.

Rubakhin SS, Hatcher NG, Monroe EB, Heien ML, Sweedler JV (2007) Mass spectrometric imaging of the nervous system. Curr Pharm Des 13:3325–3334.

Sakanaka M, Shibasaki T, Lederis K (1987) Corticotropin releasing factor-like immunoreactivity in the rat brain as revealed by a modified cobalt-glucose oxidase-diaminobenzidine method. J Comp Neurol 260:256–298.

Seidah NG, Chretien M (1997) Eukaryotic protein processing: Endoproteolysis of precursor proteins. Curr Opin Biotechnol 8:602–607.

Shibdas B, Shyamalava M (2012) Electrospray ionization mass spectrometry: A technique to access the information beyond the molecular weight of the analyte. Int J Anal Chem Article ID 282574.

Smith D, Hanley A (1997) Purification of synthetic peptides by high performance liquid chromatography. In: Neuropeptide Protocols (Irvine CB, Williams CH,eds), pp 75–87. Totowa, New Jersey: Humana Press.

Steele KM, Lunte CE (1995) Microdialysis sampling coupled to on-line microbore liquid chromatography for pharmacokinetic studies. J Pharm Biomed Anal 13:149–154.

Steiner DF (1998) The proprotein convertases. Curr Opin Chem Biol 2:31–39.

Strand FL (1999) Neuropeptides: Regulators of physiological processes. Cambridge, Massachusetts: The MIT Press.

Sturm R, Greer T, Chen R, Hensen B, Li L (2013) Comparison of NIMS and MALDI platforms for neuropeptide and lipid mass spectrometric imaging in *C. borealis* brain tissue. Anal Methods 5:1623–1628.

Svensson M, Sköld K, Nilsson A, Fälth M, Svenningsson P, Andrén P (2007) Neuropeptidomics: Expanding proteomics downwards. Biochem Soc Trans 35:588–593.

Sweedler JV, Li L, Alexeeva V, Rubakhin SS, Dembrow NC, Dowling O, Jing J, Weiss KR, Vilim FS (2002) Identification and characterization of the feeding circuit-activating peptides, a novel neuropeptide family of *Aplysia*. J Neurosci 22:7797–7808.

Tanaka K, Waki H, Ido Y, Akita S, Yoshida Y, Yoshida T (1988) Protein and polymer analyses up to m/z 100,000 by laser ionization time-of-flight mass spectrometry. Rapid Commun Mass Spectrom 2:151–153.

Wallingford RA, Ewing AG (1988) Capillary zone electrophoresis with electrochemical detection in 12.7 microns diameter columns. Anal Chem 60:1972–1975.

Watson S, Akil H, Fischli W, Goldstein A, Zimmerman E, Nilaver G, van wimersma Griedanus T (1982) Dynorphin and vasopressin: Common localization in magnocellular neurons. Science 216:85–87.

Yalow RS, Berson SA (1960) Immunoassay of endogenous plasma insulin in man. J Clin Invest 39:1157–1175.

Yan W, Li Y, Zhao L, Lin JM (2009) Determination of estrogens and bisphenol A in bovine milk by automated on-line C30 solid-phase extraction coupled with high-performance liquid chromatography mass spectrometry. J Chromatogr 1216:7539–7545.

CHAPTER 16

ADVANCING CHRONIC INTRACORTICAL ELECTRODE RECORDING FUNCTION

Lohitash Karumbaiah[*], Tarun Saxena[†], and Ravi Bellamkonda[†]

[*]The University of Georgia,
[†]Georgia Institute of Technology

16.1 INTRODUCTION

Brain electrode implants have been used since the mid-1900s for electrical stimulation and measurement of neuronal action potentials in different regions of the brain (Heath and Norman, 1946; Clark and Ward, 1948a; Hayne et al., 1949; Delgado, 1952; Bradley and Elkes, 1953). More recently, implantable intracortical electrodes have been used as part of more complex brain–computer interfaces (BCIs) to control external prosthetics by interfacing with local neuronal populations in the brain (Jarosiewicz et al., 2008; Moritz et al., 2008; Velliste et al., 2008; Hochberg et al., 2012). Recent advances have demonstrated the tremendous promise of this technology in controlling assistive prosthetics to restore the independence of patients suffering from long-term paralysis (Hochberg et al., 2012). Although promising, long-term (chronic) applications of this technology could be hindered due to the formation of astroglial scar tissue, which results from the brain tissue's reaction to the chronic presence of intracortical electrode implants (Biran et al., 2007; McConnell et al., 2009b; Winslow et al., 2010; Woolley et al., 2013). Astroglial scar tissue temporally isolates the electrode implant from the surrounding neural tissue, ultimately leading to loss of recording function and resulting in device failure. In addition to scar tissue formation, persistent micromotion of brain tissue and the resulting inflammatory milieu surrounding chronic implants has also been suggested to lead to the degeneration of surrounding neurons (Gilletti and Muthuswamy, 2006b; McConnell et al., 2009b).

This chapter discusses the differences between various types of cortical electrodes used to record neural activity, recent advances in our understanding of the inflammatory response to invasive intracortical electrode implants, and methods of assessment.

16.2 CORTICAL ELECTRODES FOR REAL-TIME MONITORING OF NEURAL ACTIVITY

Cortical electrodes used for recording neural activity from the brain can be classified into non-invasive scalp electrodes, minimally invasive brain surface electrodes, and invasive intracortical electrodes. Hans Berger was the first to use extradural surface electrodes to record neural activity in the brain (Berger, 1929). Electroencephalography (EEG) — a technique pioneered by Berger — showed for the first time that electrical oscillations could indeed be recorded directly from the intact brain, and that changes in EEG signals could be used to explain epileptic seizures and aberrant neural activity associated with lesions to the brain. Hans Berger's discovery of the brain's electrical activity should also be credited for having spawned the field of clinical neurophysiology, and for providing impetus to more advanced studies involving the use of these signals for control and communication using BCIs.

16.2.1 NON-INVASIVE SCALP ELECTRODES FOR MEASUREMENT OF EEG ACTIVITY

EEG activity from the brain is acquired *via* numerous non-invasively positioned electrodes on the scalp (Figure 16.1). BCIs using EEG activity for movement control rely heavily on the ability of the individual to voluntarily induce changes in EEG activity. EEG activity for motor imagery applications is obtained from the sensorimotor cortex after filtering it through the skull and scalp, and complex signal processing to facilitate accurate cursor control (Wolpaw et al., 1991; Pfurtscheller et al., 1994; McFarland et al., 1997a, b). Although the use of motor imagery and steady-state visual evoked potentials (SSVEPs) for volitional control of a prosthesis can be achieved using EEG activity (Muller-Putz and Pfurtscheller, 2008; Kaiser et al., 2013), the localized electrical activity required for more complex movements of the prosthesis can only be achieved *via* minimally invasive subdural brain surface electrodes used to record electrocortcographic (ECoG) activity, and invasive intracortical electrode implants used to record single unit activity (Schwartz, 2004).

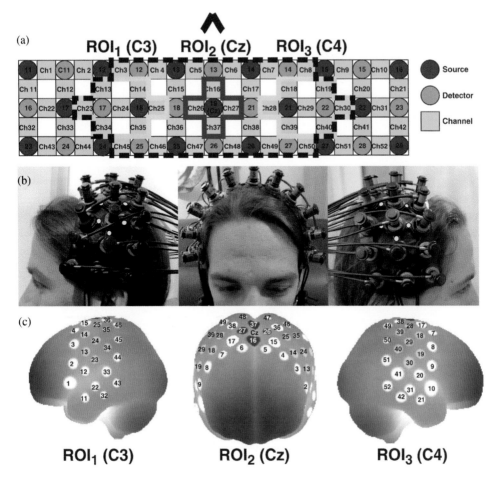

Figure 16.1 (a) Illustration of a 52 channel functional near infrared spectroscopy (fNIRS) array. Dotted line represents channels used for analysis; (b) scalp electrodes positioned on an individual's sensorimotor cortex for measurement of EEG signals; (c) representation of the position of fNIRS arrays on the brain. Reprinted with permission from Kaiser et al., 2013.

16.2.2 MINIMALLY INVASIVE SUBDURAL ELECTRODES FOR MEASUREMENT OF ECoG ACTIVITY

ECoG activity from the motor cortex is measured *via* the minimally invasive placement of brain surface electrodes between the dura and the pia (Figure 16.2). ECoG activity-based BCIs require lesser patient training times, provide higher spatial resolution and lower electromyographic (EMG) artifacts (Freeman et al., 2003; Ball et al., 2009), and provide the high amplitude neural frequencies (>40Hz) responsible for control of speech and motor control in humans when compared to EEG signals (Leuthardt et al., 2004, 2006, 2011). Although a majority of ECoG-based BCIs have mainly focused on the development of speech and communication tools till date, recent studies have demonstrated long-term stability

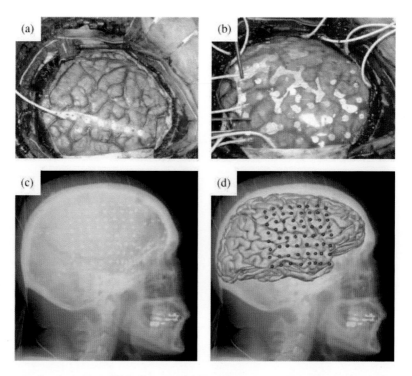

Figure 16.2 Implantation of ECoG arrays. (a) Exposed brain surface after craniotomy and durotomy; (b) placement of subdural ECoG arrays; (c) X-ray image of implanted ECoG arrays; (d) position of ECoG arrays on the brain. Reprinted with permission from Schalk et al., 2008.

and the ability to effectively decode high degree of freedom arm kinematics from ECoG signals obtained from monkeys implanted with subdural ECoG electrodes (Chao et al., 2010). Although promising, neither scalp nor subdural electrodes are currently able to provide the high spatiotemporal signal resolution required for the real-time control of assistive prosthetics such as a robotic arm (Nicolelis, 2001; Schwartz, 2004).

16.2.3 INVASIVE INTRACORTICAL ELECTRODES FOR MEASUREMENT OF SINGLE UNIT ACTIVITY

Invasive intracortical microwire or silicon-based intracortical electrode implants penetrate brain tissue and interface with local populations of neurons (Figure 16.3). These electrodes are either directly bonded to the external connector (headcap), or are connected to the headcap *via* a flexible cable (tether) that is aimed at prolonging electrode function by reducing the foreign body response (FBR) triggered in response to the implant. By virtue of their close proximity to neuronal

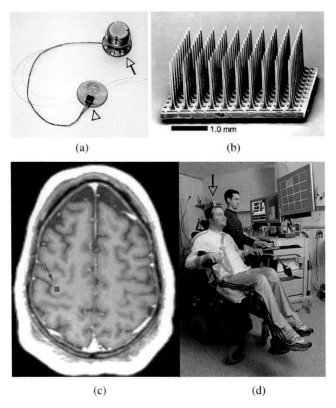

Figure 16.3 (a) Image of a high-density intracortical Utah Array and attached CerePort™ connector (Blackrock Microsystems, UT); (b) scanning electron micrograph (SEM) of a 100 electrode Utah Array; (c) box indicates location of placement of the intracortical Utah electrode array in the motor cortex; (d) patient implanted with an intracortical Utah electrode array, and connected to a BCI is seen controlling a cursor on the monitor. Reprinted with permission from Hochberg et al., 2006.

populations, intracortical electrodes are able to record single unit neural activity with high signal-to-noise ratios, and thereby provide fine control of assistive prosthetics with high degrees of freedom. Intracortical electrode implants have been used with tremendous success in recording motor intent from paralyzed patients (Hochberg et al., 2006; Kim et al., 2008; Chadwick et al., 2011; Simeral et al., 2011; Ajiboye et al., 2012; Hochberg et al., 2012). In studies conducted in both monkeys (Moritz et al., 2008; Velliste et al., 2008) and humans (Hochberg et al., 2012; Collinger et al., 2013) implanted with these electrodes, neural recordings have been shown to have sufficient spatiotemporal resolution to control a robotic arm, and to perform specific tasks. However, in order for these devices to be clinically viable, they are required to perform reliably for at least a decade or longer (Perge et al., 2013). Although recent studies have demonstrated significant device functionality chronically (Hochberg et al., 2012), in-depth studies to address issues related to electrode material and design, as well as the triggering of an FBR that can eventually degrade neural signal quality and lead to device failure need to be undertaken.

16.3 FBR TO INTRACORTICAL ELECTRODES

The immunological response of the body to the presence of a foreign material is termed the FBR, and it dictates biocompatibility of the implanted material. FBRs are aimed to eliminate the foreign material and shield the body from any adverse reactions that may occur as a result of the presence of the material (Anderson et al., 2008). To this end, upon sensing the presence of foreign material, immune cells mediate a concerted cascade of events to eliminate the foreign body. The brain is usually considered to be an immune privileged zone. Nevertheless, intracortical electrode implants, by virtue of their invasive placement procedures, do elicit a strong FBR in the brain (Polikov et al., 2005; Reichert et al., 2008). The FBR of the brain to intracortical electrodes has been implicated in the progressive loss of functionality of these devices over time (Zhong et al., 2001; Reichert et al., 2008; Grill et al., 2009; McConnell et al., 2009b). The presence of intracortical electrodes leads to the activation of resting glial cells (astrocytes and microglia) in a process termed as "reactive gliosis", which leads to the formation of an astroglial scar. The astroglial scar forms a glial sheath that walls off the intracortical electrode from the surrounding neurons, purportedly leading to loss of signal from neurons (Polikov et al., 2005; Schwartz et al., 2006). However, there is a temporal mismatch between these events as the scar stabilizes well before loss of signal fidelity. Recent reports suggest that chronic inflammation, which results due to the intracortical electrodes, leads to neurotoxicity and neurodegeneration that ultimately causes recording failure of intracortical electrodes (Karumbaiah et al., 2012, 2013; Saxena et al., 2013).

16.3.1 BRAIN RESPONSE TO INTRACORTICAL ELECTRODE IMPLANTS

The FBR of the brain can be divided into two temporal phases: (i) the acute phase that lasts from hours to days; and the (ii) chronic response, which lasts from weeks to years. These phases are described below. Excellent reviews on the subject are referenced below and the interested reader is directed to them for an in depth understanding of the FBR of the brain.

16.3.1.1 Acute response

Upon intracortical electrode insertion, acute responses are triggered due to the mechanical trauma of insertion. Insertion trauma leads to the death of neurons physically. Blood vessels and capillaries are disrupted releasing erythrocytes, clotting factors, excitotoxins, neurotoxins, and inflammatory factors. This activates resting glia and begins the process of glial scarring. Upon activation, resting astrocytes become hypertrophic, proliferate, and migrate toward the site of injury in order to limit the extent of neurotoxins and inflammatory species. Activated microglia also undergo a morphological change from amoeboid to ramified, and begin to phagocytose debris around the implant wound (Zhong et al., 2001; Polikov et al., 2005; Reichert et al., 2008; Grill et al., 2009). Insertion trauma also leads to the disruption of the blood–brain barrier (BBB), a physiological barrier that actively monitors and prevents blood-borne moieties from entering the brain parenchyma (Abbott et al., 2006), leading to circulating monocytes and macrophages, and blood serum proteins entering the brain (Biran et al., 2005; Liu et al., 2012; Saxena et al., 2013). There is also a build-up of edema that causes neuronal death. The edema however subsides within 3–7 days (Hatashita and Hoff, 1990). Thus, the acute FBR, mediated primarily by mechanical damage, excitotoxicity, and neurotoxicity, lasts for up to a week, and sets in motion a cascading set of molecular events that lead to the chronic FBR.

16.3.1.2 Chronic response

Over a period of 1–3 weeks, reactive glia form a dense sheath around the intracortical electrode termed the astroglial scar. The astroglial scar is a capsule consisting predominantly of hypertrophied astrocytes and reactive microglia. It is known that astroglial scar can act as a diffusion barrier to cytokines (Roitbak and Sykova, 1999; Saxena et al., 2012), and is perceived to be a neuroprotective mechanism to prevent damage to cells surrounding the injury zone (Fawcett and Asher, 1999; Karumbaiah and Bellamkonda, 2013). Further, the BBB is open around the implant site, leading to chronic inflammation causing neuronal death and degeneration as time progresses (McConnell et al., 2009b; Bellamkonda et al., 2013; Saxena et al., 2013). The dura mater (a meningeal sheath covering the cortex) that is ruptured due to electrode insertion reseals itself as meningeal fibroblasts migrate to the wound site and form a fibrotic scar. There is continual remodeling of the matrix around the implant site as new vasculature begins to form. Electrode micromotion can cause the wound site to remain "fresh" and exacerbate inflammation as well (Gilletti and Muthuswamy, 2006a; Karumbaiah et al., 2012).

16.3.2 CURRENT METHODS OF ASSESSMENT

16.3.2.1 Histology

Conventional methods of assessment of the brain's FBR to intracortical electrodes employ histological techniques to probe the presence of reactive glia, neurons, BBB leakage, infiltrating fibroblasts, and other byproducts in the intracortical electrode–brain neural interface. Traditionally, animals are implanted with intracortical electrodes and the state of the FBR is assessed at various times post-implant (Kim et al., 2004; Biran et al., 2005; Potter et al., 2012; Karumbaiah et al., 2013). Quantification of fluorescence-based histology is often performed by fitting empirical functions to the fluorescence intensity as it varies with distance from the implant site, and estimating statistical differences from the parameters of the function (Gutowski et al., 2013; Karumbaiah et al., 2013). In some cases, where cell counts are necessary, total number of cells in a given area can be estimated. However, histological techniques can only provide a snapshot in time, and are not nearly sensitive enough to assess fine molecular details in the dynamic wound site. Further, histology is an end-point assay and cannot provide any real-time information on the intracortical electrode–tissue neural interface (Schwartz et al., 2006; Karumbaiah et al., 2013). For intracortical electrodes to be a mainstream technology, real-time monitoring methods, amenable to humans, need to be employed. Two techniques, impedance spectroscopy and cyclic voltammetry (CV), are routinely used in animal experimental models and are briefly discussed below.

16.3.2.2 Impedance spectroscopy

Tissue impedance spectroscopy (TIS) is a technique that measures the impedance of tissues as a function of frequency. The impedance of tissues arises from the cellular components and the acellular components of the extracellular space. Hence, change in cellular structures and matrix remodeling can manifest themselves in changing tissue impedance (Williams et al., 2007; Grill et al., 2009; McConnell et al., 2009a). Specific to intracortical electrodes, it is known that glial cells become hypertrophic and proliferate at sites of injury. The astroglial scar has been shown to have increased tortuosity and poses a diffusion barrier to small molecules. These changes can contribute to changes in measured impedance. Williams et al. probed brain tissue impedance at a frequency of 1 kHz (as neurons typically fire at 1 ms), and demonstrated a general trend of increased impedance with astroglial scarring upon intracortical electrode implantation (Williams et al., 2007). Similar results were obtained by McConnell et al. (2009a) who performed brain TIS in a range of 100 Hz–1 MHz. These studies and others

(Frampton et al., 2010; Prasad et al., 2012) have demonstrated a correlation of astroglial scarring with tissue impedance. In general, it has been observed from animal models of intracortical electrode implantation that impedance values drop upon implantation, correlating with edema buildup, and start increasing as time progresses, correlating with astroglial scar formation (Polikov et al., 2005; McConnell et al., 2009a; Prasad et al., 2012). However, insulation damage to the electrode shanks and corrosion of the electrodes are factors that cannot be discounted as contributors to measured impedance (Prasad et al., 2012). Thus, noninvasive, real-time methods using impedance spectroscopy have been shown to detect the degree of gliosis to some extent but are lacking in resolution and specificity (Schwartz et al., 2006).

16.3.2.3 Cyclic voltammetry

Neurotransmitters such as dopamine and serotonin are electroactive and respond to electrical fields (Dale et al., 2005). When a potential is applied, electroactive species undergo oxidation causing electron release, resulting in a measurable current. This current is directly proportional to the concentration of the electroactive species. CV is a technique that identifies the presence of electrochemical reactions and can provide useful information on the quantity of electroactive material on the electrode (Bockris and Reddy, 1973). In a typical CV experiment, potential is ramped and cycled across a set of electrodes in solution while measuring the current. Individual species of electroactive molecules have unique oxidation potentials and as this potential is reached, an increase in current (proportional to concentration of the species) is observed, helping to identify the species (Kawagoe et al., 1993; Strong et al., 2001). Traditionally, CV has been used to study neurotransmitter dynamics and concentrations (Phillips et al., 2003; Heien et al., 2005). However, CV can also help identify corrosion and stability of the intracortical electrode, for example, since injury sites usually produce acidic environments, leading to corrosion of the electrode (Cogan, 2008; Prasad et al., 2012). As biomolecules and ionic species adsorb and penetrate into the intracortical electrode coatings, it is expected that the voltammograms change over time. Indeed, Cogan (2008) showed that in electrodes implanted *in vivo* for a period of six weeks; there was a distinct change in the charge storage capacity of intracortical electrodes as time progressed. This was attributed to both astroglial scar formation and ionic changes occurring at the electrode insulation layer. Glial cells encapsulating the electrode produce a variety of cytokines (Karumbaiah et al., 2013; Saxena et al., 2013) and CV holds the potential to monitor these cytokines in real-time (Fillenz, 2005; Hashemi et al., 2011). Direct real-time measurements of neurochemicals in the brain are extremely valuable when combined with simultaneous recording of neuronal activity. This can be especially useful to link chemical signaling to physiologically observable behaviors and combining CV with intracortical electrodes can help realize this goal.

16.3.3 ADVANCED MOLECULAR METHODS

Although the invasive intracortical electrodes have been studied for almost 65 years, our understanding of the brain's FBR has remained limited since early implantation studies in cat and primate brains by Clark and Ward (1948b), and Delgado (1952). Recently, however, we (Karumbaiah et al., 2012, 2013; Bellamkonda et al., 2013; Saxena et al., 2013) have elucidated fine molecular details of the brain's FBR to intracortical electrodes employing non-invasive imaging and quantitative gene expression techniques, as discussed below.

16.3.3.1 Non-invasive imaging of the BBB

An open BBB leads to the influx of various neurotoxic serum proteins (Bell et al., 2012; Davalos et al., 2012), proinflammatory cells, and potentiates expression of cytokines that can lead to neurodegeneration and cell death (Hassel et al., 1994; Fitch and Silver, 1997; Matz et al., 2001; Gorter et al., 2003; Banks, 2005; Abbott et al., 2006; van Vliet et al., 2007; Zlokovic, 2008; Banks and Erickson, 2010; Liu et al., 2012). Indeed, in neurodegenerative disorders such as multiple-sclerosis and Alzheimer's disease, a highly permeable BBB contributes actively to the disease pathology and active neurodegeneration (Zlokovic, 2008). Therefore, BBB permeability is a critical factor that can predict electrode performance, and the non-invasive monitoring of the BBB can help facilitate chronic electrode function *via* means that can modulate BBB permeability (Saxena et al., 2013). There are various techniques that can assist with this. However, the metallic components associated with commercially available intracortical electrodes are incompatible with conventional non-invasive imaging techniques. Fluorescence molecular tomography (FMT) is a powerful preclinical tool that allows quantitative 3-D deep tissue fluorescence imaging (Weissleder, 2001) by using near-infrared (NIR) fluorophores. To assess BBB breach using NIR fluorescence techniques, a serum protein or particle is labeled with an NIR fluorophore and injected into the blood stream (Hama et al., 2007; Klohs et al., 2009; Saxena et al., 2013). Under normal conditions, the particle or serum proteins cannot cross the BBB. However, if the BBB is breached, the particle or protein extravasates from the blood stream and accumulates at breached sites, enabling detection. While NIR techniques are mostly pre-clinical, adaptation of electrode materials to suit conventional imaging methods might bridge this gap (Dunn et al., 2009).

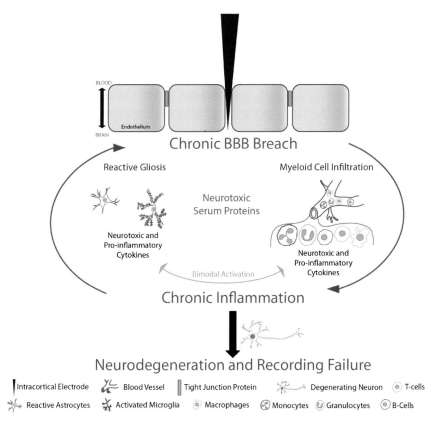

Figure 16.4 Schematic illustration of the FBR triggered in response to intracortical electrode implantation, and putative inflammatory mechanisms leading to chronic neurotoxicity and neurodegeneration. Reprinted with permission from Saxena et al., 2013.

It is surprising, that for brain–machine interfaces, the mechanistic link between the physiological response and intracortical electrode failure has been studied little. In a recent study, we identified the integrity of the BBB around the intracortical electrode implant site to be the key mediator of neuronal health, and ultimately intracortical electrode failure (Saxena et al., 2013). We showed that intracortical electrodes induce a chronic breach of the BBB leading, in a positive feedback loop, to chronic inflammation, culminating in neurodegeneration and electrode failure (Figure 16.4). We had previously observed the presence of degenerating neurons around chronic electrode implants (McConnell et al., 2009b). Drawing from the literature on neurodegenerative disorders, where BBB disruption underlies many neurodegenerative pathologies (Abbott et al., 2006; Zlokovic, 2008; Banks and Erickson, 2010; Liu et al., 2012), we investigated the contribution of BBB breach and the consequent infiltration of neurotoxic factors and proinflammatory myeloid cells to neuronal health (Hatashita and Hoff, 1990; Hassel et al., 1994; Seiffert et al., 2004; Davalos et al., 2012) and the failure of chronically implanted intracortical electrodes. We observed an inverse correlation between chronic intracortical electrode functionality and BBB breach: electrodes that fail chronically have a highly permeable BBB leading to an increased presence of active inflammatory cells and neurotoxic factors, whereas electrodes that perform better have significantly better healing of the wound and a lesser BBB breach (Saxena et al., 2013).

16.3.3.2 Inflammatory cytokine signaling

Reactive microglia are major producers of neurotoxic cytokines in the injured brain (Stoll et al., 2002; Block and Hong, 2005). The occurrence of reactive gliosis around intracortical electrode implants, along with the chronic breach of the BBB and influx of blood-borne myeloid cells results in the temporal accumulation of proinflammatory and neurotoxic cytokines such as IL6, IL1, IL17, and TNF among others that can exacerbate BBB breach, and significantly diminish neuronal viability. Chronic intracortical electrode function therefore depends heavily on the ability to preserve neuronal health and viability, and to mitigate proinflammatory and neurotoxic cytokine accumulation.

Histological evaluation of inflammatory markers around intracortical electrode implants has been the conventional method of investigating the tissue response to intracortical electrode implants. Although useful to visually assess the overall state of the tissue response to chronic electrode implants, histology does not provide detailed quantitative information regarding the prevalence of inflammatory or neurotoxic cytokines in the local milieu. Further, since only a few histological markers can be evaluated per tissue section, the technique needs to be combined with other high-throughput screening methods such as quantitative mass spectrometry and gene expression analyses to obtain a comprehensive understanding of the prevailing inflammatory milieu. In recent studies, we have used fluorescently tagged rat IgG to selectively stain brain tissue sections for regions of BBB breach, and excised this tissue from the remaining tissue using laser capture microdissection (LCM) (Karumbaiah et al., 2012, 2013; Saxena et al., 2013). The excised tissue was then used to conduct sensitive quantitative gene expression analyses of inflammatory and neurotoxic cytokine-encoding transcripts in a high-throughput manner. Through these types of analyses, we were able to detect temporal changes in wound healing and BBB breach-inducing factors, which may have contributed to differences in recording function between silicon-based Michigan and metallic microwire electrodes (Saxena et al., 2013). In a subsequent, more extensive study comparing electrodes of different shape, size, and tethering variations, we detected significant differences in the chronic expression levels of inflammatory and neurotoxic cytokine transcripts in animals implanted with electrodes of different designs (Karumbaiah et al., 2013). These results are the first to show that cytokine gene expression analysis is a reliable predictor of intracortical electrode recording function. These results also indicate that intracortical electrode implants of different design variations induce significant differences in the expression of inflammatory and neurotoxic cytokine transcripts, and therefore needs to be taken into consideration for the rational design of chronically functional intracortical electrodes.

16.4 CONCLUSIONS

BCIs and assistive neural prosthetics have the tremendous potential of providing paralyzed patients with the capacity to communicate with and control external neuroprosthetics. Intracortical electrode-based BCIs that interact directly with local populations of neurons in the brain provide the required high resolution of neural signals to control and manipulate assistive prosthetics with high degrees of freedom when compared to other EEG or ECoG-based BCIs. While the FBR triggered by the immune system in response to implanted intracortical electrodes is a significant challenge, recent advances in our understanding of the relationship between electrode design, FBR, and chronic recording stability of the neuron–electrode interface could help inform better electrode design and function. Future application of this technology will depend heavily on intracortical electrode design alterations to minimize FBR and facilitate chronic recording function.

REFERENCES

Abbott NJ, Ronnback L, Hansson E (2006) Astrocyte–endothelial interactions at the blood–brain barrier. Nat Rev Neurosci 7:41–53.

Ajiboye AB, Simeral JD, Donoghue JP, Hochberg LR, Kirsch RF (2012) Prediction of imagined single-joint movements in a person with high-level tetraplegia. IEEE Trans Biomed Eng 59:2755–2765.

Anderson JM, Rodriguez A, Chang DT (2008) Foreign body reaction to biomaterials. Semin Immunol 20:86–100.

Ball T, Kern M, Mutschler I, Aertsen A, Schulze-Bonhage A (2009) Signal quality of simultaneously recorded invasive and non-invasive EEG. Neuro Image 46:708–716.

Banks WA (2005) Blood–brain barrier transport of cytokines: A mechanism for neuropathology. Curr Pharm Des 11:973–984.

Banks WA, Erickson MA (2010) The blood–brain barrier and immune function and dysfunction. Neurobiol Dis 37:26–32.

Bell RD, Winkler EA, Singh I, Sagare AP, Deane R, Wu Z, Holtzman DM, Betsholtz C, Armulik A, Sallstrom J (2012) Apolipoprotein E controls cerebrovascular integrity *via* cyclophilin A. Nature 485:512–516.

Bellamkonda R, Karumbaiah L, Saxena T, Wang Q, Stanley G (2013) Is the extent of blood–brain-barrier breach predictive of intracortical electrode performance? Biophys J 104:376a–376a.

Berger H (1929) Uber das Electrenkephalogramm des Menchen. Arch Psychiat Nervenkr 87:527–570.

Biran R, Martin DC, Tresco PA (2005) Neuronal cell loss accompanies the brain tissue response to chronically implanted silicon microelectrode arrays. Exp Neurol 195:115–126.

Biran R, Martin DC, Tresco PA (2007) The brain tissue response to implanted silicon microelectrode arrays is increased when the device is tethered to the skull. J Biomed Mater Res A 82:169–178.

Block ML, Hong JS (2005) Microglia and inflammation-mediated neurodegeneration: Multiple triggers with a common mechanism. Prog Neurobiol 76:77–98.

Bockris JOM, Reddy AK (1973) Modern electrochemistry: An introduction to an interdisciplinary area. Springer.

Bradley PB, Elkes J (1953) A technique for recording the electrical activity of the brain in the conscious animal. Electroen Clin Neuro 5:451–456.

Chadwick EK, Blana D, Simeral JD, Lambrecht J, Kim SP, Cornwell AS, Taylor DM, Hochberg LR, Donoghue JP, Kirsch RF (2011) Continuous neuronal ensemble control of simulated arm reaching by a human with tetraplegia. J Neural Eng 8.

Chao ZC, Nagasaka Y, Fujii N (2010) Long-term asynchronous decoding of arm motion using electrocorticographic signals in monkeys. Frontiers in neuroengineering 3:3.

Clark G, Ward JW (1948a) Responses elicited from the cortex of monkeys by electrical stimulation through fixed electrodes. Brain 71:332–342.

Clark G, Ward JW (1948b) Responses elicited from the cortex of monkeys by electrical stimulation through fixed electrodes. Brain 71:332–342.

Cogan SF (2008) Neural stimulation and recording electrodes. Annu Rev Biomed Eng 10:275–309.

Collinger JL, Wodlinger B, Downey JE, Wang W, Tyler-Kabara EC, Weber DJ, McMorland AJC, Velliste M, Boninger ML, Schwartz AB (2013) High-performance neuroprosthetic control by an individual with tetraplegia. Lancet 381:557–564.

Dale N, Hatz S, Tian F, Llaudet E (2005) Listening to the brain: Microelectrode biosensors for neurochemicals. Trends Biotechnol 23:420–428.

Davalos D, Ryu JK, Merlini M, Baeten KM, Le Moan N, Petersen MA, Deerinck TJ, Smirnoff DS, Bedard C, Hakozaki H, Gonias Murray S, Ling JB, Lassmann H, Degen JL, Ellisman MH, Akassoglou K (2012) Fibrinogen-induced perivascular microglial clustering is required for the development of axonal damage in neuroinflammation. Nat Commun 3:1227.

Delgado JMR (1952) Permanent implantation of multilead electrodes in the brain. Yale J Biol Med 24:351–.

Dunn JF, Tuor UI, Kmech J, Young NA, Henderson AK, Jackson JC, Valentine PA, Teskey GC (2009) Functional brain mapping at 9.4T using a new MRI-compatible electrode chronically implanted in rats. Magn Reson Med 61:222–228.

Fawcett JW, Asher RA (1999) The glial scar and central nervous system repair. Brain Res Bull 49:377–391.

Fillenz M (2005) *In vivo* neurochemical monitoring and the study of behaviour. Neurosci Biobehav R 29:949–962.

Fitch MT, Silver J (1997) Activated macrophages and the blood–brain barrier: Inflammation after CNS injury leads to increases in putative inhibitory molecules. Exp Neurol 148:587–603.

Frampton JP, Hynd MR, Shuler ML, Shain W (2010) Effects of glial cells on electrode impedance recorded from neuralprosthetic devices *in vitro*. Ann Biomed Eng 38:1031–1047.

Freeman WJ, Holmes MD, Burke BC, Vanhatalo S (2003) Spatial spectra of scalp EEG and EMG from awake humans. Clin Neurophysiol 114:1053–1068.

Gilletti A, Muthuswamy J (2006a) Brain micromotion around implants in the rodent somatosensory cortex. J Neural Eng 3:189.

Gilletti A, Muthuswamy J (2006b) Brain micromotion around implants in the rodent somatosensory cortex. J Neural Eng 3:189–195.

Gorter JA, Goncalves Pereira PM, van Vliet EA, Aronica E, Lopes da Silva FH, Lucassen PJ (2003) Neuronal cell death in a rat model for mesial temporal lobe epilepsy is induced by the initial status epilepticus and not by later repeated spontaneous seizures. Epilepsia 44:647–658.

Grill WM, Norman SE, Bellamkonda RV (2009) Implanted neural interfaces: Biochallenges and engineered solutions. Annu Rev Biomed Eng 11:1–24.

Gutowski SM, Templeman KL, South AB, Gaulding JC, Shoemaker JT, LaPlaca MC, Bellamkonda RV, Lyon LA, García AJ (2013) Host response to microgel coatings on neural electrodes implanted in the brain. J Biomed Mater Res A.

Hama Y, Koyama Y, Choyke PL, Kobayashi H (2007) Two-color *in vivo* dynamic contrast-enhanced pharmacokinetic imaging. J Biomed Opt 12.

Hashemi P, Walsh PL, Guillot TS, Gras-Najjar J, Takmakov P, Crews FT, Wightman RM (2011) Chronically implanted, nafion-coated Ag/AgCl reference electrodes for neurochemical applications. ACS Chem Neurosci 2:658–666.

Hassel B, Iversen EG, Fonnum F (1994) Neurotoxicity of albumin *in vivo*. Neurosci Lett 167:29–32.

Hatashita S, Hoff JT (1990) Brain edema and cerebrovascular permeability during cerebral ischemia in rats. Stroke 21:582–588.

Hayne RA, Belinson L, Gibbs FA (1949) Electrical activity of subcortical areas in epilepsy. Electroen Clin Neuro 1:437–445.

Heath RG, Norman EC (1946) Electroshock therapy by stimulation of discrete cortical sites with small electrodes. Proc Soc Exp Biol Med 63:496–502.

Heien MLAV, Khan AS, Ariansen JL, Cheer JF, Phillips PEM, Wassum KM, Wightman RM (2005) Real-time measurement of dopamine fluctuations after cocaine in the brain of behaving rats. Proc Natl Acad Sci USA 102:10023–10028.

Hochberg LR, Serruya MD, Friehs GM, Mukand JA, Saleh M, Caplan AH, Branner A, Chen D, Penn RD, Donoghue JP (2006) Neuronal ensemble control of prosthetic devices by a human with tetraplegia. Nature 442:164–171.

Hochberg LR, Bacher D, Jarosiewicz B, Masse NY, Simeral JD, Vogel J, Haddadin S, Liu J, Cash SS, van der Smagt P, Donoghue JP (2012) Reach and grasp by people with tetraplegia using a neurally controlled robotic arm. Nature 485:372–U121.

Jarosiewicz B, Chase SM, Fraser GW, Velliste M, Kass RE, Schwartz AB (2008) Functional network reorganization during learning in a brain–computer interface paradigm. Proc Natl Acad Sci USA 105:19486–19491.

Kaiser V, Bauernfeind G, Kreilinger A, Kaufmann T, Kubler A, Neuper C, Muller-Putz GR (2013) Cortical effects of user training in a motor imagery based brain–computer interface measured by fNIRS and EEG. NeuroImage.

Karumbaiah L, Bellamkonda R (2013) Neural tissue engineering. In: Neural engineering, pp 765–794. US: Springer.

Karumbaiah L, Norman SE, Rajan NB, Anand S, Saxena T, Betancur M, Patkar R, Bellamkonda RV (2012) The upregulation of specific interleukin (IL) receptor antagonists and paradoxical enhancement of neuronal apoptosis due to electrode induced strain and brain micromotion. Biomaterials.

Karumbaiah L, Saxena T, Carlson D, Patil K, Patkar R, Gaupp EA, Betancur M, Stanley GB, Carin L, Bellamkonda RV (2013) Relationship between intracortical electrode design and chronic recording function. Biomaterials 34:8061–8074.

Kawagoe KT, Zimmerman JB, Wightman RM (1993) Principles of voltammetry and microelectrode surface-states. J Neurosci Methods 48:225–240.

Kim SP, Simeral JD, Hochberg LR, Donoghue JP, Black MJ (2008) Neural control of computer cursor velocity by decoding motor cortical spiking activity in humans with tetraplegia. J Neural Eng 5:455–476.

Kim YT, Hitchcock RW, Bridge MJ, Tresco PA (2004) Chronic response of adult rat brain tissue to implants anchored to the skull. Biomaterials 25:2229–2237.

Klohs J, Steinbrink J, Bourayou R, Mueller S, Cordell R, Licha K, Schirner M, Dirnagl U, Lindauer U, Wunder A (2009) Near-infrared fluorescence imaging with fluorescently labeled albumin: A novel method for non-invasive optical imaging of blood–brain barrier impairment after focal cerebral ischemia in mice. J Neurosci Methods 180:126–132.

Leuthardt EC, Schalk G, Wolpaw JR, Ojemann JG, Moran DW (2004) A brain–computer interface using electrocorticographic signals in humans. J Neural Eng 1:63–71.

Leuthardt EC, Miller KJ, Schalk G, Rao RP, Ojemann JG (2006) Electrocorticography-based brain–computer interface — the Seattle experience. IEEE Trans Neural Syst Rehabil Eng: A publication of the IEEE Engineering in Medicine and Biology Society 14:194–198.

Leuthardt EC, Gaona C, Sharma M, Szrama N, Roland J, Freudenberg Z, Solis J, Breshears J, Schalk G (2011) Using the electrocorticographic speech network to control a brain–computer interface in humans. J Neural Eng 8.

Liu JY, Thom M, Catarino CB, Martinian L, Figarella-Branger D, Bartolomei F, Koepp M, Sisodiya SM (2012) Neuropathology of the blood–brain barrier and pharmacoresistance in human epilepsy. Brain 135:3115–3133.

Matz PG, Lewén A, Chan PH (2001) Neuronal, but not microglial, accumulation of extravasated serum proteins after intracerebral hemolysate exposure is accompanied by cytochrome c release and DNA fragmentation. J Cereb Blood Flow Metab 21:921–928.

McConnell GC, Butera RJ, Bellamkonda RV (2009a) Bioimpedance modeling to monitor astrocytic response to chronically implanted electrodes. J Neural Eng 6:055005.

McConnell GC, Rees HD, Levey AI, Gutekunst CA, Gross RE, Bellamkonda RV (2009b) Implanted neural electrodes cause chronic, local inflammation that is correlated with local neurodegeneration. J Neural Eng 6:056003.

McFarland DJ, Lefkowicz AT, Wolpaw JR (1997a) Design and operation of an EEG-based brain–computer interface with digital signal processing technology. Behav Res Meth Ins C 29:337–345.

McFarland DJ, McCane LM, David SV, Wolpaw JR (1997b) Spatial filter selection for EEG-based communication. Electroen Clin Neuro 103:386–394.

Moritz CT, Perlmutter SI, Fetz EE (2008) Direct control of paralysed muscles by cortical neurons. Nature 456:639–U663.

Muller-Putz GR, Pfurtscheller G (2008) Control of an electrical prosthesis with an SSVEP-based BCI. IEEE Trans Biomed Eng 55:361–364.

Nicolelis MAL (2001) Actions from thoughts. Nature 409:403–407.

Perge JA, Homer ML, Malik WQ, Cash S, Eskandar E, Friehs G, Donoghue JP, Hochberg LR (2013) Intra-day signal instabilities affect decoding performance in an intracortical neural interface system. J Neural Eng 10.

Pfurtscheller G, Flotzinger D, Neuper C (1994) Differentiation between Finger, Toe and Tongue Movement in Man Based on 40 Hz EEG. Electroen Clin Neuro 90:456–460.

Phillips PEM, Stuber GD, Heien MLAV, Wightman RM, Carelli RM (2003) Subsecond dopamine release promotes cocaine seeking. Nature 422:614–618.

Polikov VS, Tresco PA, Reichert WM (2005) Response of brain tissue to chronically implanted neural electrodes. J Neurosci Methods 148:1–18.

Potter KA, Buck AC, Self WK, Capadona JR (2012) Stab injury and device implantation within the brain results in inversely multiphasic neuroinflammatory and neurodegenerative responses. J Neural Eng 9:046020.

Prasad A, Xue QS, Sankar V, Nishida T, Shaw G, Streit WJ, Sanchez JC (2012) Comprehensive characterization and failure modes of tungsten microwire arrays in chronic neural implants. J Neural Eng 9:056015.

Reichert WM, He W, Bellamkonda RV (2008) A molecular perspective on understanding and modulating the performance of chronic central nervous system (CNS) recording electrodes.

Roitbak T, Sykova E (1999) Diffusion barriers evoked in the rat cortex by reactive astrogliosis. Glia 28:40–48.

Saxena T, Gilbert J, Stelzner D, Hasenwinkel J (2012) Mechanical characterization of the injured spinal cord after lateral spinal hemisection injury in the rat. J Neurotrauma 29:1747–1757.

Saxena T, Karumbaiah L, Gaupp EA, Patkar R, Patil K, Betancur M, Stanley GB, Bellamkonda RV (2013) The impact of chronic blood–brain barrier breach on intracortical electrode function. Biomaterials 34:4703–4713.

Schalk G, Miller KJ, Anderson NR, Wilson JA, Smyth MD, Ojemann JG, Moran DW, Wolpaw JR, Leuthardt EC (2008) Two-dimensional movement control using electrocorticographic signals in humans. J Neural Eng 5:75–84.

Schwartz AB (2004) Cortical neural prosthetics. Annu Rev Neurosci 27:487–507.

Schwartz AB, Cui XT, Weber DJ, Moran DW (2006) Brain-controlled interfaces: Movement restoration with neural prosthetics. Neuron 52:205–220.

Seiffert E, Dreier JP, Ivens S, Bechmann I, Tomkins O, Heinemann U, Friedman A (2004) Lasting blood–brain barrier disruption induces epileptic focus in the rat somatosensory cortex. J Neurosci 24:7829–7836.

Simeral JD, Kim SP, Black MJ, Donoghue JP, Hochberg LR (2011) Neural control of cursor trajectory and click by a human with tetraplegia 1000 days after implant of an intracortical microelectrode array. J Neural Eng 8.

Stoll G, Jander S, Schroeter M (2002) Detrimental and beneficial effects of injury-induced inflammation and cytokine expression in the nervous system. Mol Cell Biol Neuroprot CNS 513:87–113.

Strong TD, Cantor HC, Brown RB (2001) A microelectrode array for real-time neurochemical and neuroelectrical recording *in vitro*. Sensor Actuat A-Phys 91:357–362.

van Vliet EA, da Costa Araujo S, Redeker S, van Schaik R, Aronica E, Gorter JA (2007) Blood–brain barrier leakage may lead to progression of temporal lobe epilepsy. Brain 130:521–534.

Velliste M, Perel S, Spalding MC, Whitford AS, Schwartz AB (2008) Cortical control of a prosthetic arm for self-feeding. Nature 453:1098–1101.

Weissleder R (2001) A clearer vision for *in vivo* imaging. Nat Biotechnol 19:316–317.

Williams JC, Hippensteel JA, Dilgen J, Shain W, Kipke DR (2007) Complex impedance spectroscopy for monitoring tissue responses to inserted neural implants. J Neural Eng 4:410.

Winslow BD, Christensen MB, Yang WK, Solzbacher F, Tresco PA (2010) A comparison of the tissue response to chronically implanted parylene-C-coated and uncoated planar silicon microelectrode arrays in rat cortex. Biomaterials 31:9163–9172.

Wolpaw JR, Mcfarland DJ, Neat GW, Forneris CA (1991) An EEG-based brain–computer interface for cursor control. Electroen Clin Neuro 78:252–259.

Woolley AJ, Desai HA, Otto KJ (2013) Chronic intracortical microelectrode arrays induce non-uniform, depth-related tissue responses. J Neural Eng 10:026007.

Zhong Y, Yu X, Gilbert R, Bellamkonda RV (2001) Stabilizing electrode–host interfaces: A tissue engineering approach. J Rehabil Res Dev 38:627–632.

Zlokovic BV (2008) The blood–brain barrier in health and chronic neurodegenerative disorders. Neuron 57:178–201.

CHAPTER 17

MEASUREMENT OF CYTOKINES IN THE BRAIN

Julie A. Stenken and Michael Elkins

University of Arkansas

17.1 INTRODUCTION

Cellular communication within the brain involves chemical and electrical information being passed through the extracellular fluid space (ECS) (Cooper et al., 1991). The historical interest in creating methods for detection of different neurotransmitters lies in wanting to elucidate the role bioactive chemicals play in behavior or disease. Chemical analysis of neurochemicals allows correlation between neurochemical transmission and the onset of either physical or pharmacological treatments. Monitoring chemical communication through the ECS has historic roots for the classical small-molecule neurotransmitters such as dopamine, glutamate, serotonin, and other molecules (Michael and Borland, 2007).

When creating new chemical analysis methods, analytical chemists typically go through a thought process combining knowledge of available transduction methods with the analyte's chemistry, i.e., an analyte that can be oxidized or reduced might be quantified using an electrochemistry technique. For elucidating neurochemical networks, electrochemical methods have been the workhorse since the catecholamines (e.g., dopamine, norepinephrine, and serotonin) are redox active and oxidase enzymes exist for glucose, glutamate, and lactate. An additional advantage that has fostered the use of *in vivo* electrochemical measurements is the ability to create low micron-sized electrodes in the range of 5 to 10 μm.

As neurochemistry knowledge increased, there became more interest in other small molecules as well as interest in how neuropeptides are networked and work in concert with classical neurotransmitters and even other neuropeptides. Some neuropeptides are co-localized in vesicles with different neurotransmitters (Merighi et al., 2011). This demonstrates the importance of being able to perform measurements across classes of analytes whether they are low molecular weight neurotransmitters or larger neuropeptides. Unlike the catecholamine neurotransmitters, which can be measured *in situ* in brain tissue using small diameter (5–10 μm) electrodes, larger molecules such as peptides and proteins are typically accessed using either microdialysis sampling or push–pull perfusion approaches (Myers et al., 1998). The collected perfusates are then subjected to an appropriate analysis scheme, typically one that involves either a chromatographic or electrophoretic separation scheme to allow for measurement of many different components (Nandi and Lunte, 2009). Readers are referred to several reviews describing neuropeptide measurements and their inherent challenges from microdialysis samples (Wotjak et al., 2008; Van Eeckhaut et al., 2011; Mabrouk and Kennedy, 2013; Schmerberg and Li, 2013).

As neuroscience discoveries evolved, so did the repertoire of chemical compounds that were discovered to act as chemical signals through the brain ECS. Bioactive proteins including cytokines and growth factors are suspected to be networked with the many different small molecule signaling systems. Cytokines have interactions with the endocannabinoid system (Molina-Holgado and Molina-Holgado, 2010), glutaminergic and GABAergic systems (Galic et al., 2012), serotinonergic system (Mueller and Schwarz, 2007), and the dopaminergic system (Capuron and Miller, 2011). This chapter gives a brief overview of cytokine proteins, their neurochemical source, methods of measurement, challenges associated with their *in vivo* chemical analysis, and data interpretation.

17.2 CYTOKINES AS SIGNALING PROTEINS

Cytokines and chemokines (*chemoattractant* cyto*kines*) are bioactive proteins with a wide variety of molecular weight, tertiary structure, and function. These proteins act as immunoregulators and neuroregulators and are considered a third neurotransmitter/neuromodulator system (Turrin and Plata-Salaman, 2000; Vitkovic et al., 2000; Adler and Rogers, 2005; Ransohoff and Benveniste, 2006). Cytokines are implicated in a wide variety of neurological processes (appetite, memory, and sleep) as well as different inflammatory disease states including alcoholism, depression, epilepsy, fever, multiple sclerosis, and various psychiatric disorders (Lampron et al., 2013). These proteins are integrated into networks that combine neuropeptides and neurotransmitters through highly complex neurochemical-signaling pathways that are still being elucidated (Johnson et al., 2005). For chemokines, the suggested nomenclature that focuses on the type of chemokine, with two consecutive cysteines (CC) or with amino acids in between the cysteines (CXC), will be used with common names.

Cytokine proteins are known to operate in networks that include a broad and overlapping range of bioactivity (Balkwill and Burke, 1989; Capuron and Miller, 2011). The term cytokine network describes the highly complex interactions within the immune system involving different cytokines. Cytokines can induce or decrease their own synthesis from any immune-derived cell in a process that is called autocrine regulation. Cytokines can also induce or decrease the synthesis of other cytokines released from different types of immune cells in what is termed a paracrine interaction (cell-to-cell). A single cytokine can also produce a cascade of cytokine activity. Cytokine-encoding genes are also pleiotrophic meaning that a cytokine gene can regulate different phenotypes of a single cytokine leading to different outcomes. The overall network activity can work to be protective or highly damaging depending upon the microenvironmental context and the cytokine cues elicited. Obtaining information about the different cytokine concentrations within a suspected network can be more informative than concentrations of a single cytokine. When performing measurements, it is necessary to carefully consider the cytokine network rather than a single cytokine due to the considerable redundancy and pleiotropism that exists among cytokines.

17.3 CLINICAL INTERESTS IN CYTOKINES IN THE BRAIN

Many central nervous system (CNS) diseases have an inflammatory component to their etiology. Chemokines are known to be involved in neurological injury and repair (Jaerve and Mueller, 2012). Numerous disease states are now believed to involve some aspect of cytokine signaling. Cytokines such as IL-1β, IL-6, and TNF-α are known actors in the CNS (Arvin et al., 1996; Turrin and Plata-Salaman, 2000; Vezzani et al., 2008; Galic et al., 2012). This leads to reasons to measure cytokines within the CNS and especially their signaling through the ECS to develop pharmaceutical agents that modulate cytokine profiles toward a desired outcome (Helmy et al., 2007).

Diseases affecting the CNS with a suspected cytokine signaling component include: alcoholism/addiction (Crews et al., 2011), Alzheimer's disease (Rubio-Perez and Morillas-Ruiz, 2012), epilepsy (Vezzani et al., 2008), pain (White and Miller, 2010), Parkinson's disease (Panaro and Cianciulli, 2012), traumatic brain injury (Frugier et al., 2010), and various psychiatric disorders (Capuron and Miller, 2011). Given this extensive listing, it is understandable why measurement of cytokines within the brain is of clinical importance.

There has been significant progress in the creation of imaging agents combined with positron emission tomography (PET) for imaging of different binding and uptake processes associated with various aspects of classical neurotransmitters in humans. However, the field of imaging cytokine receptors using PET is in its infancy. There is research demonstrating imaging of activated microglial cells by targeting a non-cytokine receptor, translocator protein 18 kDa (TSPO), using PET (Lavisse et al., 2012). The chemokine, CXCL12 (also known as stromal-derived factor-1, SDF-1), serves to recruit neural stem cells and therefore is of significant clinical interest. It binds several receptors including CXCR4. The CXCL12/CXCR4 signaling axis and its involvement with recruitment of neural stem cells is of significant clinical interest in the area of Alzheimer's, glioblastoma, and stroke treatments. Therefore, there is significant interest in drugs against (Oishi and Fujii, 2012; Peled et al., 2012), as well as imaging agents for, the CXCR4 receptor which binds SDF-1/CXCL12 (De et al., 2011; Kuil et al., 2012).

17.4 OVERVIEW OF COMMON METHODS FOR CYTOKINE MEASUREMENTS AND TISSUE MAPPING

Measurements of cytokines in the clinic are typically performed in blood or cerebral spinal fluid (CSF) samples (Kwon et al., 2010). It is becoming more common to see suggestions of using blood and/or CSF cytokine analyses for disease management or disease detection strategies for alcoholism (Achur et al., 2010), Alzheimer's (Olson and Humpel, 2010), multiple sclerosis (Stangel et al., 2013), and stroke (Lambertsen et al., 2012). The common methods for measuring cytokine proteins are either standard ELISA or multiplexed bead-based immunoassays that allow for measurements of many cytokines in a single low μL sample (Vignali, 2000). The difficulty with all of these methods is that blood and CSF fluid is far removed from the actual sites of cytokine action in the brain. Furthermore, cytokines in the blood are indicative of inflammation, but the source of this inflammation can be from anywhere in the body. Finally, blood and whole tissue levels of cytokines in the brain after exposure to the endotoxin, lipopolysaccharide (LPS), have been shown to be divergent for many cytokines (Erickson and Banks, 2011).

Much of what is known about cytokines in the brain does not come from direct measurements of proteins found in the ECS, but rather indirectly from different types of measurements (protein, immunohistochemistry, or RNA measurements), from collected tissue, or from CSF. For this reason, it is challenging to make inferences regarding the role of cytokines in different neurochemical pathways as the direct transmission of these molecules through the ECS is not easy to map. A significant amount of knowledge about cytokines has come from mRNA analyses. However, there have also been whole protein measurements performed and the different procedures for extracting cytokines and growth factors from the brain have been described (Matalka et al., 2005).

Researchers have extensively used mRNA for many studies related to mapping cytokines to different disease states in the brain (Minami et al., 1991; Fan et al., 1995; Hausmann et al., 1998; Quan et al., 1999; Churchill et al., 2006). Recently, it has been found that mRNA analysis accounts for approximately 40–60% of the overall proteome change (de Sousa Abreu et al., 2009; Vogel and Marcotte, 2012). Appropriate methods for normalization and overall assessment of total mRNA in samples have been reviewed (Guenin et al., 2009).

A significant concern with mRNA analyses is that few correlations have been demonstrated between protein levels and mRNA levels (Maier et al., 2009). The kinetics of mRNA production and protein removal are different for different proteins. Some proteins continue on to be post-translationally modified. Proteins have different binding kinetics to receptors. Proteins have different enzymatic removal processes. Thus, there is not a direct correlation between levels of mRNA and actual levels of protein in tissues as has been observed for Brain Derived Neurotrophic Factor (BDNF) (Nawa et al., 1995). Reviews are available related to mRNA analysis as it pertains to cytokines (Giulietti et al., 2001; Overbergh et al., 2003).

17.5 CELLULAR SOURCES OF CYTOKINES

Performing sampling or measurements of molecules within an *in vivo* setting necessitates understanding analyte source and sinks. Neurotransmitter cellular sources, metabolism, synthesis pathways, receptor thermodynamics and kinetics, release rates, and transporter kinetics are well-studied. While single-cell analysis of cytokines have been reported (Han et al., 2010), cytokine production from different neuronal cellular sources and cytokine removal processes are not as well documented. Outlined below is a brief synopsis of the current understanding of cytokine sources in the brain.

Cytokines are produced from cells within the brain as well as immune cells that pass the blood–brain barrier (BBB) in response to injury (Table 17.1). Originally, it was believed that the only way for cytokines to enter brain tissue would be through a breach of the BBB (Ransohoff and Benveniste, 2006). This perception came from the knowledge that cytokines are immune-cell derived and that the BBB provides immune privilege to the brain, so only through a disruption of the BBB would immune-derived and cytokine-releasing cells enter the brain. While there is still significant interest in understanding the role that the BBB plays in the transport of cytokines and immune cells (Banks, 2005), it has become apparent that nearly all the cell types in the brain are capable of producing cytokines.

Cytokines have been found to exist in all CNS-based cells — astrocytes, glial cells, and neurons (Bruce-Keller, 1999). Chemokines are a subclass of cytokines and their function in the CNS has been reviewed (Reaux-Le et al., 2013).

Table 17.1 Selected cytokines and their sources.

Cytokine	Cellular Source and Other Comments
CCL2/MCP-1 (monocyte chemoattractant protein-1)	Microglial cells. Little evidence for neuronal release. Neuronal release appears to occur during pathological conditions.
CXCL12/SDF-1	Astrocytes; microglia; neurons. Studied extensively in adult brain. Localization within the brain known.
CX3CL1 (fractalkine)	Primarily neurons. Other evidence suggests possible involvement of astrocytes and endothelial cells.
IL-1β	Most brain cells. Astrocytes; endothelial; glia; neurons.
IL-6	Astrocytes; microglia.

Chemokines are typically released first and then there is a cascade of different production levels after injury. For some cytokines, their receptors have been mapped out in different brain regions using either whole protein measurements, immunohistochemistry, or mRNA analyses. The list of known cytokines that influence brain function continues to grow, but in the brain, the major cytokines and growth factors known and their roles are briefly described below. A general overview of the interactions of different chemokines and cytokines along with their receptors is shown in Figure 17.1.

17.5.1 CCL2 (MCP-1)

Monocyte chemoattractant protein-1 (CCL2) is a chemokine that is associated with different inflammatory responses. CCL2 has been described in the neuroscience literature as being involved with different neurodegenerative diseases (Bose and Cho; Mahad and Ransohoff, 2003; Semple et al., 2010). Its locations throughout the brain along with its receptor, CCR2, have been mapped as shown in Figure 17.2 (Conductier et al., 2010). Among the cytokines that have been collected from the ECS in either rats or humans using microdialysis sampling, this chemokine frequently exhibits high pg/mL concentrations. CCL2 has been collected from the brain using microdialysis sampling by a few groups (Helmy et al., 2011b; Herbaugh and Stenken, 2011; Vasicek et al., 2013).

17.5.2 CXCL12 (SDF-1)

Stromal cell-derived factor-1/CXCL12 is another important chemokine that has significant effects in the brain (Skrzydelski et al., 2007). CXCL12 is primarily involved in recruitment of stem cells to regenerate cells in the brain (Callewaere

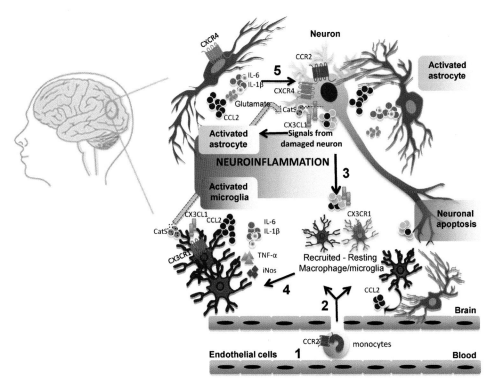

Figure 17.1 An overview of the cytokine/chemokine system in the brain. Reprinted from Reaux-Le et al., 2013 with permission from Elsevier.

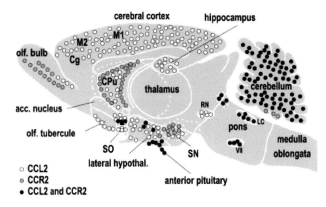

Figure 17.2 Locations of CCL2 production and its receptor, CCR2, in the brain. Reprinted from Conductier et al., 2010 with permission from Elsevier.

et al., 2006; Tiveron and Cremer, 2008; Terasaki et al., 2011; Li et al., 2012). Interestingly, there are no reports of collecting this protein using microdialysis. It is not clear why this is the case. CXCL12 is known to bind heparin and other glycosaminoglycans (GAGs) which may preclude it from being easily sampled from the ECS (De Paz et al., 2007).

17.5.3 CX3CL1

Fractalkine (CX3CL1) is a chemokine that is released from neurons and is critically important to neuroprotection (Berangere Re and Przedborski, 2006). This chemokine signals through the microglia with a known neuron/microglial signaling component. CX3CL1 is upregulated during many different neurological disease states. CX3CL1 has been collected using microdialysis sampling in human brain (Helmy et al., 2012).

17.5.4 IL-1β

Interleukin-1β (IL-1β) is a cytokine that has been extensively studied in the CNS (Yan et al., 1992; Vitkovic et al., 2000; Rijkers et al., 2009; An et al., 2011). It has significant implications in neuronal injury (Figure 17.3). IL-1β is expressed from several brain cells — astrocytes, endothelial cells, glia, and neurons. In response to stimuli, its mRNA can be upregulated within minutes (Allan et al., 2005). The

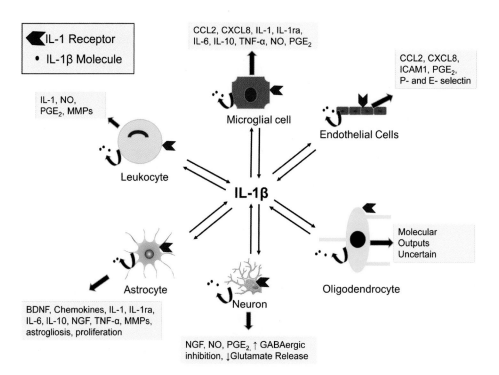

Figure 17.3 Overview of IL-1β interactions redrawn from (Allan et al., 2005). Reprinted by permission from Macmillan Publishers Ltd: [Nature Reviews Immunology] (5:629-640), 2005.

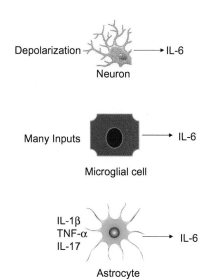

Figure 17.4 The role of IL-6 in different aspects of brain chemistry and disease. Redrawn with permission from Erta M, Quintana A, Hidalgo J. Interleukin-6, a Major Cytokine in the Central Nervous System. Int J Biol Sci 2012; 8(9):1254–1266 (Erta et al., 2012).

receptors for IL-1β exist on different cells in the brain which are believed to regulate the release of different bioactive agents including other cytokines, nitric oxide, and prostaglandin E_2 (PGE_2).

17.5.5 IL-6

Interleukin-6 (IL-6) is a pleotropic cytokine that has been involved in many neurological disorders (Van and Benveniste, 1999; Spooren et al., 2011; Erta et al., 2012). IL-6 was demonstrated to be produced in rat astrocytes as early as 1990. Like IL-1β, IL-6 is believed to be involved in many different pathways within the brain. IL-6 has important implication in brain injury as well as many different disease states as outlined in Figure 17.4.

17.6 CYTOKINE COLLECTION FROM BRAIN ECS

In vivo chemical collection of cytokines has been achieved using different sampling methods since it is important for the sampling method to be non-selective to collect different cytokines residing within the sampling milieu. To date, there are only three methods that allow for cytokine sampling within living systems and only two have been applied for the collection of cytokines from brain ECS. These three methods are microdialysis sampling (Ao and Stenken, 2006), open-flow microperfusion (Pachler et al., 2007), and push–pull perfusion (Myers et al., 1998). Among these three sampling methods, only microdialysis sampling and push–pull perfusion have been used in neurochemical sampling of cytokines.

Interpretation of cytokine concentrations collected from *in vivo* sampling devices such as microdialysis sampling or push–pull perfusion requires understanding the analyte mass transport mechanisms through the tissue space and into the device as well as the different kinetic processes that either add or remove analyte from the sampling space. Another important issue is calibration of the devices between *in vitro* settings and *in vivo*. For cytokine collections, it is important to remember that these are not small hydrophilic molecules. Cytokine proteins have an extensive range of molecular weight and hydrodynamic radii with molecular weights between 6,000 and 70,000 Da. Extrapolating data obtained using small hydrophilic molecules may not be appropriate for these larger solutes. This is problematic as it is known that larger molecules incur different diffusivity in the brain relative to smaller molecules (Sykova and Nicholson, 2008).

17.6.1 SAMPLING CONSIDERATIONS: OVERVIEW OF MASS TRANSPORT, BINDING, AND UPTAKE PROCESSES

Any solute passing through the brain ECS possesses its own unique diffusivity through the brain space (D_{ECS}) and is influenced by the volume fraction in the vicinity of the collection device (Φ_{ECS}), removal processes (k_{ECS}), and the rate at which it is introduced into the surrounding space or its generation rate (G_{ECS}) (Sykova and Nicholson, 2008). Note that G is used rather than the standard engineering Q since Q is used to denote flow rates through microdialysis probes. In order for solutes to be collected *via* either push–pull perfusion or microdialysis sampling, it must be transported to the tissue implant site. Differences in any of the different parameters denoted above will affect solute transport through the ECS. In addition to understanding the mass transport theories that are used for the brain, researchers performing *in vivo* methods would be wise to review the components of the extracellular matrix in the brain (Novak and Kaye, 2000).

The diffusion process through tissue ECS can be considered to be a random-walk process with imposed barriers (Wolak and Thorne, 2013). For any solute, its tortuosity through the ECS is defined by the ratio of its diffusion coefficient in an aqueous solution to its diffusion coefficient through the ECS (Eq. 17.1). For small molecules, this tortuosity has been extensively studied in the brain using different electrochemical techniques and has been well-described with an approximate value of 2.25 for τ (Sykova and Nicholson, 2008). Larger molecules such as bioactive proteins have their tortuosity values increase since their ability to traverse the narrow tortuous pathways within the ECS is diminished. This leads to larger than expected values for the tortuosity and the necessity of the molecule to require deformation in order to diffuse through the space as demonstrated for dextrans with molecular weights of greater than >200 kDa (Xiao et al., 2008).

$$\tau = \frac{D_{(aq)}}{D_{(ECS)}},\qquad \text{Eq. 17.1}$$

For most solutes, alterations in volume fraction, Φ_{ECS}, and removal processes would be expected to significantly affect collected solute amounts. The standard accepted value for Φ_{ECS} under normal conditions is 0.2. A larger volume fraction, which could occur due to edema, would provide a larger space for molecules to diffuse and would reduce their tortuosity. How different sampling conditions affect the local value of Φ_{ECS} has not been thoroughly studied. One example in the microdialysis literature for small molecules showed that Φ_{ECS} had to be adjusted to a value of 0.4 for mathematically modeled data to fit experimental data (Dykstra et al., 1992). Alternatively, the volume fraction can be diminished in situations such as ischemia where there would be less volume and perhaps a larger concentration of material to collect. Changes in the volume fraction and tortuosity have been recently examined under disease conditions in cortical dysplasia (Vargova et al., 2011).

For cytokines, the primary tissue binding processes would be to either receptors or other tissue components such as the GAGs. Chemokines, some cytokines, and many growth factors bind to GAGs, complex sugars, lining the tissue surface. GAGs such as heparan sulfate and chondroitin sulfate line the ECS within the brain (Maeda et al., 2011; Bartus et al., 2012). These complex carbohydrates include negatively charged sulfate groups that lead to a space within the tissue resembling an ion-exchange chromatography column. Chemokines bind to GAGs and this binding is hypothesized to set up cellular migration gradients throughout the tissue space. The interactions of chemokine binding to GAGs are well-documented in the literature (Gibbs, 2003; Handel et al., 2005; Gandhi and Mancera, 2008; Rot, 2010).

Chemokines bind to GAGs with nM binding constants. Interestingly, the kinetics of chemokine/GAG are believed to be more rapid than for protein/protein interactions as demonstrated by surface plasmon resonance experiments (Cochran et al., 2009). What this means for cytokine collection *via* any sampling method is that chemokines or cytokines bound to GAGs in the ECS might be in high concentration in the tissue, but cannot be sampled since they are in equilibrium with tissue GAGs. Thus, in addition to the fraction that is collected from the sampling process due to mass transport restrictions, the solute amount collected is further reduced since any sampling technique allows the collection of only unbound solutes.

Cytokines are removed from the ECS by several different pathways. The binding interaction with cytokine receptors will remove cytokines from the ECS. Proteins in the ECS are typically removed by the bulk flow through the ECS leading to the CSF (Abbott, 2004; Iliff et al., 2012). Cytokines would be expected to be degraded *via* proteolytic activity (Zhao et al., 2005) since there are known trypsin and trypsin-like proteases in the brain (Wang et al., 2008). Cytokine activity is also reduced by binding to their own soluble receptors (Jones and Rose-John, 2002). For cytokine removal from the ECS, which mechanisms are most significant for cytokine removal and how these removal processes may be influenced under different neurological states is poorly understood.

17.6.2 PUSH–PULL PERFUSION

Prior to the use of microdialysis sampling, push–pull perfusion had been a widely used method for sampling from the brain. The technique is straightforward and involves inserting a cannula into the brain site of interest, slowly passing an isotonic saline solution through it, waiting for equilibration with the infused fluid and then pulling back on the fluid lines. The push–pull perfusion technique has been used for proteins since it does not have the diffusion and mass transport restrictions that are associated with microdialysis sampling. Proteins being collected only have to diffuse into the perfusion fluids being passed into the brain (Bayon et al., 1986). Cytokines have been collected in the brain using push–pull perfusion (Klir et al., 1994; Kakizaki et al., 1999; Watanobe and Hayakawa, 2003). While there have been recent reports of miniaturizing and significantly reducing the perfusion flow rates for push–pull perfusion (Kottegoda et al., 2002; Rupert et al., 2013; Slaney et al., 2013), these newly reported improvements in the push–pull perfusion technique have not been used for collection of cytokines. However, they may provide some benefit over the microdialysis sampling approach in animal studies since it may be possible to collect high concentrations and possibly more cytokines due to the lack of membrane restriction to mass transport.

17.6.2.1 Device calibration

It is interesting to note that relatively little has been written about push–pull perfusion device calibration. We contend this has a significant amount to do with researchers being generally satisfied with actually collecting intended solutes. This approach is not much different with standard microdialysis approaches where only absolute values of solutes in the collected perfusion fluids are generally reported despite significant possibilities for alterations in tissue mass transfer and volume fraction due to different treatments given. Such alterations would be expected to significantly affect device calibration.

In the early push–pull perfusion literature, there are relatively few citations for the mass transport properties to consider for this device. This was confirmed with searches in SciFinder Scholar for push–pull perfusion (>1,300 citations) and a cross reference with mathematical modeling (0 citations) or mass transport (0 citations). Model experiments with contact area show that to be more important than flow rate using quinine (Szerb, 1967). Experiments performed by the Myers group demonstrated, with ^{3}H-norepinephrine and ^{14}C-dopamine in the hypothalamus, that recovery of these radiolabeled neurotransmitters ranged between 85 and 88%. Both of these catecholamines are known to innervate the hypothalamus. Since both dopaminergic and norepinephrine pathways and neurons are reported in this region, it would be expected that an infusion of either neurotransmitter would result in loss through transporter removal, enzymatic removal, or receptor attachment. Thus, it seems the push–pull perfusion technique at least for these two neurotransmitters was able to reach high recovery values. While there may be information in the literature describing recovery rates from push–pull devices for other solutes including higher molecular weight proteins, these works are not easily searchable and may be in original articles focused on different protein collections.

17.6.3 MICRODIALYSIS SAMPLING

Microdialysis sampling is derived from the push–pull perfusion concept as a means to reduce the suspected damage that is incurred during push–pull perfusion (Ungerstedt, 1991). Of particular concern was that pushing in and pulling out the fluid caused significant damage and edema surrounding the device. This would suggest that sampling was not occurring near normal tissue, but rather a "lake" of fluid induced by the action of the push–pull mechanism. There are significant works available describing many aspects of microdialysis sampling as it relates to neuroscience and other applications (Robinson et al., 1991; Westerink and Cremers, 2007; Müller, 2013).

Microdialysis sampling applications for cytokine collection as well as other peptides or proteins were pale in comparison to the large body of work describing applications of collection of hydrophilic small molecules such as glucose and various neurotransmitters. The literature related to microdialysis sampling of larger molecules such as the cytokines has been reviewed (Clough, 2005; Ao and Stenken, 2006). As will be further delineated below in the calibration section, the major drawbacks for using microdialysis sampling to collect proteins is the significant reduction in recovery caused by microdialysis sampling being a diffusion-based separation technique and non-specific adsorption issues onto the materials used with the technique.

17.6.3.1 Microdialysis sampling device calibration

Microdialysis sampling is a diffusion-based collection method. For any solute to be collected, it has to diffuse through the tissue space and traverse the extensive tortuosity within the space, then pass through the membrane material, and finally diffuse into the perfusion fluid passing through the dialysis membrane inner lumen. For any membrane technique, the recovery of material can be determined using a mass transport coefficient approach resulting in the overall equation simply relating what is collected to a product of that mass transfer coefficient and the membrane surface area (Jacobson et al., 1985). This is illustrated in Eq. 17.2 where C is the concentration collected in the dialysate, C_o is the concentration in which the dialysis probe is immersed, k is the mass transfer coefficient, A is the membrane surface area, and Q is the volumetric flow rate through the membrane lumen. For microdialysis sampling, k is unique for every solute and represents a composite of all the tissue mass transport processes that would affect solute recovery. With cytokines, the value for the mass transport coefficient would be expected to be primarily due to diffusion through the tissue space. Diffusion coefficients for larger molecules are lower than for small molecules. Additionally, effects of tortuosity would also be expected to be a contributing factor to diffusion of large molecules through the tissue space.

$$C = C_o(1 - e^{-kA/Q})$$

For these reasons, microdialysis recoveries of cytokines are typically poor (Waelgaard et al., 2006). However, for human cytokines, the levels are much higher for recovery as the membranes are longer (10 mm *versus* 4 mm in animals for brain cannulas) and the flow rates used are far lower at 0.3 μL/minute rather than the 1 μL/minute flow rates used in animal studies.

Mechanistic models of the mass transport processes occurring during microdialysis sampling in the brain have been described by different authors and include relevant diffusion and kinetic processes (Chen et al., 2002; Bungay et al., 2011). The aim of these models is to help researchers be able to predict outflowing dialysate concentrations of the targeted analyte knowing the different mass transport parameters in the tissue space. However, it is difficult to apply these models to many aspects of microdialysis sampling since the knowledge of the different fundamental parameters (diffusion coefficients *in vivo*, production and removal rates, receptor density and receptor binding affinity, and any other processes that may serve to add or remove cytokines) are not well documented. It is important to note that differences between protein diffusivity and smaller solute diffusivity (e.g., dopamine or small ions) have been extensively described by Nicholson and Sykova (Hrabetova and Nicholson, 2007; Sykova and Nicholson, 2008).

Another significant problem with the collection of cytokines and other proteins is the inability to obtain even an estimate of microdialysis probe calibration or recovery. While many techniques for *in vivo* recovery calibration during microdialysis sampling have been reviewed (de Lange, 2013), none are appropriate for cytokine collection since no group has demonstrated a steady-state concentration for cytokines during microdialysis sampling. For these reasons, most groups only report absolute dialysate concentrations of cytokines.

17.6.4 SELECTED APPLICATIONS OF CYTOKINE COLLECTION

Both push–pull perfusion and microdialysis sampling have been widely used for collection of cytokines from mammalian brain. Push–pull perfusion has been more widely used for basic research studies involving cytokines in rodents and has not been used in clinical research. On the other hand, microdialysis sampling in the clinic for use in traumatic-brain injured patients has been used by neurosurgeons in Sweden since the early 1990s principally for elucidating glucose and other energy metabolites activity and flux in the injured portion of the brain (Hillered et al., 2005). This long clinical acceptance of the microdialysis sampling technique in humans has led to its use for collection of other biomarkers from the injured brain space.

17.6.4.1 Push–pull perfusion applications for cytokine collection

Cytokines produced during pathophysiological events have been collected using push–pull perfusion. There has been a particular interest in the role of cytokines as they pertain to different effects related to fever (Roth et al., 1993; Jansky et al., 1995; Kakizaki et al., 1999). Research performed in the early 1990s focused on collection of cytokines after giving an LPS dose that was sufficient to induce a significant immune response. LPS, one of the endotoxins associated with *E.coli*, is commonly used in cytokine research since it reliably upregulates the immune system into an extensive inflammatory response resulting in significant increases in cytokine concentrations.

Kluger's research group (Klir et al., 1993) has performed push–pull perfusion in the hypothalamus of Sprague–Dawley rats treated with LPS (20 μg/kg). Both

IL-6 and TNF-α were collected and subjected to appropriate bioassays for both cytokines. It was interesting to note that slight increases in IL-6 bioactivity appeared in the data for measurements out to three hours. However, the authors did not denote the statistical significance levels for this data and only compared the treated groups to the controls. For the treated groups, the levels of the inflammatory cytokines increased to significantly high levels compared to control.

17.6.4.2 Microdialysis applications for cytokine collection

The first description of microdialysis sampling in human brain was reported by Lars Hillered and colleagues in 1990 where microdialysis was used to monitor glucose or energy metabolism in injured brain. The technique has been widely used to elucidate dysfunction in energy metabolism in injured brain by leading groups in Sweden and in the UK (Hutchinson et al., 2002; Marklund et al., 2009; Mellergard et al., 2012). The microdialysis sampling probe is connected in a bolt that contains the pressure and oxygen sensors which are significantly larger than the ~600 μm clinical brain probes. For this reason, human microdialysis sampling is frequently applied to patients who have had major brain surgery due to either a traumatic brain injury (motorcycle accident or other high-impact head injury) or other major surgery involving tissue removal in either oncology or epilepsy patients. Like other applications of microdialysis sampling, it soon became apparent that other molecules besides hydrophilic solutes (e.g., glucose or catecholamines) would be of interest to sample. For cytokines in the brain, the major goal is to elucidate the particular cytokines that are involved in disease or injury and then determine if different therapies can alter the cytokines present and their concentrations at the site of action and thus improve clinical outcomes.

The first microdialysis sampling application for cytokine collection in the human brain was demonstrated by the group at Southampton with Professor Geraldine Clough and Martin Church where IL-1β, IL-6, and nerve growth factor (NGF) were collected from a human who had a traumatic brain injury (Winter et al., 2002). Here, a plasmapheresis membrane was used that has a significantly higher MWCO of 3,000 kDa as compared to the CMA Microdialysis (now μ dialysis) probes with 100 kDa MWCO. The use of the higher MWCO membrane was to allow for higher recovery of the cytokines since the 100 kDa MWCO membranes restrict diffusion. Many other research groups have reported on various aspects of human cytokine collection and Table 17.2 lists a summary of concentrations found among different groups. The addition of multiplex bead-based immunoassay platforms such as those produced by Luminex and BD Biosciences have allowed for measurements of greater than 15 cytokines in single 25 μL samples (Vignali, 2000). Some of this information can be found in far more detail for cytokines involved in traumatic brain injury in depth by a review by Helmy et al. (2011a).

Table 17.2 shows a brief overview of the different cytokine measurements that have been reported for human brain. Note that due to the difficulties of defining a relative recovery for cytokines *in vivo*, these values reflect absolute dialysate concentrations. Of particular interest to note are the variable levels observed. Also of

Table 17.2 Summary of cytokine microdialysis collections in humans.

Work Cited	Clinical Diagnosis (Patient #s)	Measured Cytokines (Concentration pg/mL range)	Comments (Probe Type, Flow Rates)
(Winter et al., 2002)	TBI (3)	IL-1β (66) IL-6 (1,000) NGF (1,100)†	3,000 kDa membrane †Max levels
(Hillman et al., 2007)	TBI (7) SAH (7)	IL-1β (3–35) IL-6 (80–1,378)	CMA 100 kDa 0.3 μL/min 24–168 hour collection
(Hutchinson et al., 2007)	TBI (15)	IL-1α (6±15)† IL-1β (10±15) IL-1ra (2,800±2,900)	CMA 71 100 kDa 0.3 μL/minute † Mean of means.
(Hanafy et al., 2010)	SAH (14)	TNF-α (750)	CMA 71,100 kDa
(Marcus et al., 2010)†	Tumor Resection Surgery (9)	IL-1α (2–8) [2–6] IL-1β (1–13) [1–15] IL-1ra (170–2,175) [90–2,150] IL-6 (6–1,540) [70–1,325] IL-8 (80–6,910) [40–1,775] TGF-α (5–100) [5–30] EGF (0.5–40) [0.5–15] VEGF (2–1,290) [2–210]	CMA71,100 kDa membrane, 0.3 μL/minute † Measured glioma tissue side and peri-glioma tissue [brackets].
(Helmy et al., 2011b)†	TBI (12)	IL-1α (80); IL-1β (5); IL-1ra (75); IL-2 (1); IL-3 (25); IL-4 (1) IL-5 (1); IL-6 (570); IL-7 (6); IL-8 (340); IL-9 (0.5); IL-10 (7), IL-12p40 (4); IL-12p70 (1.5), IL-13 (0.25), IL-15 (2); IL-17 (1); IP-10 (2835); MIP-1α (40); MIP-1β (70); RANTES (22); TGF-α (18); TNF-α (1); TNF-α (1); VEGF (12)	CMA71 100 kDa membrane, 0.3 μL/min †Median concentrations.

(Continued)

Table 17.2 (*Continued*)

Work Cited	Clinical Diagnosis (Patient #s)	Measured Cytokines (Concentration pg/mL range)	Comments (Probe Type, Flow Rates)
(Perez-Barcena et al., 2011)	TBI (16)	IL-1β (0–15) IL-6 (70–2,725) IL-8 (4–2,385) IL-10 (0–2) IL-12p70 (0–3) TNF-α (0–2)	CMA71 100 kDa membrane, 0.3 μL/min
(Mellergard et al., 2008)†	ICU brain injury patients (38)	IL-1β IL-6 IL-8 IL-10 MIP-1β RANTES VEGF FGF-2	CMA71 100 kDa membrane, 0.3 μL/min † Difficult to extract exact data for a table. See the paper for more detailed information.
(Mellergard et al., 2010)†	SAH TBI (145)	FGF-2 (50) Day 2 Mean VEGF (120) Day 2 Mean	† Collected over many days. See the paper for more details.
(Mellergard et al., 2012)†	TBI (69)	IL-1β (15) IL-6 (830) IL-8 (1350) IL-10 (20) MIP-1β (270) RANTES (95) FGF-2 (10) VEGF (60)	† Data placed into groups based on patient age. Median data is shown is for 25–44 year old. See the paper for more details.

Abbreviations: TBI: traumatic brain injury; SAH: subarachnoid hemorrhage.

interest is that certain cytokines have much higher concentrations than others such as IL-6 and IL-8, which are both inflammatory cytokines.

For human studies, the probes are often placed in the penumbra region (Engstrom et al., 2005). Occasionally, two separate microdialysis probes may be used with a second probe placed in a non-injured brain region to compare concentrations (typically energy metabolites such as glucose, lactate, and pyruvate).

The use of microdialysis sampling for cytokine collection in animals has fewer reports than humans. The main reason for this is the technical challenge with trying to collect cytokines through smaller membrane lengths (4 mm *versus* 10 mm in humans) at higher flow rates. These two conditions lead to lower recoveries with cytokine concentrations hovering at or quite near the limits of detection

(LOD) which is the pg/mL for the ELISA and bead-based immunoassays used for cytokine analysis.

The earliest report in animal models for microdialysis of cytokines from the brain was in 1991 related to a traumatic brain injury (Woodroofe et al., 1991). Interestingly, this work actually used the microdialysis probe as the means to create the stab wound injury and then followed the kinetics/dynamics of IL-1β and IL-6 and saw changes in these cytokines immediately after insertion of the dialysis probe. Indeed, this group collected the microglia from the probe implant site and clearly demonstrated that the microglia were capable of producing the inflammatory cytokines, IL-1β and IL-6. This work of course begs a question that our group has recently sought to answer. What levels of cytokines are observed after probe insertion and how might these levels change as a function of probe implantation procedure or protocol (Vasicek et al., 2013)?. The other issue is *how can one elucidate the production of cytokines as being from a potential disease state versus from the insertion of the dialysis probe?* Yet, it is a critically vital and important question to answer if microdialysis sampling of cytokines in the brain is to have a beneficial clinical potential.

As mentioned above, in human studies, the microdialysis probe is commonly placed in the penumbra region. A second microdialysis probe can also be placed in a non-injured region far away from the injury or diseased site for biochemical comparison as is typically performed for energy metabolite analysis. While it is likely that severe injury to the brain causes significant changes in cytokine levels compared to probe implantation, the possibility of confounding information caused by cytokines released due to mechanical injury *versus* cytokines present due to disease or brain trauma cannot be easily eliminated or elucidated. For example, platelets are a rich source of chemokines and inflammatory cytokines (Gear and Camerini, 2003; Galliera et al., 2012). Thus, probe insertion may release high levels of chemokines derived from platelets.

Since cytokines are redundant, the common inflammatory cytokines would be expected to be present both from a mechanical injury due to probe implantation *versus* those arising from the disease or brain injury. The only way to answer this vital question is to map the cytokine network relative to mechanical injury caused by the probe implantation *versus* the network activated during various disease states.

A common problem with microdialysis sampling is that nearly every research group has their own protocols that are followed. Some groups implant cannulas for a week and then sample. Others implant cannulas and wait overnight while slowly perfusing the microdialysis sampling probe. Cytokine concentrations in other contexts are known to change temporally. Thus, the microdialysis probe

Table 17.3 Summary of cytokine data for different implantation procedures.

Cytokine	Day 0 Animals with Cytokines Quantified (Out of 4)	Day 7 Animals with Cytokines Quantified (Out of 4)	Concentration Range in Day 0 Dialysates (pg/mL)	Concentration Range in Day 7 Dialysates (pg/mL)
CCL2/MCP-1	1	4	<LOD*–35	<LOD–275
CCL3/MIP-1α	4	4	5–70	<LOD–130
CCL4/MIP-2	4	4	>LOD–150	<LOD–1700
CCL5/RANTES	2	3	<LOD–17	<LOD–35
IL-1β	2	4	<LOD–75	<LOD–55
IL-6	1	3	<LOD–100	<LOD–160
IL-10	4	4	<LOD–23	<LOD–45
KC/GRO	4	4	<LOD–600	<LOD–650

*<LOD means there were samples that had concentrations lower than the assay limit of detection (LOD).

implantation protocols followed may give different cytokine concentrations. We have compared the cytokine levels obtained between implanting the microdialysis probe and immediately sampling *versus* implanting a guide cannula and waiting one week prior to sampling in the work of Vasicek et al. (2013). What is particularly interesting to note in Table 17.3 are the high levels of different cytokines at seven days post-implantation. Additionally, nearly all animals at Day 7 had quantifiable cytokine concentrations as compared to Day 0. Additional work is necessary to see how these values compare to overnight perfusions that are used by many researchers.

17.7 CYTOKINE PRODUCTION DURING BRAIN INJURIES AND OTHER DEVICE IMPLANTATION PROCEDURES

A common concern with any *in vivo* invasive chemical analysis is whether that analysis influences the measurements or production of the targeted solutes. This is a particular concern with measurements of cytokines since they are produced by so many different cell types and elucidating the exact origin requires careful experimental planning.

A recent study with a cortical impact procedure found that within four hours many of the inflammatory cytokines increase above controls (Dalgard et al., 2012). This information is provided in Table 17.4 although the full paper can be consulted for information about Days 3 and 7 post-impact. What is clearly observed is that inflammatory cytokines such as CCL2 and KC/GRO were elevated to high levels. Interestingly, IL-6 was not measured in these studies.

Table 17.4 Cytokine levels after traumatic brain injury.*

Cytokine	Control (pg/mg protein)	4 Hours (pg/mg protein)	12 Hours (pg/mg protein)	24 Hours (pg/mg protein)
KC/GRO	<5	~175	~25	<5
IFN-γ	N.D.	17.5	5	N.D.
TNF-α	~5	~60	~15	~10
IL-4, IL-5, IL-13	Low†	Low	Low	Low
CCL2 (MCP-1)	<25	~900	~1300	~500
CCL20 (MIP-3α)	~2	~15	~15	~50
IL-1β	~3	~17	~12	~17

*Table data derived from (Dalgard et al., 2012).
†These concentrations were not different between controls and treated animals and were less than 50 pg/mg protein.

A significant body of literature exists that has been focused on improving the functional longevity of brain implants. This is an important field that researchers using microdialysis sampling in the brain should consult. This field has been focused on primarily long-term implants using electrodes for various stimulation therapies particularly related to brain–computer interface field and methods to control prosthetic limbs (Jackson and Zimmermann, 2012; Ordonez et al., 2012).

All implants will elicit a foreign body reaction and the reactions to brain implants have been reviewed (Polikov et al., 2005; Tresco and Winslow, 2011). The foreign body reaction involves activation of microglial cells that have a defined role in creating a glial scar around the implanted device to wall it off from healthy tissue (Grand et al., 2010; Ereifej et al., 2011). An extensive amount of research has focused on using different materials (e.g., polymer, metal, ceramic), coatings, and controlled release strategies for neural implants, as a means to reduce and/or control the inflammatory response as much as possible (Seymour and Kipke, 2007; Leprince et al., 2010; Capadona et al., 2012; Chen and Allen, 2012; Forcelli et al., 2012; Kolarcik et al., 2012; Yue et al., 2013). The topic is of such interest that even visualized protocols are available to study the glial scar formation around an implant (Woolley et al., 2013).

Different device factors seem to be important. Blood–brain barrier breach can affect cortical implants (Saxena et al., 2013). Interestingly, an array of implants has stronger inflammatory reactions at the edges nearest the tissue rather than internal to the array (Lind et al., 2012). Different inflammatory proteins are expressed just with implant and there is some concern that localized micromotion may also affect the inflammatory response (Karumbaiah et al., 2012). The response to the probe implantation and release of neurotrophins exhibits regional variations (Humpel et al., 1995).

Upon implantation of a microdialysis probe as a means to provide trauma into rat striatum, quantifiable levels of IL-1β were found (200 pg/mL/hour) within the first hour (Fassbender et al., 2000). The IL-1β concentrations continued to rise to 600 pg/mL/hour two days after probe implantation and then declined to ~250 pg/mL/hour at three days post-probe implantation. Different inhibitors of IL-1β synthesis were used demonstrating the IL-1β to be produced on demand rather than residing as a basal level. This work is critically important as it demonstrated that the source of IL-1β was due to reactions to the probe implantation process and that nearly non-detectable levels of IL-1β was present as a basal concentration at the microdialysis probe implant site.

17.8 CONCLUSIONS AND FUTURE PROSPECTS

Microdialysis sampling has been successfully used for collection of different cytokines from rodent and human brain. Significant challenges remain for measurement of cytokines from low μL volume samples. However, the availability of bead-based immunoassays has ameliorated this analysis problem with microdialysis.

Of particular importance to address in works now focused on cytokine signaling networks in the brain is the issue of determining the cytokine source. It is clear that implantation of devices into the brain causes an inflammatory response related to the implantation procedure. This raises concerns about interpretation of the cytokine source, i.e., is the cytokine related to the actual disease state or is it related to microdialysis probe implantation processes. For some cytokines, it may be possible to locally infuse inhibitors of cytokine synthesis as described for IL-1β collections in the brain (Fassbender et al., 2000). Other ways would be to determine differences between collections from control animals *versus* treated animals. Of particular concern are studies from human brain as there are not easy ways to obtain controls and using different cocktails of different agents to promote or inhibit synthesis of cytokines will likely be difficult to be approved by the institutional review boards (IRB) at a particular site.

The human study of most relevance with respect to elucidating cytokine source is that denoted in Table 17.2 by (Marcus et al., 2010). Comparing the data side-by-side shows that only a few cytokines exhibit much wider concentration ranges between the probes in the glioma *versus* those in surrounding tissue. These cytokines were IL-8, TGF-α, and VEGF. IL-8 is a known inflammatory chemokine. TGF-α is a known tumor growth factor and would be expected to be in higher concentrations in a tumor. VEGF is also known to be highly upregulated in tumors due to its role in angiogenesis. It is interesting that cytokines known to be involved with the foreign body reaction such as IL-1β, IL-1ra, and IL-6 exhibited similar concentrations between the two sites suggesting that probe insertion may very well cause an unexpected or unwanted foreign body reaction.

Another significant need is the creation of bioinformatics approaches that can be used in combination with cytokine and other brain neurotransmitter measurements. Cytokines work in networks with other cytokines and in concert with different neurotransmitters. Being able to measure the combination of both and perform appropriate data analyses is necessary to fully elucidate chemical communication processes in the brain.

ACKNOWLEDGMENTS

The authors acknowledge NIH grants NS075874 and EB014404 for funding the work related to cytokine microdialysis sampling.

REFERENCES

Abbott NJ (2004) Evidence for bulk flow of brain interstitial fluid: Significance for physiology and pathology. Neurochem Int 45:545–552.

Achur RN, Freeman WM, Vrana KE (2010) Circulating cytokines as biomarkers of alcohol abuse and alcoholism. J Neuroimmune Pharmacol 5:83–91.

Adler MW, Rogers TJ (2005) Are chemokines the third major system in the brain? J Leukocyte Biol 78:1204–1209.

Allan SM, Tyrrell PJ, Rothwell NJ (2005) Interleukin-1 and neuronal injury. Nat Rev Immunol 5:629–640.

An Y, Chen Q, Quan N (2011) Interleukin-1 exerts distinct actions on different cell types of the brain in vitro. J Inflammation Res 4:11–20.

Ao X, Stenken JA (2006) Microdialysis sampling of cytokines. Methods 38:331–341.

Arvin B, Neville LF, Barone FC, Feuerstein GZ (1996) The role of inflammation and cytokines in brain injury. 20:445–452.

Balkwill FR, Burke F (1989) The cytokine network. Immunol Today 10:299–304.

Banks WA (2005) Blood–brain barrier transport of cytokines: A mechanism for neuropathology. Curr Pharm Des 11:973–984.

Bartus K, James ND, Bosch KD, Bradbury EJ (2012) Chondroitin sulphate proteoglycans: Key modulators of spinal cord and brain plasticity. Exp Neurol 235:5–17.

Bayon A, Anton B, Leff P, Solano S (1986) Release of proteins, enzymes, and the neuroactive peptides, enkephalins, from the striatum of the freely moving rat. Ann NY Acad Sci 473:401–417.

Berangere Re D, Przedborski S (2006) Fractalkine: Moving from chemotaxis to neuroprotection. Nat Neurosci 9:859–861.

Bose S, Cho J Role of chemokine CCL2 and its receptor CCR2 in neurodegenerative diseases. Arch Pharmacal Res: Ahead of Print.

Bruce-Keller AJ (1999) Microglial–neuronal interactions in synaptic damage and recovery. J Neurosci Res 58:191–201.

Bungay PM, Sumbria RK, Bickel U (2011) Unifying the mathematical modeling of *in vivo* and *in vitro* microdialysis. J Pharm Biomed Anal 55:54–63.

Callewaere C, Banisadr G, Desarmenien MG, Mechighel P, Kitabgi P, Rostene WH, Parsadaniantz SM (2006) The chemokine SDF-1/CXCL12 modulates the firing pattern of vasopressin neurons and counteracts induced vasopressin release through CXCR4. Proc Natl Acad Sci USA 103:8221–8226.

Capadona JR, Tyler DJ, Zorman CA, Rowan SJ, Weder C (2012) Mechanically adaptive nanocomposites for neural interfacing. MRS Bull 37:581–589.

Capuron L, Miller AH (2011) Immune system to brain signaling: Neuropsychopharmacological implications. Pharmacol Ther 130:226–238.

Chen KC, Hoistad M, Kehr J, Fuxe K, Nicholson C (2002) Theory relating *in vitro* and *in vivo* microdialysis with one or two probes. J Neurochem 81:108–121.

Chen S, Allen MG (2012) Extracellular matrix-based materials for neural interfacing. MRS Bull 37:606–613.

Churchill L, Taishi P, Wang M, Brandt J, Cearley C, Rehman A, Krueger JM (2006) Brain distribution of cytokine mRNA induced by systemic administration of interleukin-1β or tumor necrosis factor α. Brain Res 1120:64–73.

Clough GF (2005) Microdialysis of large molecules. AAPS J 7:E686–E692.

Cochran S, Li CP, Ferro V (2009) A surface plasmon resonance-based solution affinity assay for heparan sulfate-binding proteins. Glycoconjugate J 26:577–587.

Conductier G, Blondeau N, Guyon A, Nahon J-L, Rovere C (2010) The role of monocyte chemoattractant protein MCP1/CCL2 in neuroinflammatory diseases. J Neuroimmunol 224:93–100.

Cooper JR, Bloom FE, Roth RH (1991) The biochemical basis of neuropharmacology, 6th Edition. Oxford: Oxford University Press.

Crews FT, Zou J, Qin L (2011) Induction of innate immune genes in brain create the neurobiology of addiction. Brain Behav Immun 25:S4–S12.

Dalgard CL, Cole JT, Kean WS, Lucky JJ, Sukumar G, McMullen DC, Pollard HB, Watson WD (2012) The cytokine temporal profile in rat cortex after controlled cortical impact. Front Mol Neurosci 5:6.

de Lange ECM (2013) Recovery and calibration techniques: Toward quantitative microdialysis. In: Microdialysis in drug development (Müller M, ed), pp 13–33. New York: Springer.

De Paz JL, Moseman EA, Noti C, Polito L, Von Andrian UH, Seeberger PH (2007) Profiling heparin–chemokine interactions using synthetic tools. ACS Chem Biol 2:735–744.

de Sousa Abreu R, Penalva LO, Marcotte EM, Vogel C (2009) Global signatures of protein and mRNA expression levels. Mol BioSyst 5:1512–1526.

De SRA, Peyre K, Pullambhatla M, Fox JJ, Pomper MG, Nimmagadda S (2011) Imaging CXCR4 expression in human cancer xenografts: Evaluation of monocyclam ^{64}Cu-AMD3465. J Nucl Med 52:986–993.

Dykstra KH, Hsiao JK, Morrison PF, Bungay PM, Mefford IN, Scully MM, Dedrick RL (1992) Quantitative examination of tissue concentration profiles associated with microdialysis. J Neurochem 58:931–940.

Engstrom M, Polito A, Reinstrup P, Romner B, Ryding E, Ungerstedt U, Nordstrom C-H (2005) Intracerebral microdialysis in severe brain trauma: The importance of catheter location. J Neurosurg 102:460–469.

Ereifej ES, Khan S, Newaz G, Zhang J, Auner GW, Vande VPJ (2011) Characterization of astrocyte reactivity and gene expression on biomaterials for neural electrodes. J Biomed Mater Res, Part A 99A:141–150.

Erickson MA, Banks WA (2011) Cytokine and chemokine responses in serum and brain after single and repeated injections of lipopolysaccharide: Multiplex quantification with path analysis. Brain Behav Immun 25:1637–1648.

Erta M, Quintana A, Hidalgo J (2012) Interleukin-6, a major cytokine in the central nervous system. Int J Biol Sci 8:1254–1266.

Ewen C, Baca-Estrada ME (2001) Evaluation of interleukin-4 concentration by ELISA is influenced by the consumption of IL-4 by cultured cells. J Interferon Cytokine Res 21:39–43.

Fan L, Young PR, Barone FC, Feuerstein GZ, Smith DH, McIntosh TK (1995) Experimental brain injury induces expression of interleukin-1β mRNA in the rat brain. Brain Res Mol Brain Res 30:125–130.

Fassbender K, Schneider S, Bertsch T, Schlueter D, Fatar M, Ragoschke A, Kuhl S, Kischka U, Hennerici M (2000) Temporal profile of release of interleukin-1 beta in neurotrauma. Neurosci Lett 284.

Forcelli PA, Sweeney CT, Kammerich AD, Lee BCW, Rubinson LH, Kayinamura YP, Gale K, Rubinson JF (2012) Histocompatibility and *in vivo* signal throughput for PEDOT, PEDOP, P3MT, and polycarbazole electrodes. J Biomed Mater Res, Part A 100A:3455–3462.

Frugier T, Morganti-Kossmann MC, O'Reilly D, McLean CA (2010) *In situ* detection of inflammatory mediators in post mortem human brain tissue after traumatic injury. J Neurotrauma 27:497–507.

Galic MA, Riazi K, Pittman QJ (2012) Cytokines and brain excitability. Front Neuroendocrinol 33:116–125.

Galliera E, Corsi MM, Banfi G (2012) Platelet rich plasma therapy: Inflammatory molecules involved in tissue healing. J Biol Regul Homeost Agents 26:35S–42S.

Gandhi NS, Mancera RL (2008) The structure of glycosaminoglycans and their interactions with proteins. Chem Biol Drug Des 72:455–482.

Gear ARL, Camerini D (2003) Platelet chemokines and chemokine receptors: Linking hemostasis, inflammation, and host defense. Microcirculation 10:335–350.

Gibbs RV (2003) Cytokines and glycosaminoglycans (GAGs). Adv Exp Med Biol 535:125–143.

Giulietti A, Overbergh L, Valckx D, Decallonne B, Bouillon R, Mathieu C (2001) An overview of real-time quantitative PCR: Applications to quantify cytokine gene expression. Methods 25:386–401.

Grand L, Wittner L, Herwik S, Goethelid E, Ruther P, Oscarsson S, Neves H, Dombovari B, Csercsa R, Karmos G, Ulbert I (2010) Short and long term biocompatibility of NeuroProbes silicon probes. J Neurosci Methods 189:216–229.

Guenin S, Mauriat M, Pelloux J, Van WO, Bellini C, Gutierrez L (2009) Normalization of qRT-PCR data: The necessity of adopting a systematic, experimental conditions-specific, validation of references. J Exp Bot 60:487–493.

Han Q, Bradshaw EM, Nilsson B, Hafler DA, Love JC (2010) Multidimensional analysis of the frequencies and rates of cytokine secretion from single cells by quantitative microengraving. Lab Chip 10:1391–1400.

Hanafy KA, Grobelny B, Fernandez L, Kurtz P, Connolly ES, Mayer SA, Schindler C, Badjatia N (2010) Brain interstitial fluid TNF-α after subarachnoid hemorrhage. J Neurol Sci 291:69–73.

Handel TM, Johnson Z, Crown SE, Lau EK, Sweeney M, Proudfoot AE (2005) Regulation of protein function by glycosaminoglycans — as exemplified by chemokines. Annu Rev Biochem 74:385–410.

Hausmann EH, Berman NE, Wang YY, Meara JB, Wood GW, Klein RM (1998) Selective chemokine mRNA expression following brain injury. Brain Res 788:49–59.

Helmy A, Carpenter KLH, Hutchinson PJ (2007) Microdialysis in the human brain and its potential role in the development and clinical assessment of drugs. Curr Med Chem 14:1525–1537.

Helmy A, De SM-G, Guilfoyle MR, Carpenter KLH, Hutchinson PJ (2011a) Cytokines and innate inflammation in the pathogenesis of human traumatic brain injury. Prog Neurobiol 95:352–372.

Helmy A, Carpenter KLH, Menon DK, Pickard JD, Hutchinson PJA (2011b) The cytokine response to human traumatic brain injury: Temporal profiles and evidence for cerebral parenchymal production. J Cereb Blood Flow Metab 31:658–670.

Helmy A, Antoniades CA, Guilfoyle MR, Carpenter KLH, Hutchinson PJ (2012) Principal component analysis of the cytokine and chemokine response to human traumatic brain injury. PLoS One 7:e39677.

Herbaugh AW, Stenken JA (2011) Antibody-enhanced microdialysis collection of CCL2 from rat brain. J Neurosci Methods 202:124–127.

Hillered L, Vespa PM, Hovda DA (2005) Translational neurochemical research in acute human brain injury: The current status and potential future for cerebral microdialysis. J Neurotrauma 22:3–41.

Hillman J, Aaneman O, Persson M, Andersson C, Dabrosin C, Mellergaard P (2007) Variations in the response of interleukins in neurosurgical intensive care patients monitored using intracerebral microdialysis. J Neurosurg 106:820–825.

Hrabetova S, Nicholson C (2007) Biophysical properties of brain extracellular space explored with ion-selective microelectrodes, integrative optical imaging and related techniques. In: Electrochemical methods for neuroscience (Michael AC, Borland LM, eds), pp 167–204. Boca Raton, FL: CRC Press.

Humpel C, Lindqvist E, Soederstroem S, Kylberg A, Ebendal T, Olson L (1995) Monitoring release of neurotrophic activity in the brains of awake rats. Science 269:552–554.

Hutchinson PJ, O'Connell MT, Kirkpatrick PJ, Pickard JD (2002) How can we measure substrate, metabolite and neurotransmitter concentrations in the human brain? Physiol Meas 23:R75–R109.

Hutchinson PJ, O'Connell MT, Rothwell NJ, Hopkins SJ, Nortje J, Carpenter KLH, Timofeev I, Al-Rawi PG, Menon DK, Pickard JD (2007) Inflammation in human brain injury: Intracerebral concentrations of IL-1alpha, IL-1beta, and their endogenous inhibitor IL-1ra. J Neurotrauma 24:1545–1557.

Iliff JJ, Wang M, Liao Y, Plogg BA, Peng W, Gundersen GA, Benveniste H, Vates GE, Deane R, Goldman SA, Nagelhus EA, Nedergaard M (2012) A paravascular pathway facilitates CSF flow through the brain parenchyma and the clearance of interstitial solutes, including amyloid β. Sci Transl Med 4:ra111, 112 pp.

Jackson A, Zimmermann JB (2012) Neural interfaces for the brain and spinal cord-restoring motor function. Nat Rev Neurol 8:690–699.

Jacobson I, Sandberg M, Hamberger A (1985) Mass transfer in brain dialysis devices — a new method for the estimation of extracellular amino acids concentration. J Neurosci Methods 15:263–268.

Jaerve A, Mueller HW (2012) Chemokines in CNS injury and repair. Cell Tissue Res 349:229–248.

Jansky L, Vybiral S, Pospisilova D, Roth J, Dornand J, Zeisberger E, Kaminkova J (1995) Production of systemic and hypothalamic cytokines during the early phase of endotoxin fever. Neuroendocrinology 62:55–61.

Johnson JD, Campisi J, Sharkey CM, Kennedy SL, Nickerson M, Greenwood BN, Fleshner M (2005) Catecholamines mediate stress-induced increases in peripheral and central inflammatory cytokines. Neuroscience 135:1295–1307.

Jones SA, Rose-John S (2002) The role of soluble receptors in cytokine biology: The agonistic properties of the sIL-6R/IL-6 complex. Biochim Biophys Acta, Mol Cell Res 1592:251–263.

Kakizaki Y, Watanobe H, Kohsaka A, Suda T (1999) Temporal profiles of interleukin-1β, interleukin-6, and tumor necrosis factor-α in the plasma and hypothalamic paraventricular nucleus after intravenous or intraperitoneal administration of lipopolysaccharide in the rat: Estimation by push–pull perfusion. Endocr J 46:487–496.

Karumbaiah L, Norman SE, Rajan NB, Anand S, Saxena T, Betancur M, Patkar R, Bellamkonda RV (2012) The upregulation of specific interleukin (IL) receptor antagonists and paradoxical enhancement of neuronal apoptosis due to electrode induced strain and brain micromotion. Biomaterials 33:5983–5996.

Klir JJ, McClellan JL, Kluger MJ (1994) Interleukin-1β causes the increase in anterior hypothalamic interleukin-6 during LPS-induced fever in rats. Am J Physiol 266:R1845–R1848.

Klir JJ, Roth J, Szelenyi Z, McClellan JL, Kluger MJ (1993) Role of hypothalamic interleukin-6 and tumor necrosis factor-α in LPS fever in rat. Am J Physiol 265:R512–R517.

Kolarcik CL, Bourbeau D, Azemi E, Rost E, Zhang L, Lagenaur CF, Weber DJ, Cui XT (2012) *In vivo* effects of L1 coating on inflammation and neuronal health at the electrode–tissue interface in rat spinal cord and dorsal root ganglion. Acta Biomater 8:3561–3575.

Kottegoda S, Shaik I, Shippy SA (2002) Demonstration of low flow push–pull perfusion. J Neurosci Methods 121:93–101.

Kuil J, Buckle T, van LFWB (2012) Imaging agents for the chemokine receptor 4 (CXCR4). Chem Soc Rev 41:5239–5261.

Kwon BK, Stammers AMT, Belanger LM, Bernardo A, Chan D, Bishop CM, Slobogean GP, Zhang H, Umedaly H, Giffin M, Street J, Boyd MC, Paquette SJ, Fisher CG, Dvorak MF (2010) Cerebrospinal fluid inflammatory cytokines and biomarkers of injury severity in acute human spinal cord injury. J Neurotrauma 27:669–682.

Lambertsen KL, Biber K, Finsen B (2012) Inflammatory cytokines in experimental and human stroke. J Cereb Blood Flow Metab 32:1677–1698.

Lampron A, Elali A, Rivest S (2013) Innate immunity in the CNS: Redefining the relationship between the CNS and its environment. Neuron 78:214–232.

Lavisse S, Guillermier M, Herard A-S, Petit F, Delahaye M, Van CN, Ben HL, Lebon V, Remy P, Dolle F, Delzescaux T, Bonvento G, Hantraye P, Escartin C (2012) Reactive astrocytes overexpress TSPO and are detected by TSPO positron emission tomography imaging. J Neurosci 32:10809–10818.

Leprince L, Dogimont A, Magnin D, Demoustier-Champagne S (2010) Dexamethasone electrically controlled release from polypyrrole-coated nanostructured electrodes. J Mater Sci: Mater Med 21:925–930.

Li M, Hale JS, Rich JN, Ransohoff RM, Lathia JD (2012) Chemokine CXCL12 in neurodegenerative diseases: An SOS signal for stem cell-based repair. Trends Neurosci 35:619–628.

Lind G, Gaellentoft L, Danielsen N, Schouenborg J, Pettersson LME (2012) Multiple implants do not aggravate the tissue reaction in rat brain. PLoS One 7:e47509.

Mabrouk OS, Kennedy RT (2013) Measurement of neuropeptides in dialysate by LC-MS. Neuromethods 75:249–259.

Maeda N, Ishii M, Nishimura K, Kamimura K (2011) Functions of chondroitin sulfate and heparan sulfate in the developing brain. Neurochem Res 36:1228–1240.

Mahad DJ, Ransohoff RM (2003) The role of MCP-1 (CCL2) and CCR2 in multiple sclerosis and experimental autoimmune encephalomyelitis (EAE). Semin Immunol 15:23–32.

Maier T, Gueell M, Serrano L (2009) Correlation of mRNA and protein in complex biological samples. FEBS Letters 583:3966–3973.

Marcus HJ, Carpenter KLH, Price SJ, Hutchinson PJ (2010) *In vivo* assessment of high-grade glioma biochemistry using microdialysis: A study of energy-related molecules, growth factors and cytokines. J Neuro-Oncol 97:11–23.

Marklund N, Blennow K, Zetterberg H, Ronne-Engstroem E, Enblad P, Hillered L (2009) Monitoring of brain interstitial total tau and beta amyloid proteins by microdialysis in patients with traumatic brain injury. J Neurosurg 110:1227–1237.

Matalka KZ, Tutunji MF, Abu-Baker M, Abu-Baker Y (2005) Measurement of protein cytokines in tissue extracts by enzyme-linked immunosorbent assays: Application to lipopolysaccharide-induced differential milieu of cytokines. Neuroendocrinol Lett 26:231–236.

Mellergard P, Sjogren F, Hillman J (2010) Release of VEGF and FGF in the extracellular space following severe subarachnoidal haemorrhage or traumatic head injury in humans. Br J Neurosurg 24:261–267.

Mellergard P, Sjogren F, Hillman J (2012) The cerebral extracellular release of glycerol, glutamate, and FGF2 is increased in older patients following severe traumatic brain injury. J Neurotrauma 29:112–118.

Mellergard P, Aneman O, Sjogren F, Pettersson P, Hillman J (2008) Changes in extracellular concentrations of some cytokines, chemokines, and neurotrophic factors after insertion of intracerebral microdialysis catheters in neurosurgical patients. Neurosurgery 62:151–157; discussion 157–158.

Merighi A, Salio C, Ferrini F, Lossi L (2011) Neuromodulatory function of neuropeptides in the normal CNS. J Chem Neuroanat 42:276–287.

Michael AC, Borland LM, eds (2007) Electrochemical methods for neuroscience. Boca Raton: CRC Press.

Minami M, Kuraishi Y, Satoh M (1991) Effects of kainic acid on messenger RNA levels of IL-1β, IL-6, TNFα and LIF in the rat brain. Biochem Biophys Res Commun 176:593–598.

Molina-Holgado E, Molina-Holgado F (2010) Mending the broken brain: Neuroimmune interactions in neurogenesis. J Neurochem 114:1277–1290.

Mueller N, Schwarz MJ (2007) The immune-mediated alteration of serotonin and glutamate: Towards an integrated view of depression. Mol Psychiatry 12:988–1000.

Myers RD, Adell A, Lankford MF (1998) Simultaneous comparison of cerebral dialysis and push–pull perfusion in the brain of rats: A critical review. Neurosci Biobehav Rev 22:371–387.

Müller M, ed (2013) Microdialysis in drug development. New York: AAPS Springer.

Nandi P, Lunte SM (2009) Recent trends in microdialysis sampling integrated with conventional and microanalytical systems for monitoring biological events: A review. Anal Chim Acta 651:1–14.

Nawa H, Carnahan J, Gall C (1995) BDNF protein measured by a novel enzyme immunoassay in normal brain and after seizure: Partial disagreement with mRNA levels. Eur J Neurosci 7:1527–1535.

Novak U, Kaye AH (2000) Extracellular matrix and the brain: Components and function. J Clin Neurosci 7:280–290.

Oishi S, Fujii N (2012) Peptide and peptidomimetic ligands for CXC chemokine receptor 4 (CXCR4). Org Biomol Chem 10:5720–5731.

Olson L, Humpel C (2010) Growth factors and cytokines/chemokines as surrogate biomarkers in cerebrospinal fluid and blood for diagnosing Alzheimer's disease and mild cognitive impairment. Exp Gerontol 45:41–46.

Ordonez J, Schuettler M, Boehler C, Boretius T, Stieglitz T (2012) Thin films and microelectrode arrays for neuroprosthetics. MRS Bull 37:590–598.

Overbergh L, Giulietti A, Valckx D, Decallonne R, Bouillon R, Mathieu C (2003) The use of real-time reverse transcriptase PCR for the quantification of cytokine gene expression. J Biomol Tech 14:33–43.

Pachler C, Ikeoka D, Plank J, Weinhandl H, Suppan M, Mader JK, Bodenlenz M, Regittnig W, Mangge H, Pieber TR, Ellmerer M (2007) Subcutaneous adipose tissue exerts proinflammatory cytokines after minimal trauma in humans. Am J Physiol 293:E690–E696.

Panaro MA, Cianciulli A (2012) Current opinions and perspectives on the role of immune system in the pathogenesis of Parkinson's disease. Curr Pharm Des 18:200–208.

Peled A, Wald O, Burger J (2012) Development of novel CXCR4-based therapeutics. Expert Opin Invest Drugs 21:341–353.

Perez-Barcena J, Ibanez J, Brell M, Crespi C, Frontera G, Llompart-Pou JA, Homar J, Abadal JM (2011) Lack of correlation among intracerebral cytokines, intracranial pressure, and brain tissue oxygenation in patients with traumatic brain injury and diffuse lesions. Crit Care Med 39:533–540.

Polikov VS, Tresco PA, Reichert WM (2005) Response of brain tissue to chronically implanted neural electrodes. J Neurosci Methods 148:1–18.

Quan N, Stern EL, Whiteside MB, Herkenham M (1999) Induction of pro-inflammatory cytokine mRNAs in the brain after peripheral injection of subseptic doses of lipopolysaccharide in the rat. J Neuroimmunol 93:72–80.

Ransohoff RM, Benveniste EN, eds (2006) Cytokines and the CNS, 2nd Edition. Boca Raton, FL: CRC Press, Taylor & Francis.

Reaux-Le GA, Van SJ, Rostene W, Melik PS (2013) Current status of chemokines in the adult CNS. Prog Neurobiol (Oxford, UK) 104:67–92.

Rijkers K, Majoie HJ, Hoogland G, Kenis G, De BM, Vles JS (2009) The role of interleukin-1 in seizures and epilepsy: A critical review. Exp Neurol 216:258–271.

Robinson TE, Justice JB, Jr., eds (1991) Microdialysis in the neurosciences. Amsterdam: Elsevier.

Rot A (2010) Chemokine patterning by glycosaminoglycans and interceptors. Front Biosci, Landmark Ed 15:645–660.

Roth J, Conn CA, Kluger MJ, Zeisberger E (1993) Kinetics of systemic and intrahypothalamic IL-6 and tumor necrosis factor during endotoxin fever in guinea pigs. Am J Physiol 265:R653–R658.

Rubio-Perez JM, Morillas-Ruiz JM (2012) A review: Inflammatory process in Alzheimer's disease, role of cytokines. Scientific World Journal 2012:756357.

Rupert AE, Ou Y, Sandberg M, Weber SG (2013) Electroosmotic push–pull perfusion: Description and application to qualitative analysis of the hydrolysis of exogenous galanin in organotypic hippocampal slice cultures. ACS Chem Neurosci 4:838–848.

Saxena T, Karumbaiah L, Gaupp EA, Patkar R, Patil K, Betancur M, Stanley GB, Bellamkonda RV (2013) The impact of chronic blood–brain barrier breach on intracortical electrode function. Biomaterials 34:4703–4713.

Schmerberg CM, Li L (2013) Function-driven discovery of neuropeptides with mass spectrometry-based tools. Protein Pept Lett 20:681–694.

Semple BD, Kossmann T, Morganti-Kossmann MC (2010) Role of chemokines in CNS health and pathology: A focus on the CCL2/CCR2 and CXCL8/CXCR2 networks. J Cereb Blood Flow Metab 30:459–473.

Seymour JP, Kipke DR (2007) Neural probe design for reduced tissue encapsulation in CNS. Biomaterials 28:3594–3607.

Skrzydelski D, Guyon A, Dauge V, Rovere C, Apartis E, Kitabgi P, Nahon Jl, Rostene W, Melik-Parsadaniantz S (2007) The chemokine stromal cell-derived factor-1/CXCL12 activates the nigrostriatal dopamine system. J Neurochem 102:1175–1183.

Slaney TR, Mabrouk OS, Porter-Stransky KA, Aragona BJ, Kennedy RT (2013) Chemical gradients within brain extracellular space measured using low flow push–pull perfusion sampling *in vivo*. ACS Chem Neurosci 2:321–329.

Spooren A, Kolmus K, Laureys G, Clinckers R, De KJ, Haegeman G, Gerlo S (2011) Interleukin-6, a mental cytokine. Brain Res Rev 67:157–183.

Stangel M, Fredrikson S, Meinl E, Petzold A, Stueve O, Tumani H (2013) The utility of cerebrospinal fluid analysis in patients with multiple sclerosis. Nat Rev Neurol 9:267–276.

Sykova E, Nicholson C (2008) Diffusion in brain extracellular space. Physiol Rev 88:1277–1340.

Szerb JC (1967) Model experiments with Gaddum's push–pull cannulas. Can J Physiol Pharmacol 45:613–620.

Terasaki M, Sugita Y, Arakawa F, Okada Y, Ohshima K, Shigemori M (2011) CXCL12/CXCR4 signaling in malignant brain tumors: A potential pharmacological therapeutic target. Brain Tumor Pathol 28:89–97.

Tiveron M-C, Cremer H (2008) CXCL12/CXCR4 signalling in neuronal cell migration. Curr Opin Neurobiol 18:237–244.

Tresco PA, Winslow BD (2011) The challenge of integrating devices into the central nervous system. Crit Rev Biomed Eng 39:29–44.

Turrin NP, Plata-Salaman CR (2000) Cytokine–cytokine interactions and the brain. Brain Res Bull 51:3–9.

Ungerstedt U (1991) Microdialysis — principles and applications for studies in animals and man. J Intern Med 230:365–373.

Van Eeckhaut A, Maes K, Aourz N, Smolders I, Michotte Y (2011) The absolute quantification of endogenous levels of brain neuropeptides *in vivo* using LC-MS/MS. Bioanalysis 3:1271–1285.

Van WNJ, Benveniste EN (1999) Interleukin-6 expression and regulation in astrocytes. J Neuroimmunol 100:124–139.

Vargova L, Homola A, Cicanic M, Kuncova K, Krsek P, Marusic P, Sykova E, Zamecnik J (2011) The diffusion parameters of the extracellular space are altered in focal cortical dysplasias. Neurosci Lett 499:19–23.

Vasicek TW, Jackson MR, Poseno TM, Stenken JA (2013) *In vivo* microdialysis sampling of cytokines from rat hippocampus: Comparison of cannula implantation procedures. ACS Chem Neurosci 4:737–746.

Vezzani A, Balosso S, Ravizza T (2008) The role of cytokines in the pathophysiology of epilepsy. Brain Behav Immun 22:797–803.

Vignali DA (2000) Multiplexed particle-based flow cytometric assays. J Immunol Methods 243:243–255.

Vitkovic L, Bockaert J, Jacque C (2000) "Inflammatory" cytokines: Neuromodulators in normal brain? J Neurochem 74:457–471.

Vogel C, Marcotte EM (2012) Insights into the regulation of protein abundance from proteomic and transcriptomic analyses. Nat Rev Genet 13:227–232.

Waelgaard L, Pharo A, Tonnessen TI, Mollnes TE (2006) Microdialysis for monitoring inflammation: Efficient recovery of cytokines and anaphylotoxins provided optimal catheter pore size and fluid velocity conditions. Scand J Immunol 64:345–352.

Wang Y, Luo W, Reiser G (2008) Trypsin and trypsin-like proteases in the brain: Proteolysis and cellular functions. Cell Mol Life Sci 65:237–252.

Watanobe H, Hayakawa Y (2003) Hypothalamic interleukin-1β and tumor necrosis factor-α, but not interleukin-6, mediate the endotoxin-induced suppression of the reproductive axis in rats. Endocrinology 144:4868–4875.

Westerink BHC, Cremers TIFH, eds (2007) Handbook of microdialysis sampling: Methods, applications, and clinical aspects. Amsterdam: Academic Press.

White FA, Miller RJ (2010) Insights into the regulation of chemokine receptors by molecular signaling pathways: Functional roles in neuropathic pain. Brain Behav Immun 24:859–865.

Winter CD, Iannotti F, Pringle AK, Trikkas C, Clough GF, Church MK (2002) A microdialysis method for the recovery of IL-1β, IL-6 and nerve growth factor from human brain *in vivo*. J Neurosci Methods 119:45–50.

Wolak DJ, Thorne RG (2013) Diffusion of macromolecules in the brain: Implications for drug delivery. Mol Pharmaceutics 10:1492–1504.

Woodroofe MN, Sarna GS, Wadhwa M, Hayes GM, Loughlin AJ, Tinker A, Cuzner ML (1991) Detection of interleukin-1 and interleukin-6 in adult rat brain, following mechanical injury, by *in vivo* microdialysis: Evidence of a role for microglia in cytokine production. J Neuroimmunol 33:227–236.

Woolley AJ, Desai HA, Gaire J, Ready AL, Otto KJ (2013) Intact histological characterization of brain-implanted microdevices and surrounding tissue. J Visualized Exp:e50126.

Wotjak CT, Landgraf R, Engelmann M (2008) Listening to neuropeptides by microdialysis: Echoes and new sounds? Pharmacol Biochem Behav 90:125–134.

Xiao F, Nicholson C, Hrabe J, Hrabetova S (2008) Diffusion of flexible random-coil dextran polymers measured in anisotropic brain extracellular space by integrative optical imaging. Biophys J 95:1382–1392.

Yan HQ, Banos MA, Herregodts P, Hooghe R, Hooghe-Peters EL (1992) Expression of interleukin (IL)-1β, IL-6 and their respective receptors in the normal rat brain and after injury. Eur J Immunol 22:2963–2971.

Yue Z, Moulton SE, Cook M, O'Leary S, Wallace GG (2013) Controlled delivery for neuro-bionic devices. Adv Drug Delivery Rev 65:559–569.

Zhao W, Oskeritzian CA, Pozez AL, Schwartz LB (2005) Cytokine production by skin-derived mast cells: Endogenous proteases are responsible for degradation of cytokines. J Immunol 175:2635–2642.

INDEX

A_1 receptor, 81, 82, 89, 99
A_1 receptor agonist, 89
A_{2A} receptor, 66, 81, 88
A_{2B} receptor, 81
$A_3 \cdot A_1$ receptor, 81
AAV, choice of, 210–211
AAV serotype(s), 2, 210–212
AAV-based ChR2 constructs, 210
AAV-packaged vectors, 210
AAV5-packaged ChR2 injections, 211
Acetylcholine (ACh), 29, 38, 138, 142–144
Acetylcholinesterase (AChE), 31
Acetylcholinesterase inhibitors, 29
ACh fluctuations, 144
ACh-driven DA release, 216–217
ACh-evoked transients, 32
AChE inhibition, 31
Actin, 192
Actin cytoskeleton, 192
Action potentials, 15
Activity-dependent adenosine release, 62, 64, 65, 69, 90, 92, 96–98
Adenosine (ADO), 4, 6, 12, 13–15, 19, 45, 47, 60–62, 65–69, 79–84, 86, 88–93, 95, 97–99, 225 .
Adenosine A1 receptor agonists, 15
Adenosine biosensing principles, 59–60
Adenosine biosensors, 14, 15, 60–62, 66, 68

Adenosine characterization, 83, 84, 95
Adenosine deaminase (AD), 14, 46, 59, 60, 69, 91, 96
Adenosine detection using FSCV, 83, 84, 87
Adenosine diphosphate (ADP), 45
Adenosine in hypercapnia, 84
Adenosine kinase (ADK), 46, 48, 88, 99–101
Adenosine kinase (ADK) inhibitor, 88
Adenosine modulates neurotransmission, 83
Adenosine monophosphate (AMP), 48, 80
Adenosine/purine biosensor, 60
Adenosine receptor manipulation, 82
Adenosine receptors, 47–48, 80–81
Adenosine release, 49, 59, 66, 67
Adenosine release, activity dependence of, 60–65
Adenosine transport inhibitor, 66
Adenosine triphosphate (ATP), 45, 79, 319
Adenylate kinase (AK), 46
Adrenal chromaffin cells, 183
Adrenocorticotropin, 336
Adsorption, 141
Alcoholism, 370, 371
Alkaline phosphatase (AP), 46, 48
α-Amino-3-hydroxy-5-methyl-4-isoxazolepropionic acid (AMPA), 90

AMPA receptors, 63, 64
α-Melanocyte-stimulating hormone (MSH), 336
α-Methyl-p-tyrosine (αMPT), 321
Alzheimer's treatment, 371, 377
AM630, 253
Amfonelic acid, 302
Amperometric biosensors, 139
Amperometric enzyme biosensors, 84, 99
Amperometric measured transmitter release, 188
Amperometric measurements, 183, 188, 196
Amperometric spikes, 186, 197
Amperometric technique, 186
Amperometry, 183, 193–197, 271–272
AMP hydrolyzes, 92
AMPH action, mechanism of, 288, 297, 302
AMPH effects, 299
AMPH-induced dopamine efflux, 298
Amphetamine (AMPH), 287
AMpT, 301
Anandamide, 256, 259
Anatomical localization of peptides, 337
Anesthesia, 119, 124
Anterior associative cortices, 28
Antidepressants, 269–270, 276, 278
Anti-Parkinsonian effects, 227
Antipsychotics, 269
Aplysia californica bag cell neuron, 338, 339
Appetite, 370
2-Arachidonoylglycerol, 256, 259, 260, 261, 262
Arc (activity regulated, cytoskeletal-associated), 296, 297
Arcuate nucleus (ARC), 54
Artificial cerebral spinal fluid (aCSF) perfusate, 321
Artificial cerebrospinal fluid (aCSF), 101

Ascorbic acid, 271
Astrocyte neuron lactate shuttle (ANLS) hypothesis, 15–18
Astrocytes, 15, 81, 92, 101, 373, 374, 376
Astroglial scar, 351, 356, 357, 358
ATP antagonists, 52
ATP biosensing principles, 50–51
ATP biosensor, 50–52, 57, 59, 68
ATP metabolism, 79, 80, 90, 99
ATP receptors, 46, 49, 59, 68
ATP signaling in chemosensing, 51–57
ATP signaling in triggering eye development, 58
ATP-release, signaling and functions, 50–59
ATP-sensitive potassium (K-ATP) channel, 55
Autism, 269
Autocrine regulation, 371
Axonal DA release, 202

Bacterial artificial chromosome (BAC) technology, 208
Basal forebrain (BFB), 101
Basal synaptic adenosine level, 100
Basolateral amygdala, 144, 146
BBB leakage, 358
β-Endorphin, 336
β-Lipotropin, 336
β-MSH, 336
Bi-directional ENT, 83
Bilipidic membranes, 189
Bioactive proteins, 370, 378
Biosensor, 1, 3, 5–7, 79, 84, 97, 102
Biosensor studies during sleep, 6–7
Blood, 372
Blood–brain barrier (BBB), 357, 373, 389
Blood-oxygen-level-dependent (BOLD) signal, 20
Bluetooth transmission chip, 235

Boron-doped diamond microelectrodes, 87
Bovine serum albumin (BSA), 118
Brain, extracellular space of, 113
Brain Derived Neurotrophic Factor (BDNF), 373
Brain electrode implants, 351
Brain glucose concentration fluctuations, 152
Brain implants, 389
Brain response to intracortical electrode implants, 356
Brain stimulation, response latency, 257, 258, 260
Brain–computer interfaces (BCI), 351

^{14}C-dopamine, 381
CAG repeat, 311, 315
Calibration, 378, 380–381, 382
CamK2a-Cre mouse line, 205
Cannabinoids, 251, 252, 253, 254
Cannabis, 251
Capillary electrophoresis (CE), 336
Carbon materials, 147
Carbon-fiber, 183
Carbon-fiber microbiosensors, 149
Carbon-fiber microelectrode (CFM), 84, 86–88, 91, 92, 95, 96, 104, 137, 147–152, 202, 229, 287
CA1 region of hippocampus, 90, 91
Catecholamine, 138, 185, 186, 197, 369, 370, 381, 384
Catecholamine amounts in PC12 cells, 185
Caudate putamen (CPu), 88, 89, 90, 91, 153, 205
CB1 receptors, 252, 253, 255
CB2 receptors, 251, 252
CCL2/MCP-1 (monocyte chemoattractant protein-1), 374, 388
CCL20 (MIP-3α), 389
CCL2 production, 375
CCR2 (receptor), 374, 375
CE-based methods, 337
CE–ESI–MS, 337
Central CO_2 chemoreception, 51–54
Central nervous system, adenosine regulation in, 79, 80, 82
Central nervous system (CNS), 114, 311, 377
Centrally programmed fatigue, 68
Ceramic alumina, 115
Ceramic-based MEAs, 115, 122, 125, 126
Cerebellum, 61, 63, 96, 97, 98
Cerebral spinal fluid (CSF), 372
CFM photo-activation, 214
CGS 21680, 88
Channel Rhodospin2 (ChR2), 205
Chemical synapse, 182
Chemical vapor deposition (CVD), 87
Chemokine, 370–374, 376, 379, 387
Chemosensitive properties of tanycytes, 55
Chemosensory signal, 54
ChETA, 216
ChI firing, 203
Chitosan, 148
Choline, 31, 32, 142
Choline acetyltransferase positive (CHAT+) neurons, 205
Cholinergic basal forebrain (BFB), 65
Cholinergic interneurons (ChIs), 202, 204–207, 213
Cholinergic projections, 143
Cholinergic synapses, 31
Cholinergic system, 27, 28, 30, 37
Cholinergic transients, 28, 31–33, 35–36, 37, 39
Choline transporter (CHT), 32
Chondroitin sulfate, 379

ChR2, 205, 208–211, 216
ChR2 expression, 208, 209, 211–213
ChR2 "reporter" lines, 209
ChR2 vectors, 209, 210
ChR2-expressing neurons, 205
ChR2-expressing striatal ChIs, 205
ChR2-expression ChIs, 213
ChR2-positive thalamic fibers, 213
Chromatographic separation, 370
Chronic implantation, 124–128
Chronic microdialysis, 114
Chronoamperometry, 272
"Classical" neuropeptides, 336
Cluster of differentiation 39 (CD39), 48
Cluster of differentiation 73 (CD73), 49, 98
CNS biocompatibility, 125
CNS diseases, 371
Cocaine, 298, 300, 303
CO_2 chemoreception, 51–54
Cognitive dysfunction, 288, 289, 293
Co-localization of peptides, 337
Concentrative nucleoside transporters, 82
Conformal front/back MEA, 117
Constant or fixed potential amperometry (FPA), 225
Controlled iontophoresis, 171, 172
Conventional amperometry, 150, 151
Cortical cholinergic input system, 28–29
Cortical dysplasia, 379
Cortical electrodes, 352, 354
Corticotropin-releasing hormone, 337
Coupling controlled iontophoresis, 172
Coupling of FCV and optogenetics, 207–208
cPTIO, 66
Cre-loxP optogenetic, 208
Cre-loxP system, 208, 210
Cre-recombinase, 203, 205, 209, 210
Cross-linked choline oxidase, 142–144
Cued-appetitive response task (CART), 32
CX3CL1 (fractalkine), 374, 376
CXCL12/SDF-1 (stromal derived factor-1), 372, 374–375
CXCR4, 372
CXCR4 receptor, 372
Cyclic voltammogram (CV), 86, 317, 358, 359
Cystine/glutamate antiporter inhibitor (S)-4-carboxyphenylglycine (CPG), 121
Cystine/glutamate exchanger, 121
Cytokine, 369–390
Cytokine network, 371, 387
Cytosine-adenine-guanine (CAG), 311
C18 ZipTip pipette tip, 341, 343

D, l-Threo-β-benzyloxyaspartate (TBOA), 121
D-Amphetamine (AMPH) sulfate, 321
D1 receptor, 314
D_1 receptors modulated, 89
D1 MSNs, 163, 170, 172, 173
D1 receptor-expressing striatonigral efferent neurons, 296
D2-autoreceptors, 163, 172
D2 MSNs, 163, 170, 172, 173
DA and ADO recording, 239
DA axon terminals, 202, 204–207
DA detection with FCV, 202
DA neurons, 201–202, 204
DA release, 202, 204, 206–207, 213, 214
DA re-uptake inhibition, 229
DA transmission, 203–204
DBS surgery, 226, 228, 239
Deep Brain Stimulation (DBS), 116, 225
Depolarization of cell membrane, 181
Depression, 138, 144, 155, 269, 370
Dextran, 379
Diamond microelectrodes, 84, 87, 88

Diffuse neuromodulator, 38
Diffusion coefficient, 378, 382
Diffusivity (*DECS*), 378, 383
3,4-Dihydroxyphenylacetic acid (DOPAC), 118, 229, 271
Disrupted sleep, 1
Dopamine (DA), 138, 161–165, 166, 170, 201, 225, 271, 273, 274, 378, 383
Dopamine degradative enzyme, 300
Dopamine neurotransmission, 161–163, 168
Dopamine receptors, 163, 168
Dopamine signaling, 287–296
Dopamine terminals form synapses, 162
Dopamine therapies, 95
Dopamine transporter (DAT), 276
Dopamine-dependent, 167
Dopamine-depleted brain, dopamine signaling in, 289–296
Dopamine-depleted conditions of Parkinson's disease (PD), 287
Dopamine-depleted striatum, 290
Dopaminergic axons, 202
Dopaminergic neurons, 161, 166
Dorsal caudate putamen, 90, 91
Dorsal raphe nucleus (DRN), 273
Dorsal STN, 236
Dual-enzyme detection strategy, 143
Dura mater, 357
Dynamin, 190
Dynasore, 191
Dynorphin peptide, 337

ECoG activity, 352, 353
ECoG arrays, implantation of, 354
Ecto 5'-nucleotidase (e5'-NT), 46, 49
Ectonucleotidases, 48–49
Edema, 379, 381
EEG activity, 352
Efflux, 298, 302–304

EGF, 385
Electrical stimulation *versus* optogenetics, 203
Electroactive molecule, 139
Electrochemical biosensors, 138–139
Electrochemical cytometry, 183–186
Electrochemical detection (CE-EC), 184
Electrochemical detection of adenosine, 79–105
Electrochemical methods, 369
Electrochemical monitoring techniques, 225
Electrochemical recording electrode lead, 235
Electrochemical techniques, 181
Electrochemistry in neuroscience, 137
Electrochemistry with electrophysiology, combining, 167–170
Electrocortcographic (ECoG) activity, 352
Electroencephalographic sleep measurement, 1, 2–4, 5, 9, 11, 13, 18
Electroencephalography (EEG), 352
Electromyograph muscle activity (EMG), 4
Electromyography (EMG), 353
Electro-osmosis, 171
Electro-osmotic flow, 171, 172
Electrophoresis, 184
Electrophoretic separation, 369
Electrophysiological techniques, 144
Electrospray ionization (ESI), 336
ELISA, 372, 387
Encapsulation, 140, 148
Endocannabinoid, 251, 256, 370
Endocannabinoid system, 255–261
Endothelial cells, 374, 376
Energy homeostasis, 54, 55
Enhanced dopamine diffusion, 290

Enhanced yellow fluorescent protein (eYFP), 205
ENT inhibitor, 88
Entrapment, 140
Enzyme encapsulation within chitosan by electrodeposition, 148
Enzyme immobilization, 140, 141
Epilepsy, 82, 84, 100–101, 370, 371, 384
Epileptic episodes, 82
Epinephrine, 138
Equilibrative nucleoside transporters (ENT), 46, 47, 63, 82
Equilibrative transporters, 82
Equilibrative-insensitive, 82
Equilibrative-sensitive, 82
Erythro-9-(2-hydroxy-3-nonyl) adenine (EHNA), 96
Escitalopram, 277, 278
ESI. *see* electrospray ionization (ESI)
ESI–MS, 337, 342
Essential tremor, 226, 243
Ethylene diamine tetra acetic acid (EDTA), 90
Ethylene glycol tetra acetic acid (EGTA), 90
Event related potential (ERP), 36
Excessive glutamatergic neurotransmission, 144
Excitatory neurotransmission, 82, 101
Exitotoxicity, 82
Exocytosis, 181, 183, 186, 191
Exocytotic, 61, 62, 64
Exocytotic flux of neurotransmitters, 188
Exogenous adenosine, 65, 68
Exogenous ATP, 68
Extended kiss and run, 183
Extracellular adenosine, 80–82, 100–101
Extracellular dopamine, 161, 164
Extracellular glutamate, 12

Extracellular space (ECS), 369
Eye-field transcription factors (EFTF), 57

FAAH, 256, 259
FAAH inhibitor URB597, 259
Fast Analytical Sensing Technology (FAST16mkIII), 114
Fast ENT-mediated, 64
Fast scan cyclic voltammetry (FSCV), 14, 79, 147, 162, 225, 252, 272, 287
Fatal Familial Insomnia, 2
Fatigue, 269
FBR to intracortical electrodes, 356
FDA (Food and Drug Administration), 115
Fever, 370
FGF-2, 386
Field potential analysis, 36
Flexible shank electrodes, 116
Fluoexetine, 231, 279
Fluorescence-based histology, 358
Fluorescence microscopy, 337
Fluorescence molecular tomography (FMT), 360
Fluorescent tag, choice of, 212
Food and Drug Administration (FDA), 226
Forced swim test (FST), 278
Force-plate actometer, 324, 325, 326
Force sensing runway, 324, 326
Forebrain cholinergic system, 27, 29
Foreign body response (FBR), 354
Fragile X syndrome (FXS), 341
Fusion pore, 186, 190–193
Fusion pore potential energy, 191

GABA/DA co-release, 216
GABAergic antagonists, 170
GABAergic system, 370

Gamma-aminobutyric acid (GABA), 39, 138, 162, 227, 314
Gamma-aminobutyric acid (GABA) ergic, 209
Gamma-aminobutyric acid (GABA) terminals, 255
γ-MSH, 336
Genetically-engineered HD model rodents, 314
Gerhardt group, real-time monitoring technique, 31
GFAP staining, 126
Glial cells, 373
Glial fibrillary acidic protein (GFAP)-positive astrocytes, 101
Glial fibrillary acid protein (GFAP), 125
Glioblastoma treatment, 372
Glioma tissue, 385
Globus pallidus external segment (GPe), 314
Globus pallidus internal segment (GPi), 314
Glucose, 7–9, 146–147, 152–153, 369
Glucosensing, 54–57
Glucosensing in hypothalamic tanycytes, 56
Glucose oxidase (GOx) enzyme, 140, 146
Glucose transporter 1 (Glut1), 8
GluOx, 118, 121, 122, 128
Glutamate, 4, 6, 11–13, 15, 118–119, 121–122, 124–125, 144–146, 369
Glutamate/DA co-release, 216
Glutamate levels, 116, 119–122
Glutamate MEAs, 119–122, 124
Glutamate oxidase (GluOx), 117, 118, 122
Glutamate transients, 144–146
Glutaminergic system, 370
Glycerol, 50, 51
Glycerol kinase, 50

Glycerol-3-phosphate oxidase, 5
Glycosaminoglycans (GAGs), 375
GO_x, 146, 148, 150, 153
GO_x/chitosan-modified electrode, 153
GO_x/chitosan-modified microelectrodes, 150, 151
G-protein coupled receptors, 47, 80, 336
G-protein linked dopamine receptors, 163
Group II metabotropic glutamate receptors (mGluR2/3), 120
Growth factors, 370, 372, 374
GTPase activity, 191, 192
GTPase mechanochemical enzyme, 190

^{3}H-norepinephrine, 381
5-HT$_{1B}$ receptor, 275
6-Hydroxydopamine (6-OHDA), 229
6-Hydroxydopamine (6-OHDA)-lesioned rat model of PD, 289–290
5-Hydroxytryptamine (5-HT), 202
Heparan sulfate, 379
Heparin, 375
High frequency stimulations (HFS), 88
High performance liquid chromatography (HPLC), 82, 121
Hippocampus, 63, 64, 67
Hippocampus dentate gyrus (DG), 124
Histamine receptors, 275
Homovanillic acid (HVA), 229
Human neurochemical recordings, 226, 238, 243
Huntingtin gene (*htt*), 311
Huntington's disease (HD), 82, 311
Hydrogen peroxide (H_2O_2), 139
Hydrogen peroxide (H_2O_2) plus α-ketoglutarate, 118
Hypercapnia, 79, 84, 100, 102
Hyperdopamine, 288, 297
Hypocapnia, 100
Hypothalamic neuronal circuitry, 56

Hypothalamic tanycytes, 54
Hypothalamus, 381, 383
Hypothalamus defense area (HDA), 100
Hypoxia, 79, 98, 99, 102, 104

IFN-γ, 389
IL-3, 385
IL-4, 385, 389
IL-5, 380, 385
IL-8, 385
IL-10, 385
IL-12, 385
IL-13, 385
IL-1α, 385
IL-1β (Interleukin-1β), 376
IL-6 (Interleukin-6), 377
IL-1ra, 385
Imaging agents, 371, 372
Immobilized GluOx, 118, 128
Immunohistochemistry, 337, 372, 374
Immunoregulators, 370
Immunotoxin 192 IgG-saporin, 28
Impedance spectroscopy, 358–359
Implantable glucose sensor, 140
Implantable intracortical electrodes, 351
Implantation of ECoG arrays, 354
Inducible nitric oxide synthase (iNOs), 102
Inducible nitric oxide synthase (iNOS) inhibitor, 14
Inflammation, 372
Infra-red differential interference contrast (IR-DIC) optics, 212
Injury, 371, 373, 374, 384–385, 387, 389
iNOS inhibitor, 66
Insertion trauma, 357
Intra- and extracellular formation of adenosine, 80
Intracortical electrode implants, 351, 352, 354, 355–357

Intracranial self stimulation (ICSS), 161
Invasive intracortical electrodes, 354
Ionized calcium binding adaptor 1(Iba1), 125
Ionotropic glutamate receptor activation, 97
Ionotropic receptors, 49
Iontophoresis, 171
IP-10, 385
Irreversible mitochondrial complex II inhibitor, 315
Ischemia, 79, 82–84, 99–100, 102, 104
Ischemic attack, 82, 99
IT15 gene, 311

JWH133, 253
JZL184, 259, 260, 261

KC/GRO, 388, 389
KCl-stimulated synaptoneurosomal preparation, 339

Lactate, 15–17, 369, 386
Large dense core vesicle (LDCV), 184
Laser, 212–214
Laser capture microdissection (LCM), 362
Laser induced fluorescence (LIF), 336
Latrunculin A treatment, 193
LC-ESI-MS, 337
LC–MALDI–MS, 339, 340
L-3,4-dihydroxyphenylalanine (L-DOPA), 95, 186, 227, 232, 233
LED illumination, 213, 214
LED-based system, 212
L-glutamate, 114–115, 118, 128
L-histidine, 88, 89
Limit of detection (LOD), 87
Lipids, 190, 197
Lipopolysaccharide (LPS), 372

Liquid chromatography (LC), 336
Lithium-polymer battery, 235
L-lactate oxidase, 5
L-NAME, 66
Local field potential (LFP), 36
L-type calcium channel blocker, 215
LY341495, 121
LY379268, 121
Lysing process, 198

MAGL inhibitor JZL184, 259
Magnetic resonance imaging (MRI), 226
MALDI–MS, 337–339, 341, 343
MALDI–TOF–MS, 338, 341
Mass spectrometry (MS), 336
Mass transport, 378–382
Matrix-assisted laser desorption/ionization (MALDI), 336
Mayo Investigational Neuromodulation Control System (MINCS), 225–226, 236–237
Mayo Neural Engineering Laboratory, 225
MEA implantation, 125–126, 127
MEA photolithographic mass fabrication process, 116
MEA recordings, 123
MEA technology, 114
Medial forebrain bundle (MFB), 230, 273, 277
Medial prefrontal cortex (mPFC), 32
Medium spiny neurons (MSN), 162
Membrane, 382, 384, 385, 386
Membrane cytoskeleton, 189, 190
Memory, 370, 377
Metabotropic glutamate (mGlu) receptor, 66
Metabotropic receptor inhibition, 97
Metabotropic receptors (P2Y), 46

Methamphetamine (METH)-induced neurotoxicity, 287
Methylphenidate, 300, 302
1-Methyl-4-phenyl-1,2,3,6-tetrahydropyridine (MPTP), 292
mGLUR$_{2/3}$ antagonist, 121
mGluR subtypes, 122
Michaelis–Menten models, 276
Microcircuitry, 167
Microdialysis, 4–6, 12, 66, 82–83, 95, 102, 121, 252–253, 256, 270–271, 289, 302, 316, 336, 342–344, 370, 374, 377–378, 380, 381–388
Microdialysis applications, 384
Microdialysis measurements, 271
Microdialysis sampling, 381–383
Microelectrode array (MEA) technology, 114
Microelectrode biosensors, 50
Microelectrode for human neurochemical recording, 238
Microelectrode methods, 114
Microglia, 125, 356, 357, 372, 374, 376, 389
Microglial cells, 372, 374
Micro-processor, 235
Midbrain DA neuron activity, 202
MINCS. see Mayo Investigational Neuromodulation Control System (MINCS)
MINCS–WINCS hardware, 236
MIP-1β, 385
Modulation, 275
Molecular weight cutoff (MWCO), 384
Monoacylglycerol lipase (MAGL), 256, 259
Monoamine oxidase (MAO), 274, 300, 301
Monocyte chemoattractant protein-1 (CCL2), 374

Motoric behavior, PD role in dopamine tone, 289
mPFC cholinergic transients, 33, 35
mPFC signaling, 34
MRI in PD patients, 293
mRNA, 372–374, 376
MSN firing, 170
MSN populations, 163
Multiple sclerosis, 370, 372, 377
Multiplexed bead-based immunoassays, 372
Muscarinic acetylcholine receptor (mAChR), 29
Muscarinic receptor antagonists, 29
Muscarinic receptors (mAChRs), 202

3-Nitropropionic acid (3-NP), 315
N6-cyclopentyladenosine (CPA), 89
NAc core, 162, 172
NAC glutamatergic–dopaminergic mechanisms, 30
NAChR. *see* nicotinic receptors (nAChRs)
NAc shell, 162
NAc subregions, 162
Nafion®, 118, 142–146
Nafion® layer, 273
Nafion® overlayer, 142
Nafion®/polypyrrole-coated electrode, 145
N-arachidonoylaminophenol (AM404), 256
Naylor study, 17, 18
Nerve growth factor (NGF), 384
Neurochemical changes evoked by DBS, 243
Neurodegenerative diseases, 374
Neuroinflammation, 82
Neurological injury, 371
Neurological repair, 371
Neuromodulation, 79, 83, 95
Neuromodulator, 202, 269, 370
Neuronal injury, 376
Neurons, 162
Neuropeptide, 335, 336, 369–370
Neuropeptide release from *ex vivo* brain slice containing SCN, 341
Neuropeptide synthesis, 335, 336
Neuroprotection, 79, 82, 83, 376
Neuroregulators, 370
Neuroscience, electrochemistry in, 137
Neurotoxicity, methamphetamine (METH)-induced, 287–289, 294, 295
Neurotransmission, 161, 269–279
Neurotransmitter, 45, 49, 114, 119, 120–124, 138, 269, 275, 359, 369, 370, 373, 381
Neurotransmitter modulation, 8
Neurotransmitter monitoring systems, 235–237
Nicotinic acetylcholine receptors (nAChR), 29
Nicotinic receptors (nAChRs), 202
Nifedipine, 215
Nigrostriatal dopamine neurons, 289
NIR fluorophore, 360
Nitric oxide, 377
Nitric oxide (NO) signaling, 66
Nitrobenzylthioinosine (NBTI), 66, 82
NKA (Na^+–K^+ ATPase), 63–65
N-methyl-D-aspartate (NMDA), 64, 90, 170
Nomifensine, 229, 231, 302
"Non-classical" neuropeptides, 336
Non-electroactive neurotransmitters, 114
"Nonfaradaic" response, 116
Non-rapid eye movement (NREM) sleep, 3, 8
Non-selective glutamate transporter inhibitor (D,L-threo-β-benzyloxyaspartate (TBOA), 121

Non-SERT serotonin reuptake
   mechanisms, 276
Non-specific adsorption, 382
Noradrenaline (desipramine) re-uptake
   inhibitors, 231
Norepinephrine, 138, 202, 369, 381
Norepinephrine transporters (NET), 276
NOS scavenger, 66
NREM sleep epochs, 3, 7, 8
NREM to awake transitions, 10
NT5E inhibitor, 90
NTPDase2, 48, 57, 58
Nucleoside transporters, 81, 82
Nucleoside triphosphate
   dihydrophosphorylase (NTPDase)
   family, 48
Nucleoside triphosphate
   diphosphohydrolase (NTPDase)
   inhibitor, 90
Nucleus accumbens (NAc), 30, 161, 205,
   274
Nucleus accumbens shell, 90, 91
Nucleus tractus solitarii (NTS), 99
Null sensor, 51

Olivocerebellar fibers, 337
OMDM-2, 256
Ω-Conotoxin, 120
Open-and-closed exocytosis, 181, 183,
   190–193, 196
Open-flow microperfusion, 377–378
Opsin expression, 208, 211, 212
Optogenetic method, 39
Optogenetics, 203
Organic cation transporters (OCT), 276
Oxidase enzymes, 369

Pain, 371
Pancreatic β cell, 55
Pannexin-1 hemichannels, 47

Paracrine, 371
Parafascicular nucleus (Pf), 205
Paraventricular nucleus (PVN), 54
Pargyline, 301
Parkinson's disease (PD), 82, 92, 95, 104,
   116, 226, 287, 371, 377
Partial pressure of carbon dioxide (PCO_2),
   51–52
"Passive Stabilization" model, 289
Passive stabilization of tonic dopamine
   signaling, 289
Patch-amperometry, 183
Patch-clamp, 182, 183
PC12 cells, 185, 192, 193, 196
Pedunculopontine tegmental nucleus
   (PPT), 229
Penumbra region, 386, 387
Peptides, 335–336
Peptidomics, 338
Performance-associated ACh levels, 30
Perfusates, 370
Peri-glioma tissue, 385
Pharmacological methods, cholinergic
   activity manipulation, 29
Phasic activity of DA neurons, 201
Phasic dopamine signaling, 288–289,
   293–299
Phasic release of neurotransmitters, 123
Phasic *versus* tonic DA signaling, 296
Physiological control of sleep, 2
Physiological firing frequencies, 205,
   216
Physiological relevance, 207, 215
Phytocannabinoids, 251
Plasma membrane monoamine
   transporters (PMAT), 276
Platelets, 387
Platinum (Pt)-based biosensors, 139
Pleiotropism, 371
p-Methoxyamphetamine (PMA), 278

Polydimethylsiloxane (PDMS)-based microfluidic chamber, 184
Polyglutamine (polyQ), 311
Poly-(meta-phenylenediamine) (mPD), 118
Pore formation, 189, 190
Positron emission tomography (PET), 227, 371
Post exocytosis foot, 183
Post-hypoxia purine efflux (PPE), 98
Post-spike feet, 193, 195, 196, 197
Postsynaptic gene expression, deficits in, 295
Post-synaptic processing, 15
Post-traumatic stress disorder (PTSD), 269
Prefrontal cortex, 90–91
Pre-prohormone, 335
Preprotachykinin (*ppt*), basal expression of, 296
Pre-spike foot, 186, 193, 195
Presynaptic autoreceptors, 39
Progenitor cells, 57, 59
Prohormone, 335
Proinflammatory and neurotoxic cytokines, 362
Pro-opiomelanocortin prohormone, 336
Prostaglandin E_2 (PGE_2), 377
Prostatic acid phosphatase (PAP), 92
Proton–ATPase pump, 274
Psychiatric disorders, 370, 371
Psychostimulant amphetamine, 186
Purine nucleoside phosphorylase (PNP), 59, 61
Purinergic control of motor pattern generation, 69
Purinergic signaling, 45–49
Purines, 45
Push-pull perfusion, 370, 378, 380, 381, 383–384
Putative sleep regulatory substance, 9, 19

P2Y1 receptors, 55, 57, 58
P2Y subtypes as ATP/ADP receptors, 47

Radial glial cells, 57
Radioimmunoassay, 337
RANTES, 385, 386, 388
Rapid adenosine monitoring, 104
Rapid eye movement (REM) sleep, 3, 8, 101
Rat, 377, 390
Reactive gliosis, 356, 362
Reactive microglia, 362
Real-time serotonin monitoring, 270–273
Recording arrangement, 56, 69
Redox active, 369
Redox polymers, 140
Reference electrode lead, 235
Research, sleep, 1
Retaliatory metabolite, 83, 104
Retina, 57, 59
Retinal pigment epithelium (RPE), 59
"Reward prediction error," 293
Rimonabant, 254, 255
R6/1 mice, 315–318, 323
R6/2 mice, 315–323, 326
Root mean square (RMS) noise, 196

S-adenosylhomocysteine (SAH), 80
"San Vitus." *see* Huntington's disease (HD)
Scanning electron microscopy of human FSCV, 239
SCH23390, 89
Schaffer collaterals, 63, 64
Schizophrenia, 269
Secretory vesicle, 183, 184
Seizure, 79, 100
Selective attention, 36
Selective serotonin reuptake inhibitors (SSRI), 276

Self-referencing, 114, 116–119
Self-referencing amperometric biosensors, 141
Self-referencing subtraction, 117–119, 122
Self-referencing technique, 143
Serotonin, 231, 269–278, 369
Serotonin modulation, 275
Serotonin transporters (SERT), 272
Single-bead SPE, 339
Sleep, 1–7, 10, 14–15, 370
Sleep deprivation, harms of, 19
Sleep deprivation (SD), 66
Sleep homeostatic process, 1, 11, 18, 19
Sleep homoeostasis, 65
Sleep onset, 4, 10, 17
Sleep pressure, 2, 8, 10
Sleep-related fluctuations, 4
Sleep/wake cycles, 8, 15, 82, 84, 101–102
Sleep-wake status, 66
Slower astrocyte mediated release, 64
Slow wave (SW) sleep, 65, 101
SNARE complex formation, 101
SNARE proteins, 190
Solid phase extraction (SPE), 338
Somnogen, 65
SPE–LC–MS, 342
Spinal lamina II neurons, 92, 94
Steady-state visual evoked potentials (SSVEP), 352
Stem cells, 374
Stereotaxic surgery, 208, 209, 214
STN DBS, 227, 229, 231, 232, 236, 239
STN stimulation, 227, 231
STN stimulation evoked striatal DA release, 234
Striatal acetylcholine (ACh), 201–202
Striatal ChIs, 203–205
Striatal DA, 229–232, 236
Striatal degeneration, 319

Striatal vesicle neurotransmitter, pharmacological manipulation of, 187
Striatum, 314, 318, 319, 323, 326
Stroke, 82, 83, 100, 372
Subarachnoid Hemorrhage, 386
Substantia nigracompacta (SNc), 229
Substantia nigra pars compacta, 201, 314
Substantianigra pars reticulata (SNR), 273
Subthalamic nucleus (STN), 227
Sulphonylurea (SUR) subunit, 55
Suprachiasmatic nucleus (SCN), 341
Sustained Attention Task (SAT), 27
Synapse, 181, 182, 194
Synapsin, 274
Synthetic cannabinoid WIN55, 212–254

Tanycyte glucosensing, 55
Tanycytes, 54–55
Temporal limitations of microdialysis, 271
Tetrabenazine (TBZ), 321
Tetrodotoxin (TTX), 120
TGF-α, 385, 390
"That disorder." see Huntington's disease (HD)
Tissue impedance spectroscopy (TIS), 358
Tissue-nonspecific alkaline phosphatase (TNAP), 92
TNF-α, 77, 371
Tonic dopamine signaling, 288–291
Tortuosity, 378–379
Transgenic HD (HDtg) model rat line, 315
Translocator protein (TSPO), 372
Traumatic brain injury (TBI), 116, 371, 384, 385, 387, 389
Tremor arrest, 243
Tri-synaptic pathway, 124
Trypsin proteases, 380

TTX, 61
Tyrosine hydroxylase activity, 229

Unwanted stimulation side effects, 228
Up-regulated dopamine synthesis, 290
Uridine diphosphate (UDP), 47
Uridine triphosphate (UTP), 47

VDM11, 256
VEGF, 385, 386, 390
Ventral root activity, simultaneous recordings of, 69
Ventral surface of medulla oblongata (VMS), 52
Ventral tegmental area (VTA), 162, 201, 276
Ventral tegmental area (VTA) dopamine neurons, 255
Ventrolateral medulla (VLM), 99
Ventrolateralpreoptic area (VLPO), 66
Ventromedial hypothalamic nucleus (VMH), 54
Vesicular acetylcholine transporter (VAChT), 210
Vesicular dopamine release, 298, 300–303

Vesicular monoamine transporter (VMAT), 274
Vesicular monoamine transporter (VMAT) inhibitor, 90
Vesicular nucleotide transporter (V-NUT), 47
Voltage-gated calcium channels (VGCC), 319
Voltammetric biosensors, 147
Volume fraction (ΦECS), 379

Wild type (WT), 316
WIN 35428, 302
WIN55, 212–255
WINCS, 225, 235–237
WINCS electrode for human application, 238
Wireless Instantaneous Neurochemical Concentration Sensing system (WINCS), 225
WT synaptoneurosomes, 340

Xanthine oxidase (XO), 59, 61

Yeast artificial chromosome (YAC), 315